STATISTICAL PHYSICS

Statistical Physics

Invited Lectures from STATPHYS 16,
the 16th International Conference on
Thermodynamics and Statistical Mechanics,
Boston University, August 11–15, 1986

Editor:

H. Eugene Stanley

Center for Polymer Physics
and Department of Physics
Boston University
Boston, MA 02215, USA

1986

NORTH-HOLLAND

Amsterdam · Oxford · New York · Tokyo

ISBN 0 444 87041 5
LCC No. 86-310 93

Published by:
NORTH-HOLLAND PHYSICS PUBLISHING
a division of Elsevier Science Publishers B.V.
P.O. Box 103
1000 AC Amsterdam, The Netherlands

SOLE DISTRIBUTORS FOR THE USA AND CANADA:
Elsevier Science Publishing Company Inc.
52, Vanderbilt Avenue
New York, N.Y. 10017, USA

REPRINTED FROM PHYSICA A VOL. 140A (1986) Nos. 1 & 2

Printed in The Netherlands

PREFACE

This volume contains a collection of 52 papers based on the principal invited talks presented at the 16th IUPAP Conference on Thermodynamics and Statistical Mechanics ('STATPHYS-16'). The authors have taken care to present their recent advances in a fashion that is unusually readable for a conference proceedings volume. Accordingly, it is envisaged that these papers will form a concise but coherent summary of the "state-of-the-art" of statistical physics in the year 1986. Publication as a free-standing paperback volume is designed to encourage use of this material to supplement existing texts on thermodynamics and statistical mechanics.

The order of the presentations is the same as in the conference itself. In structuring the program of both invited and contributed presentations, the Local Committee found it helpful to divide the field into six broad "topical groups." Since a list of those groups may be useful to scholars and students everywhere, we reproduce it here:

Topical group 1

A. Conformal invariance
   Exact solutions
B. Computational physics
C. Cellular automata (memory, vision)
   Artificial Intelligence: applications in stat. mech.

Topical group 2

D. Liquids
E. Strongly coupled Coulomb gases
   Transport phenomena in dense plasma
F. Reaction kinetics
   Chemical kinetics

Topical group 3

G. Wetting, surfaces
H. Colloids
   Microemulsions
   Micelles
   Polymers
   Sol-gel transition
I. Random surfaces
J. Liquid crystals (see group 4 also)

Topical group 4

K. Glasses, metastability
   Spin glasses
L. Systems with competing periods
   Quasi-crystalline phases
M. Liquid crystals (see group 3 also)

Topical group 5

N. Physics of disorder
   Random fields
   Physics of porous media
   Transport in porous media
O. Equilibrium and kinetic aggregation processes
   Chaos and turbulence
   Non-linear driven oscillators
   Pattern formation (snowflake, spinodal decomposition)
   Fractals

Topical group 6

P. 2d electron gas in a field
   Quantum Hall effect
Q. Heavy fermion systems
R. Disorder and transport in electronic systems
   Localisation

In addition to the main invited talks, the program consisted of roughly 500 poster presentations and 250 twelve-minute talks. The former were scheduled in a plenary session, taking place at the end of the morning session in order that discussions initiated at the poster session could continue over lunch. The lunch break itself was rather generous, in an effort to further foster the sort of informal interactions that are all too often in short supply at large international conferences. I wish to thank all who presented contributed papers for their efforts, and especially to thank that subset of contributed posters and contributed talks that the Organizing Committee invited to be "up-graded" to the status of invited poster and invited talk. Those in the former category include A. Compagner, J.A.M.S. Duarte, H. Eicke, P. Hänggi, H.-M. Huang, B. Jancovici, K. Kakurai, K. Kaneko, J. Kertèsz, D. Kim, H.R. Krishna-Murthy, J. Machta, P. Mausbach (with A. Geiger), P. Pearce, R. Pindak, C.J. Thompson (with C. Pisani and E.R. Smith), M. Velarde and J.C. Wheeler, while those in the latter category include D. Andelman (with F. Brochard, P.G. de Gennes and J.F. Joanny), H. Au-Yang (with J.H.H. Perk), M. Azbel (with P. Bak and P. Chaikin), D. Bedeaux, M.J. de Oliveira (with S.R. Salinas and C.S.O. Yokoi), V. Dohm (with W. Huhn and R. Schloms), F. Dowell, P. Egelstaff, J. Feder (with T. Jøssang, K.J. Måløy and U. Oxaal), J.M. Gordon (with A.M. Goldman, J. Maps, D. Costello, R. Tiberio and B. Whitehead), A.B. Harris (with T.C. Lubensky), R.M. Hornreich (with S. Shtrikman), M.V. Jarić (with D.R. Nelson and U. Mohanty), M. Kardar, M. Kerszberg (with A. Bergman), D. Kessler (with J. Koplik and H. Levine), S. Milošević (with S. Elezović and M. Knezević), J. Pankert (with V. Dohm), J.H.H. Perk (with F.Y. Wu), S.A. Safran (with D. Roux, D. Andelman and M. Cates), B. Shapiro, T. Vicsek (with F. Family, J. Kertész and D. Platt), J.L. Viovy (with W.M. Gelbart and A. Ben-Shaul), J. Wallenborn (with V. Zehnlé and B. Bernu), D.A. Weitz (with H.M. Lindsay, S.K. Sinha, P. Dimon, Z. Chen, P. Sheng and R. Pynn) and R.K.P. Zia.

Finally, those who served as discussion leaders rose to the challenge admirably: P. Argyrakis, J. Bernasconi, D.D. Betts, A.R. Bug, A.D. Bruce, A. Bunde, H.W. Capel, J.W.D. Connolly, H. Craubner, C. Domb, M. Droz, J.W. Dufty, J. Essam, M. Fontana, R.B. Griffiths, S. Havlin, B.D. Hughes, S. Ichimaru, N. Jan, R. Kubo, D.P. Landau, J.M.H. Levelt Sengers, T.C. Lubensky, M.A. Moore, R. Mountain, M. Nauenberg, R.J. Speedy, M. Suzuki, P. Szépfalusy, M.M. Telo da Gama, H. Thomas, A.M.S. Tremblay, J.M.J. van Leeuwen, T.A. Witten and M.J. Zuckermann.

It is a privilege and an honor to thank all who gave so generously of their time and energy to make this meeting the success that it was generally acknowledged to have been. Chief among these individuals were the members of the international and local scientific organizing committees, who for12 months were barraged with questions and proposals that required measured and timely responses. The chairmen of the "topical groups" put in extra time in an effort to achieve that elusive goal of a balanced program: G.B. Benedek, A.N. Berker, R.J. Birgeneau, B.I. Halperin, A.M. Jaffe, P.C. Martin, I. Oppenheim and F.Y. Wu.

N. Andreopoulos, J. Morrow and E. Shuman each devoted many patient hours to the Boston University organization. Perhaps the most impressive feature of the "organic organization" (as one attendee described it) was the conference staff itself - a team of 14 *undergraduate* students! These individuals, ranging in age from 17 to 21, functioned as a remarkably responsible and creative team - organizing almost single-handedly every aspect of the conference organization. Thus I speak for everyone when I especially thank Daniel S. Aronson, William Chisholm, Marcelo Colon, Delia Evans, Brenda M. Gray, Craig D. Howard, William Karlon, Tira Maloney, Jose Mercado, Dulce Nordlund, Tim Raczkowski, Sheila J. Roberts, Mark Shneyder and Bryan Taylor.

This list would not be complete if it did not identify those individuals who spontaneously gave of their ideas and energy at moments when those precious commodities were most needed: Robin Blumberg-Selinger, Antonio Coniglio, Thomas N. Earnest, Cynthia Goh, Jenny Green, William Klein, Sava Milošević, Sidney Redner, Janet Nittmann, Savely Rabinovitch, Paul Roepe, Jonathan Selinger, Robin Speedy, Idahlia Stanley, Michael Stanley, and Rachel Stanley.

To bring together 1000 of the leading workers in a field from 43 countries and over 700 universities, laboratories, and industrial firms requires considerable financial resources. Roughly 500 individuals (a substantial fraction of which were graduate students) received partial support from resources kindly provided by the President of Boston University, John R. Silber, the National Science Foundation, the Office of Naval Research, the IBM Corporation, and our sponsors, the IUPAP.

No list of acknowledgements for any book would be complete without an expression of the gratitude an editor feels for the support of the publishers themselves. This statement is particularly true in this case. From the initial interactions with Peter de Châtel, Pieter Bolman and Joost Kircz, it became clear that we were working with a firm of integrity and imagination. Plans were laid to achieve two "firsts" in publication of conference proceedings: A record fast publication time (3½ months after the conference, including time for type-setting and proof correction), and a record low price ($9 pre-publication and $12 thereafter). The first goal was realized only because of the gargantuan efforts of North-Holland editor Peter Tel, who laboured throughout the week of the conference massaging manuscripts into such a form that typesetting could begin immediately: imagine authors' surprise when the first page proofs were received the week after the meeting ended! The second goal is based on the optimism of the editor that these lectures will serve to supplement the standard texts of a substantial number of undergraduate and graduate courses in statistical mechanics and thermodynamics, none of which describes even a fraction of the existing new developments presented herein.

So . . . STATPHYS-16 is now "history". The scientific and non-scientific exchanges that transpired are the testimony to the ultimate value of this meeting. And this little volume is a modest step in the direction of sharing with the entire community some representative fraction of the many scientific developments that were exposed in Boston. Last but not least, I wish to thank the authors of these papers for their timely efforts to produce work of the highest scientific and pedagogical quality under the time pressures mentioned above and I wish to thank the 1000 participants themselves who seemed to understand that the conference had expected only 300–400 and so forgave the many imperfections of the organization: "Boston is not Zurich and the USA is not Switzerland!"

H. Eugene Stanley
Boston, September 1986

THE BOLTZMANN AWARD 1986

Presentation by Joel L. Lebowitz, Chairman, IUPAP Commission on Thermodynamics and Statistical Mechanics, DAVID RUELLE and YASHA G. SINAI: BOLTZMANN MEDALISTS 1986.

The award of the Boltzmann Medal to David Ruelle and Yasha G. Sinai recognizes their contributions to statistical mechanics: contributions both fundamental and practical. They have given us not only many new exact results, but also novel ways of thinking about statistical mechanics, thus revealing its breadth and great inner beauty.

I still find it amazing that the foundations laid by Boltzmann, Gibbs, Einstein, and others have held firm as statistical mechanics has extended to become an essential ingredient of all sciences, from pure mathematics to applied biology. The works of Ruelle and Sinai have certainly contributed greatly to both the crown and the root of our beautiful subject.

I cannot and will not try to enumerate all the important theorems proven by David Ruelle and Yasha Sinai. You can find one aspect of their work described in their joint article in this volume. This relates at a very deep level the time invariant measures of certain dynamical systems to Gibbs measures of particles on a one dimensional lattice (with translation playing the role of time shifts). These beautiful results (which are truly joint ones despite the fact that this is their first joint paper) have their origin in the incisive characterization of Gibbs states provided by the Dobrushin–Lanford–Ruelle equations and of (good) dynamical systems by the Kolmogorov–Sinai entropy. The lecture was given by Ruelle since Sinai was very regretably unable to come to the meeting. I will add to that article by mentioning briefly a few other topics to which our Boltzmann medalists have made deep contributions.

Ruelle and Takens' paper, showing that the onset of turbulence can occur directly as a transition from quasi-periodic motion with as few as three frequencies, opened the way for a new era in the theoretical and experimental study of this subject. The problem of fully developed turbulence is, however, still very much with us – an important sign post on the way is Ruelle's upper bound on the density of positive Lyapunov exponents.

The work of Sinai which I would like to mention is his proof of ergodicity and K-character of the billiard system – appropriately named Sinai's Billiards. Sinai's work in this field includes also the stretched exponential bound with Bunimovich on the decay of the velocity autocorrelation function with the number of collisions and his upper bound on the Kolmogorov–Sinai entropy/unit volume of a system of hard spheres. In a completely different area is the very beautiful work of Sinai and Pirogov on the phase diagram of general lattice systems at low temperatures. That theory, still under very active development, provides the only rigorous approach for systems without general symmetry. It is, in my opinion, far too little known among physicists and engineers – so do look it up.

Response by David Ruelle:

It is difficult for me to answer the very kind words of Joel Lebowitz. He is the nicest, and anything I say after him will appear clumsy. So let me just tell you how happy and honored I am to be here this evening and to be added to the prestigious list of former recipients of the Boltzmann Medal. I am also particularly happy to have this occasion with Yasha Sinai. He is surely with us in spirit, but since he is not here in person, I may be allowed to express my admiration for a man whom I consider to be one of the truly great scientists of our time.

In the course of my scientific wanderings I have visited several of the regions which lie across physics and mathematics: equilibrium statistical mechanics, ergodic theory, differential dynamics, and a few more. Each of these regions is like a valley, large or small, with its own character and its towering peaks of learnedness which are the local experts. Going from one valley to another, the landscape and the people change, the experts are different. Yet, I have found that in these various regions across physics and mathematics which I visited, Yasha Sinai was part of the landscape and had made fundamental contributions. It is not that he has written a myriad of articles on different topics. Rather, his work appears like a high mountain which can be seen from far away, and not just from one particular valley. It is one of the satisfactions of science that one can see some exceptional minds at work. I consider myself fortunate in having been able to follow part of the work of Yasha Sinai and in having interacted on occasion with him.

IUPAP COMMISSION ON STATISTICAL PHYSICS

J.L. Lebowitz, USA, Chairman; A. Corciovci, Romania, Secretary; M.J. Buckingham, Australia; I.M. Khalatnikov, USSR; O. Penrose, UK; G. Prosperi, Italy; A. Sjölander, Sweden; M. Suzuki, Japan; H. Thomas, Switzerland; J. van Kranendonk, Canada; J.M.J. van Leeuwen, The Netherlands; G. Vojta, E. Germany

INTERNATIONAL SCIENTIFIC ORGANIZING COMMITTEE

S. Alexander, Israel; P.W. Anderson, USA; C.A. Angell, USA; R.J. Baxter, Australia; H. Benoit, France; B.J. Berne, USA; D.D. Betts, Canada; K. Binder, W. Germany; S. Chandrasekhar, India; M.H. Cohen, USA; R.A. Cowley, UK; P.G. de Gennes, France; C. Domb, Israel; K. Dusek, Czechoslovakia; I. Dzyaloshinskii, USSR; S.F. Edwards, UK; J. Feder, Norway; M.E. Fisher, USA; M. Fixman, USA; P.J. Flory, USA (*dec.*); J. Fröhlich, Switzerland; W.I. Goldburg, USA; R.B. Griffiths, USA; J.D. Gunton, USA; E. Guyon, France; J.P. Hansen, France; Hao Bai-lin, China; A.J. Heeger, USA; L.P. Kadanoff, USA; K. Kawasaki, Japan; D. Kim, S. Korea; R. Kubo, Japan; R. Landauer, USA; J.L. Lebowitz, USA; E.H. Lieb, USA; S. Milosevic, Jugoslavia; K.A. Müller, Switzerland; J. Mydosh, Netherlands; R. Orbach, USA; G. Parisi, Italy; P. Pincus, USA; V. Pokrovsky, USSR; I. Prigogine, Belgium; A. Rahman, USA; S.A. Rice, USA; F. Rondelez, France; D.W. Schaefer, USA; J.R. Schrieffer, USA; M. Suzuki, Japan; C.J. Thompson, Australia; G. Toulouse, France; C. Tsallis, Brazil; M.G. Velarde, Spain; J. Villain, W. Germany; D.J. Wallace, UK; B. Widom, USA; K.G. Wilson, USA; S. Wolfram, USA; M.J. Zuckermann, Canada; R. Zwanzig, USA

LOCAL SCIENTIFIC ORGANIZING COMMITTEE

A. Aharony, R. Bansil, G.B. Benedek, A.N. Berker, R.J. Birgeneau, S.H. Chen, A. Coniglio, J.M. Deutch, R.L. Devaney, F. Family, C.W. Garland, R.J. Glauber, H. Gould, E.P. Gross, B.I. Halperin, J. Herzfeld, A.M. Jaffe, J. Jose, G.J. Kalman, T. Keyes, W. Klein, C. Lobb, B.B. Mandelbrot, P.C. Martin, R.B. Meyer, D.R. Nelson, M.A. Novotny, I. Oppenheim, P.S. Pershan, S. Redner, D. Ronis, R.J. Silbey, H.E. Stanley, T. Tanaka, L. Tisza, G. Vichniac, F.Y. Wu, S. Yip

CONTENTS

Physica **140A** (1986) 1–8
North-Holland, Amsterdam

FROM DYNAMICAL SYSTEMS TO STATISTICAL MECHANICS AND BACK

D. RUELLE
Institut des Hautes Etudes Scientifiques 91440, Bures-sur-Yvette, France

and

Ya.G. SINAI
Landau Institute of Theoretical Physics, Moscow, USSR

The relation between statistical mechanics and dynamical systems established by Boltzmann has led to surprising later developments. Some recent issues are reviewed here.

1. Introduction

In order to lay the foundations of equilibrium statistical mechanics, Boltzmann invoked the *ergodic hypothesis*. The precise formulation, and then the analysis of this hypothesis, have proved extraordinarily fruitful but also difficult, and many of the fundamental questions in the field remain open to this day.

The ergodic hypothesis makes use of dynamical concepts to help equilibrium statistical mechanics. Conversely, statistical mechanics has come to be an important tool in the study of dynamical systems. This has occurred in subtle ways. In particular, the entropy concept of statistical mechanics, borrowed by Shannon for information theory, has been turned into the measure theoretic entropy or the Kolmogorov–Sinai invariant of ergodic theory, which has then been applied to the study of differentiable dynamical systems, leading finally to the understanding of "chaotic" dynamical systems in physics (Ruelle–Takens, etc.).

We shall now return to some of these questions in more detail, following a logical rather than chronological order. Sections 2–5 are written by Ruelle, sections 6 and 7 by Sinai.

2. Symbolic dynamics

A differentiable dynamical system is a differentiable time evolution $(x, t) \mapsto f^t x$ with $x \in$ manifold M, t integer or real. We restrict attention to a set Λ of recurrent (nonwandering) points of M. Time evolution will stretch or contract distances; there is a class of nice systems (hyperbolic, or Axiom A systems) where one can define continuously an expanding and a contracting direction on Λ. For such systems one can almost replace the set Λ by the set of configurations of a 1-dimensional classical lattice spin system. For simplicity we describe this symbolic dynamics in the discrete time case, i.e., $t \in$ integers.

Let J be a finite set of *symbols* (= spin values), and (t_{ij}) a *transition matrix* with elements 0 or 1. We define the *configuration space* $\Omega = \{(\xi_i)_{i \in \mathbf{Z}} : t_{\xi_i \xi_{i+1}} = 1\}$. The existence of symbolic dynamics (Sinai, Bowen) is the assertion that one can choose J, (t_{ij}), and a map π: $\Omega \mapsto \Lambda$ with the following properties: π is continuous, onto Λ, finite-to-one, in fact one-to-one on a dense residual set, and such that $f \circ \pi = \pi \circ \tau$ where τ is the *shift* (lattice translation) of Ω.

One virtue of this symbolic dynamics is to replace the study of probability measures on a manifold, invariant under a diffeomorphism f, by translationally invariant probability distributions for a spin system. Equilibrium statistical mechanics provides a powerful theory of such probability distributions: the theory of *Gibbs measures*. In particular, one can define the Gibbs states either by "local" properties (DLR equations) in terms of conditional probabilities, or "global" properties by a variational principle (Dobrushin, Lanford–Ruelle)*.

3. Gibbs measures on hyperbolic attractors as a description of "chaos"

There is a natural measure on hyperbolic (Axiom A) attractors which corresponds to time averages in the presence of a small amount of noise. This natural measure is a Gibbs measure (Sinai, Ruelle, Bowen, Kifer). For *strange* attractors it describes a *chaotic* time evolution.

4. Chaos in physics and the nature of turbulence

Stretching distances by time evolution corresponds to the phenomenon of *sensitive dependence on initial condition*, corresponding to *strange attractors*, or *chaos*. This phenomenon was recognized by Hadamard and its philosophical implications (for unpredictability and the justification of *chance*) were understood by Duhem and Poincaré. A modern, specific and lucid application is due to Lorenz. The general assertion that strange attractors are characteristic of hydro-

*Among the mathematical applications of symbolic dynamics and the thermodynamic formalism let us mention the study by Parry and Pollicott[1]) of the distribution of periods of an Axiom A flow by the methods of the prime number theorem. Also the proof that the Fourier transforms of time correlation functions for Axiom A systems are meromorphic in a strip (Pollicott, Ruelle).

dynamic turbulence is a physical statement, open to experimental verification. The verification (Ahlers, Gollub–Swinney, etc.) has been positive. It is now known that the onset of hydrodynamic turbulence corresponds to the appearance of strange attractors of lower dimension. The same is true for a variety of other chaotic physico–chemical systems.

5. Ergodic theory of differentiable dynamical systems and diagnosis of chaos

The early discussion of physical chaos was largely based on a geometrical reconstruction of attractors, which is possible only in low dimension (i.e. at the onset of chaos).

For moderately excited systems, other tools are available from the ergodic theory of differentiable dynamical systems. To an ergodic measure ρ, various parameters are associated:

–*characteristic exponents* $\lambda_1 \geqslant \lambda_2 \geqslant \cdots$ (also called Liapunov exponents) from the multiplicative ergodic theorem (Oseledec). The λ_i give the rate of exponential divergence of nearby orbits of the dynamical system.

–*entropy* $h(\rho)$: this is the mean rate of creation of information by the system, or Kolmogorov–Sinai invariant.

–*information dimension* $\dim_H \rho$: smallest Hausdorff dimension of a set S such that $\rho(S) = 1$.

Following an idea of Grassberger and Procaccia (also Young) one can readily obtain the information dimension from an experimental time series. In principle, entropy is also accessible, and characteristic exponents are now being obtained. For a detailed discussion and references we can refer the reader to the review article of Eckmann and Ruelle[2]). Using these methods one can now test all kinds of time series (from economics, electroencephalograms, variable stars) to see whether (or to what extent) they behave like deterministic chaotic systems. The game has only just started, one has to wait a little bit for interesting results.

We have seen above how the ideas of equilibrium statistical mechanics were put to use in the study of differentiable dynamical systems (including turbulence and chaos). Now we come to the use of the methods of dynamical systems in equilibrium statistical mechanics, and specifically in the study of the "ergodic hypothesis" for the billiard problem.

6. Dispersed billiards

The billiard system is the dynamical system generated by the uniform motion of a point inside a domain Q with elastic reflections at the boundary ∂Q. We will always assume that Q lies either in Euclidean space R^d or in the d-dimensional torus Tor^d. The motion takes place in the phase space M consisting of pairs $x = (q, v)$ where $q \in Q$ and v is the velocity. Without any loss of generality we assume that $\|v\| = 1$.

The ergodic properties of the corresponding flow $\{S^t\}$ depend crucially on the properties of the boundary ∂Q. If $d = 2$ and Q is smooth and convex then $\{S^t\}$ is non-ergodic. The phase space contains a set of positive measure consisting of caustics, i.e. unit tangent vectors to convex closed curves inside Q such that any tangent vector to the curve after the reflection transforms into a vector tangent to the same curve. This statement was proven by Lazutkin[3]) using KAM-theory. Its multidimensional analogues seem to be unknown.

Quite a different situation arises when the boundary ∂Q is concave. To make the definition more precise, denote by $n(q)$ the normal unit vector to ∂Q directed inside Q. Having the field of normals $n(q)$ one can introduce the self-adjoint operator $\mathscr{F}(q)$ of the second fundamental form which acts in the tangent plane $\mathscr{F}_q$ to ∂Q at the point q.

Definition. The billiard system $\{S^t\}$ is called dispersed if $\mathscr{F}(q) > 0$ in all regular points of ∂Q. Dispersed billiards have very strong instability properties, almost

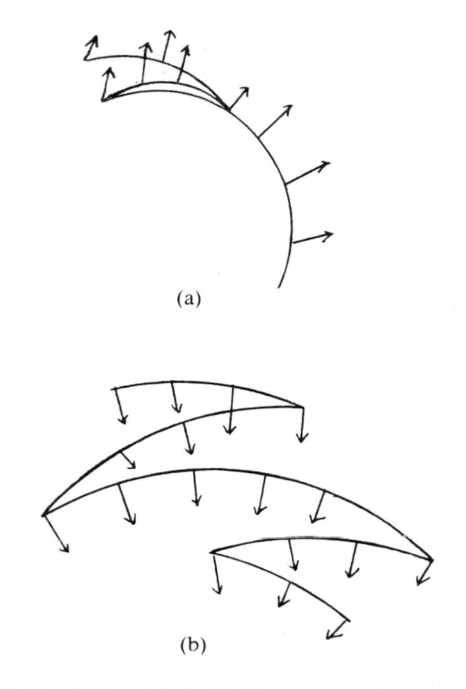

Fig. 1. a) Unstable manifolds; b) stable manifolds.

as strong as in Anosov systems. Apparently Krylov[4]) was the first to notice this fact. An essential difference is the discontinuity which is always present in dispersed billiard systems. We shall see that the discontinuity implies some memory leading to a slower decay of autocorrelation functions.

The first consequence of the instability which follows more or less directly from the properties of dynamics is the existence of stable and unstable manifolds (s.m. and u.m.) for typical points. This statement might look surprising to non-experts. The form of these manifolds for $d = 2$ is drawn in the figure. The cusp singularities correspond to the trajectories, shown in fig. 1, which are tangent to the boundary.

In a paper by Chernoff, more general conditions on ∂Q were found which guarantee the existence of s.m. and u.m. In particular, the results by Chernoff concern the so-called semi-dispersed billiards where $\mathscr{F}(q) \geqslant 0$.

A global s.m. or u.m. is a submanifold with singularities imbedded in the phase space. Following the general strategy, one can introduce "individual Gibbs measure" on a s.m. $\Gamma^{(s)}(x)$ (or on a u.m. $\Gamma^{(u)}(x)$). Denote by $B^{(s)}(x)$ the operator of the second fundamental form of $\tilde{\Gamma}^{(s)}(x) = \pi\Gamma^{(s)}(x)$ where π is the natural projection of M to Q. In the dispersed billiards and many semi-dispersed billiards $B^{(s)}(x) < 0$.

Put $\alpha(x) = \operatorname{tr} B^{(s)}(x)$ and for any two points $y_1, y_2 \in \Gamma^{(s)}(x)$ define

$$a(y_1, y_2) = \int_0^\infty \left[\alpha\left(S^t y_1\right) - \alpha\left(S^t y_2\right)\right] \mathrm{d}t.$$

The integral is finite because the semi-trajectories $\{S^t y_1, t \geqslant 0\}$, $\{S^t y_2, t \geqslant 0\}$ converge exponentially fast to each other as $t \to \infty$. Define the measure on $\Gamma^{(s)}(x)$ whose density with respect to the Riemannian volume on $\Gamma^{(s)}(x)$ is given by $\exp a(x, y)$ with an arbitrary fixed point y.

Assume that a partition ξ of M is given whose elements $C_\xi(x)$ are compact subsets of $\Gamma^{(s)}$.

Lemma. The Liouville measure μ induces on elements $C_\xi(x)$ conditional measures whose densities with respect to the Riemannian volume are given by

$$\frac{\exp\{a(x, y)\}}{\int_{C_\xi(x)} \exp\{a(x, z)\}\,\mathrm{d}z}.$$

Any element $C_\xi(x)$ might be considered as a "configuration space" of a spin system on a half-line under boundary conditions on another half-line. Different elements $C_\xi(x)$ correspond to different boundary conditions. The denominator is an analogue of the partition function. In spite of the fact that the lemma is more or less trivial it presents a deep reconstruction of the initial Liouville measure to the form of conditional Gibbs distribution.

Using DLR-conditions one can say that the Liouville measure μ is the Gibbs state with respect to the potential $a(x, y)$. One should emphasize that the whole reconstruction becomes possible because we have stable and unstable manifolds of the dimension $d - 1$. Now the problem of ergodicity is formulated in the spirit of the theory of phase transitions. Namely, ergodicity means that the limit Gibbs state corresponding to the potential a is unique. A natural expectation of the affirmative answer is based upon the fact that from the point of view of equilibrium statistical mechanics we have a one-dimensional system and the result might follow from an estimation of "the decay of interaction".

However direct arguments give only that, for dispersed billiards or for semi-dispersed billiards where $B(x) > 0$ a.e., all ergodic components have positive measure and on each component the induced transformation of the boundary is a K-automorphism.

The proof of the fact that the ergodic component is unique needs a more heavy geometrical technique. The corresponding statement is known as "the main theorem of theory of dispersed billiards". Now there exist at least four different proofs of it (Sinai, Bunimovich–Sinai, Sinai, Chernoff) based on various but similar ideas. The main difficulty lies in the fact that smooth components of s.m. or u.m. can be arbitrarily small and the usual Hopf argument does not work. The main theorem says that if we begin with a small s.m. component then with a uniformly large probability there are smooth components of the u.m. intersecting the initial component of the s.m. which are approximately of the same order in size as the s.m..

In Bunimovich–Sinai, Markov partitions for two-dimensional dispersed billiards were constructed. Using these partitions one can describe in a more explicit way the representation of the induced transformation as the shift in the space of configurations of a spin system where spin variables take infinitely many values. The complicated nature of elements of the Markov partition is connected with a discontinuous character of dynamics.

The first application of the technique based on Markov partitions is the estimation of auto-correlation functions. Consider the two-dimensional Lorentz gas where a particle moves on a plane between periodically distributed scatterers and reflects from their boundaries. If $v(n)$ is the velocity after n collisions then $|\langle v(n), v(0)\rangle| \leqslant \text{const} \exp\{-n^\gamma\}$ for $\gamma > 0$. For the Lorentz gas with a finite horizon this was shown in Bunimovich–Sinai and was later generalized by Bunimovich.

Billiard systems in concave domains are not the only ones which have strong instability and statistical properties. There exist convex boundaries where the instability manifests itself in a divergence of small convex beams of trajectories due to the fact that, during the dynamics, convex beams transform into concave beams, and the fraction of time when the beam is convex is larger than the fraction of time when it is concave. The first important example of this type was discovered by Bunimovich. We have in mind billiards in a stadium, i.e. in the domain whose boundary consists of two parallel straight segments and two half-circles. In his paper Bunimovich[5]) constructed a big class of such domains. Several other examples were added in[6]).

It turns out that this quasi-exponential asymptotics appears when the configuration of scatterers is highly ordered, i.e. purely periodic. There have been several numerical studies of the exponent γ (Casati, Kubo, Machta and Zwanzig, Bouchaud and Le Doussal) which apparently show that $\gamma < 1$. One can mention that quasi-exponential decay of correlations appears now in such other cases as spin glasses and chemical reactions.

If a configuration of scatterers is random then we have a usual Lorentz gas. Markov partitions are not constructed so far. If it were so then we would have a representation of the Lorentz gas as a random walk in a random environment. In this case it is natural to expect the usual asymptotics $\overline{\langle v(n), v(0)\rangle} \sim \text{const}\, n^{-(d/2)-1}$ to appear.

7. Mechanical models of Brownian motion

One of the main features of dynamical systems of statistical mechanics is the presence of long tails of time-correlation functions and the corresponding divergencies in the expression for transport coefficients. General belief is that such slow decay is connected with recollisions of pairs of particles and the induced memory in the dynamics. Lebowitz suggested several models where one can see this mechanism in a more explicit form. One of them is a mechanical system modelling one-dimensional Brownian motion (Sinai–Soloveichik[7])). Assume that we have a massive particle (m.p.) of mass M, $0 < M < \infty$, moving on the line under reflections with an ideal gas of point particles of mass $m = 1$ distributed according to the equilibrium Gibbs distribution with the inverse temperature β and the density ρ. If $Q(t)$ is the displacement of the m.p., then as was shown in[7,8])

$$0 < \mathscr{D}_0(\beta, \rho) \leqslant \varliminf_{t\to\infty} \frac{\overline{(Q(t))^2}}{t} \leqslant \varlimsup_{t\to\infty} \frac{\overline{(Q(t))^2}}{t} \leqslant \mathscr{D}_1(\beta, \rho) < \infty .$$

Surprisingly enough, $\mathscr{D}_0(\beta, \rho)$, $\mathscr{D}_1(\beta, \rho)$ do not depend on M. Numerical studies (Hasin) show that $\lim_{t \to \infty} \overline{(Q(t))^2}/t$ really depends on M. If it were independent of M then it would follow from the results of Szasz–Toth that $Q(t)/\sqrt{t}$ has in the limit $t \to \infty$ the Gaussian probability distribution. However, the same numerical studies by Hasin show that this is not the case and it is unlikely that the probability distribution for $Q(st)/\sqrt{ts}$, $0 \leqslant s \leqslant 1$, converges as $t \to \infty$ to the Wiener measure. This result can be explained only by the essential role of recollisions. We see that mechanical models of Brownian motion have a more complicated nature than the usual Wiener measure.

References

1) W. Parry and M. Pollicott, An analogue of the prime number theorem for closed orbits of Axiom A flows, Ann. Math. **118** (1983) 573–591.
2) J.-P. Eckmann and D. Ruelle, Ergodic theory of chaos and strange attractors, Rev. Mod. Phys. **57** (1985) 617–656.
3) V.F. Lazutkin, The existence of caustics for the billiard problem in a convex domain, Math. USSR Izv. **37** (1973) 186–216.
4) N.S. Krylov, Works on the Foundation of Statistical Physics (Princeton Univ. Press, Princeton, 1979).
5) L.A. Bunimovich, On the ergodic properties of some billiards, Funct. Anal. Appl. **8** (1974) 73–74; On the ergodic properties of nowhere dispersing billiards, Commun. Math. Phys. **65** (1979) 295–312.
6) M. Wojtkowski, Principles for the design of billiards with nonvanishing Liapunov exponents, Commun. Math. Phys. **105** (1986) 391–414.
7) Ya.G. Sinai and M.R. Soloveichik, One dimensional classical massive particles in the ideal gas, Commun. Math. Phys. **104** (1986) 423–443.
8) S. Szasz and B. Toth, Bounds for the limiting variance of the "heavy particle" in R^1, Commun. Math. Phys. **104** (1986) 445–455.

Physica 140A (1986) 9–25
North-Holland, Amsterdam

TOWARDS A SCALING THEORY OF DRAG REDUCTION

P.G. DE GENNES
Collège de France, 75231 Paris Cedex 05, France

Flexible polymers in dilute solution enhance the viscosity in slow flows. But in strong, rapidly varying, shear fields, they behave *elastically*. A turbulent cascade (from large to small scales) should thus be deeply modified when the elastic stresses become comparable to the Reynold's stress. A (tentative) scaling picture for these effects has been proposed by M. Tabor and the present author[1]): it involves one unknown exponent n relating polymer deformation (λ) and spatial scales (r) in the cascade. We now show that, depending on the control parameters (turbulent power; polymer concentration and molecular weight) the cascade may proceed according to two "scenarios". In the first scenario[1]) the smallest Kolmogorov eddy occurs when the chains are only partly stretched. In the second scenario, the smallest eddies display nearly full chain extension: polymer degradation is expected to be much more serious in the latter case.

We also transpose these ideas to wall turbulence, in the first scenario. At a distance y from the wall, the smallest eddy scale available $r^{**}(y)$ is a *decreasing* function of y (very different from the classical Lumley picture, where it is an increasing function of y). The overall result is again an increase of the buffer layer, provided that the polymer concentration exceeds a – very low but finite – threshold.

We point out finally that the elastic effects discussed here could be present in many other systems: one amusing (although impractical) example is a binary mixture near a consolute point: the concentration fluctuations should be very similar, in this effects, to a polymer solution with coil size ξ (the correlation length), at the overlap concentration c^*.

1. Introduction

Flexible polymers in dilute solution can reduce turbulent losses very significantly[1]). The main (tentative) interpretation of this effect is due to Lumley[2]). He emphasized that remarkable visco-elastic effects can occur only when certain hydrodynamic frequencies become higher than the relaxation rate of one coil $1/T_z$ (the "time criterion"). He then proposed a crucial assumption: namely that, in regions of turbulent flow, the solution behaves as a fluid of *strongly enhanced viscosity* – presumably via regions of elongational flow. On the other hand, Lumley noticed that – for turbulent flow near a wall – the viscosity in the laminar sublayer near the wall should remain *low*: this last observation does agree with the viscometric data on dilute, linear polymers in good solvents, which show shear thinning[3]). Starting from the above assumptions, and performing a careful

0378-4371/86/$03.50

matching of velocities and stresses beyond the laminar layer, Lumley was able to argue that the overall losses in pipe flow should be reduced.

This explanation has been rather generally accepted. However, it is now open to some question: in recent experiments with polymer injection at the center of a pipe, one finds drag reduction in conditions where wall effects are not involved[4,5]).

This observation prompted Tabor and the present author to try a completely different approach[6]): namely to discuss first the properties of homogeneous, isotropic, 3 dimensional turbulence *without any wall effect*, in the presence of polymer additives. This "cascade theory" is described in section 2. The central idea is that polymer effects at small scales (high frequencies) are not described by a viscosity, but by an *elastic modulus*. The general notion of elastic behavior at high frequencies is classical for molten, entangled chains[3]). We claim that it is also important for dilute polymers. In our approach the viscosity effects are mostly trivial, and we do not even discriminate between solvent viscosity and solution viscosity.

In section 3 we return to wall turbulence, and try to set up a modified version of the Lumley approach, where, at each distance y from the wall, we have a cascade, but it is truncated elastically. This gives a law for the minimum eddy size r^{**} versus distance y which is qualitatively different from Lumley's viscous effect. But the net result is still an enhancement of the intermediate "buffer layer". We expect drag reduction from this, although we have not carried out the detailed analog of Lumley's matching.

In section 4 we list some more general systems which can show drag reduction on turbulent flow. Some of the systems are dominantly elastic while others are probably dominantly viscous.

Our whole discussion is very qualitative. But, even at this modest level, it leads to a surprisingly rich classification of possible cascades and flows. For instance, in bulk turbulence, we have three control parameters: a) the dissipation per unit mass ε; b) the polymer chain length, or equivalently the number of monomers per chain N; c) the monomer concentration c (or the number of coils/cm^3 $c_p = c/N$).

This 3 dimensional parameter space can be split into regions where different "scenarios" for the cascade should occur. The identification of these scenarios is a natural aim for future experimental research.

2. The cascade theory

2.1. *The time criterion*

Our starting point is the classical view of Kolmogorov[7]) for homogeneous, isotropic, 3 dimensional turbulence.

At each spatial scale (r) there is a characteristic fluctuating velocity $U(r)$, related to (r) by the condition

$$\frac{U^3(r)}{r} = \varepsilon = \text{constant}. \tag{2.1}$$

We must compare the characteristic frequencies $U(r)/r$ to the Zimm relaxation rate of one coil[8])

$$\frac{1}{T_z} \cong \frac{kT}{\eta_0 R^3}, \tag{2.2}$$

where k is the Boltzmann constant, T the temperature, η_0 the solvent viscosity, and R the gyration radius of the coil at rest. We focus our attention on linear, flexible, neutral polymers in good solvents, where the Flory law holds[9,10])

$$R \cong N^{3/5} a, \tag{2.3}$$

a being a monomer size, and N the number of monomers per coil.

At large scales r, the hydrodynamic frequency U/r is smaller than $1/T_z$. But, if we go down in scale, we may reach a value $r = r^*$ where the two frequencies become equal. Thus

$$U(r^*)T_z = r^*. \tag{2.4}$$

Solving the coupled equations (2.1), (2.4) we arrive at

$$r^* = \left(\varepsilon T_z^3\right)^{1/2} \qquad (\sim N^{2.7}\varepsilon^{1/2}), \tag{2.5}$$

$$U^* = \left(\varepsilon T_z\right)^{1/2} \qquad (\sim N^{0.9}\varepsilon^{1/2}). \tag{2.6}$$

Note that r^* (and U^*) depend on molecular weight, but not on concentration.

Another parameter of interest for our discussion will be the Reynolds number computed *at the scale* r^*, namely

$$\mathrm{Re}^* = \frac{U^* r^*}{\nu} = \frac{\varepsilon T_z^2}{\nu} \qquad (\sim N^{3.6}\varepsilon), \tag{2.7}$$

where $\nu = \eta_0/\rho$ is the kinematic viscosity, ρ being the fluid density.

The condition (2.4) defining r^* is the natural expression of Lumley's time criterion[2]). Most interesting viscoelastic effects will occur only at frequencies higher than $1/T_z$, or equivalently at scales $r < r^*$.

2.2. *The passive range*

If our solute macromolecules are very dilute, their reaction on the flow pattern is weak. Thus we expect that there exists a certain interval $r^* > r > r^{**}$ where eddies of size r are still described by the Kolmogorov cascade, but where the polymer begins to undergo strong distortions.

2.2.1. *Information from laminar flows*

Let us concentrate first on *elongational flows*. Two regimes have been probed in some detail.

(i) *Constant shear rate* $\dot{\gamma}$ (fig. 1). Here one expects that the coils are essentially unperturbed when $\dot{\gamma} < 1/T_z$ and that they are strongly elongated when $\dot{\gamma} > 1/T_z$[11,12,10]). This sharp coil–stretch transition has been observed in an important series of experiments by Keller and coworkers[13,14]). We might think at first sight that this transition should show up in turbulent flows and bring in some important nonlinear effects. We shall argue, however, that this is not correct: for the situations of interest, where $\dot{\gamma}$ (as seen by the molecule) is rapidly varying in time, the coil stretch–transition disappears completely.

(ii) *Variable shear rates*. Two examples are shown on figs. 2a, b: in fig. 2a we have a duct with a periodic modulation of the cross section. In fig. 2b we consider the converging flow towards a very thin (~ 500 Å) capillary. The main conclusion, obtained first from detailed calculations by Daoudi[15]) for a coil under periodic modulations, is the following: whenever the modulation frequency is higher than the Zimm relaxation rate, the coil *follows passively the deformations of the local volume element*. The dimensionless elongation λ of the coil is entirely fixed by the flow.

Of course there are still some local modulations in the coil shape: more specifically, if we call ω the modulation frequency, we can define subunits of p monomers such that

$$\frac{1}{T_z(p)} = \omega, \qquad T_z(p) = \frac{\eta_0 a^3 p^{1.8}}{kT}, \tag{2.8}$$

where $T_z(p)$ is the Zimm type of the subunit. Inside each subunit we still have some relaxation, but at larger scales the coil deforms affinely.

These considerations have been transposed long ago[15]) to the converging flow of fig. 2b. Here the shear rate at a distance r from the entrance point is of order

$$\dot{\gamma}(r) = \text{const.}\, r^{-3}. \tag{2.9}$$

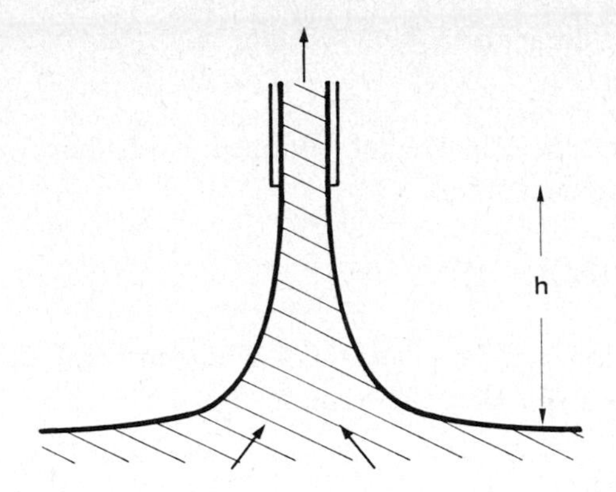

Fig. 1. The tubeless siphon: a dilute solution of long, flexible polymers can be sucked up over large intervals h (~ 20 cm). This shows the dramatic effects of the polymer induced stresses in longitudinal shear flows.

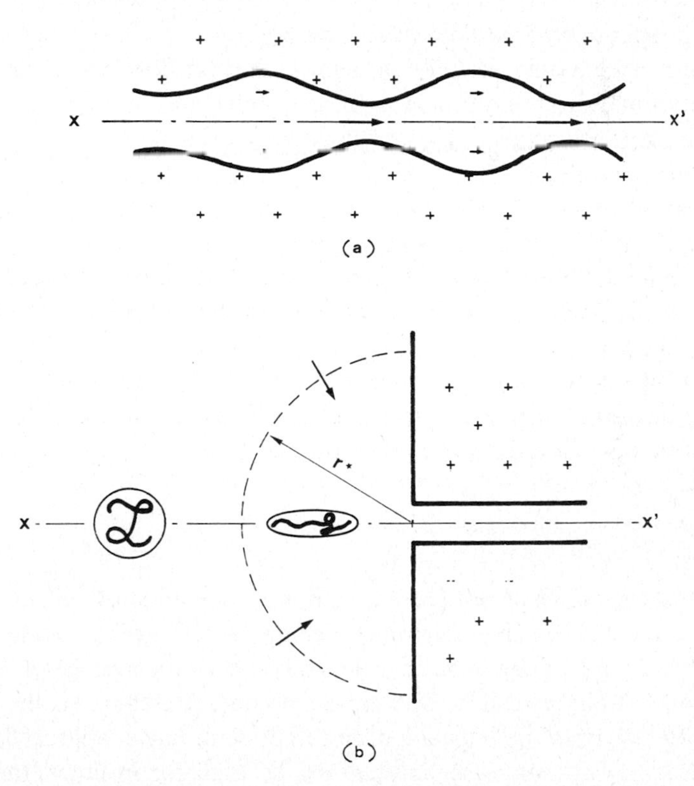

Fig. 2. Two (approximate) examples of longitudinal shear flows. a) Tube with periodic constriction; b) entry of a capillary. In both cases the molecules which lie exactly on the axis of symmetry (xx') experience a purely longitudinal shear.

Viscoelastic effects occur for $r < r^*$, where:

$$\dot{\gamma}(r^*) = 1/T_z. \tag{2.10}$$

At distances $r > r^*$ the coils are not deformed. At distances $r < r^*$ they deform affinely, and their elongation is

$$\lambda = \left(\frac{r^*}{r}\right)^2 \quad (3d). \tag{2.11a}$$

A similar discussion can be given for a 2 dimensional flow, where the fluid converges towards a slit. Here the result is

$$\lambda = \frac{r^*}{r} \quad (2d). \tag{2.11b}$$

Thus for a simple longitudinal flow, there is always a simple power law relating the striction parameter r^*/r and the polymer elongation.

2.2.2. *Transposition to the Kolmogorov cascade*[6])

We now make a bold assumption: namely that for flows which are admixtures of longitudional shear and simple shear, and which are turbulent, there remains a power law between polymer elongation and spatial scale

$$\lambda(r) = \left(\frac{r^*}{r}\right)^n, \tag{2.12}$$

where n is an unknown exponent. The extreme case quoted in eq. (2.11) suggests that $n < 2$. In practical discussions we shall attempt to use $n = 1$ and $n = 2$ as possible values.

It may be worthwhile to return here to the definition of molecular elongation. For one particular coil, the dimensionless elongation $\lambda_{(1)}$ can be constructed from the radius of gyration $\tilde{R}$ in the distorted state

$$\lambda_{(1)}^2 = \frac{\tilde{R}^2}{R^2}, \tag{2.13}$$

where R is the radius at rest [eq. (2.3)]. For an ensemble of coils in a turbulent flow, we should select the coils which belong, in real space, to eddies of size r (and which do not belong to any smaller eddy). Then the average of $\lambda_{(1)}$ over this population is what we call λ. One immediate question concerns the distribution of $\lambda_{(1)}$ values within the population. In the present, naive, approach, we assume that this distribution is reasonably *narrow*, so that, for instance, the average of the squares is given by

$$\langle \lambda_{(1)}^2 \rangle = k_2 \lambda^2 \tag{2.14}$$

with a constant k_2 which is independent of r, and of order unity. It may well be that this assumption of narrow distributions is not satisfactory, and that independent exponents $n_1 n_2 \ldots$ would be required to describe the successive moments of the elongation. But, at our present level of ignorance, we shall omit this complication.

To summarize: we expect that any coil, located in eddies of size $r < r^*$, will follow passively the surrounding volume element, and will deform accordingly. We postulate a scaling law describing this effect [eq. (2.12)] in terms of a single exponent n.

This simple behavior, with affine deformation, and without significant reactions of the coils on the flow, will hold in a finite interval of spatial scales $r^* > r > r^{**}$. We call this interval the *passive range*.

2.3. *The first scenario: semi-stretched chains*

2.3.1. *Stresses in a partly stretched state*

Let us now consider the reaction of the polymers on the flow: when our coils are stretched by a factor λ, a certain elastic energy is stored in each of them. We shall, for the moment, assume that the coils are significantly stretched ($\lambda \gg 1$) but that they are still far from full extension: the deformed size $\tilde{R}$ is still much smaller than the contour length Na: or returning to eqs. (2.3–13):

$$1 \ll \lambda \ll N^{2/5}. \tag{2.15}$$

Since $N \sim 10^4$–10^5 in typical experiments, we may go up to $\lambda \sim 100$.

What is the elastic energy of a coil in this regime? In the harmonic approximation, it would be proportional to $(\lambda - 1)^2$ (or equivalently to λ^2, since $\lambda \gg 1$). However, for coils in good solvents, the harmonic approximation is not very good: the shape, and the monomer repulsions, change with λ. This has been analyzed by Pincus[16]).

The final result is an anharmonic energy

$$F_1 \cong kT\lambda^{5/2} \quad (1 \ll \lambda \ll N^{2/5}) \tag{2.16a}$$

or a free energy per unit volume

$$F_{\mathrm{el}} = \frac{c}{N} kT\lambda^{5/2} = G\lambda^{5/2}, \tag{2.16b}$$

where G has the dimensions of one elastic modulus, and is linear in concentration. Equivalently the restoring force on one spring is

$$f_1 \cong \frac{kT}{R}\lambda^{3/2} \tag{2.17a}$$

and the stress due to c/N springs/cm^3 is

$$\tau \cong \frac{c}{N} f_1 \lambda R \cong F_{\mathrm{el}}. \tag{2.17b}$$

2.3.2. *The elastic limit* r_1^{**}

Whenever τ is much smaller than the Reynolds stresses ρU^2, the reaction of the polymer on the flow is negligible. If we go towards smaller and smaller scales, λ and τ increase, while the local Reynolds stress $\rho U^2(r)$ decreases. Thus, at a certain scale r_1^{**}, the two stresses become equal.

$$G[\lambda(r_1^{**})]^{5/2} = \rho U^2(r_1^{**}). \tag{2.18}$$

Using the Kolmogorov formula (2.1) this leads to

$$\frac{r_1^{**}}{r^*} = X^{v}, \quad v = \left(\frac{5n}{2} + \frac{2}{3}\right)^{-1}, \tag{2.19}$$

where we have introduced a dimensionless parameter

$$X \equiv \frac{G}{\rho U^{*2}} (\sim cN^{-2.8}\varepsilon^{-1}). \tag{2.20}$$

X is a natural measure of concentration effects. Consider a typical case with $U^* = 1$ m/s, $\rho = 1$ g/cm^3, $ca^3 = 10^{-4}$, $a = 2$ Å $T = 300$ K, $N = 10^4$. Then $G = 60$ erg/cm^3 and $X = 6 \times 10^{-3}$. Depending on our choice of n, the exponent v might be in range $\frac{1}{3}$–$\frac{1}{6}$.

2.3.3. *Comparison with the Kolmogorov limit*

Can we effectively go down to the elastic limit r_1^{**}, retaining the inertial cascade all the time? The Kolmogorov scheme is always truncated by viscous dissipation at a scale r_κ defined by[7]):

$$\frac{r^*}{r_\kappa} = (\mathrm{Re}^*)^{3/4}. \tag{2.21}$$

The elastic limit is observable only if $r_1^{**} > r_\kappa$. Comparing (2.19) and (2.21) we find that this condition is equivalent to

$$X > (\mathrm{Re}^*)^{-3/4v}. \tag{2.22}$$

The condition (2.22) has no counterpart in the Lumley scheme, where drag reduction was expected to occur at arbitrarily low polymer concentrations. Here, we do find a minimum X, or equivalently, a concentration threshold c_m, below which the polymer should have no visible effect. Using eqs. (2.20) for X and (2.7) for Re*, we can find how the threshold concentration c_m depends on N and ε:

$$c_m \sim N^{2.8-2.7/v}\varepsilon^{1-3/4v}. \tag{2.23}$$

Since $1/v$ is expected to be in the range 3–6, c_m should be a strongly decreasing function of N. It may well be that for long chains, c_m is extremely small and practically invisible. But systematic experiments at variable N might detect the threshold c_m.

2.3.4. *Ultimate fate of the turbulent energy*

At the scales $r = r^{**}$, the liquid should behave like a strongly distorted rubber, carrying elastic waves (longitudinal and transverse) with comparable kinetic and elastic energies.

At the scales $r < r^{**}$ inertial nonlinearities are not, by themselves able to generate smaller structures. But the *elasticity* is also nonlinear, and may have the ingredients required to produce shock waves in the "rubber". This process may imply a further thinning of scales – down to the natural width of a shock front. This scheme is completely conjectural: we may deal with a sea of rarefaction waves and shock waves, with very peculiar couplings between them: Thus, we do not know the ultimate fate of the turbulent energy.

On the whole, it is tempting to assume that the formation of new eddies is strongly restricted for $r < r^{**}$. Using the description of eqs. (2, 18–23), this would then lead to a truncation in the cascade at $r = r_1^{**}$ (the index 1 stands for the "first scenario" with partly elongated chains, which was the only one discussed above).

The result is

$$r_1^{**} = r^* X^v, \tag{2.24}$$

$$r_1^{**} \sim N^{2.7-2.8v}\varepsilon^{1/2-v}c^v. \tag{2.25}$$

We expect at last a qualitative change, and possibly a truncation, of the cascade at $r = r_1^{**}$. Note that r_1^{**} should increase rapidly with N.

Recent experiments on pipe turbulence or with planar mixing layers[17]), using strophometry or laser Doppler anemometry, do suggest that polymer additives can suppress certain small scales. But detailed proposals, such as (2.25), for the truncation, remain to be checked.

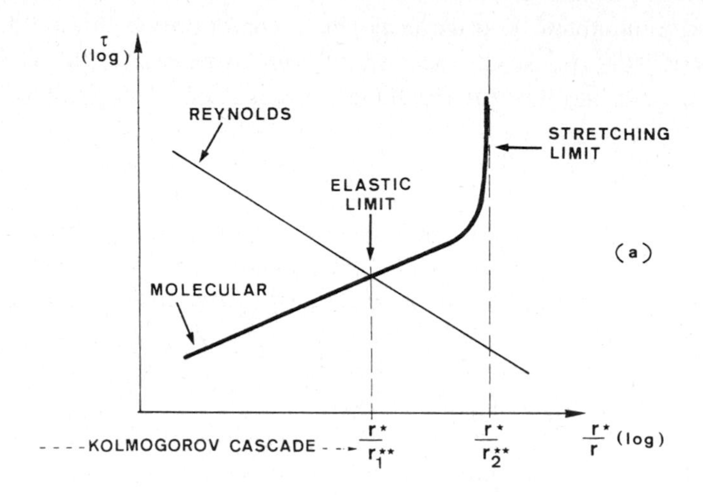

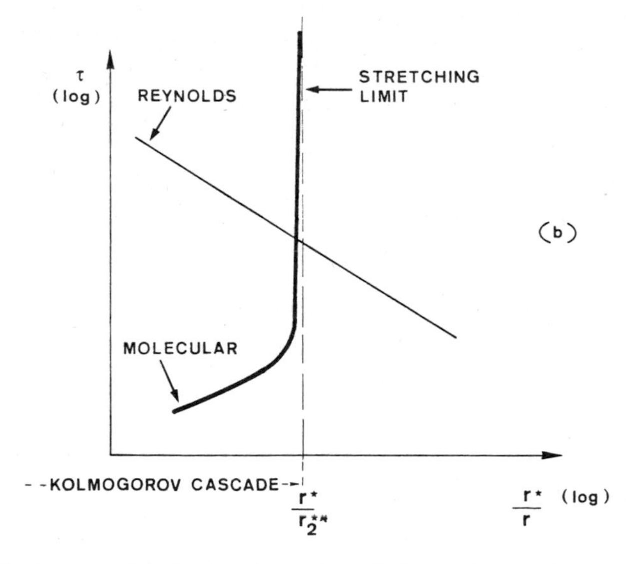

Fig. 3. Reynolds stresses and molecular stresses in a Kolmogorov cascade with a dilute polymer solution. a) In the first scenario, the Kolmogorov cascade is truncated at a certain scale r_1^{**}: at this scale, the chains are only partly stretched; b) in the second scenario, the trunction occurs (at a scale r_2^{**}) where the chains are totally stretched.

2.4. *The second scenario: strongly stretched chains*

Let us return to the passive regime, and assume that we can reach very small scales r, so that the polymer under flow may become fully elongated. Returning to (2.15), we see that this corresponds to $\lambda = \lambda_{max} \cong N^{2/5}$. Inserting this value into the scaling law (2.12), we arrive at a certain characteristic scale

$$r_2^{**} = R^* N^{-2/(5n)} . \tag{2.26}$$

We call r_2^{**} the stretching limit. The meaning of r_2^{**} is made more apparent by the construction of fig. 3 which is a log–log plot of the stresses as a function of the striction ratio r^*/r. When λ gets very close to λ_{max}, the stresses tend to diverge (detailed laws for this have been constructed on simple molecular models): this fixes r_2^{**}.

In this second scenario the behavior at scales r smaller than r^{**} is probably rather different: viscous forces are very important when we have rod particules immersed in a fluid; it may well be that viscosity takes over at $r = r_2^{**}$.

3. Flow near a wall

3.1. *A reminder on pure fluids*[7])

Wall turbulence is characterised by a velocity U_τ such that the wall stress is equal to ρU_τ^2. The average velocity profile for a pure fluid is represented in fig. 4:

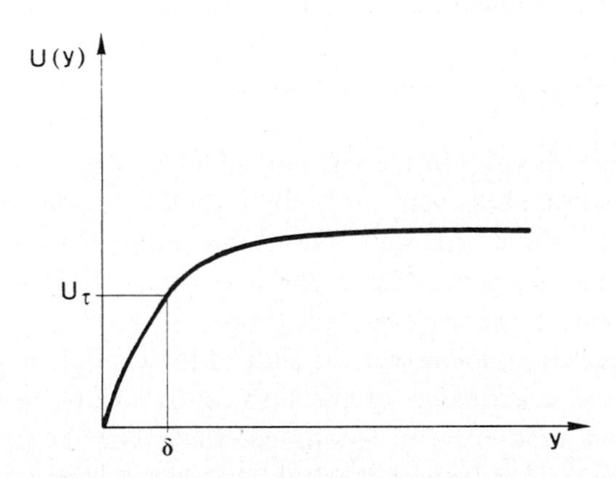

Fig. 4. Average velocity profile in the wall turbulence of a pure fluid (qualitative). y is the distance from the wall, and $\delta = \nu/U_\tau$ is the thickness of the laminar sublayer. At $y \gg \delta$ the profile is logarithmic.

it is linear very close to the wall, in a laminar sublayer of size

$$\delta \cong \frac{\nu}{U_\tau} \tag{3.1}$$

then, at long distances from the wall, it is logarithmic. At any distance y ($\gg \delta$) from the wall we find eddies with waves vectors k lying between two limits: the largest size (or the smallest wave vector $k_{\min}^{-1}$) is given by the distance y itself; the smallest size (or the largest wave vector $k_{\max}^{-1}$) is given by a Kolmogorov limit r_κ similar to eq. (2.21). The main difference is that now the dissipation ε is a function of the distance to the wall.

$$\varepsilon \cong U_\tau^3/y. \tag{3.2}$$

A formula, equivalent to (2.21, for $r_\kappa(y)$ is

$$r_\kappa = \left(\frac{\nu^3}{\varepsilon}\right)^{1/4} = \delta^{3/4} y^{1/4}. \tag{3.3}$$

Thus the eddies expected around the observation point (y) have wave vectors in the range

$$\frac{1}{y} < k < \frac{1}{\delta^{3/4} y^{1/4}}. \tag{3.4}$$

We call the lower limit the geometrical limit, and the upper limit the viscous limit. The range of turbulent eddies is shown on fig. 5.

3.2. *The Lumley model for polymer solutions* [2])

Lumley kept the viscosity in the laminar sublayer at its Newtonian value, but he assumed an increased viscosity η_t in the turbulent regions. This implied that, at any distance y from the wall, the Kolmogorov limit $r_\kappa(y)$ was shifted upwards; in the log/log plot of fig. 5, the new viscous limit is represented by a dotted line, parallel to the original limit (slope -4 according to eq. (3.3)). The higher the concentration, the higher the shift of the viscous limit.

The net result is a shrinkage of the turbulent domain: beyond the laminar layer, we now find a buffer layer, of size increasing with the polymer concentration. It is then natural to expect that the turbulent losses be *reduced*; Lumley gave a detailed argument to show this[2]). Note that the whole effect occurs at arbitrarily low c: as soon as we add some polymer, the viscous limit shifts.

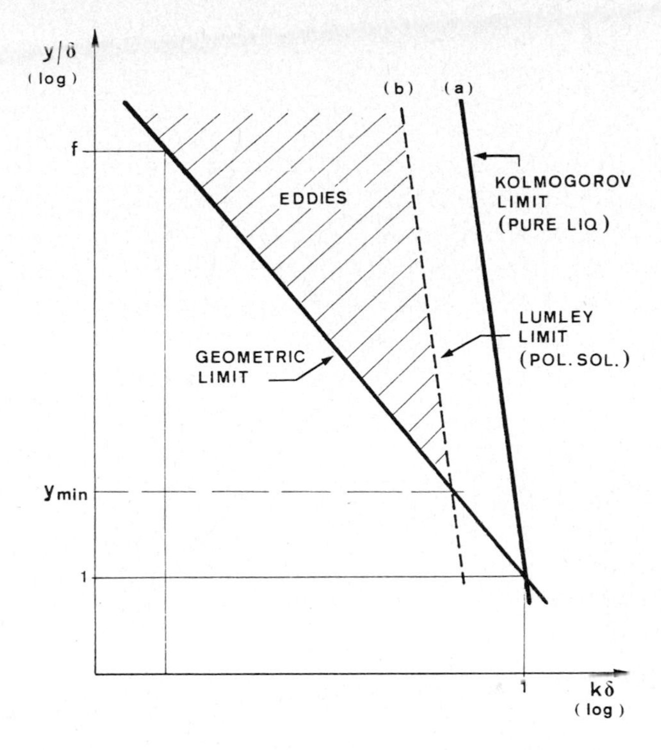

Fig. 5. The distribution of eddy wave vectors k at various distances y from the wall. a) Pure fluid: the smallest eddies are defined by the viscous limit of Kolmogorov. b) Polymer solution in the Lumley scheme: the limit is shifted to the left but the slope of the limiting line remains the same (-4).

3.3. *A modified version in the first scenario*

Let us assume that the discussion of section (2.3) holds: the chains are partly stretched, but never fully extended, and there is an elastic limit r_1^{**}, given, in terms of the local energy dissipation ε, by eqs. (2.21, 24, 25). Inserting eq. (3.2) for $\varepsilon(y)$, this gives

$$r_1^{**}(y) \sim U_\tau^{3(1/2-v)} y^{v-1/2} N^{2.7-2.8v} c^v. \tag{3.5}$$

Let us further assume that no singular dissipation occurs at scales smaller than the elastic limit. Then we are led to a modified Lumley construction, shown in fig. 6. There is an elastic limit, described by the dotted line, and the slope of this line is reversed in sign. At very low c, the new limiting line intersects the geometrical limit at thickness y smaller than the laminar sublayer δ: in this regime, we expect no macroscopic effect. But, beyond a certain threshold c_o (o stands for: onset),

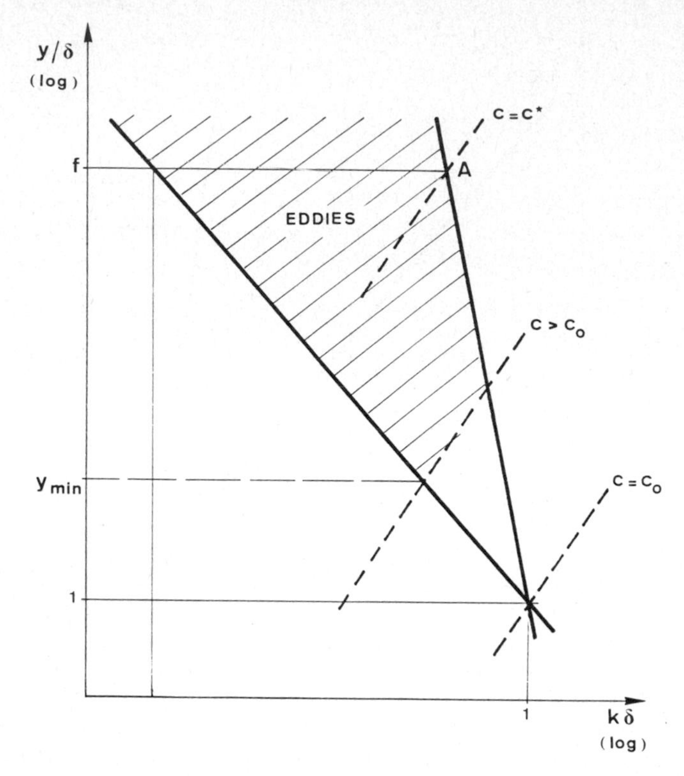

Fig. 6. The plot of fig. 5, transposed to the present model, in the first scenario. The maximum wave vector of turbulent eddies can depend upon an elastic limit: note the difference in slope with fig. 5. At very low polymer concentrations $c < c_{\mathrm{o}}$, this limit is never relevant. Above $c = c_{\mathrm{o}}$, it becomes relevant. When the elastic limit reaches point A the polymer concentration just corresponds to coils in contact ($c = c^*$).

the turbulent domain is actually truncated. The scaling structure of c_{o} can be extracted from (2.5), writing $r_1^{**}(\delta) = \delta$. c_{o} is the wall analog of the concentration c_{m} introduced in our discussion of homogeneous turbulence. In many practical cases, c_{o} will be very small. But, conceptually, the existence of c_{o} reveals a significant difference with the Lumley model.

If we increase c beyond c_{o}, the elastic limit shifts upward in fig. 6, and we again find a buffer layer: we conjecture that, in this regime, dissipation is reduced.

At a certain higher concentration, the intersection of the elastic limit and the Kolmogorov limit reaches point A. At this moment, the largest eddies cease to satisfy the time criterion. We expect that any addition of polymer beyond this point will be less effective.

It turns out that the polymer concentration associated with point A is a familiar object of solution theory: it is the concentration.

$$c^* = N/R^3 = a^{-3}N^{-3/4} \tag{3.6}$$

at which neighboring coils begin to overlap. Thus, in the interval $c_o < c < c^*$ we expect a significant drag reduction, increasing steadily with c. At concentrations beyond c^*, the behavior is more complex, because some eddies are operating in a Newtonian regime, and the trivial increase of Newtonian viscosity due to the polymer may become the leading feature.

4. Conclusions and perspectives

4.1. *Elasticity versus viscosity*

The main idea of this paper is that flexible coils, even in the dilute regime, behave elastically at high frequencies: a description of small eddies in terms of a renormalised (real) viscosity is not entirely adequate.

A Kolmogorov cascade remains unaltered by polymer additives only down to a certain limit r^{**} where the polymer stresses balance the Reynolds stresses. In the second scenario, this occurs at full stretching. Chemical degradation is probably severe in this last case.

The fate of the turbulent energy below the limiting scale r^{**} is unclear: in the first scenario, it might result from a delicate balance between elastic shock waves and rarefaction waves. In the second scenario, viscous effects may be immediately dominant.

4.2. *Other elastic systems*

a) Apart from linear polymers, many other systems may show an elastic behavior at high frequencies. One obvious example is *branched* polymers, which may show an improved resistance to degradation. However, the regime of extreme deformation (the analog of the second scenario) is reached at much small elongations λ: we need a special discussion of branched systems, and the details of the branching statistics (e.g. the number of internal loops) will be important.

b) Another family of interesting systems is obtained with *binary fluid mixtures* near a consolute point. Ruiz and Nelson[18]) have studied situations where, at the starting point, the two fluids are not fully mixed. They pointed out that the resulting concentration gradients induced elastic stresses in the system, and that these stresses *react on the turbulent field*.

c) We may also consider a modified version of the Nelson–Ruiz problem, where the binary mixture is macroscopically homogeneous; there remains an effect due to fluctuations of concentration, with a characteristic size ξ (the correlation length). These fluctuations are remarkably similar to polymer coils: ξ is the analog of the unperturbed polymer size R. There is a characteristic time (first introduced long ago by Ferrell and Kawazaki) which is the exact analog of the Zimm time (eq. (2.2)) with $R \to \xi$. There are elastic tensions (although, to our knowledge, the analog of the nonlinear Pincus formula, eq. (2.16), has not been worked out). The fluctuations span all space: this is the analog of a polymer system at the concentration of first overlap $c^* = N/R^3$. In practice, the main limitation is that, to reach long times T_z (needed to satisfy the time criterion) we need a large ξ, i.e. a good temperature stabilisation, in the one phase region, near the critical point.

d) Similar effects may exist in the 2-phase region, where small droplets are constantly broken (and coalesce) in turbulent flow: here, the elasticity of the droplets can be expressed in terms of one interfacial tension, following the classic papers of Taylor. Very near to the critical point, however, we should return to a microscopic description: the basic formulas are described by Aronowitz and Nelson[19]).

4.3. *The third scenario*

All our discussions have used, as a starting point, the "time criterion" of Lumley: spatial effects have been ignored because the coil size R is usually much smaller than the smallest eddy size. However, after stretching, the situation may be different. A deformed coil of length λR, with $\lambda \gg 1$, may become comparable to the eddy size r: this may lead to a third scenario – an idea first suggested to us by E. Siggia. It is easy to construct the scale r_3^{**} at which we would meet this effect. It is far more difficult to perceive what would happen at even smaller scales. In practice, the third scenario should occur only if we achieve turbulence even at very small scales, with extremely large Reynold's numbers; it will probably be associated with strong degradation.

4.4. *Open problems*

All our discussions are extremely conjectural a) The very existence of a single exponent n characterizing the elongation at different scales (eq. (2.12)) is unproven. (Among other things, we might need a family of exponents to give separate scaling law for the various moments λ^m.) b) The behavior of the cascade beyond the elastic threshold is entirely unclear. c) The intermittency features which are omitted in the Kolmogorov description of the cascade may be much more significant for polymer solutions than for pure liquids.

However, we feel that the proposed scheme, the classification of scenarios, and even the tentative scaling laws which we propose, should be of some use to guide future experiments.

Apart from the elastic systems discussed here, there also exist some interesting questions with rigid rods: they do have an orientational entropy, which leads to some analog of an elastic energy in aligning flows: but they cannot follow the local deformation affinely, as done by the coils. Viscous dissipation is thus much stronger. It may be that the Lumley scheme holds for rods.

Acknowledgments

This work originated from an unforgettable discussion with M. Tabor: my only regret is that geographical distance did not allow us to continue it together. I have also greatly benefited from written exchanges with E. Siggia and D. Nelson, (which are reflected in section 4) and with Y. Rabin, A number of earlier discussions with A. Ambari and A. Keller are also gratefully acknowledged

References

1) J. Lumley, Ann. Rev. Fluid Mech. (1969), p. 367.
2) J. Lumley, J. Polymer Sci. **7** (1973) 263.
3) W. Graessley, Adv. Polymer Sci. **16** (1974) 1.
4) W. McComb and L. Rabie, Phys. Fluids (USA) **22** (1979) 183.
5) H.W. Bewersdorff, Rheol. Acta **21** (1982) 587, **23** (1984) 552.
6) M. Tabor and P.G. de Gennes, Europhys. Lett. (to be published).
7) H. Tennekes and J. Lumley, A First Course in Turbulence (MIT press, Cambridge MA, 1972).
8) W. Stockmayer, in Fluides Moléculaires, (R. Balian and G. Weill, eds.) (Gordon and Breach, New York, 1976). P.G. de Gennes, J. Chimie Phys. **76** (1979) 763.
9) P. Flory, Principles of Polymer Chemistry (Cornell U.P., Ithaca, 1971).
10) P.G. de Gennes, Scaling Concepts in Polymer Physics (Cornell U.P., Ithaca, 1984), 2nd printing.
11) J. Hinch, Phys. Fluids **20** (1977) S22.
12) P.G. de Gennes, J. Chem. Phys. **60** (1974) 5030.
13) D. Pope and A. Keller, Coll. Pol. Sci. **256** (1978) 751.
14) J. Odell and A. Keller, in Polymer-Flow Interactions, Y. Rabin, ed. (AIP Press, New York, 1986), p. 33.
15) S. Daoudi and F. Brochard, Macromolecules **11** (1978) 751.
16) P. Pincus, Macromolecules **9** (1976) 386.
17) A. Ambari and E. Guyon, Ann. NY Acad. Sci. **404** (1982) 87. R. Scharf, Rheol. Acta **24** (1985) 272.
18) R. Ruiz and D. Nelson, Phys. Rev. A **24** (1981) 2727.
19) J. Aronovitz and D. Nelson, Phys. Rev. A **29** (1984) 2012.

Physica **140A** (1986) 26–34
North-Holland, Amsterdam

HAMILTONIAN AND PHENOMENOLOGICAL MODELS OF MICROEMULSIONS

B. WIDOM, K.A. DAWSON and M.D. LIPKIN
Department of Chemistry, Cornell University, Ithaca, NY 14853, USA

We review briefly a phenomenological microemulsion model, its phase diagram, and its interfacial tensions. We then describe a lattice model of a microemulsion, based on a prescribed Hamiltonian equivalent to that of an Ising model with competing nearest- and further-neighbor interactions. Its phase diagram and interfacial tensions are compared with those in the phenomenological model.

A microemulsion is a homogeneous solution of an oil and water, rendered mutually soluble by a third component, a surfactant, which is amphiphilic: partly hydrophilic and partly hydrophobic (lipophilic). Among the most striking properties of microemulsions, both in the laboratory and in the theoretical models described here, are that they can coexist in three-phase equilibrium with two other phases, one of which is almost pure oil and the other almost pure water, and that the tensions of the interfaces between pairs of these coexisting phases is "ultralow": typically the small fraction 10^{-4} or 10^{-5} of the tension of the oil–water interface without surfactant.

Balbuena et al.[1]) recently described the phase equilibria in a phenomenological model of such a microemulsion[2]), that model having evolved from earlier ones of Talmon and Prager[3]) and of de Gennes et al.[4,5]). Borzi et al.[6]) calculated the interfacial tensions in the same model.

One imagines, in the model, that the space is divided into cubical cells, each of edge-length ξ, and each filled with either oil or water, at random, subject only to the restriction that a prescribed fraction ϕ be oil-filled and $1 - \phi$ water-filled. There are ρ surfactant molecules per unit volume, but they are confined to the microscopic interfaces between oil-filled and water-filled cells (the "surfactant film"). The model is further defined by its free-energy density $F(\phi, \rho, \xi)$,

$$\begin{aligned} F(\phi, \rho, \xi) &= kT[\phi \ln \phi + (1 - \phi) \ln (1 - \phi)]/\xi^3 + 6\sigma_0\phi(1 - \phi)/\xi \\ &\quad - \rho kT \ln [\phi(1 - \phi)/\xi\rho] + \rho D[1 - \Lambda(2\phi - 1)\xi]/2\xi^2, \qquad \xi > a, \\ F(\phi, \rho, \xi) &= \infty, \qquad \xi < a, \end{aligned} \tag{1}$$

0378-4371/86/$03.50 © Elsevier Science Publishers B.V.
(North-Holland Physics Publishing Division)

where a (> 0) may be thought of as a microscopic cutoff enforcing upon ξ a lower limit that is of molecular size; σ_0 (> 0) as the oil–water interaction energy per unit area of contact; D (> 0) as a parameter that determines the free energy of curvature of the surfactant film (this free energy increasing with decreasing ξ); and Λ as a curvature-bias parameter that determines a favored curvature: a flat film when $\Lambda(2\phi - 1) \leqslant 0$, a film that is concave toward water and convex toward oil when $\Lambda > 0$ and $\phi > \frac{1}{2}$, and the opposite when $\Lambda < 0$ and $\phi < \frac{1}{2}$.

For any chemical composition ϕ, ρ the equilibrium cell size ξ is that which minimizes $F(\phi, \rho, \xi)$ in (1). A phase in which the minimum F is achieved with $\xi > a$ is identified as the microemulsion. A phase in which the minimum F is achieved only at $\xi = a$ is a normal, non-microemulsion phase; such are the oil- and water-rich phases with which a microemulsion may be in equilibrium in this model.

When $\Lambda = 0$, and when, in addition, the oil and water activities are equal (as they would be when $\Lambda = 0$ and $\phi = \frac{1}{2}$), the oil and water components play symmetrical rôles. When a three-phase equilibrium occurs in such a state, the microemulsion is a middle phase β, in which the oil fraction $\phi_\beta = \frac{1}{2}$, and is symmetrically related to an upper phase α in which the oil fraction $\phi_\alpha > \frac{1}{2}$ (oil-rich phase) and a lower phase γ in which the oil fraction $\phi_\gamma = 1 - \phi_\alpha < \frac{1}{2}$ (water-rich phase). In this symmetrical phase equilibrium the tensions $\sigma_{\alpha\gamma}$, $\sigma_{\alpha\beta}$, and $\sigma_{\beta\gamma}$ of the $\alpha\gamma$, $\alpha\beta$, and $\beta\gamma$ interfaces are found[6]) to be

$$\sigma_{\alpha\gamma} \cong 2\sigma_{\alpha\beta} = 2\sigma_{\beta\gamma} \cong 10^{-3} kT/a^2; \tag{2}$$

i.e., they are ultralow, corresponding well to experimental observation.

For these symmetrical states ($\Lambda = 0$ and equal oil and water activities), the phase diagram[1]) in the plane of surfactant chemical potential μ_s and dimensionless curvature-energy parameter $q = D/a^2kT$, at fixed $\sigma_0 a^2/kT$, is as shown schematically in fig. 1. The phase labeled $\alpha\gamma$ is a homogeneous solution of oil and water with a high concentration of surfactant (μ_s large) and with $\xi = a$. The region $\alpha + \gamma$, at lower μ_s, is that of coexisting oil-rich and water-rich phases α and γ, in each of which, again, $\xi = a$. The line CC′ is the line of α, γ critical points. When $q = 0$ these are the only possible equilibrium phases; there is no three-phase equilibrium. Likewise, at $q = 0$, the $\alpha\gamma$ tension is not low, except near the critical point C. What is remarkable about the ultralow tensions of the interfaces in microemulsion phase equilibria is that they are so low even when the phases, judged by their chemical compositions, are far from a critical point. Thus, if in fig. 1 we look only at $q = 0$, there is not yet an identifiable microemulsion.

At large enough q ($q \approx 1$) a new phase, β, appears, which is the microemulsion. Like $\alpha\gamma$, it is a homogeneous oil–water solution, but is distinguished from

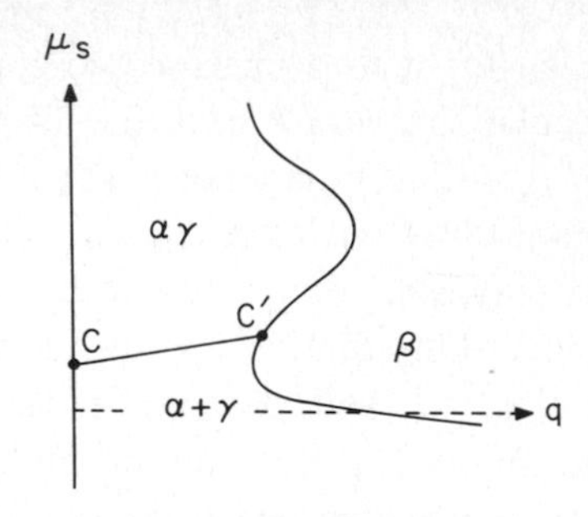

Fig. 1. Phase diagram of the phenomenological model in the plane of surfactant chemical potential μ_s and curvature-energy parameter $q = D/a^2kT$. CC' is a line of critical points; C' is a critical endpoint. The phases $\alpha\gamma$, $\alpha + \gamma$, and β are a homogeneous oil–water solution, phase-separated oil- and water-rich solutions, and the microemulsion.

$\alpha\gamma$ by a larger structural parameter, $\xi > a$, and by its transition to $\alpha + \gamma$ being of first rather than second order, so that it can coexist with α and γ. Far enough along the three-phase line, i.e., for large enough q (which still need not be much greater than 1), the α and γ phases are nearly pure oil and water. The tensions of the interfaces are even then ultralow.

Note the crucial rôle played in this story by the curvature-energy parameter q: without it, and until it is about 1, there is no microemulsion. Note also that C′ in fig. 1 is a critical endpoint: the line of β, $\alpha\gamma$ transitions (which are typically of first order near C′ but can be second order) and the three-phase line have the same slope at C′, but different from that of the critical line CC′. These aspects of the phase diagram will be recalled when we obtain the analogous diagram for a model system defined by a spin Hamiltonian, to which we now turn.

Consider a spin-$\frac{1}{2}$ Ising model with a spin variable $s_i = \pm 1$ at each site (pictorially, + or −; fig. 2). Every nearest-neighbor ++ pair may now be thought of as a water molecule, every −− pair as an oil molecule, and every +− pair as an amphiphile. Each lattice site belongs simultaneously to c such molecules, where c is the coordination number of the lattice: $c = 4$ for the square lattice in two dimensions (fig. 2) and $c = 6$ for the simple-cubic lattice in three dimensions. Thus, only like ends of molecules (only + ends or only − ends) are present at any lattice site. The + ends of amphiphiles are therefore oriented only toward water molecules or toward the + ends of other amphiphiles, while the − ends are oriented only toward oil or toward the − ends of other amphiphiles. That is what is responsible for their behavior as amphiphiles[7,8]).

If the model were simply that, then it would be equivalent to the nearest-neighbor, spin-$\frac{1}{2}$ Ising model, with a dictionary of translations[7]) as given in table I. The spin–spin interaction parameter and magnetic field in the former arc J and H, in conventional notation, while θ is the temperature. We define

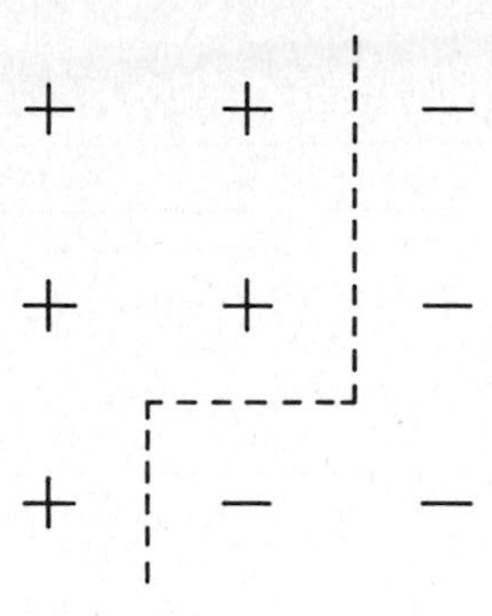

Fig. 2. A spin configuration of the Ising model. The dashed line separates regions of (+) spin from regions of (−) spin.

TABLE I

Equivalences between the spin-$\frac{1}{2}$ Ising model with nearest-neighbor interactions and the model oil–water–amphiphile mixture.

Ising	Oil–water–amphiphile
$j\ (=J/k\theta)$	$\frac{1}{2}\ln[(z_o z_w)^{1/2}/z_a]$
$h\ (=H/k\theta)$	$\frac{1}{4}c\ \ln(z_w/z_o)$

$j = J/k\theta$ and $h = H/k\theta$. The coordination number of the lattice is again c, while z_o, z_w, and z_a are the activities of the oil, water, and amphiphile in the equivalent mixture model. The only possible phases are a homogeneous oil–water mixture with a high concentration of the amphiphile [large $z_a/(z_o z_w)^{1/2}$, equivalent to high temperature θ in the Ising model], this phase being the disordered Ising paramagnet, which we now call $\alpha\gamma$, and separate oil-rich and water-rich phases at low concentration of the surfactant [low $z_a/(z_o z_w)^{1/2}$, equivalent to low θ], which are the ordered ferromagnetic phases of the Ising model and which we call $\alpha + \gamma$. These are analogous to what one sees along the $q = 0$ axis of fig. 1, where the surfactant chemical potential μ_s is analogous to the present $\ln[z_a/(z_o z_w)^{1/2}]$ and where the critical point C is analogous to the present critical-solution point or to the Ising-model Curie point.

For the same reasons that there was no microemulsion at $q = 0$ in fig. 1, there is none, as yet, here: we are lacking the analog of the curvature-energy parameter q. We introduce that[9]) by noting that here, too, we have a surfactant film, shown as dashed lines in fig. 2, which separates water-coherent (+) from oil-coherent (−) regions, and that there is a bend in this film whenever two or more amphiphiles share the same lattice site. We thus introduce an interaction energy $Q(1 + \lambda)$ for every pair of amphiphile (+) ends that meet at the same lattice site

TABLE II
Equivalences between the extended spin-$\frac{1}{2}$ Ising model and the microemulsion model.

Ising	Microemulsion
$j\ (= J/k\theta)$	$\frac{1}{2}(c-1)q + \frac{1}{2}\ln[(z_o z_w)^{1/2}/z_a]$
$h\ (= H/k\theta)$	$\frac{1}{8}c(c-1)\lambda q + \frac{1}{4}c\ln(z_w/z_o)$
$m\ (= M/k\theta)$	$-\frac{1}{4}q$
$l\ (= L/k\theta)$	$-\frac{1}{4}\lambda q$

and $Q(1-\lambda)$ for every pair of amphiphile $(-)$ ends that do. If T is the temperature of our mixture model (distinguished from θ of the equivalent Ising model), then Q/kT, which we call q, is analogous to q in the earlier phenomenological model; while λ, which distinguishes one sense of bend from the other [$(+)$ inside and $(-)$ outside, or the opposite], is analogous to the earlier curvature-bias parameter Λ — more precisely, to the dimensionless Λa.

This more general model is again equivalent to a spin-$\frac{1}{2}$ Ising model[9]), but now, besides interaction of energy $-Hs_1$ with a magnetic field H and nearest-neighbor spin–spin interactions of energy $-Js_1s_2$, in this equivalent Ising model there are also interactions of energy $-2Ms_1s_3$, where M is another coupling constant and s_1, s_3 are any pair of spins separated by the distance $\sqrt{2}\,a$ (a = lattice spacing) on either the square or simple-cubic lattice, interactions of energy $-Ms_1s_3$ with s_1, s_3 any pair of spins separated by $2a$, and interactions of energy $-Ls_1s_2s_3$ with s_1, s_2, s_3 the spins at any connected triple of sites. Table II is the new table of equivalences. It reduces to table I when $m(= M/k\theta) = l(= L/k\theta) = 0$ (in Ising-model language) or $q = \lambda = 0$ (in microemulsion-model language).

Most of what is so far known about this extended Ising model[9–12]) is for the symmetric states $h = l = 0$ (in Ising-model language) or $\lambda = \ln(z_w/z_o) = 0$ (in microemulsion-model language), in the mean-field approximation. On the simple-cubic lattice with sites at integer x, y, z, the free energy F in this approximation, for $h = l = 0$, is given by

$$F/k\theta = \sum_{x,y,z} \left\{ \tfrac{1}{2}s\left[-6(j+5m)s - (j+12m)\,\Delta^2 s - m\,\Delta^4 s\right] + \frac{1+s}{2}\ln\frac{1+s}{2} + \frac{1-s}{2}\ln\frac{1-s}{2} \right\}, \tag{3}$$

where s is $s_{x,y,z}$, the mean spin at x, y, z; $\Delta^2 s$ is the second lattice difference $(\Delta_x^2 + \Delta_y^2 + \Delta_z^2)s_{x,y,z}$ with $\Delta_x^2 s_{x,y,z} = s_{x+1,y,z} + s_{x-1,y,z} - 2s_{x,y,z}$, etc.; and $\Delta^4 s$ is the fourth lattice difference $(\Delta^2)^2 s$. The equilibrium spin density $s_{x,y,z}$ is

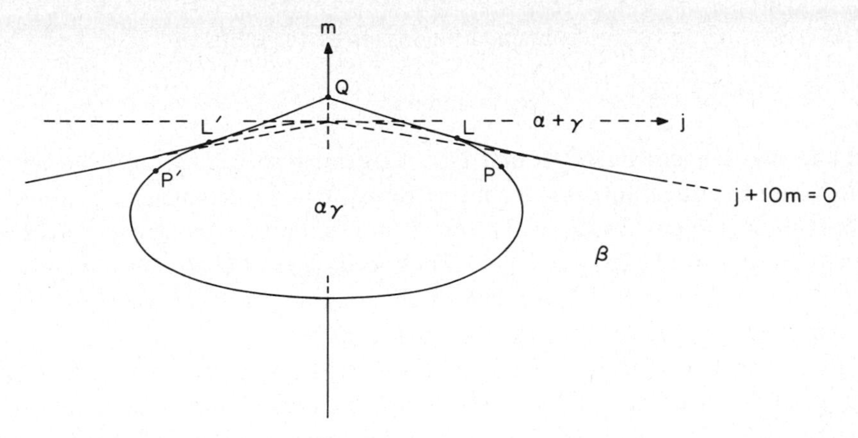

Fig. 3. Phase diagram of the extended Ising model in the j, m plane. Regions $\alpha\gamma$, $\alpha + \gamma$, and β are paramagnetic, ferromagnetic, and modulated.

that which minimizes $F/k\theta$ in (3), and so is a solution of

$$-6(j + 5m)s - (j + 12m)\,\Delta^2 s - m\,\Delta^4 s + \frac{1}{2}\ln\frac{1+s}{1-s} = 0. \qquad (4)$$

Although the solutions of (4) (and of its analog for the square lattice in two dimensions) are not yet completely catalogued, much is now known about them[10-12]) and we have begun to construct the phase diagram[10-12]) and determine the interfacial structures and tensions[10]). Both isotropic and anisotropic phases are found, with structures both commensurate and incommensurate with the lattice. Among the anisotropic phases are the same layered ones as are found in the ANNNI model[9,13-18]).

A simplified phase diagram in the m, j plane for the simple-cubic lattice, still for $h = l = 0$, is in fig. 3. Any line

$$(Y^2 - 6)m + Yj = 1 \quad \text{(critical line)} \qquad (5)$$

in the j, m plane is a critical line for periodic ordering with wavelengths p_1, p_2, p_3 in the x, y, z directions such that

$$\frac{1}{2}Y = \cos\frac{2\pi}{p_1} + \cos\frac{2\pi}{p_2} + \cos\frac{2\pi}{p_3}; \qquad (6)$$

the side of the line (5) that includes the origin $j = m = 0$ is the disordered side and the side away from the origin is the ordered side. The envelope of lines (5) for all possible Y, $-6 \leqslant Y \leqslant 6$, then outlines the disordered (paramagnetic) region of the j, m plane, which contains the origin $j = m = 0$ in its interior. This is the region $\alpha\gamma$ in fig. 3. Its boundary consists of the ellipse

$$6j^2 + (12m + 1)^2 = 1 \quad (\alpha\gamma \text{ boundary}) \tag{7}$$

and the two tangent lines QL and QL', tangent, respectively, at L and at L', which are the critical lines (5) for the extremes $Y = 6$ (ferromagnet, $\cos(2\pi/p_1) = \cos(2\pi/p_2) = \cos(2\pi/p_3) = 1$) and $Y = -6$ (antiferromagnet, $\cos(2\pi/p_1) = \cos(2\pi/p_2) = \cos(2\pi/p_3) = -1$). The coordinates of Q are $j = 0$, $m = 1/30$; of L are $j = 2/7$, $m = -1/42$; and of L' are $j = -2/7$, $m = -1/42$. In the related ANNNI model,[13-18]) L is the Lifshitz point.

The region $\alpha + \gamma$ in fig. 3 is that of the Ising ferromagnet, while region β is that of all the modulated phases with all possible periodicities, which we identify collectively as the microemulsion in the equivalent amphiphile-solution model. The boundary (first-order-transition line) between β and $\alpha + \gamma$ approaches its asymptote $j + 10m = 0$ exponentially rapidly[9,13]) as $j \to \infty$. The line $j + 10m = 0$ is that along which no work is required to increase the area of the surfactant film[9]). The β, $\alpha + \gamma$ boundary is the three-phase line, along which the microemulsion β is in equilibrium simultaneously with an oil-rich phase α and a water-rich phase γ. As $j \to \infty$ along that boundary, the fraction of bonds in the β phase that are $+-$ bonds (i.e., the mole fraction of amphiphile in the microemulsion) approaches $1/9 = 11\%$, while the α and γ phases approach pure oil and pure water.

Points P and P' are at $j = 4/11$, $m = -1/22$ and $j = -4/11$, $m = -1/22$, respectively; they are critical points for orderings with $Y = \pm 4$. Just outside the paramagnetic boundary between P and L (i.e., for $4 < Y < 6$) the stable phases consist of ferromagnetic layers ($\cos(2\pi/p_1) = \cos(2\pi/p_2) = 1$), modulated with wavelength p_3 in the direction perpendicular to the layers, such that $\cos(2\pi/p_3) = \frac{1}{2}Y - 2$. In the middle of the arc LP, where $Y = 5$, we have $p_3 = 6$; i.e., just outside the paramagnetic boundary at that point the stable phase has ferromagnetic $(+)$ layers alternating by threes with ferromagnetic $(-)$ layers. This point in the middle of the arc LP is that at which the elliptical boundary intersects the line $j + 10m = 0$, and for $0.366 < j < \infty$ along the three-phase line this phase of $+$, $-$ ferromagnetic layers alternating by threes is the β phase that coexists with $\alpha + \gamma$.

The phase diagram is symmetric about $j = 0$. States with any m and $j < 0$ are the antiferromagnetic images (alternate spins change sign) of those with the same m and $j > 0$. Only states with $j > 0$ and $m < 0$ (competing ferromagnetic nearest-neighbor and antiferromagnetic further-neighbor interactions) are relevant as microemulsions.

In the present model the closest analog of the surfactant chemical potential μ_s of the phenomenological model is $kT \ln[z_a/(z_o z_w)^{1/2}]$, which is the difference

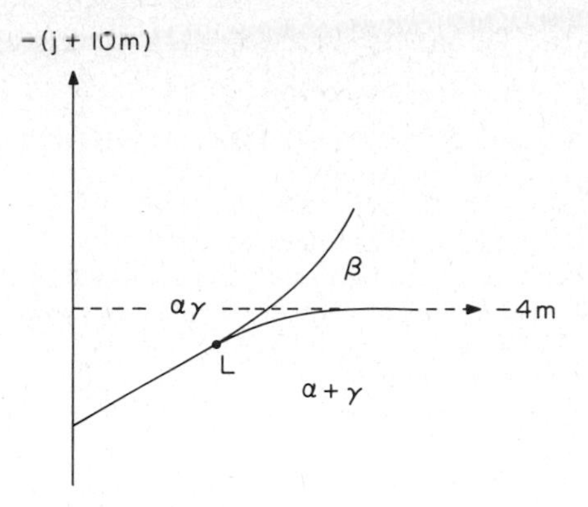

Fig. 4. Same as fig. 3, simplified, in the $-(j + 10m)$, $-4m$ plane.

between the chemical potential of the amphiphile and the average chemical potential of the oil and water. By table II, this is proportional to and increases with $-(j + 10m)$ (for the simple-cubic lattice, for which $c = 6$). The analog of the former q, as already remarked, is the present q, which by table 2 is $-4m$. To obtain the analog of fig. 1 for the present model we then replot fig. 3 in the $-(j + 10m)$, $-4m$ plane, as in fig. 4, where now only the relevant parts are shown. We see a rough analogy with fig. 1 (there the 0 of the scale of ordinates is arbitrary), although at the Lifshitz point L in fig. 4 the three-phase boundaries have a common slope[18]), while at the critical endpoint C′ in fig. 1 they do not.

Along the β, $\alpha + \gamma$ three-phase boundary the tension $\sigma_{\alpha\gamma}$ ($\approx 2\sigma_{\alpha\beta} = 2\sigma_{\beta\gamma}$) vanishes both at L and in the limit $-m$, $j \to \infty$, and so is maximal somewhere between. Dawson[10]) has calculated $a^2\sigma_{\alpha\gamma}/kT$ (together with the interfacial profiles) along the whole three-phase line and finds its maximal value to be

$$\max a^2\sigma_{\alpha\gamma}/kT = 1.8 \times 10^{-3}. \tag{8}$$

This agrees well with (2), and confirms that these models account readily for the ultralow tensions in microemulsion phase equilibria.

This and other lattice models[19-24]) appear to be promising approaches to microemulsions, as is also a continuum model[25]) with interactions resembling those here.

Acknowledgements

This work was supported by the National Science Foundation and the Cornell University Materials Science Center, and by the award of a Lindemann Trust Fellowship to K.A.D.

References

1) P. Balbuena, C. Borzi and B. Widom, Physica **138A** (1986) 55.
2) B. Widom, J. Chem. Phys. **81** (1984) 1030.
3) Y. Talmon and S. Prager, J. Chem. Phys. **69** (1978) 2984.
4) P.G. de Gennes and C. Taupin, J. Phys. Chem. **86** (1982) 2294.
5) J. Jouffroy, P. Levinson and P.G. de Gennes, J. Physique **43** (1982) 1241.
6) C. Borzi, R. Lipowsky and B. Widom, J. Chem. Soc. Faraday Trans. II **82** (1986) 1739.
7) J.C. Wheeler and B. Widom, J. Am. Chem. Soc. **90** (1968) 3064.
8) B. Widom, J. Phys. Chem. **88** (1984) 6508.
9) B. Widom, J. Chem. Phys. **84** (1986) 6943.
10) K.A. Dawson, submitted to Phys. Rev.
11) M.D. Lipkin and K.A. Dawson, to be submitted.
12) K.A. Dawson and P. Balbuena, to be submitted.
13) M.E. Fisher and W. Selke, Phys. Rev. Lett. **44** (1980) 1502.
14) W. Selke and P.M. Duxbury, Z. Phys. B **57** (1984) 49.
15) T. DeSimone and R.M. Stratt, Phys. Rev. B **32** (1985) 1537.
16) P. Bak and J. von Boehm, Phys. Rev. B **21** (1980) 5297.
17) M.H. Jensen and P. Bak, Phys. Rev. B **27** (1983) 6853.
18) C.S.O. Yokoi, M.D. Coutinho-Filho and S.R. Salinas, Phys. Rev. B **24** (1981) 4047.
19) J.C. Wheeler, ACS DCSC symposium, Chicago, September 1985, paper no. 154.
20) A. Robledo, Europhys. Lett. **1** (1986) 303.
21) C. Varea and A. Robledo, Phys. Rev. A **33** (1986) 2760.
22) A. Robledo, C. Varea, and E. Martina, J. Physique Lett. 46 (1985) 967.
23) M. Teubner, preprint; R. Pandit, C. Ebner, K. Chen and C. Jayaprakash, 6th Int. Symp. on Surfactants in Solution, New Delhi, August 1986 (K.L. Mittal, General Chairman), abstract ME 4.
24) M. Schick and W.-H. Shih, Phys. Rev. B **34** (1986) 1797.
25) M.M. Telo da Gama and J.H. Thurtell, J. Chem. Soc. Faraday Trans. II **82** (1986).

Physica **140A** (1986) 35–43
North-Holland, Amsterdam

DECAY OF METASTABLE AND UNSTABLE STATES: MECHANISMS, CONCEPTS AND OPEN PROBLEMS

K. BINDER

Institut für Physik, Universität Mainz, D-6500 Mainz, Postfach 39 80, Fed. Rep. Germany

The decay of metastable states is initiated by a nucleation process in which a free energy barrier is overcome, while in unstable states fluctuations can grow unaffected by such a barrier. The first part of this review considers the significance of the *spinodal line* separating the metastable and unstable regime. A *Ginzburg criterion* for the validity of the spinodal line, as well as for the mean-field theory of nucleation and the linear theory of spinodal decomposition will be discussed. Evidence from simulations and experiments will be mentioned.

The second part considers the scaling behavior of the structure function (and droplet or domain size distribution, etc.) in the late stages of the decay: do we understand on which factors the associated exponents depend? Is there a universality as in critical phenomena?

1. Introduction and overview

In this talk, we consider a phase diagram of a gas–fluid system or a binary mixture, as shown schematically in fig. 1, plotting the plane temperature T-order parameter ψ of the transition (in a mixture, ψ is the concentration difference $\phi_{\mathrm{crit}} - \phi$, etc.). In this plane, the two branches $\phi_{\mathrm{coex}}^{(1)}$ and $\phi_{\mathrm{coex}}^{(2)}$ of the coexistence curve meet in the critical point T_c, ϕ_{crit}; they separate the one-phase region from the two-phase region. In the former, thermal equilibrium states are homogeneous, while in the latter, thermal equilibrium requires mixed phase states of macroscopic domains which are in the states $\phi_{\mathrm{coex}}^{(1)}$ and $\phi_{\mathrm{coex}}^{(2)}$.

Suppose now a quenching experiment, where the system at time $t = 0$ suddenly is brought from an initial equilibrium state at T_0 in the one-phase region to a state inside this miscibility gap. This macroscopically homogeneous initial state now is not thermodynamically stable, however, and hence phase separation starts. According to the mean-field (Cahn–Hilliard[1])) theory of phase separation kinetics, two mechanisms are distinguished: for $\phi_{\mathrm{coex}}^{(1)} < \phi < \phi_{\mathrm{sp}}$ (cf. fig. 1), the system has to overcome the nucleation barrier associated with the formation of a critical size droplet; here the spinodal curve $\phi_{\mathrm{sp}}(T)$ is defined from the condition $\chi_T^{-1}|_{\phi_{\mathrm{sp}}} \equiv (\partial^2 F'(\phi)/\partial\phi^2)_T|_{\phi_{\mathrm{sp}}} = 0$, F' being a hypothetical free energy of the one-phase states in the two-phase region. Assuming $F'(\phi)$ to have the familiar

0378-4371/86/$03.50

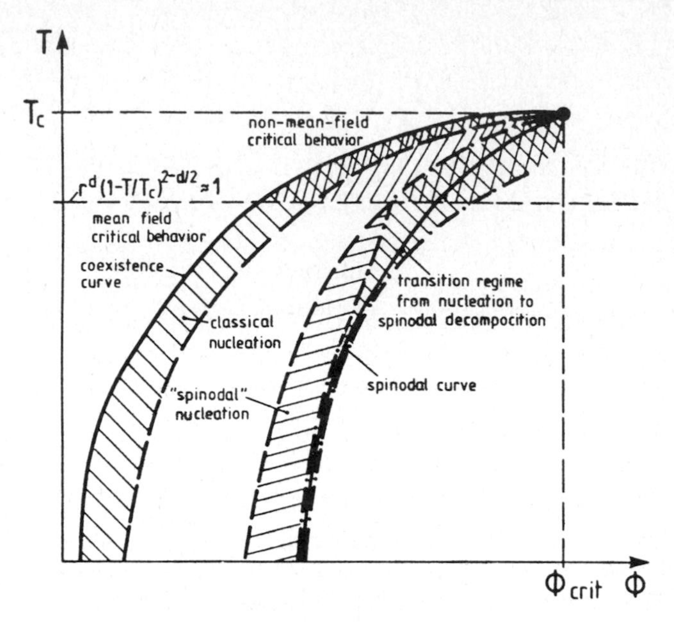

Fig. 1. Various regimes in the temperature-concentration plane of a binary mixture near T_c (only concentrations $\phi < \phi_{crit}$ are shown). The regime inside the two dash–dotted lines around the spinodal line $\phi = \phi_{sp}(T)$ is the regime where a gradual transition from nucleation to spinodal decomposition occurs. The regime between the coexistence curve and the left of the two broken curves is described by classical nucleation theory. In this regime, a crossover from mean-field-like critical behavior to non-mean-field behavior occurs for $r^d(1 - T/T_c)^{(4-d)/2} \approx 1$ (r is the interaction range). The regime between the right broken curve and the left dash–dotted curve is the regime of "spinodal nucleation" (due to ramified rather[7]) than compact droplets); it exists only in the regime of mean-field critical behavior. In the regime to the right of the right dash–dotted regime the linearized theory of spinodal decomposition[1]) holds for the early stages; this regime also exists in the region of mean-field critical behavior. From Binder[2]).

double-well form $(\partial^2 F'(\phi)/\partial\phi^2) > 0$ for $\phi_{coex}^{(1)} < \phi < \phi_{sp}$ (metastable region), while $\chi_T^{-1} = (\partial^2 F'(\phi)/\partial\phi^2)_T < 0$ for $\phi_{sp} < \phi$ (unstable region). In the latter regime, long wavelength fluctuations get spontaneously amplified as the time proceeds after the quench (*spinodal decomposition*[1])).

Now the assumption of double-well free energies $F'(\phi)$ is so common practice that many workers consider this notion of metastable and unstable branches of a free energy density as a fundamental concept of general validity. However, a thermodynamic potential such as the free energy $F(\phi)$ is well-defined in thermal equilibrium only, and must then be a convex function of ϕ: this fact excludes multiple minima! Equilibrium statistical mechanics hence should not yield any branches of $\mathscr{F}(\phi)$ lying above the double-tangent construction, and, in fact, it does not[3]): such branches really are only artefacts of mean-field approximations; the interpretation of these branches in terms of metastable and unstable states really is a kind of "folklore" lacking any fundamental justification by statistical mechanics[4-6]). However, mean-field approximations do become justifiable for systems with long range interactions; this is also true for the mean-field description of metastability. As a consequence, some remnants of the singular behavior associated with the spinodal $\{\chi_T(\phi) \to \infty$ as $\phi \to \phi_{sp}(T)\}$ do occur also in systems with long but finite range of the forces. While this fact has been understood qualitatively for a long time[3,5]), it has only recently been investigated in detail by a Ginzburg criterion[2]) as well as more detailed analytical[7]) and numerical[8,9]) calculations. This work, which also has bearing on the understanding of systems such as polymer mixtures[10,11]), will be reviewed in the first part of this talk: we shall see that several regimes with rather distinct kinetic behavior can be distinguished in the T–ϕ phase diagram, as anticipated in fig. 1.

In the second part, we shall be concerned with the behavior in the very late stages of phase separation and related processes where large ordered domains grow out of disordered initial configurations[12-14]). In particular, we present phenomenological scaling considerations for the structure factor (and related quantities, such as the order parameter distribution function, etc.) and corresponding evidence from computer simulations. It is argued that a single exponent suffices to characterize the growth behavior, unlike scaling behavior for critical phenomena.

2. Ginzburg criteria for nucleation and spinodal decomposition

According to homogeneous nucleation theory, for $\delta\phi \equiv \phi_{ms} - \phi^{(1)}_{coex} \ll \phi^{(2)}_{coex} - \phi^{(1)}_{coex}$ the creation of a d-dimensional spherical droplet out of a metastable phase at concentration ϕ_{ms} involves a free energy barrier ΔF^* [2])

$$\Delta F^* = \left[\frac{S_d}{d}\right]^d \left[\frac{d-1}{V_d}\right]^{d-1} \frac{f_{int}^d \chi_T^{d-1}}{\left[\phi^{(2)}_{coex} - \phi^{(1)}_{coex}\right]^{2(d-1)}} \left[\frac{\delta\phi}{\phi^{(2)}_{coex} - \phi^{(1)}_{coex}}\right]^{-(d-1)}, \quad (1)$$

where S_d and V_d are surface area and volume of a d-dimensional unit sphere, and f_{int} is the interfacial free energy between two coexisting bulk phases at concentrations $\phi^{(1)}_{coex}, \phi^{(2)}_{coex}$.

We now focus on a system described by a free energy functional

$$\mathscr{F}\{\phi(\boldsymbol{x})\} = \int d^d x \{ f[\phi(\boldsymbol{x})] + r^2 k_B T [\nabla\phi(\boldsymbol{x})]^2 \}, \quad (2)$$

with a large but finite interaction range r. The Ginzburg criterion[15]) tells us that for $r^d(1 - T/T_c)^{(4-d)/2} \gg 1$, mean-field critical behavior occurs $\{\chi_T =$

$\Gamma(1 - T/T_c)^{-\gamma}$, $\phi^{(2)}_{\text{coex}} - \phi^{(1)}_{\text{coex}} = B(1 - T/T_c)^{\beta}$, correlation length $\xi_{\text{coex}} = \xi_0(1 - T/T_c)^{-\nu}$ with exponents $\gamma = 1, \beta = \frac{1}{2}, \nu = \frac{1}{2}, \xi_0 \propto r, f_{\text{int}} \propto r(1 - T/T_c)^{3/2}$), and hence

$$\Delta F^*/k_B T_c \propto r^d(1 - T/T_c)^{(4-d)/2}\{\delta\phi/[\phi^{(2)}_{\text{coex}} - \phi^{(1)}_{\text{coex}}]\}^{-(d-1)}. \tag{3}$$

The prefactor $r^d(1 - T/T_c)^{(4-d)/2}$ has exactly the same form as in the Ginzburg criterion: hence in the mean-field critical region the nucleation barriers are always large, even away from $\phi^{(1)}_{\text{coex}}$: when $\delta\phi$ gets larger ΔF^* no longer depends on $\delta\phi$ in the simple power law form, Eq. (3), but the *scale* for the function $\Delta F^*/k_B T_c$, is still given by the factor $r^d(1 - T/T_c)^{(4-d)/2} \gg 1$ right up to the spinodal.

Conversely, for $r^d(1 - T/T_c)^{(4-d)/2} < 1$ non-mean-field critical behavior occurs, with $d\nu = \gamma + 2\beta$ and $f_{\text{int}} = \hat{f}(1 - T/T_c)^{(d-1)\nu}$, and hence

$$\frac{\Delta F^*}{k_B T_c} = \left(\frac{S_d}{d}\right)^d \left(\frac{d-1}{V_d}\right)^{d-1} \frac{\hat{f}\Gamma^{d-1}}{B^{2(d-1)}} \left\{\frac{\delta\phi}{\phi^{(2)}_{\text{coex}} - \phi^{(1)}_{\text{coex}}}\right\}^{-(d-1)}. \tag{4}$$

In eq. (4), there is no longer any temperature dependence and moreover the amplitude combination $\hat{f}\Gamma^{d-1}/B^{2(d-1)}$ is a universal constant of order unity. Now the nucleation barriers are large for small $\delta\phi$ only: $\Delta F^*/k_B T$ becomes of order unity long before the spinodal is reached, and thus there is a broad concentration regime where a gradual transition from nucleation to spinodal decomposition occurs.

For treating nucleation at larger $\delta\phi$, we recall the field-theoretic description of the nucleation barrier. One starts from the fact that the barrier corresponds to a saddle point in configuration space, which is interpreted as a metastable state containing just one droplet of critical size. As in the calculation of the free energy F from eq. (2),

$$F = -k_B T \ln Z = -k_B T \ln \int d\{\phi\} \exp\left[-\mathscr{F}\{\phi(\boldsymbol{x})\}/k_B T\right], \tag{5}$$

we consider a functional integral but restrict the phase space to states near a metastable minimum of the coarse-grained free energy density $f(\phi_{\text{ms}})$ and to nonuniform solutions $\phi(\boldsymbol{x})$ tending to $\phi = \phi_{\text{ms}}$, the constant concentration of the metastable phase, at large distances from the origin. Just as in the mean-field theory of stable or metastable states where one replaces the actual distribution $\exp\{-\mathscr{F}\{\phi\}/k_B T\}$ by a delta function $\delta(\phi - \bar{\phi})$, where $\bar{\phi}$ minimizes $\mathscr{F}$, the mean-field theory of nucleation amounts to replacing the functional integral by a saddle point solution of $\delta\mathscr{F}\{\phi(\rho)\}/\delta\phi(\rho) = 0$, where $\phi(\rho)$ is a spherically symmetric concentration profile of the critical droplet. In the mean-field critical region, this yields

$$\Delta F^*/k_B T_c \propto r^d (1 - T/T_c)^{(4-d)/2} \{(\phi_{ms} - \phi_{sp})/(\phi_{coex}^{(2)} - \phi_{coex}^{(1)})\}^{(6-d)/2}. \tag{6}$$

Thus for $r^d(1 - T/T_c)^{(4-d)/2} \gg 1$ the barrier $\Delta F^*/k_B T_c \gg 1$, even if one comes close to the spinodal. The limit of metastability is reached for $\Delta F^*/k_B T_c \approx 1$, i.e., the width of the region over which the spinodal is smeared out is given by

$$\delta\phi_{sp}/[\phi_{coex}^{(2)} - \phi_{coex}^{(1)}] \propto [r^d(1 - T/T_c)^{(4-d)/2}]^{-2/(6-d)}. \tag{7}$$

Computer simulations of medium-range Ising models[8]) confirm this result. Another derivation of eq. (7) starts from the condition that the mean-field approximation to the functional integral is only justified if the mean-square amplitude of fluctuations of the coarse-grained order parameter along the radial droplet concentration profile must be smaller than the square of the order parameter difference between $\rho = 0$ and $\rho = \infty$ as described by the profile itself[2])

$$\langle [\delta\phi(\boldsymbol{x})]^2 \rangle_T \ll [\phi(\rho \to \infty) - \phi(\rho = 0)]^2. \tag{8}$$

If $\phi(\boldsymbol{x})$ is defined by coarse-graining of Ising spins $S_i = \pm 1$ over a cell of linear dimension L, $\phi(\boldsymbol{x}) \equiv L^{-d}\Sigma S_i$, one finds[2])

$$\langle [\delta\phi(\boldsymbol{x})]^2 \rangle_{T,L} \propto L^{2-d} r^{-2}. \tag{9}$$

Putting $L = \xi_{coex}$, $\phi(\rho \to \infty) \approx \phi_{coex}^{(1)}$, and $\phi(\rho = 0) \approx \phi_{coex}^{(2)}$, eqs. (8) and (9) yield the standard Ginzburg criterion $r^d(1 - T/T_c)^{(4-d)/2} \gg 1$; hence, it is no surprise that this also rules the crossover from eq. (3) to eq. (4). Near the spinodal, however, $L = \xi \propto r(1 - T/T_c)^{-1/2}(\phi/\phi_{sp} - 1)^{-1/2}$, and $|\phi(\rho \to \infty) - \phi(\rho = 0)| \propto |\phi - \phi_{sp}|$, and, hence, one finds from eq. (8) that

$$1 \ll r^d(1 - T/T_c)^{(4-d)/2} |\phi/\phi_{sp} - 1|^{(6-d)/2}, \tag{10}$$

consistent with eq. (6). This treatment can also be generalized to spinodal decomposition[2]): the linearized theory holds for initial times if

$$\langle [\delta\phi(\boldsymbol{x}, t)]^2 \rangle_T \ll (\phi - \phi_{sp})^2; \tag{11}$$

introducing $L = \lambda_c$, where $\lambda_c \propto r(1 - T/T_c)^{-1/2}(1 - \phi/\phi_{sp})^{-1/2}$ is the critical wavelength that fluctuations must exceed in order to get amplified, one finds instead of eq. (10),

$$\exp[2R(2\pi/\lambda_m)t] \ll r^d(1 - T/T_c)^{(4-d)/2}(1 - \phi/\phi_{sp})^{(6-d)/2}, \tag{12}$$

λ_m being the wavelength of maximum growth and $R(2\pi/\lambda_m)$ is the amplification factor[1,2]). Thus only for $r \gg 1$ the linear theory predicting exponential growth holds initially, and even then the time range t over which it holds is rather limited ($t \propto \ln r$). Again, these results are corroborated by computer simulations on medium range Ising models[9]) and by experiments on polymer mixtures[11]). Since the coarse-grained free energy density of a polymer mixture has the Flory–Huggins form[16])

$$\frac{f_{pol}(\phi)}{k_B T} = \frac{\phi \ln \phi + (1 + \phi) \ln (1 - \phi)}{N} + \chi_{FH}\phi(1 - \phi), \tag{13}$$

N being the chain length and χ_{FH} a phenomenological parameter responsible for the unmixing tendency, one finds that eq. (13) can be mapped on the free energy of the Ising–mean-field model $f(\phi)/k_B T = \phi \ln \phi + (1 - \phi) \ln (1 - \phi) + 2(T_c^{MF}/T)\phi(1 - \phi)$ by[2,10])

$$N f_{pol}(\phi) = f(\phi), \qquad N\chi_{FH} = 2T_c^{MF}/T, \qquad r^2 \equiv N r_p^2, \tag{14}$$

where r_p is the range parameter appearing in $\mathscr{F}_{pol}$, and hence one has $\mathscr{F}_{pol}/k_B T \equiv (\mathscr{F}_{Ising}/k_B T)/N$. As a consequence, eqs. (3)–(12) carry over to the polymer problem if one replaces r^d by $r_{pol}^d N^{(d-2)/2}$.

We note in passing that a still unsolved problem with eqs. (6)–(13) is the apparent self-consistency of the spinodal for $d > 6$, though this is corroborated by renormalization group arguments[17]). While a double-well-shaped bulk free energy $F'(\phi)$ has no meaning, for the coarse-grained free energy density $f(\phi)$ the double-well shape is perfectly reasonable since for $L \lesssim \xi_{coex}$ no phase-separation within one coarse-graining cell is possible. The cell-size dependence of $f(\phi)$ has also been obtained recently by Monte Carlo methods[18]).

3. Scaling phenomena at late stage

While there are various regimes with different behavior during the initial stages of phase separation (fig. 1), less distinctions occur during the late stages: there the system is locally unmixed ($\phi \cong \phi_{coex}^{(1)}$ or $\phi \cong \phi_{coex}^{(2)}$, respectively), even if it started by spinodal decomposition. The typical linear dimensions $l(t)$ of unmixed regions grow with time according to a power law, $l(t) \propto t^x$, and the structure factor in this coarsening regime satisfies a scaling law[12,19]).

$$S(\boldsymbol{k}, t) = [l(t)]^d \tilde{S}_e\{kl(t)\}, \qquad t \to \infty,\ kl(t) \text{ finite}, \tag{15}$$

where $\tilde{S}_e$ is a scaling function which also depends on the direction of $\boldsymbol{k}$, $\boldsymbol{e} \equiv \boldsymbol{k}/|\boldsymbol{k}|$ [20]). This scaling behavior is somewhat reminiscent of the critical point scaling. In fact, if we consider a quench to $T = T_c$ we expect, instead of eq. (15) [14]),

$$S(k, t) = [l'(t)]^{2-\eta} \tilde{S}'\{kl'(t)\}, \qquad l'(t) \propto t^{1/z}. \tag{16}$$

Here the exponent z describing the growth of the time-dependent correlation length $l'(t)$ is the dynamic critical exponent[21]), and the exponent η (note also $\tilde{S}'(z \to \infty) \propto z^{-(2-\eta)}$ and hence $S(k, \infty) = S_{T_c}(k) \propto k^{-2+\eta}$) reflects the fractal structure of critical correlations. Alternative scaling descriptions of phase separation consider the number $n_s(t)$ of clusters of size s at time t after the quench,

$$n_s(t) = t^{-dx} \tilde{n}(st^{-dx}), \tag{17}$$

and the probability that an order parameter ψ $(= \phi_{\text{crit}} - \phi)$ is established in a subsystem of linear dimension L (or a finite system, respectively), $P_L(\psi, t)$. It is also possible to incorporate both the scaling of phase separation below T_c and critical point scaling in an unified description, which then properly brings out the temperature dependence of the growth laws. As an example, we consider $P_L(\psi, t)$, which scales as

$$P_L(\psi, t) = L^{\beta/\nu} \bar{P}(\psi L^{\beta/\nu}, t^{1/z}/L, \xi/L), \qquad \int_{-\infty}^{+\infty} \mathrm{d}\psi\, P_L(\psi, t) = 1. \tag{18}$$

At T_c, $\xi = \infty$ and then eq. (18) as (16) expresses the fact that L scales with $l'(t) \propto t^{1/z}$. In the static limit $(t \to \infty)$, eq. (18) reduces to standard finite size scaling. Both eqs. (16) and (18) imply that the time-dependent "susceptibility" $L^d \langle \psi^2 \rangle_t = L^d \int \mathrm{d}\psi\, \psi^2 P_L(\psi, t) = S(k = 0, t) \propto t^{(2-\eta)/z}$. In the critical region for $T < T_c$, however, eq. (18) reduces to[13])

$$P_L(\psi, t) = \xi^{\beta/\nu} \bar{\bar{P}}(\psi \xi^{\beta/\nu}, L\xi^{zx-1} t^{-x}), \tag{19}$$

from which one finds $l(t) \propto (\Omega t)^x$, $\Omega \propto \xi^{1/x-z}$. In addition, $P_L(\psi, t)$ is a simple gaussian in the variable ξ: this expresses the fact that on length scales much larger than $l(t)$ only negligible correlations have built up[13]).

Eqs. (15)–(19) not only apply to phase separation but also to order–disorder transition where a system in a disordered initial state is quenched to $T \leqslant T_c$ and ordered domains grow. Understanding of the exponent x that describes the growth of structure in phase separation and domain growth is a currently debated issue (see, e.g.[12–14,22–30]) for further references). In phase-separating solids, most workers – but not all[23]) – believe that the Lifshitz–Slyozov[31]) result $x = \frac{1}{3}$ holds, and this law was also suggested for certain order–disorder processes at constant density[14]). In this mechanism, clusters or domains grow (or shrink) by random condensation/evaporation of atoms, which then may diffuse from one cluster to the next. In order–disorder processes without conservation laws, the random motion of domain walls leads to $x = \frac{1}{2}$ [32]) and simulations[27,28]) confirm this result. Still, there are some problems: the theory[32]) does not yield the scaling temperature dependence of the rate Ω as mentioned above, but rather yields $\Omega(T = T_c) = \text{const}$; for some[30]) – but not all[13]) – models with broad domain walls $x = \frac{1}{4}$ was found[30]). If the ordering is strongly degenerate, as happens for q-state Potts models for large q, one finds $x \approx 0.41$ [24]). In order–disorder transitions for $d = 2$ at constant density, one sometimes finds $x \approx 0.2$ [25,26]): perhaps this is due to the mechanism[19]) where domains randomly diffuse and only grow when two domains of the same type meet and coalesce. Thus, it is seen that the "universality classes" for cluster or domain growth are not really understood – even the existence of universal (rather than, e.g., temperature-dependent) exponents x is questionable[22]). Analytic theories which reliably predict the scaling function $\tilde{S}$ in eq. (15) would be desirable.

In conclusion, we comment on the description of growth by one single exponent x in eq. (15) – rather than two exponents $2 - \eta, 1/z$ in eq. (16). This fact reflects the compact, rather than fractal, structure of the growing objects: the correlation function is trivial for short distances, $\langle \psi_i \psi_j \rangle \approx \psi^2$ for $|\boldsymbol{r}_i - \boldsymbol{r}_j| \ll l(t)$, and for long distances, $\langle \psi_i - \psi_j \rangle \propto \exp\{-|\boldsymbol{r}_i - \boldsymbol{r}_j|/l(t)\} \to 0$ for $|\boldsymbol{r}_i - \boldsymbol{r}_j| \gg l(t)$: it exhibits nontrivial structure only on the scale of $l(t)$ itself[13]). On the other hand, in phase separation there does occur a change of morphology from well-separated growing droplets (for ϕ near $\phi^{(1)}_{\text{coex}}$) to a percolating interconnected structure (for ϕ around ϕ_{crit}). At the percolation transition separating these two regimes, fractal clusters are observable (see also[23])) but this percolation phenomenon (and a related transient percolation, the so called "dynamical spinodal"[33]) does not seem to show up in the structure function $S(\boldsymbol{k}, t)$. Again the interplay between the geometrical description of the growing objects and the dynamical growth laws is not fully understood.

Acknowledgments

The author has benefitted from the interaction with J.D. Gunton, D.W. Heermann, K. Kaski, W. Klein, A. Milchev, A. Sadiq, and D. Stauffer.

References

1) J.W. Cahn and J.E. Hilliard, J. Chem. Phys. **28** (1958) 258, **31** (1959) 688; J.W. Cahn, Acta Metall. **9** (1961) 795.
2) K. Binder, Phys. Rev. A **29** (1984) 34.
3) O. Penrose and J.L. Lebowitz, J. Stat. Phys. **3** (1971) 211.
4) J.S. Langer, Physica **73** (1974) 61.
5) K. Binder, Phys. Rev. B **8** (1973) 3423.
6) K. Binder and D. Stauffer, Adv. Phys. **25** (1976) 343.
7) W. Klein and C. Unger, Phys. Rev. B **28** (1983) 445; C. Unger and W. Klein, Phys. Rev. B **29** (1984) 2698.
8) D.W. Heermann, W. Klein and D. Stauffer, Phys. Rev. Lett. **49** (1982) 1262.
9) D.W. Heermann, Phys. Rev. Lett. **52** (1984) 1126; Z. Phys. B **61** (1985) 311.
10) K. Binder, J. Chem. Phys. **79** (1983) 6387.
11) T. Izumitani and T. Hashimoto, J. Chem. Phys. **83** (1985) 3694.
12) K. Binder and D.W. Heermann, in Scaling Phenomena in Disordered Systems, R. Pynn and A. Skjeltorp, eds. (Plenum, New York, 1985), p. 207.
13) A. Milchev, K. Binder and D.W. Heermann, Z. Physik B **63** (1986) 521.
14) A. Sadiq and K. Binder, J. Stat. Phys. **35** (1984) 617; Phys. Rev. Lett. **51** (1983) 674.
15) V.L. Ginzburg, Sov. Phys. Solid State **2** (1960) 1824.
16) P.J. Flory, Principles of Polymer Chemistry (Cornell Univ. Press, Ithaca, 1967).
17) G. Dee, J.D. Gunton and K. Kawasaki, J. Stat. Phys. **24** (1981) 87.
18) K. Kaski, K. Binder and J.D. Gunton, Phys. Rev. B **29** (1984) 3996.
19) K. Binder, C. Billotet and P. Mirold, Z. Physik B **30** (1978) 183; K. Binder and D. Stauffer, Phys. Rev. Lett. **33** (1974) 1006; K. Binder, Phys. Rev. B **15** (1977) 4425.
20) J.R. Simon, P. Guyot and A. Ghilarducci de Salva, Phil. Mag. A **49** (1984) 151.
21) P.C. Hohenberg and B.I. Halperin, Rev. Mod. Phys. **49** (1977) 435.
22) H. Furukawa, Adv. Physics **34** (1985) 703, and references therein.
23) M. Schöbinger, S.W. Koch and F.F. Abraham, J. Stat. Phys. **42** (1986) 1071; S.W. Koch, preprint.
24) P.S. Sahni, G.S. Grest, M.P. Anderson and D.J. Srolovitz, Phys. Rev. B **28** (1983) 2705.
25) G.S. Grest and P.S. Sahni, Phys. Rev. B **30** (1984) 226.
26) J. Vinals and J.D. Gunton, Surf. Sci. **157** (1985) 473.
27) E.T. Gawlinski, M. Grant, J.D. Gunton and K. Kaski, Phys. Rev. B **31** (1985) 281.
28) P.S. Sahni, G. Dee, J.D. Gunton, M.K. Phani, J.L. Lebowitz and M.H. Kalos, Phys. Rev. B **24** (1981) 410.
29) J.L. Lebowitz, J. Marro and M.H. Kalos, Acta Metall. **30** (1982) 297.
30) O.G. Mouritsen, Phys. Rev. Lett. **56** (1986) 850.
31) I.M. Lifshitz and V.V. Slyozov, J. Phys. Chem. Solids **19** (1961) 35.
32) S.W. Allen and J.W. Cahn, Acta Metall. **27** (1979) 1085.
33) D.W. Heermann, Z. Phys. B **55** (1984) 309.

Physica **140A** (1986) 44–50
North-Holland, Amsterdam

RECENT DEVELOPMENTS IN THE THEORY OF PATTERN FORMATION

J.S. LANGER

Institute for Theoretical Physics, University of California, Santa Barbara, CA 93106, USA

The solvability mechanism for pattern selection in two-dimensional dendritic solidification and in the closely analogous Saffman–Taylor problem is described briefly. There remain several potentially very important unanswered questions.

The theory of pattern formation in nonlinear dissipative systems has taken some interesting turns in the last year or so. Most of the recent developments seem to be moving us in a forward direction, but there are also new indications that we may not yet be understanding some of the most important questions. In what follows, I shall describe briefly one of the developments that I consider to be most promising – the discovery that weak capillary forces act as singular perturbations which lead to very nearly identical selection mechanisms in both dendritic solidification and viscous fingering. I shall then point out where pieces of the puzzle seem to be missing.

Let us start with the "symmetric" model of crystal growth[1]). In the conventional thermodynamic model of the solidification of a pure substance from its melt, the fundamental rate-controlling mechanism is the diffusion of latent heat away from the interface between the liquid and solid phases. The latent heat that is released in the transformation warms the material in the neighborhood of the solidification front and must be removed before further solidification can take place. This is a morphologically unstable process which characteristically produces dendritic – i.e., snowflake-like – structures.

The dimensionless thermal diffusion field is conveniently chosen to be

$$u = \frac{T - T_\infty}{(L/c)}, \tag{1}$$

where T_∞ is the temperature of the liquid infinitely far from the growing solid, and the ratio of the latent heat L to the specific heat c is an appropriate unit of

0378-4371/86/$03.50

undercooling. The field u satisfies the diffusion equation

$$\frac{\partial u}{\partial t} = D \nabla^2 u, \tag{2}$$

where D is the thermal diffusion constant which we shall take to be the same in both liquid and solid phases. The latter condition defines the symmetry between the two phases in this version of the "symmetric" model. The remaining ingredients of the model are the boundary conditions imposed at the solidification front. First, there is heat conservation:

$$v_n = -[D\hat{\boldsymbol{n}} \cdot \nabla u], \tag{3}$$

where $\hat{\boldsymbol{n}}$ is the unit normal directed outward from the solid, v_n is the normal growth velocity, and the square brackets denote the discontinuity of the flux across the boundary. The physically more interesting boundary condition is a statement of local thermodynamic equilibrium which determines the temperature u_s at the two-phase interface:

$$u_s = \Delta - d_0 \mathscr{K}, \tag{4}$$

where

$$\Delta = \frac{T_M - T_\infty}{(L/c)} \tag{5}$$

is the dimensionless undercooling, a measure of the driving force for the processes that we are considering. The second term on the right-hand side of (4) is the Gibbs–Thomson correction for the melting temperature at a curved surface: $\mathscr{K}$ is the sum of the principal curvatures and $d_0 = \gamma c T_M / L^2$ is a length, ordinarily of order ångströms, which is proportional to the surface tension γ. The latter quantity and, accordingly, d_0 may be functions of the angle of orientation of the interface relative to the axes of symmetry of the crystal.

The hydrodynamic analog of dendritic solidification is the fingering instability that occurs when one causes an inviscid fluid to drive a viscous one through a porous medium. The two-dimensional version of this situation is a Hele–Shaw cell in which the two (immiscible) fluids are constrained to move between narrowly separated parallel plates. The invading inviscid fluid can be visualized as playing the role of the growing solid, and the more viscous fluid that is being pushed away is like the melt. The analog of the thermal field u is the pressure P, which can be taken to be constant in the "solid" and to satisfy Laplace's equation in the "melt". Here is the main difference between fingering and solidification; the Laplace equation is not the diffusion equation. The velocity of the viscous fluid in the porous medium is given by Darcy's law to be simply proportional to $-\nabla P$; thus the expression for the velocity of the interface between the two fluids is precisely the analog of the one-sided version of the conservation law (3). Finally, the interfacial tension γ causes the pressure at the interface to be reduced by an amount proportional to $\gamma\mathscr{K}$, in exact analogy to the thermodynamic boundary condition (4). There is, however, no crystalline anisotropy associated with this γ. Directional information can be provided only by the interaction between the long-ranged pressure field and the walls of the container or else by adding to the model – "by hand", so to speak – some anisotropy of the medium through which the fluids are moving.

There are sharply defined problems of pattern selection associated with both of these models. In solidification, it is known that the growth rates and shapes of dendritic tips are determined reproducibly by the undercooling Δ. In the hydrodynamic case, specifically, the two-dimensional Saffman–Taylor experiment[2,3]) in which a steady-state finger forms in a long channel, the ratio λ of the width of the finger to the width of the channel is determined uniquely by the flow speed. In both cases, surface tension appears at first glance to be a negligible perturbation; the length d_0 is orders of magnitude smaller than other characteristic lengths. However, the omission of surface tension in either problem leads to continuous families of solutions and, thus, to no explanation whatsoever of the experimentally observed selection principles. Obviously, surface tension is playing some special and subtle role in these processes.

Our current understanding of this capillary selection mechanism in the case of the two-dimensional dendrite is as follows[4-7]). Let $z = \zeta(x)$ be the position of a steady-state solidification front in a frame of reference moving at speed v in the $+z$ direction. Measure lengths in units of ρ, the radius of curvature of the tip of such a front at undercooling Δ with $d_0 = 0$. The corresponding relation between Δ, v, and ρ, originally derived by Ivantsov[8]), is

$$\Delta = 2\sqrt{p}\,\mathrm{e}^{p}\int_{\sqrt{p}}^{\infty}\mathrm{e}^{-y^2}\mathrm{d}y \approx (\pi p)^{1/2}, \tag{6}$$

where $p = v\rho/2D$ is known as the Peclet number, the final approximation is valid for $p \ll 1$, and the steady-state front itself is $\zeta_{\mathrm{I}v} = -x^2/2$. For values of d_0 which are small but nonzero, it is legitimate to study the linearized equation for the shape correction $\zeta_1 = \zeta - \zeta_{\mathrm{I}v}$, which turns out to be (for $p \ll 1$):

$$\sigma\frac{\mathrm{d}^2\zeta_1}{\mathrm{d}x^2} - \frac{3\sigma x}{1+x^2}\frac{\mathrm{d}\zeta_1}{\mathrm{d}x} - \frac{(1+x^2)^{3/2}}{2\pi A(x)} \times\int_{-\infty}^{\infty}\mathrm{d}x'\,\frac{(x+x')\left[\zeta_1(x)-\zeta_1(x')\right]}{(x-x')\left[1+\frac{1}{4}(x+x')^2\right]} = \sigma. \tag{7}$$

Here, $\sigma = 2Dd_0/v\rho^2$ is the same dimensionless group of parameters that has appeared in linear stability analyses[9]), and $A(x)$ is a function associated with the anisotropy of the surface tension. In particular, for a surface tension proportional to $1 - \alpha\cos(4\theta)$, where θ is the angle between the normal to the interface and the z axis,

$$A(x) = 1 - \alpha + \frac{8\alpha x^2}{(1+x^2)^2}. \tag{8}$$

The single most important feature of eq. (7) is that the first two terms on the left-hand side constitute a singular perturbation proportional to the capillary parameter σ. In the absence of these terms, (7) has physically acceptable solutions for any value of σ; but, when these derivatives of ζ_1 are present with nonvanishing coefficients, (7) is solvable only for special values of σ, say, $\sigma^*(\alpha)$. A necessary condition for the existence of solutions of (7) is that the inhomogeneous term on the right-hand side, here just σ, be orthogonal to the null space of the (non–self-adjoint) linear operator on the left. The null eigenvector of the adjoint of this operator may be computed by a WKB approximation for small values of σ, in which limit the solvability condition can be written in the form

$$\Lambda(\sigma,\alpha) = \int_{-\infty}^{\infty} \mathrm{d}x\,\sigma R(x)\exp\left[\frac{\mathrm{i}}{\sqrt{\sigma}}\psi(x)\right] = 0, \tag{9}$$

where

$$\psi(x) = \int_0^x \mathrm{d}x'\,\frac{(1-\mathrm{i}x')^{1/4}(1+\mathrm{i}x')^{3/4}}{A^{1/2}(x')} \tag{10}$$

and $R(x)$ is the slowly varying σ-independent prefactor of the WKB solution whose details are of no interest for present purposes. For small σ, (9) can be evaluated by performing a steepest-descent calculation in the neighborhood of $x = \mathrm{i}$, where the phase $\psi(x)$ is stationary. At vanishing anisotropy, $\alpha = 0$, $A = 1$,

$$\Lambda(\sigma,0) \approx \sigma^{-1/28}\exp\left(-\frac{a}{\sqrt{\sigma}}\right), \tag{11}$$

where

$$a = \int_0^1 \mathrm{d}u\,(1+u)^{1/4}(1-u)^{3/4} \cong 0.6159\ldots. \tag{12}$$

The result (11) demonstrates both the singular nature of the perturbation σ and the fact that (9) has no solutions other than $\sigma^* = 0$. For $\alpha > 0$, on the other hand, the factor $A^{-1/2}$ in (10) produces a new branch point in $\psi(x)$ which causes Λ to oscillate rapidly at small σ. The solvability condition (9) now has a denumerably infinite set of solutions σ^*. It turns out that only the largest of these, for which $\sigma^* \propto \alpha^{7/4}$, can correspond to a stable tip of a dendrite.

The solvability mechanism for the Saffman–Taylor problem[10–12]) is strikingly similar to the above results for the dendrite. The parameter σ is replaced by the dimensionless group σ_{ST},

$$\sigma_{ST} = \frac{\gamma b^2 \pi^2}{12 \mu v w^2 (1-\lambda)^2}, \tag{13}$$

where γ is the surface tension, b the spacing between the plates, μ the viscosity, v the speed of the finger, w the width of the channel, and λw the width of the finger. All other ingredients of the solvability function Λ_{ST} defined in analogy to eq. (9) are the same except that the function $A^{1/2}$ in (10) is replaced by

$$A_{ST}^{1/2}(x) = x^2 + \left(\frac{1-\lambda}{\lambda}\right)^2. \tag{14}$$

[The factor $R(x)$ in (9) also is different but remains uninteresting.] The WKB phase function $\psi_{ST}(x)$ now has a logarithmic branch point on the imaginary x axis which coincides with the point of stationary phase at $x = \mathrm{i}$ when $\lambda = \frac{1}{2}$. Thus, the boundary-related quantity $\lambda - \frac{1}{2}$ is playing a role in this problem that is closely analogous to that played by the anisotropy strength α for the dendrite. Λ_{ST} oscillates for small values of σ_{ST}, and the physically meaningful solution of the solvability condition has the form $\sigma_{ST}^* \propto (\lambda - \frac{1}{2})^{3/2}$. The convergence of λ to the value $\frac{1}{2}$ at small σ_{ST} (large v) is consistent with experiment.

The picture summarized above – an apparently exact analytic description of a mechanism for selecting steady-state configurations in two different nonlinear dissipative systems – seems elegant and compelling. There are, however, several recent developments that lead me to believe that this picture is very far from being complete.

First, there is a new phenomenon discovered experimentally by Couder and coworkers[13,14]) which indicates that anisotropy might not be so essential an ingredient of pattern-forming models as we have been led to believe by earlier results. There is already some hint of this possibility in the above analysis. In both of the above models, a nonzero solvability function Λ may be interpreted (probably) as being proportional to the magnitude of a discontinuity in the slope $d\zeta/dx$ at the tip $x = 0$. For the versions of these models discussed so far, physically acceptable solutions require $\Lambda = 0$. However, if one were able to perturb the system in such a way as to fix Λ at some nonzero value, then the mathematics tells us that solutions should exist in the absence of anisotropy. Couder et al. have produced such perturbations by attaching small bubbles to the tips of viscous fingers, and in this way have succeeded in generating Saffman–Taylor fingers with λ appreciably less than $\frac{1}{2}$. Moreover, in the circularly symmetric geometry where radial fingers ordinarily suffer tip-splitting instabilities, they have shown that fingers with bubbles at their tips behave very much like dynamically stable dendrites. A detailed theory of these phenomena has yet to be worked out.

Second, there is the question of dendrites in three dimensions. Experiments[15]) indicate that the parameter σ is indeed a Δ-independent constant, but that it may depend only weakly, if at all, on the strength of the crystalline anisotropy[16]). My initial expectation was that the basic structure of the theory would look very different in three dimensions than it did in two, and this expectation seemed to be borne out by the appearance of new mathematical features in the early stages of the analysis. However, calculations[7]) have now been completed for the situation in which the anisotropy remains cylindrically symmetric around the growth axis, and the result is that the solvability condition miraculously reduces to the same formula that we obtained in two dimensions. The resulting $\alpha^{7/4}$ law is almost certainly inconsistent with experiment. It therefore seems that some axial anisotropy, which ostensibly should play little role in setting the growth direction or controlling tip-splitting instabilities, must make a major difference.

Finally, there is the question of side-branching. Until quite recently, I had assumed that the tip of a real three-dimensional dendrite must be weakly – perhaps marginally – unstable against some oscillatory mode of deformation, and that this oscillation must generate the train of side-branches that seems always to be observed in these systems. Couder's fingers with bubbles at their tips, when driven fast enough, quite definitely do oscillate and emit side-branches. On the other hand, theory now tells us that no such oscillatory instability exists in either ordinary fingers[17]) (without bubbles) or two-dimensional dendritic tips[18]). There is no clear discrepancy between theory and experiment implied here; we do not yet know enough about real two-dimensional solidification to be able to tell whether side-branching is as common there as it seems to be in three dimensions. There is also the possibility that, even in three dimensions, dendritic side-branches may be generated by the selective amplification of noise[19,20]). My point is simply that the search for the missing pieces of the puzzle – the selection mechanism and side-branching dynamics in three dimensions – may still require fundamentally new ideas.

Acknowledgments

This research was supported by U.S. Department of Energy Grant no. DE-FG-03-84ER45108 and by National Science Foundation Grant no. PHY 82-17853, supplemented by funds from the National Aeronautics and Space Administration.

References

1) J.S. Langer, Rev. Mod. Phys. **52** (1980) 1.
2) P.G. Saffman and G.I. Taylor, Proc. Roy. Soc. A **245** (1958) 312.
3) J.W. McLean and P.G. Saffman, J. Fluid Mech. **102** (1981) 455.
4) P. Pelce and Y. Pomeau, Dendrite in the small undercooling limit, Stud. Appl. Math., to appear.
5) M. Ben Amar and Y. Pomeau, Theory of dendritic growth in a weakly undercooled melt, preprint.
6) D. Kessler, J. Koplik and H. Levine, Pattern formation far from equilibrium: the free space dendritic crystal, to appear in Proc. NATO A.R.W. on Patterns, Defects, and Microstructures in Non-Equilibrium Systems, Austin, Texas, March 1986.
7) A. Barbieri, D.C. Hong and J.S. Langer, Velocity selection in the symmetric model of dendritic crystal growth, ITP Preprint 86-65, submitted to Phys. Rev. A.
8) G.P. Ivantsov, Dokl. Akad. Nauk SSSR **58** (1947) 567.
9) J.S. Langer and H. Müller-Krumbhaar, Acta Metall. **26** (1978) 1681, 1689, 1697; H. Müller-Krumbhaar and J.S. Langer, Acta Metall. **29** (1981) 145.
10) B.I. Shraiman, Phys. Rev. Lett. **56** (1986) 2028.
11) R. Combescot, T. Dombre, V. Hakim, Y. Pomeau and A. Pumir, Phys. Rev. Lett. **56** (1986) 2036.
12) D.C. Hong and J.S. Langer, Phys. Rev. Lett. **56** (1986) 2032. Although the results obtained in this paper agree with those of refs. 10 and 11 and apparently are correct, the presentation is, at best, misleading. A more satisfactory derivation of the solvability condition for the Saffman–Taylor problem can be constructed using the methods described in ref. 7 and in the present paper, which are essentially those discussed by Shraiman in ref. 10.
13) Y. Couder, N. Gerard and M. Rabaud, The observation of the missing fingers of the Saffman–Taylor instability, submitted to Phys. Rev. Lett.
14) Y. Couder, O. Cardoso, D. Dupuy, P. Tavernier and W. Thom, Dendritic growth in the Saffman–Taylor experiment, preprint.
15) M.E. Glicksman, R.J. Shaefer and J.D. Ayers, Metall. Trans. A **7** (1976) 1747; S.C. Huang and M.E. Glicksman, Acta Metall. **29** (1981) 701, 717.
16) M.E. Glicksman, private communication.
17) D. Kessler and H. Levine, Discrete set selection of Saffman–Taylor fingers, preprint.
18) D. Kessler and H. Levine, Stability of dendritic crystals, preprint, submitted to Phys. Rev. Lett.
19) R. Deissler, J. Stat. Phys. **40** (1985) 371.
20) R. Pieters and J.S. Langer, Phys. Rev. Lett. **56** (1986) 1948.

Physica 140A (1986) 51–61
North-Holland, Amsterdam

MULTIFRACTAL STRUCTURE OF CLUSTERS AND GROWING AGGREGATES

Antonio CONIGLIO

Center for Polymer Studies and Department of Physics, Boston University, Boston, MA 02215, USA
and GNSM, Dipartimento di Fisica, Università di Napoli, Mostra d'Oltremare Pad. 19, 80125 Napoli, Italy

Various phenomena on a given fractal object have different critical behavior. This is related to the underlying multifractal structure of the aggregate. Here we discuss the problem with particular emphasis on a scaling approach which leads to a simple real space renormalization group. A mechanism which generates multifractality, based on a multiplicative process, is also illustrated. The multifractal structure of the incipient infinite cluster in percolation and diffusion limited aggregation is discussed in detail.

1. Introduction

The concept of fractals has been extremely useful to describe many problems in physics[1]). Fractals are objects whose mass M scales with their linear size L with a power law $M \sim L^{d_f}$. More recently it has been recognized that by studying a phenomenon on a fractal, not all the elements of the fractal play the same role. For example in a dilute Ising model at the percolation threshold only the single-connected bonds of the backbone contribute to the correlation length exponent[2]). In a dilute Heisenberg model instead, a larger set of bonds plays the dominant role, the same set which contributes to the resistance in a random resistor network. As a consequence, a different correlation length exponent is found, related to the resistivity exponent of the random resistor network[2]).

Similarly, a fractal object, such as a protein in a solution or a branched polymer, exhibits a strong screening with respect to a diffusive particle. It was recognized that only the "unscreened" sites, a subset of the total surface sites, are exposed to the external probe[3]). In growth phenomena, like diffusion limited aggregation (DLA)[4]), where screening plays an important role in the growth mechanism, the sites which are most likely to grow are at the tips of the aggregate; these sites are a subset with a fractal dimension less than the fractal dimension of the entire aggregate. Multifractality is related to the decomposition of the fractal object in many fractal sets, each one with its own fractal dimension.

0378-4371/86/$03.50 © Elsevier Science Publishers B.V.
(North-Holland Physics Publishing Division)

Different sets can play a dominant role in different problems, giving rise to different critical exponents. Multifractality was first discovered in connection with turbulence by Mandelbrot[5]). Further work for the specific problem of turbulence can be found in refs. 6 and 7. It was subsequently discovered in percolation[8-11]), growth phenomena[12-14]), dynamical systems[15]), localization problems[16]) and polymer absorption[17]). This talk will be restricted to multifractality in percolation and growth phenomena, although the scaling results and the underlying mechanism are rather general.

2. Multifractality in percolation

Consider a d-dimensional hypercubic lattice of size L. Suppose that each bond has a probability p of being active and $1-p$ of being nonactive. From percolation theory we know that in the limit of infinite system size a percolation threshold p_c exists above which an infinite cluster of active bonds is present. Right at p_c, the bonds in the spanning configurations, as noted by Stanley[18]) in 1977, may be partitioned in dangling bonds that do not contribute to the electrical resistance and the remaining backbone bonds. The backbone bonds may also be divided into links ("red" bonds) which are singly-connected and the remaining multiply-connected bonds ("blobs"). This partition has the great advantage of being very general and can be done for any aggregate. The next major problem is to determine whether the links or the blobs or both are critical quantities. In fact various models were proposed for the percolation cluster, some of which emphasized only the links and others only the blobs. It was only four years later that it was proved that links and blobs are both relevant, and it was proposed that the backbone of the incipient infinite cluster is made of links and blobs in a *self-similar* way[2]). The number of links L_1 diverges as $L_1 \sim L^{1/\nu}$ in any dimension where ν is the connectedness length exponent. These results were later confirmed by computer simulations[19]).

In the random resistor network we associate a unit resistance to each active bond. If we apply a unit voltage at the opposite boundary of the cell connected by the percolating cluster, we characterize each bond by the absolute value of the voltage V across it. $V_{\max}$ occurs in the links. Since these bonds carry the total current I passing through the network, $V_{\max} = I = G$, where G is the conductance of the network, $V_{\max} \sim L^{-\tilde{\zeta}}$, where $\tilde{\zeta}$ is the conductance exponent. Thus the links form a subset of bonds with a fractal dimension $1/\nu$ and singularity $\tilde{\zeta}$. To partition all the other backbone bonds into different subsets, we introduce the following moments and their critical behavior at p_c[8]):

$$M(k) = \sum n(V) V^k \sim L^{-\tilde{p}(k)}, \tag{1}$$

where $n(V)$ is the number of bonds with an absolute value of the voltage drop V across them. de Arcangelis et al.[8]) and Rammal et al.[9]) independently found that $\tilde{p}(k)$ form an infinite set of independent exponents. This situation is rather different from ordinary critical phenomena as in an Ising model where the moments of the order parameter scale with the gap exponent, namely with an exponent linear with k.

The sum in (1) can be evaluated by the steepest descent method. If $V^* = V(k)$ is the value of V, which maximizes the summand in (1), we assume for large values of L

$$V(k) = A(k)L^{-\alpha(k)}, \tag{2a}$$

$$n(V(k)) = B(k)L^{f(k)}, \tag{2b}$$

where $A(k)$ and $B(k)$ are amplitudes. From (1) we obtain

$$\tilde{p}(k) = k\alpha(k) - f(k), \tag{3a}$$

while the condition that V^* is maximum implies that

$$\frac{\mathrm{d}}{\mathrm{d}k}\tilde{p}(k) = \alpha(k). \tag{3b}$$

Therefore, given $\tilde{p}(k)$ one can calculate $\alpha(k)$ and $f(k)$ and vice versa. Eliminating k from (3) we obtain the new function $f(\alpha) = f(k(\alpha))$. From (2), using α as an independent parameter, we have

$$n(V) = C(\alpha)L^{f(\alpha)}, \tag{4}$$

where α is related to V by

$$\alpha = -\frac{\log V}{\log L} \tag{5}$$

and we have neglected terms of the order $1/\log L$. Eqs. (4) and (5) express the fact that the bonds divide into different fractal sets, each characterized by a given value α and a fractal dimension $f(\alpha)$, directly related to the moments exponents $\tilde{p}(k)$ (fig. 1). Note that a linear dependence in k of $\tilde{p}(k)$ (presence of gap

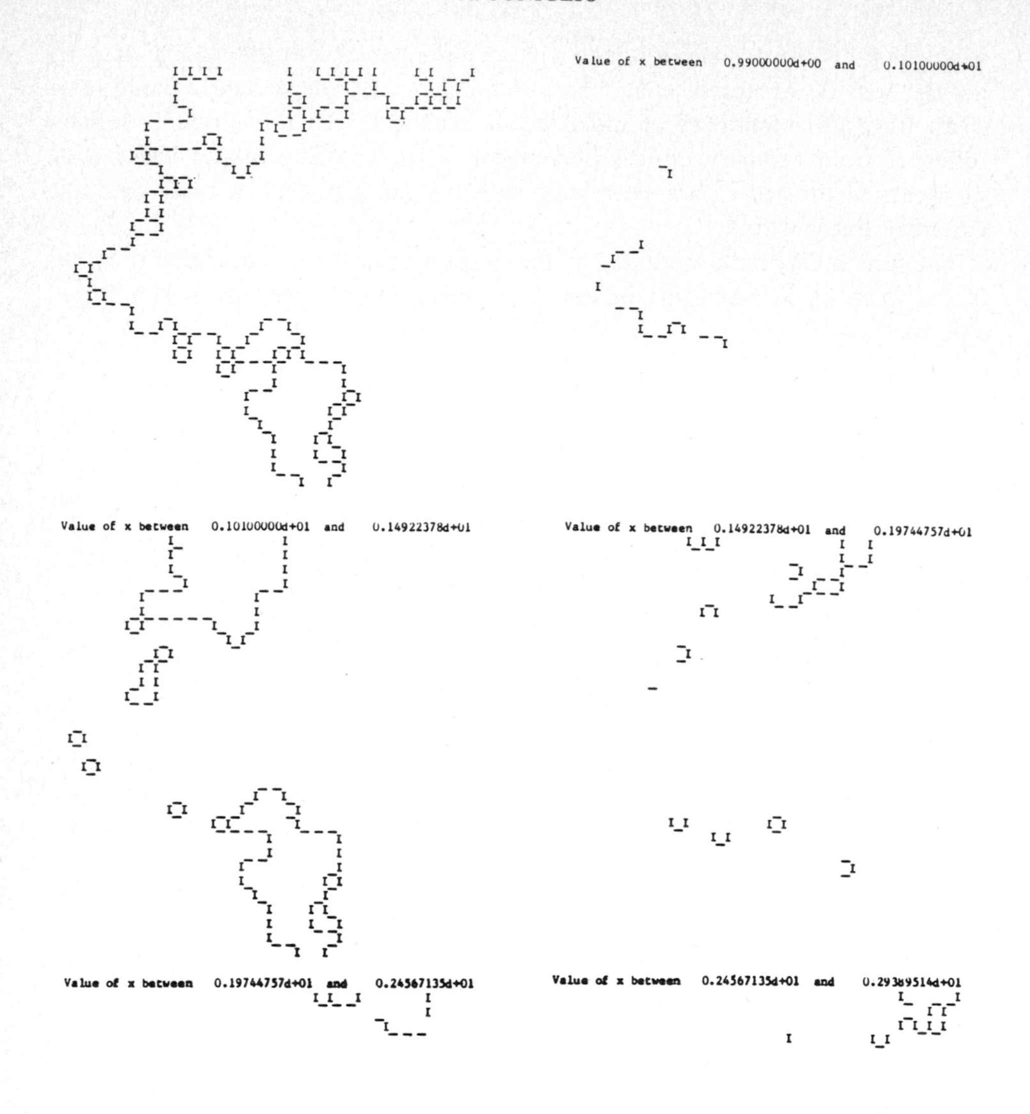

Fig. 1. A computer-generated configuration of the backbone for a system of size $L = 30$. The bonds are divided into five fractal sets corresponding to different intervals of $x \equiv \alpha = -\ln V/\ln L$.

exponent) implies only one pair of values f and α, thus one fractal dimension. Any departure from a linear behavior implies multifractality.

Exact results on a hierarchical model and computer simulations[8]) for percolation clusters showed absence of gap exponent and the anomalous scaling (5) in the voltage distribution characterized by a nonconstant critical exponent. The nonlinearity in the moment exponents $\tilde{p}(k)$ was later confirmed by ε-expansion[10]). Multifractality was also found in random superconducting networks, where for $d = 2$ the moments exponents were related by duality to those of the random resistor network[8]).

3. Mechanism for multifractality: multiplicative process

The voltage distribution can be calculated exactly on a hierarchical model. I shall illustrate the results since they will be instructive on the mechanism which produces multifractality. The model is obtained by successive iterations, substituting each bond of the previous iteration with the unit cell. Consider the general case where the bonds of the unit cell labelled $1, 2, 3, 4$ have, respectively, conductance $\sigma_1, \sigma_2, \sigma_3, \sigma_4$. If $\Delta V = 1$ is the voltage across the cell, the voltages across the bonds are V_1, V_2, V_3, V_4, obtained by solving Kirchhoff's laws. Note that from the invariance of Kirchhoff's laws under transformation $V \to \lambda V$, it follows that if the voltage across the cell is $\Delta V = \lambda$, the voltage along the bonds is λV_i $(i = 1, 2, 3, 4)$. At the next iteration $N = 2$ the network is made of four cells labeled $1, 2, 3, 4$ corresponding to the four bonds of the unit cell. Within the cell i the voltages are given by $V_iV_1, V_iV_2, V_iV_3, V_iV_4$ $(i = 1, 2, 3, 4)$ since the voltage across the cell is V_i. Therefore, the voltages at each iteration are obtained using the multiplicative rule illustrated in fig. 2. Consequently the moments are given by

$$M(k) = \left(V_1^k + V_2^k + V_3^k + V_4^k\right)^N. \tag{6}$$

For the simple case when the conductances σ_i are all equal we have $V_1 = V_2 = \frac{1}{5}$, $V_3 = V_4 = \frac{2}{5}$ and we find an infinite set of critical exponents

$$\nu\tilde{p}(k) = \frac{\log M(k)}{N \log 2} = 1 + \log\left[\left(\frac{1}{5}\right)^k + \left(\frac{2}{5}\right)^k\right] \Big/ \log 2, \tag{7}$$

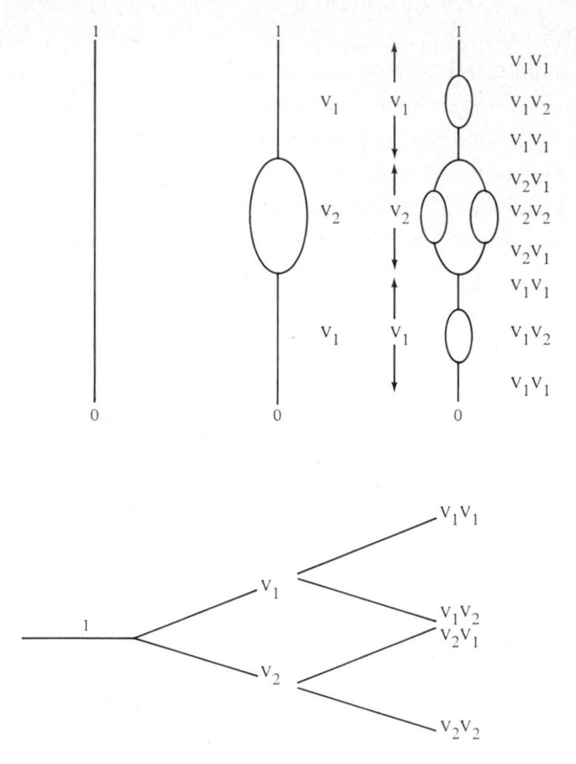

Fig. 2. Hierarchical model for the backbone of the incipient infinite cluster (ref. 8). The multiplicative rule to obtain the voltage at each interaction is indicated.

where we have used the relation $L = 2^{N\nu}$, where L is the linear dimension and 2^N is the number of links. In conclusion we see that multifractality arises in a self-similar structure where the field (in our case V) associated to each element satisfies the multiplicative rule. This rule is satisfied when the equation obeyed by the field is invariant under scaling transformation $\phi \to \lambda\phi$. The importance of the multiplicative process was also stressed in refs. 15 and 20.

Note that the multiplicative process is reminiscent of many nonlinear problems. An example is the nonlinear resistor network[21,22]) where the relation between voltage and current is given by $V = RI^{\alpha}$. In a general hierarchical model where the unit cell is characterized by m bonds with voltage $V_i(\alpha)$ ($i = 1, \ldots, m$), the conductance $G = R^{-1/\alpha}$ is given by $G = V_1(\alpha)^{1+1/\alpha} + \cdots + V_m(\alpha)^{1+1/\alpha}$. This has to be compared with the kth moment of the voltage of the linear resistor network. For a general hierarchical lattice $M(1 + 1/\alpha) = G(\alpha)$ provided that in the linear resistor network problem the conductances in the unit cell are chosen in such a way that the two sets of voltages in the unit cell coincide. The correspondence is particularly simple in the case of the hierarchical model of fig. 2[23]).

4. Multifractality in growth phenomena

The formalism of the previous section can be extended to growth phenomena. For simplicity consider the DLA model. At each step of the growth a diffusive particle is released from infinity. When the particle reaches one perimeter site it becomes part of the aggregate. In the off lattice model the resulting aggregate is a self-similar fractal. For a given aggregate of size L we can define for each site i of the perimeter, the probability that site i becomes part of the aggregate at the next step. From the growth probabilities one can define the moments and their critical exponents[12–14]):

$$M(q) = \sum n(p) p^q \sim L^{-(q-1)D(q)}, \tag{8}$$

where $n(p)$ is the number of perimeter sites characterized by a probability p. Can we expect multifractality in this case? Namely, do we expect departure from linearity in the critical exponents? The answer is yes. In fact, using the electrostatic analogy of DLA and dielectric breakdown[23]) in the continuum limit, one can show that the growth probability $p(x)$ at site x is proportional to $\boldsymbol{n} \cdot \nabla\phi(x)$, where $\boldsymbol{n}$ is the normal to the surface of the aggregate at site x and $\phi(x)$ is the electrostatic potential satisfying the Laplace equation in the exterior, under the condition that $\phi = 1$ on the aggregate and zero on a circle very far away. The field ϕ obeys the same equation as the voltage in the random resistor network, which is invariant under scaling $\phi \to \lambda\phi$. Thus from the consideration of the previous section we expect multifractal behavior, in particular nonlinear behavior, in the moments exponents, in contrast to ref. 11.

The moments have been computed by direct simulation by Meakin et al.[12]). A different method, based on the electrostatic analogy, was used by Amitrano et al.[14]) to calculate the whole spectrum of exponents. As expected, the results show the typical nonlinear behavior in the moments exponents for the DLA and dielectric breakdown model. The moments have also been calculated for other fractal and nonfractal aggregates. It was found that for nonfractal aggregates[12]), the exponents were linear in q, while for fractal aggregates deviations from linearity were observed. Nonlinearity was also observed for linear polymers using ε-expansion[17]).

From the nonlinear behavior of the moments exponents it follows that the probability distribution $n(p)$ obeys a scaling law of the type of (4) and (5) where V is replaced by p. This scaling law has been confirmed both in DLA aggregates[12]) and screen growth models[24]).

5. Scaling approach

In section 2 we have pointed out that the nonlinearity of the k-moments exponents implies an anomalous scaling law in the corresponding distribution, where the power law exponent depends on the variable field. In a similar way, in critical phenomena one can show that the linear dependence on k in the exponents of the moments of the order parameter implies the familiar generalized homogeneous function for the free energy. However, much deep understanding was obtained when the scaling laws were obtained on a more physical ground in the classical approach as developed by Kadanoff. It was this approach, in fact, that paved the way to the formulation of Wilson's renormalization group.

We want to generalize Kadanoff's approach to multifractality and indicate a way for a simple renormalization group approach. For simplicity let us focus on a DLA aggregate. The size of the aggregate, L, can be considered as the correlation length and thus as a temperature type variable. The other variable is the probability p varying in a range $p_{\min} \leqslant p \leqslant p_{\max}$. The distribution probability $n(L, p)$ plays the role of a typical thermodynamic quantity, where we have explicitly written the L dependence.

Divide the system in cells of linear size l, and associate to each cell a renormalized variable $\tilde{p} = \Sigma' p_i$, where the sum is over the sites in the cell. Due to dilation invariance, the scaling hypothesis asserts that the distribution probability for the system of cells under this transformation must be the same as for the original system apart from a scaling factor, namely

$$\tilde{L} = \frac{L}{l}, \tag{9a}$$

$$\tilde{p} = l^{x(L,p)} p, \tag{9b}$$

$$n(\tilde{L}, \tilde{p}) = l^{-y(L,p)} n(L, p). \tag{9c}$$

Eqs. (9) are the most general transformation in which the exponents x and y are considered functions of the variables L and p. In usual scaling the exponents are assumed to be constant.

We have to require now that the transformations (9) must satisfy group composition. More precisely, if T_l is the operator which transforms the site variables into cell variables we must have $T_{l_1}T_{l_2} = T_{l_1 l_2}$. The physical reason for the group constraint is that if we divide the system into cells of length l_1 and perform on the system of cells a further division in supercells of length $l_2 l_1$, all the quantities in (9) must scale in the same way as if we divide the original system directly into cells of length $l_2 l_1$. These generalized scaling laws were studied in the context of critical phenomena[25]). It was found that x and y are invariant under group transformation, namely $x(\tilde{L}, \tilde{p}) = x(L, p)$ and $y(\tilde{L}, \tilde{p}) = y(L, p)$. From these conditions it is possible to derive differential equations whose general solutions under the requirement that $\tilde{p} = 1$ for $l = L$ give

$$\tilde{p} = l^{\alpha} p, \tag{10a}$$

$$n(\tilde{L}, \tilde{p}) = l^{-f(\alpha)} n(L, p), \tag{10b}$$

where α is related to p by $\alpha = -(\ln p/\ln L)$ and terms of the order $1/\ln L$ have been neglected.

From (10) one recovers (4). Moreover, it is possible to show that the moments are given by

$$\sum n(\tilde{L}, \tilde{p})\tilde{p}^q = l^{(q-1)D(q)} \sum n(L, p) p^q = \left(\frac{L}{l}\right)^{-(q-1)D(q)}, \tag{11}$$

where $(q-1)D(q)$ is related to f and α by transformations (2) and (3). In conclusion, starting from a general scaling transformation which obeys group properties we obtain the spectrum $f(\alpha)$ with the infinite hierarchy of exponents (11). Note that in (11) one can keep L fixed and vary l, which corresponds to the box counting method definition of the fractal dimensions, or l can be fixed equal to 1, and L can be changed corresponding to the definition of the fractal dimension of the aggregate in terms of the relation between the linear dimension and the mass. Within the scaling hypothesis the two coincide, as (11) shows.

An important question regarding multifractal scaling arises: how can we adopt renormalization group ideas to calculate the fractal dimension of all fractal sets. One simple way to renormalize geometrical quantities is by mapping a cell into a single site. For example, to find the fractal dimension of the backbone in the percolation problem, one may evaluate the mass rescaling of the backbone by evaluating the average number of backbone bonds in the spanning cluster. This same procedure cannot be applied directly to each fractal set of bonds, since there are only a small number of bonds in a cell that cannot represent all the sets. A very simple alternative is to renormalize the moments in a single cell using scaling (11). Since in a single cell the renormalized probability is $\tilde{p} = \Sigma' p_i$, where the Σ' is over all the sites in the cell, we have

$$\tilde{p}^q = l^{(q-1)D(q)} \sum{}' p_i^q. \tag{12}$$

This simple renormalization procedure has been tested to evaluate the moments exponents for the voltage distribution on the hierarchical model, where it gives exact results and on a square lattice at p_c by an exact rescaling of a 4×4 into a 3×3 cell. The results are in excellent agreement with the value of the moments exponents evaluated numerically for large system size.

In summary, the existence of an infinite hierarchy of independent exponents of the moments of the voltage or probability distribution implies a new type of scaling related to the multifractal structure of the cluster or aggregate. The form of distribution can also be obtained from a generalization of Kadanoff's scaling approach to critical phenomena. From this generalized scaling, a simple real space renormalization group approach has been suggested.

Acknowledgments

I would like to thank my collaborators in this field: C. Amitrano, L. de Arcangelis, F. di Liberto, P. Meakin, S. Redner, H.E. Stanley and T. Witten. In particular, I would like to thank L. de Arcangelis for providing the computer-generated sets of fig. 1, and H. Herrmann for a critical reading of the manuscript.

References

1) B.B. Mandelbrot, The Fractal Geometry of Nature (Freeman, San Francisco, 1982), and references therein.
2) A. Coniglio, Phys. Rev. Lett. **46** (1981) 250; J. Phys. A **15** (1982) 3824.
3) A. Coniglio and H.E. Stanley, Phys. Rev. Lett. **52** (1984) 1068.
4) T.A. Witten and L.M. Sander, Phys. Rev. Lett. **47** (1981) 1400.
5) B.B. Mandelbrot, J. Fluid Mech. **62** (1974) 331.
6) See, e.g., U. Frisch and G. Parisi, in Proc. Varenna Summer School LXXXVII (1983); R. Benzi, G. Paladin, G. Parisi and A. Vulpiani, J. Phys. A **17** (1984) 3521.
7) H.G.H. Hentschel and I. Procaccia, Physica **8D** (1983) 435; P. Grassberger and I. Procaccia, Physica **13D** (1984) 34.
8) L. de Arcangelis, S. Redner and A. Coniglio, Phys. Rev. B **31** (1985) 4725; Phys. Rev. B, October 1986.
9) R. Rammal, C. Tannous, P. Breton and A.M.S. Tremblay, Phys. Rev. Lett. **54** (1985) 1718; R. Rammal, C. Tannous and A.M.S. Tremblay, Phys. Rev. A **31** (1985).
10) Y. Park, A.B. Harris and T.C. Lubensky, Phys. Rev. B, in press.
11) F.M. Bhatti and J.W. Essam, J. Phys. A **19** (1986) L519; Y. Meir, R. Blumenfeld, A. Aharony and A.B. Harris, preprint.

12) P. Meakin, H.E. Stanley, A. Coniglio and T.A. Witten, Phys. Rev. A **32** (1985) 2364; P. Meakin, A. Coniglio, H.E. Stanley and T.A. Witten, Phys. Rev. B **34** (1986) 3325.
13) T.C. Halsey, P. Meakin and I. Procaccia, Phys. Rev. Lett. **56** (1986) 854.
14) C. Amitrano, A. Coniglio and F. di Liberto, Phys. Rev. Lett. **57** (1986) 1016.
15) T.C. Halsey, M.H. Jensen, L.P. Kadanoff, I. Procaccia and B.I. Shraiman, Phys. Rev. A **33** (1986) 1141.
16) C. Castellani and L. Peliti, J. Phys. A **19** (1986) L429.
17) M.E. Cates and T.A. Witten, Phys. Rev. Lett. **56** (1986) 2497.
18) H.E. Stanley, J. Phys. A **10** (1977) L211.
19) R. Pike and H.E. Stanley, J. Phys. A **14** (1981) L169.
20) L. Pietronero and A.P. Siebesma, preprint.
21) W. Kenkel and J.P. Straley, Phys. Rev. Lett. **49** (1982) 767.
22) R. Blumenfeld and A. Aharony, J. Phys. A **18** (1985) L443; A.B. Harris, Phys. Rev. B, in press.
23) L. de Arcangelis, A. Coniglio and S. Redner, J. Phys. A **18** (1985) L805.
24) P. Meakin, Phys. Rev. A **34** (1986) 710.
25) A. Coniglio and M. Marinaro, Physica **54** (1971) 261.

Physica **140A** (1986) 62–69
North-Holland, Amsterdam

DIFFUSION LIMITED AGGREGATION AND ITS RESPONSE TO ANISOTROPY

R.C. BALL

Cavendish Laboratory, Madingley Road, Cambridge CB3 OHE, UK

The stability of diffusion limited growths with n equivalent major fingers is investigated in two dimensions using a conformal mapping. The results imply that square lattice but not hexagonal lattice bias are relevant in simple Diffusion Limited Aggregation (DLA). In general it is found that the maximum number of fingers stable with respect to finger loss by competition is given by $n_{\max} = 2 + 2/(D - 1)$, where D is the apparent fractal dimension of the fingers, at least when n is even.

Computer simulations of DLA with n-fold symmetric growth rules are shown for $n = 5$–8, 12.

Finally it is argued that (unbiased) isotropic fractal DLA should lie at the stability limit. When combined with cone angle arguments independently relating D to the effective numbers of fingers, this determines the value $D^* = 1 + \frac{1}{2}\sqrt{2} = 1.707\ldots$ for DLA in two dimensions. This corresponds to a characteristic exterior half angle at the leading tips given by $\beta^* = \pi/\sqrt{2}$.

1. Introduction

Several recent simulation studies have found that pure diffusion limited growth, as in the DLA model of Witten and Sander, is biased in its large scale morphology by local anisotropy of the growth rules. For uniaxial bias[1]) and also just a simple square lattice[2]), major fingers grow systematically along the "easy" directions; on hexagonal lattices[3,4,5]) the same effect has only been observed when the shot noise inherent in the original DLA model is strongly reduced.

These biased growths differ in their behaviour from unbiased (e.g., off-lattice) DLA in at least two ways, of which this paper directly addresses the second. The first is that they have unambiguously identifiable major fingers which are resistant to tip splitting, a feature which is presumably directly associated with the local bias systematically favouring its overall growth direction.

The second difference is that the competition between the major fingers is apparently *stable* in the sense that it does not lead to loss of fingers. This appears to defy the usual argument for DLA, that on all scales the fingers which happen to be slightly ahead will grow preferentially and eventually screen out others. It also contrasts with the classic analysis of Mullins and Sekerka[6]) applied to a spherical growth, showing that all angular harmonic perturbations above the second (i.e., ellipticity) are unstable. Note that diffusion on a lattice, and hence

0378-4371/86/$03.50

screening, grows rapidly isotropic with distance so that any bias cannot directly impede the competition between fingers.

The first objective of this paper is to understand how the Mullins–Sekerka instability is modified for a growth of $n = 2m$ symmetrically arranged fingers, at least in two dimensions. We obtain a result for stability which depends on the radius-mass exponent D of the growth, and is consistent with a series of clusters grown with deposition rules of various symmetries.

The relevance of the stability criterion to unbiased DLA will then be discussed, leading to a possible self-consistent argument as to its fractal dimension.

2. Tip growth and charge singularity

The rate of advance of a major finger, whose size is measured by the length R to its tip, is given by

$$\frac{1}{a}\frac{\mathrm{d}R}{\mathrm{d}N} = k\frac{q_{\mathrm{tip}}}{Q_{\mathrm{cluster}}}, \tag{2.1}$$

where a is a particle radius, k is a dimensionless geometrical factor (assumed equal for all fingers) and q_{tip} and Q_{cluster} are, respectively, the rates at which particles deposit on the tip and on the whole cluster[7]).

The quantities q_{tip} and Q_{cluster} can also be thought of as corresponding charges in the electrostatic analogy, where the cluster is a well isolated charged conductor. If the structure grows with $N \sim R^D$ (as for, but not necessarily, a fractal) then we have

$$\frac{q_{\mathrm{tip}}}{Q_{\mathrm{cluster}}} \cong \left(\frac{b}{R}\right)^{D-1}, \tag{2.2}$$

where b is a microscopic length. This may be interpreted[1,8,9]) in terms of a singularity in the charge distribution such that the charge $q(r)$ within distance r of the tip varies as r^{D-1}, with a cutoff length b equal to the tip radius.

It is important to note that eq. (2.2) is the key definition of D in this paper and is not intrinsically the fractal dimension (though by integration of eqs. (2) it *is* the exponent of the radius–mass relation of the cluster). Thus for example if the absorbing fingers have the geometry of a straight line one finds $D = \frac{3}{2}$ from the scaling of their tip charge singularity[1]), and in Rossi's branchless DLA model[10]) one would have $D - 1 = 0$ here since the absorbers are isolated points.

Note also that because the nature of the tip charge singularity depends on the surrounding absorber shape, it is preserved locally under smooth conformal transformations. This is a crucial motivation in the analysis below where we need to relate perturbations in the growth rates q_{tip} to perturbations in the finger lengths.

3. Multi-finger stability

We consider an even number $n = 2m$ of symmetrically disposed competing fingers in two dimensions and their stability under diffusion controlled growth with respect to finger loss. We *assume* that the fingers can preserve their overall shape and scaling properties under growth, and that they are equivalent in these respects; subject to these constraints they are of arbitrary structure. The scaling of the Mullins–Sekerka instability[6]) suggests that the most unstable perturbation in the finger lengths is that of highest wavevector, the alternating mode for even n, and we assume that it suffices to study this to establish stability with respect to finger loss. Thus we take alternate fingers to be slightly ahead and behind the mean growth with lengths $R_{\pm}$ and tip charges $q_{\pm}$ (see fig. 1), and seek the dependence of q_+/q_- on R_+/R_- when both are close to unity.

If the cluster is centered on $z = 0$ in the complex z plane and oriented as shown then the conformal mapping

$$z' = \left[z^m - \tfrac{1}{2}(R_+^m - R_-^m)\right]^{1/m}, \tag{3.1}$$

which adequately preserves the topology of the plane for small perturbations, has the property that in the complex z' plane: the tip shapes are preserved and the fingers are rendered of equal length. However the (microscopic) tip radii are different in z' because

$$b'/b = [\mathrm{d}z'/\mathrm{d}z]_{\text{tip}} = (z/z')^{m-1} \tag{3.2}$$

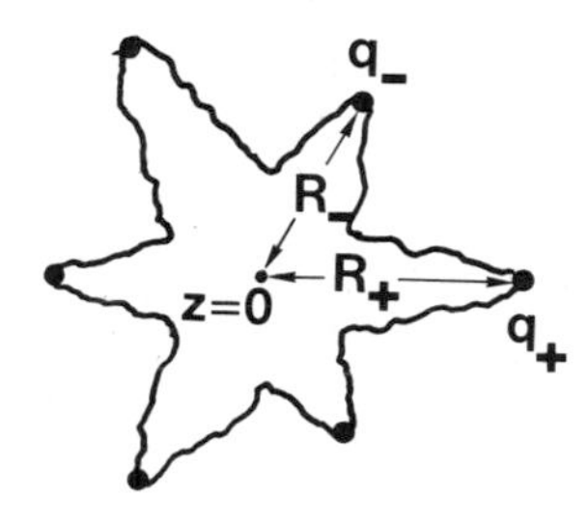

Fig. 1. Schematic view of a cluster with $n = 6$ major fingers, subject to an alternating perturbation such that their lengths are alternately R_+ and R_-. The corresponding tip advance rates are $q_{\pm}$, and the cluster is centered on the origin in the complex z plane.

from which it follows that

$$b'_+/b'_- = (R_+/R_-)^{m-1}. \tag{3.3}$$

We now make a further important assumption, that the large scale aspects of the cluster and its charge singularities are restored to full n-fold symmetry in the z' plane. Then the tip charges differ only through the tip radii and we have

$$q_+/q_- = (b'_+/b'_-)^{D-1} = [R_+/R_-]^{(D-1)(m-1)}, \tag{3.4}$$

where the charges are preserved (hence unprimed) under conformal transformation.

Eq. (3.4) gives the relative growth rates of the two classes of finger as

$$\mathrm{d}R_+/\mathrm{d}R_- = [R_+/R_-]^{(D-1)(m-1)} \tag{3.5}$$

so that the relative amplitude of the alternating perturbation grows or declines according to whether

$$(m-1)(D-1) \gtrless 1. \tag{3.6}$$

One check on the analysis is provided by the case $D = 2$, which corresponds in scaling to the perturbation of a compact circular growth, for which the critical symmetry is indeed equivalent to $m = 2$. At the opposite extreme the calculation is explicitly exact for line ($D = \frac{3}{2}$) absorbers and exact up to logarithms for point absorbers ($D = 1$). The result in the latter case indicates that in a circular geometry version of branchless DLA[10]), any number of fingers would be stable for R large enough.

It is convenient to restate the result (3.6) in terms of the maximum number of stable fingers $n_{\max}$ as

$$(\tfrac{1}{2}n_{\max} - 1)(D - 1) = 1. \tag{3.7}$$

For needles, that is unbranched fingers where – as in the uniaxial bias model – the width/length ratio of the fingers indefinitely decreases so that $D = \frac{3}{2}$ as per a straight line absorber, we have

$$n_{\max} = 6 \qquad (D = \tfrac{3}{2}), \tag{3.8}$$

whereas for fingers of fractal Witten–Sander scaling

$$n_{\max} \cong 4.9 \qquad (D = 1.7). \tag{3.9}$$

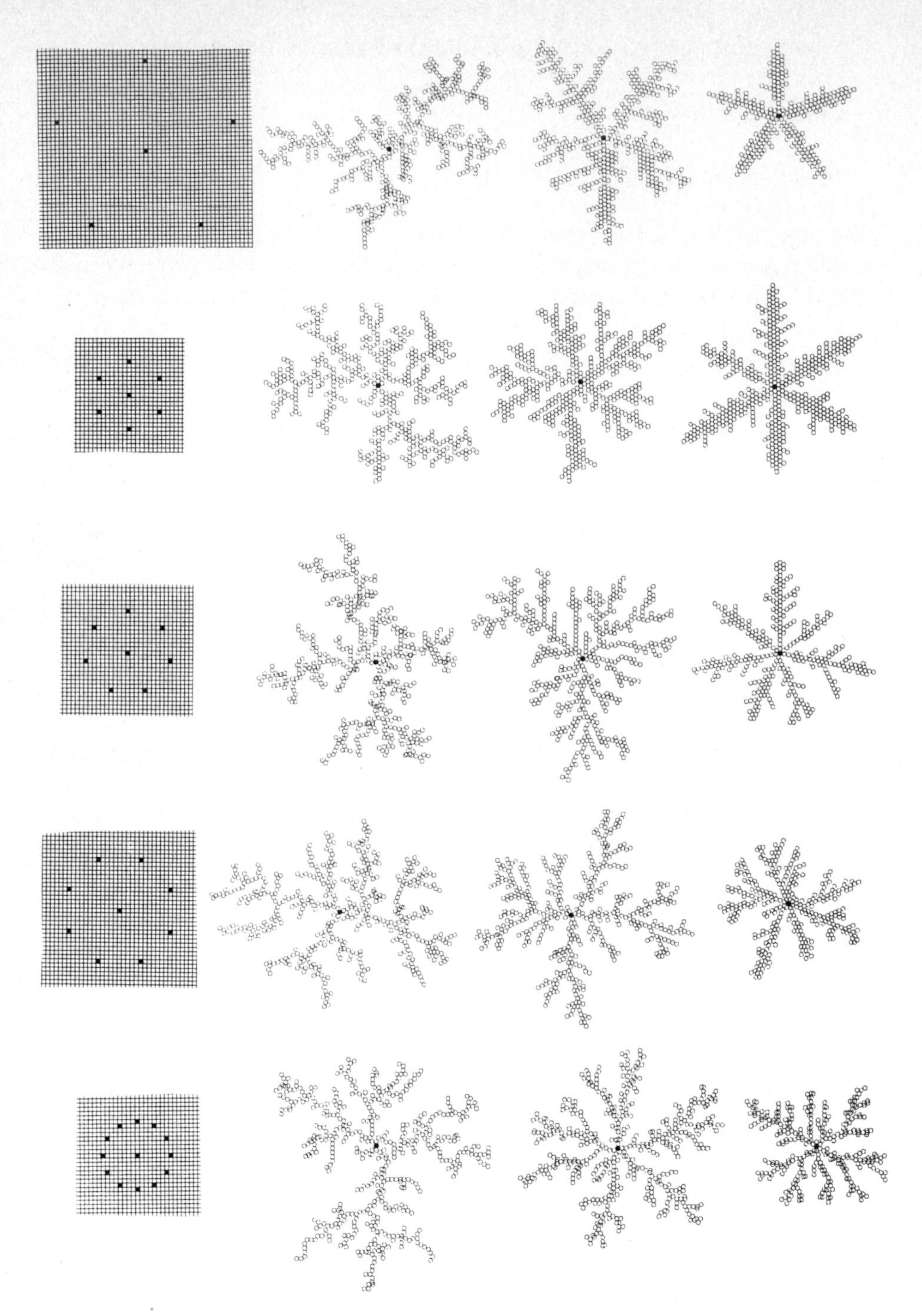

Fig. 2. DLA clusters grown with deposition rules of (approximate) n-fold symmetry for $n = 5, 6, 7, 8, 12$ from top to bottom. The noise was reduced from left to right by requiring respectively $M = 1, 10, 100$ diffusing particles absorbed ("hits") for each site grown. The deposition patterns (left) show the surrounding ring of sites which became new absorbers when the central one grew, all drawn to the same scale on the underlying square lattice used in these simulations. The clusters are drawn with particle diameter corresponding to the radius of the deposition ring, and their seed marked.

4. Simulations

The above results unambiguously support the relevance of square lattice bias in DLA simulations, in that four major fingers are *stable* for $D < 2$, and suggest that hexagonal lattice bias is irrelevant because $n_{max} < 6$ for $D > 1.5$ as in the fractal case[10]).

Some further simulation tests of these predictions are shown in fig. 2. The clusters were grown with an open sticking rule* on an underlying square lattice as shown. When a cluster site was grown, a ring of n almost equivalent sites around it became absorbers for growth. The spacing of this absorber pattern was varied with n to exploit particularly good approximations to n-fold symmetry on the lattice. Each absorber was in turn grown when it had absorbed ("been hit by") M walkers, following Tang's idea[12]) to reduce the noise in the growth process.

The values of $n = 4, 6$ corresponding to exact lattices are equivalent in symmetry to other studies[2,3-5]). As the clusters show, the competition for growth is such that other values of n do not lead to excessive densities even though simple steric hindrance is absent.

At $M = 100$ hits, which is close to noiseless in these small simulations, needle-like fingers clearly begin to grow for n up to 6, whereas for $n = 7$ there is strong sidebranching and some tip splitting, and for $n = 8, 12$ the full complement of fingers is never present. At $M = 10$ hits the full symmetry of the deposition rule is barely discernible even for $n = 5$.

These clusters are compatible with n_{max} in the range 5–7, but such other possibilities as n_{max} being simply an increasing function of M cannot be excluded. A much larger simulation study is clearly required.

5. Relevance to isotropic DLA

Although no systematic count of the number of major fingers of unbiased DLA clusters has been established, such features can nevertheless be clearly seen qualitatively. Furthermore the large fingers must have some stability to grow to their attained size without loss of identity.

It has recently been argued that their charge singularities might be modelled[1,8,9]) as the tips of conical absorbers, of full external angle 2β for which, in $d = 2$,

$$D - 1 = \pi/2\beta. \tag{5.1}$$

*I am indebted to R. Brady and E. Samulski for discussion of their results on various local sticking rules.

If in turn we adopt the assumption of a polygonal profile[8]) (so that the angle at each major tip is consistent with being defined by the neighbouring ones), then we have

$$n(2\beta - \pi) = 2\pi \tag{5.2}$$

giving a direct relation between the fractal dimension and the number of tips:

$$D - 1 = \frac{n}{n+2}. \tag{5.3}$$

We now postulate that n is further constrained to equal n_{max} given by eq. (3.5), since if $n < n_{max}$ then it will increase by underimpeded splitting of major fingers, whereas for $n > n_{max}$ the growth is unstable and will lose fingers through competition as we have shown.

In terms of n we then have flow to a stable fixed point obeying

$$\frac{n}{n+2}\left(\tfrac{1}{2}n - 1\right) = 1 \tag{5.4}$$

which has a positive root at

$$n = n^* = 2 + \sqrt{8} = 4.828\ldots \tag{5.5}$$

to which corresponds

$$D = D^* = 1 + \tfrac{1}{2}\sqrt{2} = 1.707\ldots. \tag{5.6}$$

This value of the fractal dimension is numerically quite acceptable but the meaning of non-integral n^* needs discussion. If this argument applies to fractal DLA which is self-similar, then it should apply at all large scales (i.e., above noise) within the cluster also, and not just to a simple circular geometry. Then it becomes more natural to think in terms of the selected tip (exterior half-) angle[13]),

$$\beta^* = \frac{\pi}{\sqrt{2}}\ \text{rad} = 127.3\ldots^\circ \tag{5.7}$$

and the existence of this characteristic angle in DLA becomes a crucial question to test.

6. Discussion

The stability results of this paper depend on a "weak" conformal assumption which is correct in the limits $D = 1.5$ and $D = 2$ but may only be an interpolation in between. However the conclusions of square lattice relevance and hexagonal lattice irrelevance in DLA lie out at these boundary cases and so appear beyond doubt.

Whether rigorous or not, the present analysis clearly exhibits how the multi-finger growths do have a greater angular stability than the circular case, and gives a fresh insight as to the self-consistency of fractal growth.

The values at the fixed point for fractal DLA depend also on: the cone angle model, the polygonal geometry assumption, and the extension of all arguments to non-integral n. The result for the fractal dimension is so appealing as to motivate further work to justify these.

Finally, all the ideas in this paper readily generalize from two to higher dimensions, but the conformal mapping technique used to derive the results does not (adequately). As well as trying to free the analysis of this limitation, we can also investigate the corresponding questions of lattice relevance and characteristic (solid) angles by simulation in 3 and higher dimensions. It promises to be difficult but interesting.

Acknowledgments

The author wishes to acknowledge stimulating discussions with the following: L. Sander, R. Savit, R. Brady, E. Samulski, R. Stinchcombe. He would also like to thank the organisers of Statphys 16.

References

1) R.C. Ball, R.M. Brady, G. Rossi and B.R. Thompson, Phys. Rev. Lett. **55** (1985) 1040.
2) R.C. Ball and R.M. Brady, J. Phys. A **18** (1985) L809.
B.R. Thompson, manuscript in preparation.
3) R.M. Brady and E. Samulski, Letter submitted to Nature.
4) J. Nittman and H.E. Stanley, Nature **321** (1986) 663; note that the boundary conditions used in this study differ slightly from those of DLA (see ref. 5).
5) R.C. Ball, R.M. Brady, E. Samulski and B.R. Thompson, Comment submitted to Nature.
J. Kertesz and T. Vicsek, J. Phys. A **19** (1986) L257.
6) See: J.S. Langer, Rev. Mod. Phys **52** (1980) 1.
7) R.C. Ball and T.A. Witten, Phys. Rev. A **29** (1984) 2966.
8) L.A. Turkevich and H. Scher, Phys. Rev. Lett. **55** (1985) 1026.
9) T.C. Halsey, P. Meakin and I. Procaccia, Phys. Rev. Lett. **56** (1986) 854.
10) G. Rossi, to appear in Phys. Rev. A.
11) C. Tang, Phys. Rev. A **31** (1985) 1977.
12) R. Stinchcombe, private discussion.

Physica **140A** (1986) 70–77
North-Holland, Amsterdam

SOME RIGOROUS RESULTS FOR RANDOM AND QUASI-PERIODIC POTENTIALS

Thomas SPENCER
School of Mathematics, Institute for Advanced Study, Princeton, NJ 08540, USA

We review some mathematical results and conjectures for random and quasi-periodic Schrödinger operators. The role of localization in certain nonlinear systems is also discussed. In addition we present some general arguments which show that the correlation length exponent $\nu \geqslant 2/d$ for a wide class of d dimensional random systems including the dilute Ising ferro-magnet.

1. Introduction

This talk will focus on some recent mathematical results and conjectures for the Schrödinger operator $H = -\varepsilon^2\Delta + v$, defined on the lattice $\mathbb{Z}^d$ or the continuum $\mathbb{R}^d$. The potential will be denoted by $v(x)$ or $v(j)$ for $x \in \mathbb{R}^d$ and $j \in \mathbb{Z}^d$ and the lattice laplacian is defined by

$$\Delta f(j) = \sum_{i:|i-j|=1} (f(i) - f(j)).$$

The constant ε will be specified later.

We shall be primarily interested in the behavior of solutions to the equation $H\psi = E\psi$. The analysis of this equation for the class of potentials considered below generally involves overcoming small divisor problems. Roughly speaking this is achieved by a multi-scale analysis related to KAM (Kolmogorov, Arnold, Moser) methods and to renormalization group techniques.

Four classes of potentials will be considered corresponding to different physical situations.

1) *Periodic potentials*. Here the hamiltonian describes, for example, the motion of non-interacting electrons in a perfect crystal at zero temperature.

2) *Random potentials on* $\mathbb{Z}^d$. We assume that $v(j)$ are independent random variables with a common bounded distribution density $g(v)\,\mathrm{d}v$. For example if g is uniformly distributed over the interval $[-w, w]$ then H is called the Anderson tight binding hamiltonian which he introduced and analysed to study electrons in crystals with impurities[1]).

3) *Quasi-periodic potentials on* $\mathbb{Z}^1$ *or* $\mathbb{R}^1$. We restrict ourselves to the case where there is just one incommensurate frequency α present and the potential has the form:

a) $v(x) = \cos(x) + \cos(\alpha x + \theta), \quad x, \theta \in \mathbb{R}$,
b) $v(j) = \cos 2\pi(\alpha j + \theta), \quad j \in \mathbb{Z}$.

The number α is assumed to be poorly approximated by rationals. More precisely we assume

$$|\sin n\pi\alpha| \geqslant C_0/n^2. \tag{1}$$

The set of irrationals satisfying (1) for some C_0 has measure one. Although we shall state our results for the cosine, this function can be replaced by any smooth, even, periodic function with precisely two non-degenerate critical points.

Quasi-periodic potentials naturally arise if one linearizes about a quasi-periodic orbit. See 4) below. They also arise in the study of a quantum electron moving in a perfect crystal subjected to a periodic potential whose period is incommensurate with that of the lattice.

4) *Nonlinear systems*. Let us consider for example the discrete time pendulum or standard map defined by the recursion relation

$$\varepsilon^2(\Delta x)_j \equiv \varepsilon^2(x_{j+1} + x_{j-1} - 2x_j) = \sin x_j. \tag{2}$$

This naturally defines an area preserving transformation of the torus to itself. Note that x_j depends on the initial data (x_0, x_1) hence so does the potential defined by $v(j) = \cos(x_j)$, $j \in \mathbb{Z}$. If we wish to determine the sensitivity of the orbit x_j on x_0 we differentiate (2) with respect to x_0 and obtain the equation

$$H\psi = 0, \quad \text{where } \psi(j) = \partial x_j/\partial x_0.$$

In this case we see that the Schrödinger equation arises from linearizing about a nonlinear orbit. Most of our comments on (2) will be conjectural.

We shall also present some recent results on an infinite chain of nonlinearly coupled oscillators whose hamiltonian has the form

$$\sum_i p_i^2 + w_i^2 q_i^2 + \varepsilon f(q_i - q_{i+1}), \tag{3}$$

where the w_i are assumed to be independent random variables and f is an analytic function which satisfies $f(q) = \mathcal{O}(q^4)$ for small q.

2. Results and conjectures

1) When $\varepsilon > 0$ it is well known that the spectrum for H in the periodic case is purely absolutely continuous. Moreover all generalized eigenfunctions (i.e. polynomially bounded solutions to $H\psi = E\psi$) are bloch waves

$$\psi(x) = p(x)\exp(ikx),$$

where p is a periodic function having the same period as v.

2) For the case $v(j)$ is random, let us first note that the spectrum of H when $\varepsilon = 0$ is an interval of dense pure point spectrum with probability one. In fact the eigenvalues are $v(j)$ and the eigenfunctions are delta functions on the lattice. Gold'sheid, Molchanov and Pastur[2]) were the first to prove that a similar picture holds when $d = 1$ for all ε. More precisely they proved that there is a basis of exponentially localized eigenstates, i.e. localization. Physically localization implies that under time evolution the electron's wave packet does not spread and the electron is trapped. Thus there is no conductivity or diffusion in the system. Recently Carmona, Klein and Martinelli (preprint), have extended these results to the case where the distribution is singular, e.g. $v_j = \pm 1$.

When $d \geqslant 2$ it is known that there is always an interval (near the edge of the spectrum), consisting only of localized states, provided that the density g is bounded. Furthermore if ε is small, then all states are localized. These results were independently obtained by a number of authors, Simon and Wolff[3]), Delyon, Levy and Souillard[4]), and Fröhlich, Martinelli, Scoppola and Spencer[5]). When $d = 2$ all states are believed to be localized for all ε, but there are no rigorous results of this kind. When $d \geqslant 3$ there should be a band of absolutely continuous spectrum $[E_{\mathrm{m}}, E'_{\mathrm{m}}]$ corresponding to extended states provided ε is large. However the only results of this kind are known when the lattice $\mathbb{Z}^d$ is replaced by the Bethe lattice[6]). E_{m} and E'_{m} are called mobility edges, since electrons should be mobile in this energy range and conduction should occur.

The basic estimate needed to establish the above results on localization is the exponential decay of the Green's function or equivalently when $d = 1$ the positivity of the Liapunov exponent. Let E be fixed. In one dimension we say that the Liapunov exponent $\gamma(E)$ is positive if with probability one

$$(H - E)\psi(j) = 0$$

only has solutions which grow at an exponential rate $\gamma(E)$ as j goes to + or − infinity. Equivalently the Green's function satisfies

$$|G(E,0,j)| \equiv |(E-H)^{-1}(0,j)| \leqslant C_v \exp -\gamma(E)|j|. \tag{4}$$

where C_v is a v dependent constant which is finite with probability one. In one dimension the positivity of γ is due to Furstenberg[7]), see also Borland[8]). Recently Kotani[9]) showed that $\gamma(E) > 0$ for almost all E for a wide class of "non-deterministic" potentials.

For $d \geqslant 2$ the decay of G when either ε is small or E lies in the band tail is due to Fröhlich and Spencer[10]). We refer to refs. 11 and 12 for mathematical reviews.

3) *Quasi-periodic case*. Some time ago Dinaburg and Sinai[13]) proved that in the continuum there are always high energy Bloch type eigenfunctions:

$$\psi(x) = q(x)\exp(-ik(E)x)$$

via KAM methods. Here q is a quasi-periodic function of x. On the lattice, in special case $v(j) = \cos(\alpha j + \theta)$, Aubry[14]) and Herman[15]) have very elegant proofs of the positivity of $\gamma(E)$ for all E when $\varepsilon^2 < \frac{1}{2}$ and Bellissard et al.[16]) showed that for small ε there are exponentially localized eigenstates. The following results are due to Fröhlich, Wittwer and Spencer[17]) and hold for a set of θ of measure 1.

For case 3a) there are an infinite set of low energy eigenstates which decay exponentially fast provided ε is small.

For case 3b). If ε is small there are only localized states. In other words exponentially decaying eigenstates form a basis. Furthermore, these eigenstates have 2^n peaks where $n = 0, 1, 2, \ldots$ which are *self-similar* under reflection. For example if ψ is a wave function with two peaks at 0 and l, then there is a constant C such that for $|j| \leqslant l/3$,

$$C\psi(j) = \psi(l-j) + \mathcal{O}(\exp -l). \tag{5}$$

If the wave function ψ has four peaks at $0 < l_1 < l_2 - l' < l_2$, then $l' = l_1$ and $l_2 \geqslant \exp cl_1$. Moreover

$$|v(0) - v(l_2)| \leqslant \exp(-\text{const.}\, l_2)$$

which by (1) and the eveness of cosine implies

$$|v(j) - v(l_2 - j)| \leqslant \exp -cl_2.$$

This relation implies (5) with l replaced by l_2.

Remarks. Sinai has also proved localization for case 3b. We believe that the methods for the lattice can be extended to prove that for the continuum (case a):

i) If ε is large there are only Bloch type eigenstates.

ii) If ε is small there are only localized states at low energy. These states exhibit a self-similar structure as described for case b).

4) An outstanding problem for standard map is to prove that the Liapunov exponent $\gamma(E = 0) > 0$, i.e. $\psi(j) = \partial x_j/\partial x_0$ grows exponentially fast in j, for a set of (x_0, x_1) of positive measure. Equivalently the Green's function $H^{-1}(0, j)$ decays exponentially fast. This means that there is sensitive dependence on initial data and the theory of Pesin assures us that there is an ergodic component of positive measure on which the orbit moves "stochastically". Unfortunately the only results of this kind are known for certain piecewise smooth nonlinearities as in the case where the sine in (2) is replaced by a sawtooth function. We also expect that the spectrum of H consists of dense point spectrum – (localized states) for almost all (x_0, x_1) provided ε belongs to a Cantor set of positive measure and $|\varepsilon| \ll 1$. Hence for ε in the Cantor set we expect $v(j) = \cos(x_j)$ to behave as in the random case. We remark that the Green's function G also has small divisors E_i^{-1}, since $E = 0$ is in the spectrum of H. This reflects the nonuniform hyperbolicity of the dynamical system (2). However these small divisors are of an entirely different kind from those encountered in conventional KAM situations where one is searching for integrable or quasi-periodic motions.

Next we briefly describe some results for the infinite chain of oscillators (3) done in collaboration with Fröhlich and Wayne[18]). We show that if ε is small, there are infinite dimensional, invariant tori of spacially localized, time almost periodic solutions to the equations of motion with high probability. In particular for some fixed i, $|p_i^2(t) + q_i^2(t)| \geqslant \text{Const.} > 0$ and the local energy does not go to zero. If the ω_j are all equal, then typically there are theorems which show that the local energy goes to zero since the wave packet can spread. The existence of these localized waves follows by using a variant of KAM methods. Much work remains to be done in understanding how general solutions to the equations of motion behave. Also our methods cannot yet handle nonlinearities of the form $f(q) = \mathcal{O}(q^2)$ for small q unless $f(q) = q^2$ in which case the problem is linear and equivalent to a random Schrödinger equation. See ref. 19 for related results.

3. Ideas on critical exponents

We now turn to a discussion of the correlation length ξ and the associated critical exponent ν for a class of random systems which include the dilute Ising model, percolation and the random potential. In the case of random potentials, $\xi(E)^{-1}$ is the exponential rate of decay of the Green's function $G(E)$ for $E < E_{\mathrm{m}}$ where E_{m} is the mobility edge. Let Λ be a cube centered at the origin of width $2L$ and let G_Λ be the Green's function of H restricted to Λ with Dirichlet boundary conditions. We define $\chi_L(E, v)$ to be the characteristic function of the event

$$\left\{ v:\ \max_{|y| \leqslant L/2} \sum_{x \in \partial\Lambda} \left| G_\Lambda(E, y, x) \right| > \tfrac{1}{2} \right\}.$$

Then

$$P_L(E) = \int \chi(E, v) \prod_{i \in \Lambda} g(v_i)\, \mathrm{d}v_i$$

is the probability that the event occurs and measures the sensitivity of G_Λ near the origin to the boundary conditions.

Theorem. There is a $c_0 > 0$ such that if $P_L(E) \leqslant c_0$ for some L, then the Green's function $G(E)$ decays exponentially fast with probability one and the correlation length $\xi(E) \leqslant$ Const. $L(E)$ where $P_{L(E)}(E) \cong c_0$. See refs. 10 and 11 for details.

The same result holds for the dilute Ising ferromagnet with E_{m} replaced by β_{c} (the inverse critical temperature) and $|G(E, y, x)|$ is replaced by $\langle \sigma_y, \sigma_x \rangle(\beta)$.

We now describe work done in collaboration with J. and L. Chayes and D. Fisher[20]). It suggests that $\nu \geqslant 2/d$ for a wide class of random systems. J. and L. Chayes have already proven this fact for percolation. There are three basic ingredients for the argument:

i) $P_L(E) = P_L(E \pm \varepsilon L^{-d/2}) + \mathcal{O}(\varepsilon)$.

Proof. Since $\chi(E + \delta, v + \delta) = \chi(E, v)$ and $\chi \leqslant 1$ we have

$$P_L(E + \delta) = \int \chi(E, v) \prod_{i \in \Lambda} g(v_i + \delta)\, \mathrm{d}v_i$$

and

$$\tfrac{1}{2}|P_L(E + \delta) - P_L(E)| \leqslant \int \left| \int_0^\delta \mathrm{d}s \frac{\mathrm{d}}{\mathrm{d}s} \prod_{i \in \Lambda} g(v_i + s) \right| \prod_{i \in \Lambda} \mathrm{d}v_i .$$

By the Schwarz inequality,

$$\tfrac{1}{2}|P_L(E+\delta)-P_L(E)| \leqslant \int_0^\delta \mathrm{d}s \int \left| \sum_{i\in\Lambda} \frac{g'(v_i+s)}{g(v_i+s)} \right| \prod_{i\in\Lambda} g(v_i+s)\,\mathrm{d}v_i$$

$$\leqslant \delta \left[\int \sum_{i,j\in\Lambda} \frac{g'(v_j)}{g(v_i)} \frac{g'(v_j)}{g(v_j)} \prod_{k\in\Lambda} g(v_p)\,\mathrm{d}v_p \right]^{1/2}$$

$$\leqslant \delta|\Lambda|^{1/2} \int \frac{g'(v_0)^2}{g(v_0)}\,\mathrm{d}v_0 \leqslant \delta L^{d/2}\,\text{const.}$$

In the last line we used the fact that $i \neq j$ terms give 0 contribution. i) now follows easily for a wide class of smooth g.

ii) There is a $c_1 \geqslant c_0$ such that

$$P_L(E_\mathrm{m}) \geqslant c_1 \quad \text{for all } L.$$

This result is an easy consequence of the theorem, for if $P_L(E_\mathrm{m}) < c_0$ for some L then $G(E_\mathrm{m}+\varepsilon)$ for some small ε decays exponentially. This contradicts the definition of the mobility edge.

iii) const.$L(E)/\log L(E) \leqslant \xi(E) \leqslant$ const.$L(E)$. The upper bound follows from the theorem. The lower bound can also be established for percolation and the dilute Ising ferromagnet. It has not yet been established for random potentials. However $L(E)$ is often used as a working definition of the correlation length.

To prove that $\nu \geqslant 2/d$ choose $c_0 \leqslant \frac{1}{2}c_1$ and note that

$$P_{L(E)}(E_\mathrm{m}) - P_{L(E)}(E) \geqslant c_1 - c_0 \geqslant c_0 > 0.$$

Hence by i)

$$|E_\mathrm{m} - E| \geqslant \varepsilon L(E)^{-d/2}$$

for some small ε depending only on c_0. Clearly by iii) $\xi(E) \geqslant (E_\mathrm{m} - E)^{-2/d}$, modulo logarithmic corrections. Analogous arguments hold for dilute ferromagnetic Ising models and for percolation.

Remark. There is as yet no satisfactory mean field theory for localization and the upper critical dimension still remains a topic of some controversy.

References

1) P. Anderson, Phys. Rev. **109** (1958) 1492.
2) I. Gold'sheid, S. Molchanov and L. Pastur, Funct. Anal. App. **11** (1977) 1.
3) B. Simon and T. Wolff, Comm. Pure App. Math. **39** (1986) 75.
4) F. Delyon, Y. Levy and B. Souillard, Commun. Math. Phys. **100** (1985) 463.
5) J. Fröhlich, F. Martinelli, E. Scoppola and T. Spencer, Commun. Math. Phys. **101** (1985) 21.
6) H. Kunz and B. Souillard, J. Physique Lett. **44** (1983) L411.
7) H. Furstenberg, Trans. Amer. Math. Soc. **108** (1963) 377.
8) R. Borland, Proc. Roy. Soc. **A274** (1963) 529.
9) S. Kotani, Proc. Toniguchi Conf., Katata, 1982 and Proc. AMS Conf. on Random Matrices, J. Cohen, ed., 1985.
10) J. Fröhlich and T. Spencer, Commun. Math. Phys. **88** (1983) 151.
11) T. Spencer, in Proc. Les Houches Summer School 1984, K. Osterwalder and R. Stora, eds., to appear.
12) R. Carmona, Ecole d'Eté de Probabilities XIV, Saint Flour, 1984. Lecture notes in Mathematics 1180.
13) E. Dinaburg and Ya Sinai, Funct. Analysis and App. **9** (1975) 279.
14) S. Aubry, Solid State Sci. **8** (1978) 264.
15) M. Herman, Comment Math. Helvetici **58** (1983) 453.
16) J. Bellissard, R. Lima and D. Testard, Commun. Math. Phys. **88** (1983) 207.
17) J. Fröhlich, T. Spencer and P. Wittwer, to appear.
18) J. Fröhlich, T. Spencer and C. Wayne, J. Stat. Phys. **42** (1986) 247.
19) M. Vittot and J. Bellissard, to appear.
20) J. Chayes, L. Chayes, D. Fisher and T. Spencer, to appear.

Physica **140A** (1986) 78–84
North-Holland, Amsterdam

RIGOROUS RENORMALIZATION GROUP AT WORK

K. GAWĘDZKI

C.N.R.S., I.H.E.S., 91440 Bures-sur-Yvette, France

We illustrate the recent progress in rigorous renormalization group on a simple example of multiscale analysis of a first order phase transition.

The *renormalization group* (RG)[1]) has been devised as a tool to study *quantum fields* and related *critical models* of statistical mechanics, characterized by diverging *correlation length*. By iterative analysis of many distance scales, from short to long ones, it translates the properties of a (nearly) critical system into those of a flow of RG transformations around *fixed points*. The flow takes place in infinite-dimensional space of *effective Hamiltonians* and is essentially driven by a few-dimensional dynamical system. Here resides the key to the success of the RG as an approximate scheme and, at the same time, the difficulty in turning the RG into a rigorous tool.

The way toward rigorous RG was long and has not yet been fully explored. Let us list some of its milestones:

1. *phase-space cell expansion* analysis of the ϕ_3^4 field theory[2]) introducing rigorous multiscale analysis independently of the parallel development of the RG ideas;
2. RG analysis of the *ultraviolet stability* in ϕ_3^4 [3]);
3. proof of the *Kosterlitz–Thouless phase transition* in two-dimensional Coulomb gas[4]);
4. proof of *permanent confinement* in three-dimensional lattice U(1) gauge theory[5]);
5. rigorous control of dipole gas[6]);
6. construction of *massless lattice* ϕ_4^4 theory[7]);
7. construction of the first *renormalizable asymptotically free* quantum field theory: massive Gross–Neveu model in two space–time dimensions[8]).

On the side of open problems, let us mention:

8. construction of more complicated renormalizable asymptotically free quantum field theories, like *σ-models* in two dimensions and *non-abelian gauge theories* in four dimensions[9]);

0378-4371/86/$03.50 © Elsevier Science Publishers B.V.
(North-Holland Physics Publishing Division)

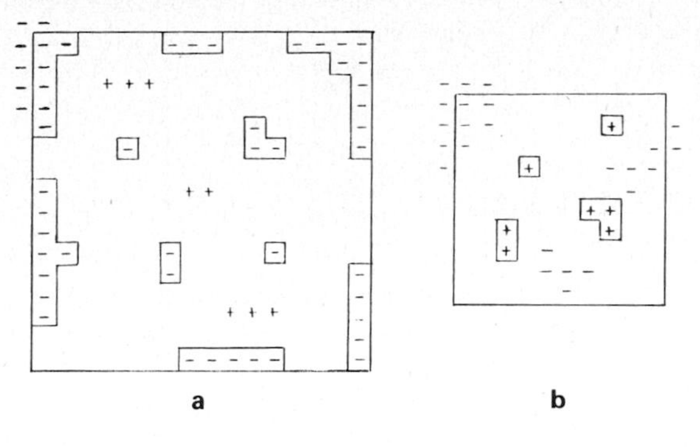

Fig. 1.

9. control of criticality governed by *non-gaussian fixed points* of the RG, as in the case of massless lattice ϕ^4 theory or Gross–Neveu quantum field in three dimensions, say, for a large number of components[10]).

In the above, we have totally omitted rigorous multiscale analyses of disordered systems, a subject in its own right; see, e.g., ref. 11.

In what follows, instead of considering any of the topics 1–9, all technically rather heavy, I shall discuss an easy application of the rigorous RG: the control of first order phase transitions at low temperatures[12]). This application shares with the others the essential feature that it combines a multiscale analysis with the convergent expansions technique used to analyse each scale.

One might think that the first order transitions constitute precisely the topic with no use for the RG analysis as the correlation length stays short in the transition region. There is however another length scale in the problem which actually diverges at the transition point. The RG is helpful in the analysis of its effects. To see how the new scale comes about, consider the low-temperature *Ising model* in a small positive *magnetic field* in two or more dimensions. Impose the negative (wrong) boundary conditions in a box Λ. If the box is big, then the typical configuration of spins looks like in fig. 1a. However, if the box is small, we shall observe a picture like fig. 1b. The size of the critical bubble in the false vacuum, that is the borderline size $\mathscr{L}$, is easily estimated by equating the energies gained and lost by the flip of signs inside the box,

$$\mathscr{L}^d h = 2\,\mathrm{d} J \mathscr{L}^{d-1}, \tag{1}$$

so that $\mathscr{L} = \mathcal{O}(J/h)$ and divergence occurs at the transition point (the correlation length is there equal to $\mathcal{O}(1/J)$.

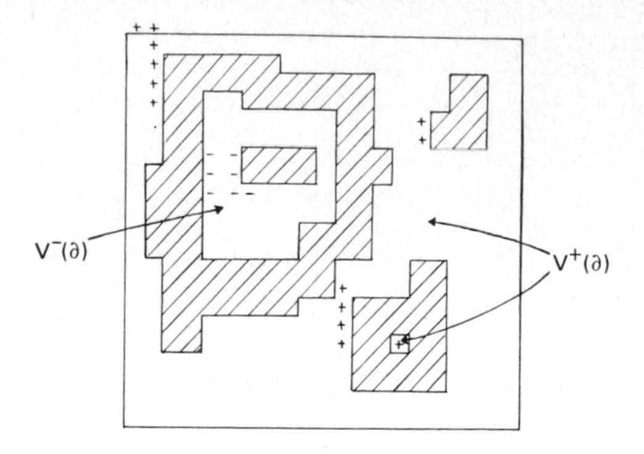

Fig. 2.

Consider a more general translation invariant Ising model with two approximate *ground states* given by the Hamiltonian

$$H(\sigma) = \tfrac{1}{4}J \sum_{\substack{x,y \\ |x-y|=1}} (\sigma_x - \sigma_y)^2 + \sum_{\substack{X \subset \mathbb{Z}^d \\ \operatorname{diam} X < R}} J_X \sigma^X, \tag{2}$$

with $J \gg 1$ and $|J_X| \ll 1$ except, possibly, the magnetic field term $J_{\{x\}}$. We would like to study the existence of the $\pm$ phases. For a spin configuration obeying one of the boundary conditions outside Λ, call a contour a connected component of the union of unit cubes centered at sites of Λ such that the configuration is not constant within radius R of them. The contours inherit from the spin configuration the signs of components of their boundary. Fix the boundary conditions outside Λ. A collection of signed contours $\partial = \{\gamma\}$ will be called compatible if it corresponds to a spin configuration obeying the boundary condition. Call then $V^{\pm}(\partial)$ the union of unit cubes centered at $\left\{\begin{smallmatrix}\text{positive}\\ \text{negative}\end{smallmatrix}\right\}$ sites of the configuration in $\Lambda \setminus \bigcup\gamma$; see fig. 2.

The partition function of the model in Λ can clearly be written as

$$Z_\Lambda^{\pm} = \sum_{\substack{\partial=\{\gamma\} \\ \text{compatible}}} \exp\left[\sum_{\pm} \sum_{x \in V^{\pm}(\partial)} h_x^{\pm}\right] \prod_{\gamma \in \partial} \zeta(\gamma), \tag{3}$$

where, away from $\partial\Lambda$,

$$h_x^{\pm} = \sum_{X \ni x} \frac{J_X}{|X|} (\pm 1)^{|X|} \equiv h^{\pm}, \tag{4}$$

and if $\pm(h^+ - h^-) > 0$, then

$$\left|\zeta(\gamma)\exp\left[-\sum_{x\in\gamma} h_x^{\pm}\right]\right| \leqslant e^{-\eta J|\gamma|} \tag{5}$$

for small η. Both h_x and $\zeta(\gamma)$ depend slightly on Λ in the vicinity of $\partial\Lambda$. We shall call expression (3) the partition function of the *contour model with external field*.

Let us consider first the case of a large external field

$$\pm(h^+ - h^-) > D \gg 1. \tag{6}$$

Take the + sign for example. Rewrite

$$Z_\Lambda^+ = \exp\left[\sum_{x\in\Lambda} h_x^+\right] \sum_{\substack{\partial=\{\gamma\}\\ \text{compatible}}} \exp\left[-\sum_{x\in V^-(\partial)} (h_x^+ - h_x^-)\right]$$

$$\times \prod_{\gamma\in\partial} \zeta(\gamma)\exp\left[-\sum_{x\in\gamma} h_x^!\right]. \tag{7}$$

Note that not only the contours but also the minus islands are now strongly suppressed. Dividing $(\cup\gamma)\cup V^-(\partial)$ into the connected components $\{X_\alpha\}$ (polymers) we easily obtain a representation

$$Z_\Lambda^+ = \exp\left[\sum_{x\in\Lambda} h_x^+\right] \sum_{\substack{\{X_\alpha\}\\ \text{disjoint}}} \prod_\alpha R(X_\alpha) \tag{8}$$

with the polymer activities

$$|R(X_\alpha)| \leqslant \exp\left(-\tfrac{1}{2}\min(\eta J, D)|X_\alpha|\right), \tag{9}$$

strongly suppressed. The polymer gas, whose partition function is given by (8) can be easily controlled in the thermodynamic limit. This leads to a construction of the + phase for our system.

Similarly, for the other sign in (6), the − phase exists. The question which will interest us more, however, is what happens for $|h^+ - h^-| \leqslant D$. The answer will be provided by an inductive RG analysis.

Let us first divide the contours into small, surrounding volumes $< L^d$ with $L = \varepsilon(J/D)$ (note the relation to scale $\mathscr{L}$), and big ones. For small contours the volume contributions from the external field inside the contour are always dominated by the damping of the contour activities. If we limit ourselves in the partition function in (3) to small contours only (distinguish the modified $Z_\Lambda^\pm$ by the tilde), then we obtain by an easy contour expansion

$$\tilde{Z}_\Lambda^\pm = \exp\left[\sum_{x\in\Lambda}\left(h_x^\pm + s_x^\pm\right)\right] \sum_{\substack{\{C_\alpha^\pm\} \\ C_\alpha^\pm \cap \Lambda \neq \varnothing \\ C_\alpha^\pm \not\subset \Lambda}} u\left(C_\alpha^\pm\right), \tag{10}$$

where $s_x^\pm$ is the free energy contribution essentially constant inside Λ except near the boundary.

$$|s_x^\pm| \leqslant \exp\left(-\tfrac{1}{2}\eta J\right). \tag{11}$$

$C_\alpha^\pm$'s are connected clusters of the expansion which decorate the boundary (surface terms)

$$\left|u\left(C_\alpha^\pm\right)\right| \leqslant \exp\left(-\tfrac{1}{2}\eta J |C_\alpha^\pm|\right). \tag{12}$$

In the original expression (3) for $Z_\Lambda^\pm$, let us fix the set of big contours $\bar{\partial}$ and let us resume the small ones in the volumes $V^\pm(\bar{\partial})$. This leads to

$$\begin{aligned} Z_\Lambda^\pm &= \sum_{\substack{\bar{\partial} \\ \text{compatible}}} \prod_{\gamma\in\bar{\partial}} \zeta(\gamma) \tilde{Z}^+_{V^+(\bar{\partial})} \tilde{Z}^-_{V^-(\bar{\partial})} \\ &= \sum_{\substack{\bar{\partial} \\ \text{compatible}}} \sum_{\substack{\{C_\alpha^\pm\} \\ C_\alpha^\pm \cap V^\pm(\bar{\partial}) \neq \varnothing \\ C_\alpha^\pm \not\subset V^\pm(\bar{\partial})}} \exp\left[\sum_\pm \sum_{x\in V^\pm(\bar{\partial})} \left(h_x^\pm + s_x^\pm\right)\right] \\ &\quad \times \prod_{\gamma\in\bar{\partial}} \zeta(\gamma) \prod_\alpha u\left(C_\alpha^\pm\right). \end{aligned} \tag{13}$$

As compared with the original expression, we have gotten rid of small contours at the price of modification of the external field and decoration of the long contours with clusters $C_\alpha^\pm$.

We are ready now to *coarse-grain* big contours. Consider all $L\times\cdots\times L$ cubes centered at points of $L\mathbb{Z}^d$ intersecting contours of $\bar{\partial}$ or clusters $C_\alpha^\pm$; see fig. 3. Let $\{L\gamma'\}$ be the collection of the connected components of their union. $\partial' = \{\gamma'\}$ naturally forms a compatible collection of contours in $\Lambda' = L^{-1}\Lambda$. Resumming in (13) $\bar{\partial}$ and $\{C_\alpha\}$ leading to fixed ∂', we obtain

$$Z_\Lambda^\pm = \sum_{\substack{\partial' \\ \text{compatible}}} \exp\left[\sum_\pm \sum_{x\in V^\pm(\partial')} h_x'^\pm\right] \prod_{\gamma'\in\partial'} \zeta'(\gamma'), \tag{14}$$

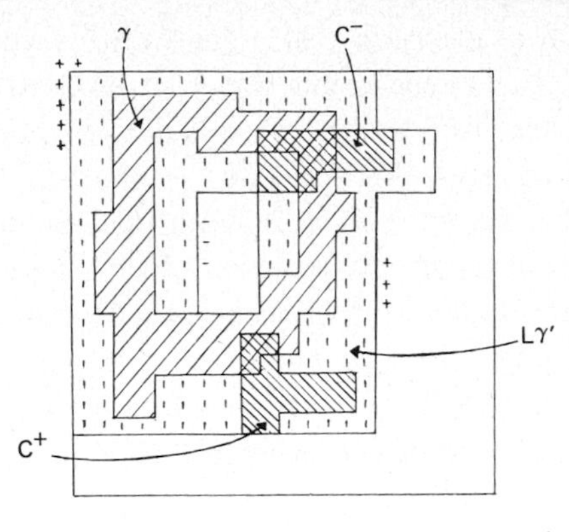

Fig. 3.

where

$$h_x'^{\pm} = \sum_{\substack{y \in L \times \cdots \times L \text{ block} \\ \text{centered at } Lx}} \left(h_x^{\pm} + s_x^{\pm}\right). \tag{15}$$

Fix γ'. Let $L\gamma'$ contain big contours Γ_i and possibly clusters $C_\beta^{\pm}$. We easily see that

$$|\gamma'| < \frac{2}{L}\left(\sum_i |\Gamma_i| + \sum_\beta |C_\beta^{\pm}|\right), \tag{16}$$

so that a bound

$$|\zeta'(\gamma')| < \exp(-L/5\eta J|\gamma'|) \equiv \exp(-\eta J'|\gamma'|) \tag{17}$$

follows for $D \ll J$.

Thus our coarse-graining RG transformation leaves us in the simple class of contour models with external field and lowers the temperature (by a factor $5/L$ at least) (note that (16) may fail for small Γ_i so that resummation of small contours was a necessary step in our RG procedure). What happens to the external field? By (15) and (11), in infinite volume,

$$h'^{\pm} = L^d\left[h^{\pm} + \mathcal{O}\left(\exp\left(-\tfrac{1}{2}\eta J\right)\right)\right], \tag{18}$$

so that the external field expands as a result of the coarse-graining ($h^+ - h^-$ is a relevant variable for the low temperature fixed point of the RG).

If $\pm(h'^+ - h'^-) > D$, then the $\pm$ phase may be constructed via the overall polymer expansion discussed above. For $|h'^+ - h'^-| \leqslant D$, which corresponds to a closed interval $I_1 \subset I_0 = \{h^+ - h^-: |h^+ - h^-| \leqslant D\}$, we coarse-grain again on a scale $L' = \varepsilon(J'/D)$. Following in this way the standard argument[13]) allowing relevant variables of the RG to be controlled, we choose the critical value $h_c \equiv (h^+ - h^-)_{\mathrm{crit}} \in \bigcap I_k$, such that for $\pm(h^+ - h^-) > h_c$ the $\pm$ phase exists. It is also easy to show that for $h^+ - h^- = h_c$ both phases coexist.

As compared to the old theory of first order phase transitions[14]), our approach has one essential virtue: it also works for complex Hamiltonians and allows one to extend the Gibbs phase rule to that case; compare with ref. 15. This is important since complex Hamiltonians arise naturally in the study of metastability or in the treatment of topological terms in lattice gauge theories.

References

1) K.G. Wilson, Rev. Mod. Phys. **55** (1983) 583.
2) J. Glimm and A. Jaffe, Fortschr. Phys. **21** (1973) 327.
3) G. Benfatto, M. Cassandro, G. Gallavotti, F. Nicoló, E. Olivieri, E. Presutti and E. Scacciatelli, Commun. Math. Phys. **71** (1980) 95.
4) J. Fröhlich and T. Spencer, Phys. Rev. Lett. **46** (1981) 1006; Commun. Math. Phys. **81** (1981) 527.
5) M. Göpfert and G. Mack, Commun. Math. Phys. **82** (1982) 545.
6) K. Gawędzki and A. Kupiainen, Ann. Phys. **147** (1983) 198.
7) K. Gawędzki and A. Kupiainen, Phys. Rev. Lett. **54** (1985) 92; Commun. Math. Phys. **99** (1985) 197; J. Feldman, J. Magnen, V. Rivasseau and R. Sénéor, Construction and Borel summability of infrared ϕ_4^4 by a phase space expansion, Ecole Polytechnique preprint, 1986.
8) K. Gawędzki and A. Kupiainen, Phys. Rev. Lett. **54** (1985) 2191; Commun. Math. Phys. **102** (1985) 1; J. Feldman, J. Magnen, V. Rivasseau and R. Sénéor, Phys. Rev. Lett. **54** (1985) 1479; Commun. Math. Phys. **103** (1986) 67.
9) K. Gawędzki and A. Kupiainen, Continuum limit of the hierarchical $O(N)$ non-linear σ-model, IHES preprint, 1986; T. Bałaban, Renormalization group approach to lattice gauge field theories, Harvard preprint, 1986.
10) G. Felder, Commun. Math. Phys. **102** (1985) 139; K. Gawędzki and A. Kupiainen, Phys. Rev. Lett. **55** (1985) 363; Nucl. Phys. B **262** (1985) 33.
11) J. Fröhlich and T. Spencer, Commun. Math. Phys. **88** (1983) 151; D. Fisher, J. Fröhlich and T. Spencer, J. Statist. Phys. **34** (1984) 863; J. Imbrie, Phys. Rev. Lett. **53** (1984) 1747; Commun. Math. Phys. **98** (1985) 145.
12) K. Gawędzki, R. Kotecký and A. Kupiainen, Coarse-graining approach to first order phase transitions, in preparation.
13) P. Bleher and Ya. Sinai, Commun. Math. Phys. **33** (1973) 23.
14) S. Pirogov and Ya. Sinai, Teor. Mat. Fiz. **25** (1975) 1185, **26** (1976) 39.
15) S. Pirogov, Teor. Mat. Fiz. **66** (1986) 331.

Physica 140A (1986) 85–95
North-Holland, Amsterdam

SYMBOLIC DYNAMICS AND SYSTEMATICS OF PERIODIC WINDOWS

Bai-lin HAO

Institute of Theoretical Physics, Academia Sinica, P.O. Box 2735, Beijing, China

Elementary symbolic dynamics of two or more letters, developed for one-dimensional mappings, may be of much help in studying ODE's. At least three periodically driven systems, including one studied in great detail by us, are known to be described in the large by symbolic dynamics of two letters. The skeleton of periodic windows in the Lorenz model is described by symbolic dynamics of three letters in close connection with the one-dimensional antisymmetric cubic map. The understanding of transitions from quasiperiodicity to chaos may also be deepened, looking at the global picture of the interrelation between the U-sequences and the Farey sequences.

1. Introduction

There are at least two fundamental levels to ask the problem of origin of stochasticity in physics. On the microscopic level there has been the long-standing problem of justifying statistical mechanics on the basis of deterministic mechanics. At the macroscopic level we face the problem of understanding various kinds of turbulence, starting from deterministic evolution equations of macro-variables. Recent progress in "chaology" has shed some light on both problems, but we shall confine ourselves to the latter.

Here one usually deals with nonlinear evolution equations of a small number of variables. Such systems may undergo sharp transitions into either of two categories of states as the control parameters vary continuously. The first category consists of states with temporal and/or spatial order and has been studied more or less thoroughly since the mid 60's under the name of nonequilibrium phase transitions, dissipative structures, synergetics and the like. The second category includes various transitions into apparently more chaotic states. These chaotic states are not simply disordered states. They should rather be compared to order without periodicity. Although known to mathematicians before in many particular cases, the chaotic transitions have only been considered as a new category of natural phenomena since the recent discovery (mainly by physicists) of universality and scaling properties associated with them and the observation of these phenomena in many computer and laboratory experiments.

A practical recipe to study bifurcation and chaos in computer experiments (and, to a certain extent, in laboratory experiments) consists of two steps. First, one should identify periodic and quasiperiodic regimes in the space of control parameters with as high as possible frequency resolution. What remains in the parameter space after separating regions of periodic and quasiperiodic regimes may correspond to chaotic motion. Then comes the second step, i.e., characterization of the chaotic attractors. This can be accomplished by studying various dimensions, entropies, and characteristic exponents, as it was explained in the review of J.-P. Eckmann and D. Ruelle[1]). We shall concentrate on the first step, emphasizing that systematics of periodic motions alone can provide useful information about the existence of quasiperiodic and/or chaotic regimes in the system.

2. Periods in chaotic sea – the U-sequences

One-dimensional mapping may be viewed as discrete time evolution equation. The most complete description of its dynamics requires the knowledge of the full set $\{x_n,\ n = 0 \text{ to } \infty\}$ of iterates starting from a given x_0. However, one can design certain coarse-grained description by ignoring the concrete numbers in the set, but retaining the essential feature of the evolution. To do so we divide the one-dimensional phase space into several segments, each corresponding to a monotonic branch of the map and separated by critical points of the map (and by end points of the map if those do not belong to critical ones). We label each monotonic branch of the map by a letter. Then every set $\{x_n\}$ may be replaced by a sequence of letters (if necessary critical points may require additional symbols). Of course, different sets $\{x_n\}$ may correspond to one and the same symbolic sequence, opening thus the possibility to introduce a classification scheme for the numerical sequences $\{x_n\}$'s. All numerical sequences, corresponding to one and the same symbolic sequence, will be considered as equivalent. It is clear that we may lose some numerical details of the iterations, but essential features of the evolution such as periodicity will be preserved.

Consider, e.g., the following maps[2]):

1) *The logistic map*

$$x_{n+1} = 1 - Ax_n^2, \qquad x_n \in (-1, 1) \tag{1}$$

which describes, among other things, the population dynamics of certain insects without generation overlap. The right hand side of eq. (1) is a symmetric function with only one maximum at the critical point $x_c = 0$, hence the name "unimodal" for this kind of maps. One can label its two monotonic branches by the letters R (Right) and L (Left).

2) *The antisymmetric cubic map*[3])

$$x_{n+1} = Ax_n^3 + (1 - A)x_n, \qquad x_n \in (-1, 1). \tag{2}$$

It has three monotonic branches and we need three letters R, M, and L to denote them. As we shall see, this map has a close relation[4]) to the celebrated Lorenz model[5]) in which one of the first strange attractors was observed.

3) *The sine-square map*

$$x_{n+1} = A \sin^2 (x_n - B), \qquad |x_n - B| \leqslant \pi \tag{3}$$

which has been used to model an optical bistable device using liquid crystal as the nonlinear medium[6]). The map (3) has two maxima and two minima. One has to use four letters to name its branches.

Unimodal maps enjoy the property that at each parameter value there exists at most one stable periodic orbit. In fact, a given period lives on a finite interval of parameter axis, called also a "periodic window". Somewhere in the middle of the interval there is a superstable period, which may be defined as the orbit, containing the critical point x_c as one of its periodic points. If we assign a letter R or L to each of the periodic points according to whether it falls to the Right or the Left of the critical point, we get a word corresponding to the period. Each period corresponds to a word made of R's and L's, but not every word describes a stable period. A word W is said to be admissible if it does correspond to a periodic orbit in the unimodal map. It is possible to introduce an ordering for all admissible words and to formulate rules to generate all the admissible words of a given period or to check the admissibility of a given word[7-9]). We skip the details and only give a few examples.

Example 1. The Feigenbaum period-doubling sequence is described by

Word:	blank	<R	<RLR	<RLRRRLR	<$\mathrm{RLR^3LRLRLR^3LR}$	< ⋯
Period:	1	2	4	8	16	⋯.

There is a *-composition rule to generate a period-doubled orbit from the preceding one[8]). Using this rule, one can present the above sequence as

blank < R < R*R < R*R*R < R*R*R*R < ⋯ .

In fact, there are infinitely many sequences of periodic windows, immerged in the chaotic sea, e.g., the period-tripling sequence

Word:	RL <	RL*RL <	RL*RL*RL <	RL*RL*RL*RL <	···
Period:	3	3^2	3^3	3^4	··· .

Example 2. There are altogether three period 5 orbits, namely,

$$RLR^2 < RL^2R < RL^3.$$

Example 3. The last of period n orbits must correspond to the word RL^{n-2}, consequently the word R represents the only period 2 orbit in unimodal mappings. All words of this type are ordered in the following way

Word:	R <	RL <	RL^2 <	RL^3 <	··· <	RL^{n-2} <	···
Period:	2	3	4	5	···	n	··· .

In the case of eq. (1) there is a one-to-one correspondence between the ordering of words and the increasing parameter value. However, the ordering of words may differ from the natural order on the parameter axis, if the height of the map does not depend on the parameter monotonically. Put in another way, although one can compile a dictionary of admissible words, the binding of the dictionary is not always of good quality: some parts of it may be bound repeatedly with the local order of words preserved or inverted.

In the case of more complicated maps such as eq. (2) or (3), there may appear basin dependence, i.e., at one and the same parameter value, there may exist more than one orbits, each with its own basin of attraction. However, one can still work out the rules to generate and order words made of 3 or 4 letters[3]). Following MSS[7]), we shall call all these ordered sequences of words U-sequences, U standing for universal.

3. Quasiperiodicity and the Farey sequences

Another class of phenomena, studied long before the "chaotic craze" has spread out, consists of frequency (or mode) locking and quasiperiodic motion whenever there are two or more competing frequencies in the system. The simplest model to exhibit quasiperiodicity is the "bare" circle mapping, i.e., the mapping of the unit circle into itself described by

$$x_{n+1} = x_n + A \qquad \mathrm{mod}(1). \tag{4}$$

If the winding number A is rational, say, $A = p/q$ with p, q being integers, we have a period of q points to be visited when making p turns around the unit circle. If A is irrational, we have quasiperiodic motion, i.e., the unit circle will be filled up by an infinite sequence of x_n without ending and repeating. Therefore, mode-locking occurs only at rational winding numbers and the totality of mode-locking regime has measure zero in the unit interval of A.

If one adds a nonlinear periodic term $g(x_n)$ to eq. (4)

$$x_{n+1} = x_n + A + Kg(x_n), \tag{5}$$

then the mode-locking regime acquires finite measure and becomes wider and wider with increasing K. These are the Arnold tongues in the A–K plane.

There are many ways to enumerate all rational numbers in the unit interval, a simple one being the Farey construction. In between any pair of rationals p/q and p'/q' there exists a rational $(p + p')/(q + q')$. Starting from 0/1 and 1/1, one can construct a Farey tree:

0/1 1/1

1/2

1/3 2/3

1/4 2/5 3/5 3/4

1/5 2/7 3/8 3/7 4/7 5/8 5/7 4/5

... ...

Eventually, any rational number will have a well-defined place in the Farey tree. Truncated at certain levels of the tree and projected to the unit interval, we get the Farey sequence, e.g.,

$$0/1 \quad 1/4 \quad 1/3 \quad 2/5 \quad 1/2 \quad 3/5 \quad 2/3 \quad 3/4 \quad 1/1.$$

It happens that in typical cases of nonlinear systems one sees mode-locking intervals ordered according to one or another Farey sequence, depending on the frequency resolution reached. Put in other words, in the quasiperiodic sea, periodic windows are ordered according to the Farey sequence. Therefore, looking at the interrelation among periods one can guess the nature of nearby aperiodic regime. This is the simple idea we are going to apply to differential equations.

4. Symbolic description of periods in ODE's

Many physical systems are modelled by ordinary differential equations. We are concerned in the first place with dissipative systems where dissipation plays a globally stabilizing role against local orbital instability and leads to the formation of strange attractors. Roughly speaking, there are three classes of ODE's to be considered: (1) Autonomous systems with 3 or more variables. (2) Non-autonomous systems with 2 or more variables. (3) Time-delayed systems with 1 or more variables. Technically, non-autonomous periodically driven systems are easier to study, since very high frequency resolution can be reached using the driving frequency at one's disposal, while time-delayed systems are too complicated to allow for a more or less thorough study at the present time[10]).

We have studied in detail the following periodically forced system[11])

$$\begin{aligned} \dot{X} &= A - (B+1)X + X^2Y + \alpha\cos(\omega t), \\ \dot{Y} &= BX - X^2Y. \end{aligned} \tag{6}$$

To present the results we use two other parameters instead of A and ω. Doing the linear stability analysis of the free oscillator, one gets two eigenvalues $\lambda_{\pm}$ for the linearized problem. Write them as

$$\lambda_{\pm} = \gamma\omega \pm i\beta\omega. \tag{7}$$

measuring the real and imaginary parts in units of the external frequency ω. In particular, β is the ratio of the limit cycle frequency in the linearized regime to the driving frequency. It happens that (γ, β) is a better plane to see mode-lockings, quasiperiodicity-to-chaos transitions and the relation between the U-sequence and the locking frequencies. Fig. 1 is the γ–β phase diagram. Not going into detailed explanation of this diagram, we indicate only that near and above the $\gamma = 0.05$ line the periods are given by the U-sequence up to period 6. As a matter of fact, we could assign letters to the observed periods and the words thus obtained are ordered strictly in accordance with the U-sequence.

Fig. 1 shows clearly the global nature important for understanding the transition from quasiperiodicity to chaos. Only in a strip of the parameter space one sees chaotic regimes separated by periodic windows described by the U-sequence, while in the lower and upper parts of fig. 1 we have mode-locking tongues, ordered according to the Farey sequences and embedded in the quasiperiodic sea. The chaotic regimes are confined to rather small regions, part of the boundary being shown in fig. 2.

Fig. 1 can be viewed as a distorted picture of the corresponding phase diagram for the circle mapping, studied by, e.g., Belair and Glass[12]) recently. It is

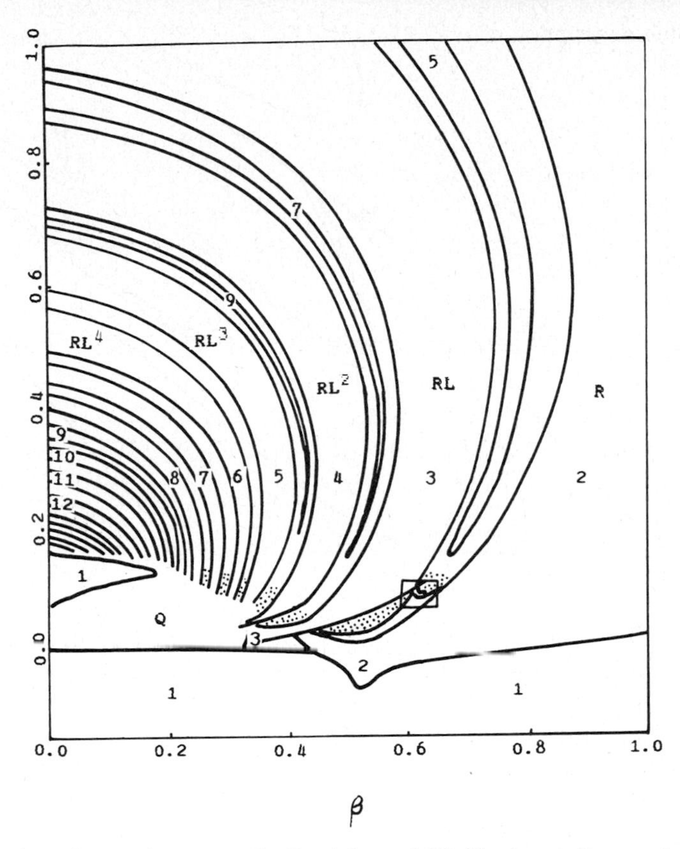

Fig. 1. A γ–β phase diagram for system (6) ($B = 1.2$, $\alpha = 0.05$). Numbers indicate periods; Q-quasi-periodic; dotted area–chaos.

interesting to note that if one keeps staying in the winding number 1/2 region, one can enter the main period-doubling sequence, passing successfully the winding numbers 2/4, 4/8, 8/16, etc. However, one can never form rational fractions with odd denominators (i.e., for periods 3, 5, 7, . . .), keeping the value of the winding number equal to 1/2. This is why the complete U-sequence as seen in fig. 1 is formed by involving bent periodic sequences originated from other winding numbers.

To our knowledge, there exist at least two other examples of similar interrelation between the U-sequence and the Farey sequence in systems of ODE's. The first one is a driven nonlinear oscillator with exact solution, studied by Gonzalez and Piro[13]). In their fig. 1, the U-sequence could be seen clearly, though it was not stated explicitly in the paper. Another example comes from an ODE model describing the yeast glycolysis[14]). Their theoretical results are shown in fig. 3. In

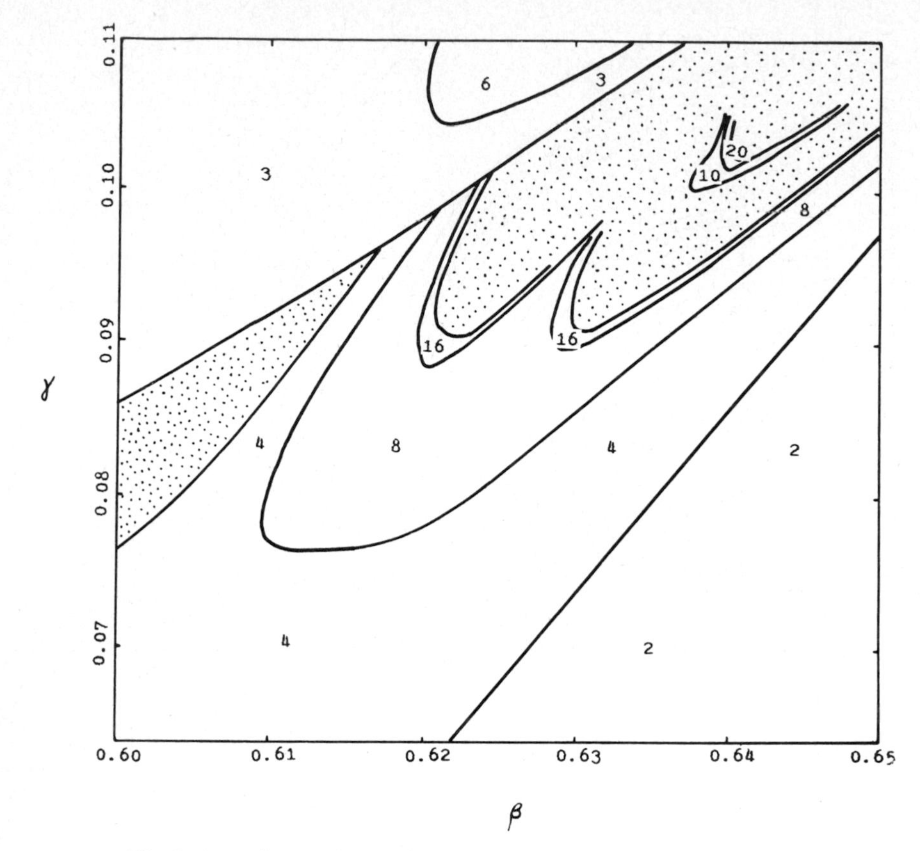

Fig. 2. A small part of fig. 1 showing a corner of chaotic zone boundary.

both cases a little more effort would reveal the nature of the ordering of the periodic windows. It seems to be dynamics of two letters too.

The systematics of periodic windows in the Lorenz model[5])

$$\begin{aligned} \dot{x} &= \sigma(y - x), \\ \dot{y} &= rx - xz - y, \\ \dot{z} &= xy - bz \end{aligned} \tag{8}$$

requires at least three letters due to the discrete anti-symmetry of system (8) with respect to $x \to -x$, $y \to -y$, which makes it closer to the antisymmetric cubic map (2). In an autonomous system like (8) the fundamental frequency of the motion may drift as the parameter varies and this seems to prevent us from associating each encountered period with an absolute name like "period 5".

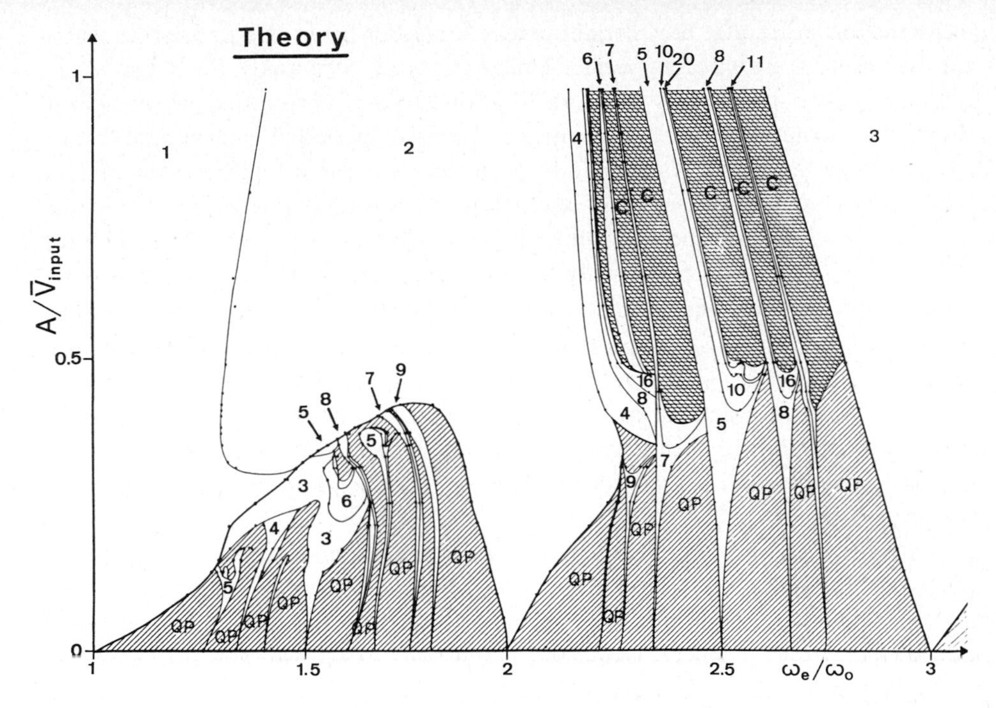

Fig. 3. The theoretical dependence of glycolyzing yeast extract on the sugar input flux frequency ω_e and amplitude A in an ODE model[14]).

Nevertheless, by carrying out extensive power spectrum analysis for all available periodic orbits, changing parameter r with σ and b fixed at the usual values, one can obtain a calibration curve for the fundamental frequency[15]). All the windows known before and newly discovered by us have been assigned a period referring to this curve. To tell the relation between different periods we devise a way to assign letters to the numerically computed Poincare map. It is a remarkable fact that 47 out from 53 prime periods fit into the "cubic" scheme, i.e., appear in the same order as they appear in the antisymmetric cubic map (2), see ref. 4 for the words thus assigned and the parameter values. Anyway, apart from a few exceptions the skeleton of the systematics for the occurrence of periodic windows in the Lorenz model obeys the "cubic" law.

5. Symbolic dynamics and the Farey sequences

When chaos and quasiperiodicity coexist in one and the same model, they must be interrelated somehow. However, U-sequence and Farey sequence are so different in nature that they cannot extend smoothly from one to another when the parameters are suitably varied. Exceptions can occur only for a few short periods, where uniqueness guarantees a one-to-one correspondence between them. For example, in the U-sequence the only period 2 and period 3 are described by the words R and RL, respectively, whereas in the Farey sequences there exists an unique period 2 (described by winding number 1/2) and an unique period 3 (1/3 and 2/3 only differ in the sense of rotation). Therefore, these periods can develop smoothly from mode-locking tongues in the quasiperiodic sea into periodic windows interspersed in the chaotic bands, as fig. 1 actually shows. It is clear that this cannot be the general rule for higher periods, e.g., for period 11 there are 93 different prime words in the U-sequence but only 5 different rational fractions in the Farey sequence. Most of the U-sequence periods must close in themselves or terminate somewhere. What was just said provides the physical reason for the folding and bending of the periodic sequences in fig. 1.

A striking feature of fig. 1 is the long period-adding sequence[16]) (resolved from period 2 to period 21). It connects the type $1/n$ winding numbers in the quasiperiodic regime and the RL^{n-2} words in the chaotic region. This is the only established case of direct connection between periods in the two regimes, including periods 2 and 3 discussed above.

In the upper part of fig. 1, embedded in between the period-adding sequence, there is a sequence of odd periods $5, 7, \ldots, 13$. They correspond to winding numbers $2/5, 2/7, \ldots, 2/13$ in the Farey tree and can be assigned numerically the words $RLRL, RL^2RL^2, \ldots, RL^5RL^5$, all inadmissible in the symbolic dynamics of two letters. The relation between type $1/n$ and $2/(2n+1)$ winding numbers on one hand, and type RL^{n-2} and $RL^{n-1}RL^{n-1}$ words on the other hand can be visualized as shown in figs. 4a and b. This opens the possibility of symbolic classification of Farey tree members.

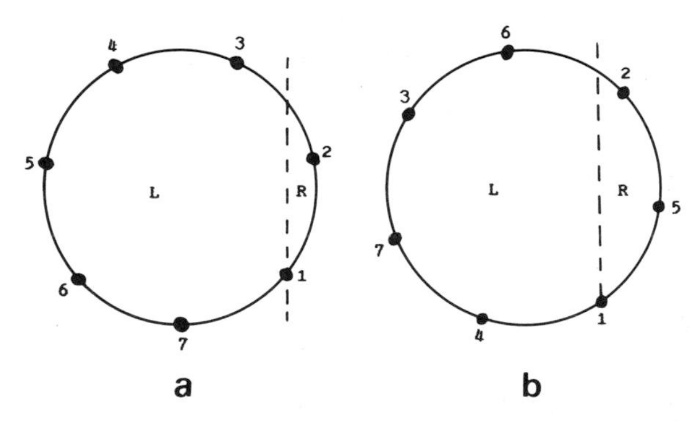

Fig. 4. The relation between winding numbers and words (schematic) (a) $A = 1/7$ RL^5; (b) $A = 2/7$ RL^2RL^2.

To conclude, we mention that U-sequences have been observed in laboratory experiments as well[17,18,14]). The deviation from U-sequence reported in[18]) was related to non-monotonic dependence of the underlying mapping on the parameter as it was discussed in section 2.

References

1) J.-P. Eckmann and D. Ruelle, Rev. Mod. Phys. **57** (1985) 617.
2) B.-L. Hao, Elementary Symbolic Dynamics, Lectures given at the ICTP Spring College on Order and Chaos in Nonlinear Physical Systems (Plenum) to be published.
3) W.-Z. Zeng, M.-Z. Ding and J.-N. Li, Chinese Phys. Lett. **2** (1985) 293; ASITP-86-013, Commun. Theor. Phys., to be published.
4) M.-Z. Ding and B.-L. Hao, Commun. Theor. Phys., to be published.
5) E.N. Lorenz, J. Atmosp. Sci. **20** (1963) 130.
6) H.-J. Zhang, J.-H. Dai, P.-Y. Wang, C.-D. Jin and B.-L. Hao, Chinese Phys. Lett. **2** (1985) 5; Commun. Theor. Phys., to be published.
7) N. Metropolis, M.L. Stein and P.R. Stein, J. Combin. Theory A **15** (1973) 25.
8) B. Derrida, A. Gervois and Y. Pomeau, Ann. Inst. Henri Poincaré **A29** (1978) 305.
9) P. Collet and J.-P. Eckmann, Iterated Maps on the Interval as Dynamical Systems (Birkhauser, Boston, 1980).
10) J.-N. Li and B.-L. Hao, ASITP-85-021, Commun. Theor. Phys., to be published.
11) For a recent review see: B.-L. Hao, Bifurcations and chaos in the periodically forced Brusselator, ASITP-86-011, to appear in a book honoring the retirement of Professor K. Tomita.
12) J. Balair and L. Glass, Physica **16D** (1985) 143.
13) D.L. Gonzalez and O. Piro, Phys. Rev. Lett. **50** (1983) 870.
14) M. Markus, S.C. Muller and B. Hess, personal communication, in preparation for publication.
15) M.-Z. Ding, B.-L. Hao and X. Hao, Chinese Phys. Lett. **2** (1985) 1.
16) K. Kaneko, Collapse of Tori and Genesis of Chaos in Dissipative Systems (World Scientific, Singapore, 1986).
17) R.H. Simoyi, A. Wolf and H.L. Swinney, Phys. Rev. Lett. **49** (1982) 245.
18) K. Cofmann, W.D. McCormick and H.L. Swinney, Phys. Rev. Lett. **56** (1986) 999.

Physica **140A** (1986) 96–104
North-Holland, Amsterdam

HOW DO OSCILLATING FRONTS FORM?

P. COLLET

Laboratoire de Physique Théorique, Ecole Polytechnique, Palaiseau, France

and

J.-P. ECKMANN

Département de Physique Théorique, Université de Genève, Switzerland

In this paper, we study a fourth order semilinear parabolic equation on the infinite real line. We show that in a certain parameter range, this equation has propagating front solutions (solutions tending to 0 at $+\infty$ and advancing to the right with a speed c) which leave behind them a *periodic* pattern in the laboratory frame. This is thus an example of spontaneous pattern formation.

1. Introduction

Despite impressive progress, the mathematical theory and the physical understanding of solidification presents a number of important problems. The most prominent are:

–do solidification fronts form?

–if yes, what is their stability?

–if there are several solutions, which one is selected "naturally?

The conjectures and ideas of Langer and Müller-Krumbhaar are among the most interesting and intriguing because they propose a mechanism of selection based on marginal stability[1]). This conjecture has only been verified for the equation

$$\partial_t u = 4\,\partial_x^2 u + u - u^3, \qquad u = u(x,t), \quad u(-\infty) = 1, \quad u(+\infty) = 0, \tag{1}$$

in a beautiful paper by Aronson and Weinberger[2]). (Their argument is based on the maximum principle which prevents a "crossing" of two solutions corresponding to different speeds.)

Despite numerous efforts, convincing answers to the selection problem are missing in other, more general cases. This had led a number of people to abandon

0378-4371/86/$03.50

the conjecture altogether, and to propose different equations for solidification, containing what could be called a "stabilizing" term. The marginal problem is thus converted into a new one, which, be it right or not, is less intriguing than the original question.

Therefore, we believe it is still interesting to study problems with marginal stability on the infinite line, and we want to study a class of problems which lead simultaneously to

–pattern formation,
–a moving front,
–marginal stability.

A typical equation which exhibits these three properties is

$$\partial_t U = \left(\varepsilon - \left(1 + \partial_x^2\right)^2\right) U - \varepsilon U^3, \qquad \varepsilon > 0. \tag{2}$$

In ref. 3 we have shown that (2) has a *two-parameter family of fronts* leaving a *periodic* pattern in the *laboratory frame*. The two parameters are the speed of the front and the wavelength of the pattern in the laboratory frame. The proof that one of the solutions is marginal is being elaborated[4]). On the other hand, we have unfortunately nothing to report on the selection problem.

Before we outline some of the highlights of our argument, we would like to point out a few basic difficulties of eq. (2):

–In order to talk about a front, we need an adequate *definition*. Several possibilities can be envisaged, and our proposal seems the most natural.

–Numerical calculations are very unstable because of the marginal stability, combined with the unwanted effects of finite volume. Our existence proof for the fronts is *constructive* and leads in principle to a *convergent expansion* in powers of $\varepsilon^{1/2}$. Thus it could be used as a constructive input to test, e.g., the marginality conjecture.

2. The definition of fronts

The basic principle of our definition is that we want to view a front as a solution connecting the zero solution (at $x = +\infty$) to a *stationary* solution, and advancing with speed $c > 0$ from left to right.

What are the stationary solutions? In the case of eq. (1), we have the stationary solution $s(x) = 1$*. (We are not interested in negative solutions.) In the case of

*By a stationary solution, we mean a solution u of the form $u(x, t) = s(x)$.

eq. (2), stationary solutions are given by

Theorem [ref. 3, Proposition 3.1]. For sufficiently small $\varepsilon > 0$ there is for every ω sufficiently close to 1 a stationary solution $U(x, t) = S(x)$ of (2) which is of the form

$$S(x) = \sum_{n=1}^{\infty} S_n \cos(\omega n x). \tag{3}$$

In fact, we require $|1 - \omega^2| < K\varepsilon$. In perturbation theory, one finds $S_n = 0$ for even n and

$$S_1 = 4/3^{1/2} + \mathcal{O}(\varepsilon),$$

$$S_n = \mathcal{O}(\varepsilon^{(n-1)/2}) \quad \text{for } n \geqslant 3.$$

(In the proofs, we get slightly worse bounds.) All the S_n are real.

What are now fronts? A front for eq. (1) is a solution u of eq. (1) which is of the form

$$u(x, t) = v(x - ct), \tag{4}$$

connecting the stationary solutions $s(x) \equiv 1$ and $s(x) \equiv 0$, i.e.,

$$\lim_{\xi \to -\infty} v(\xi) = 1, \qquad \lim_{\xi \to +\infty} v(\xi) = 0. \tag{5}$$

Our *definition* of a front for eq. (2), leaving a pattern with frequency ω in the laboratory frame is that of a solution U of eq. (2) which is of the form

$$U(x, t) = W(x, \varepsilon^{1/2} x - \varepsilon c t), \tag{6}$$

where

$$W(x, \xi) = \sum_{n \in \mathbb{Z}} W_n(\xi) e^{i\omega n \xi} \tag{7}$$

and

$$\lim_{\xi \to -\infty} 2W_n(\xi) = S_n, \tag{8}$$

$$\lim_{\xi \to +\infty} 2W_n(\xi) = 0. \tag{9}$$

Thus, we require the *amplitudes* to converge. Note that $W_n(\xi)$ is not real, but $W_n = \overline{W_{-n}}$, and contrary to what the notation suggests, the nodes of $W(x, \xi)$ need *not* be at $x = 2n\pi/\omega$, $n \in \mathbb{Z}$. They will, however, move to these positions when $t \to \infty$, i.e., when $\xi \to -\infty$.

Note further that the speed of the front, if it exists, is $\varepsilon^{1/2}c$, and that the W_n form a sort of multicomponent hull of the periodic pattern.

Main theorem. For sufficiently small $\varepsilon > 0$, for every $c > 4$ sufficiently close to 4, $|\omega^2 - 1| < \mathcal{O}(\varepsilon)$ there exists a front of the form (9) and moving with speed $\varepsilon^{1/2}c$.

In fact, perturbation theory suggests that

$$W(x, \xi) \approx \frac{4}{3^{1/2}} l(\xi) \cos(\omega x), \tag{10}$$

where $l(x)$ is a front of the problem (1), moving with speed c,

$$u(x, t) = l(x - ct), \tag{11}$$

i.e., l satisfies

$$4l'' + cl' + l - l^3 = 0, \qquad l(-\infty) = 1, \qquad l(+\infty) = 0. \tag{12}$$

3. Hints about the proof

Our first idea is to convert the problem to a fixed point problem. Substituting (7) into eq. (2) we are led to an infinite system of coupled equations

$$A_{\omega n} W_n = \varepsilon \sum_{p+q+r=n} W_p W_q W_r, \qquad n \in \mathbb{Z}. \tag{13}$$

Here,

$$A_{\omega n}(q) = \eta^2 - \left(1 - (\omega n + \eta q)^2\right)^2 + i\eta^2 cq \equiv P(q), \tag{14}$$

where $q = -\mathrm{i}\partial_x$ and $\eta = \varepsilon^{1/2}$. Basically, we would like to rewrite eq. (13) as

$$W_n = \varepsilon A_{\omega n}^{-1} \sum_{p+q+r=n} W_p W_q W_r = \varepsilon F_n(W), \qquad n \in \mathbb{Z}. \tag{13a}$$

Then, if we can define F_n, we could rewrite the existence problem as a fixed point problem for the map εF_n. To see whether this dream is possible (in fact it will be for $|n| \neq 1$), we analyze the roots q_j of $A_{\omega n}(q)$, and use the identity

$$A_{\omega n}^{-1} f(x) \approx \sum_j \int \mathrm{d}y \frac{\mathrm{e}^{\mathrm{i}q_j(x-y)}}{P'(q_j)} f(y). \tag{15}$$

This leads to a bound

$$\|A_{\omega n}^{-1} f\| \approx \sum_j \frac{1}{|\mathrm{Im}\, q_j P'(q_j)|} \|f\|.$$

For a suitable sup-norm and since the roots are located at positions as shown in fig. 1, for $n \neq 1$ (and $\mu \equiv \omega n \gg 1$), we see that

$$\eta^2 \|A_{\omega n}^{-1}\| \approx \eta^2 \frac{1}{\eta^4 \eta^{-3} \omega n} \approx \frac{\eta}{n}.$$

[The factor η^{-4} is the coefficient of q^4 in P, $|\mathrm{Im}\, q_j| \approx \mathcal{O}(\mu^{1/4}\eta^{-3/4})$ and P' is bounded by the product of differences of roots.]

Things are very different for $n = 1$ (or $n = -1$), since potentially in the ω, c region we are considering, two of the roots of P may coincide, rendering $1/P'$ unbounded.

In fact, when $n = 1$ (and, for simplicity, assuming also $\omega = 1$), we see that $P(q)$ is of the form

$$\eta^2\left\{1 - \left(2q + \eta q^2\right)^2 + \mathrm{i}cq\right\}.$$

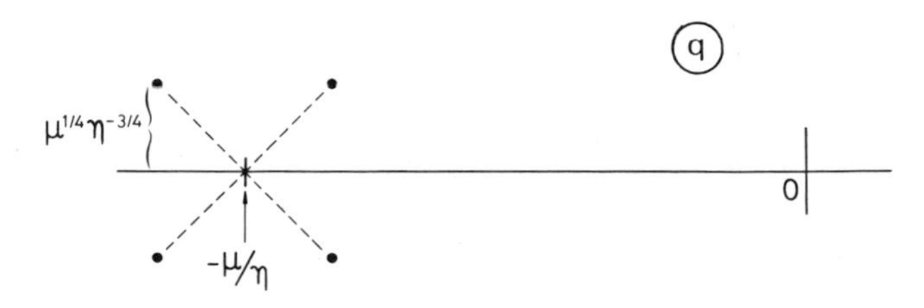

Fig. 1. Typical spectrum for $|n| \gg 1$.

Eq. (13) now becomes

$$\{1-(2q+\eta q^2)^2+\mathrm{i}cq\}W_1=\sum_{p+q+r\neq 1} W_pW_qW_r, \tag{16}$$

and when $\eta \to 0$, there are no small terms left, and the polynomial $1-4q^2+\mathrm{i}cq$ has for $c=4$ a *double root* at $q=-\mathrm{i}/2$.

But note now that this polynomial is the *same* as the one associated with the differential operator in eq. (12)! Furthermore, the sum on the r.h.s. of (16) contains a term $3W_1^2W_{-1}$. In view of the bounds on the S_n, we proceed with the analysis on the assumption that all other terms on the r.h.s. are at least $\mathcal{O}(\varepsilon)$. This assumption will turn out to be correct. Thus we rewrite (16) as

$$\{1-(2q+\eta q^2)^2+\mathrm{i}cq\}W_1-3W_1|W_1|^2=\sum_{p+q+r=1}^{*} W_pW_qW_r, \tag{16a}$$

where Σ^* excludes the three terms we have put on the other side. Clearly, the l.h.s. is, up to a trivial change of scale, the *same* as in eq. (12), apart from terms of higher order in η. Note, however, that these higher order terms are also higher order in q, i.e., in the derivatives, so that perturbation theory has to be done with care.

Since we know the existence and many properties of the solution to eq. (12), we attack the eq. (16a) by a *change of strategy*, inverting the *nonlinear* problem.

In more detail, we write $W_1=3^{-1/2}l+a+\mathrm{i}b$ and view now (16a) as a pair of coupled equations for a and b (depending of course on all other W_n). In real notation, (16a) leads to

$$L\binom{a}{b}=\Sigma^*+\eta^2(l^{\mathrm{iv}}+a^{\mathrm{iv}})+\cdots=K\binom{a}{b}+R, \tag{17}$$

with

$$L\binom{a}{b}=\begin{pmatrix}4a''+ca'+a-3al^2\\ 4b''+cb'+b-bl^2\end{pmatrix}=\begin{pmatrix}L_1a\\ L_2b\end{pmatrix}, \tag{18}$$

and all other terms have been brought to the r.h.s. of (17). Here, K is a *linear* fourth order differential operator. Note now that L_2 (and also L_1) can be explicitly inverted since

$$L_2=4\left(\partial_x+\frac{l'}{l}+\frac{c}{4}\right)\left(\partial_x-\frac{l'}{l}\right), \tag{19}$$

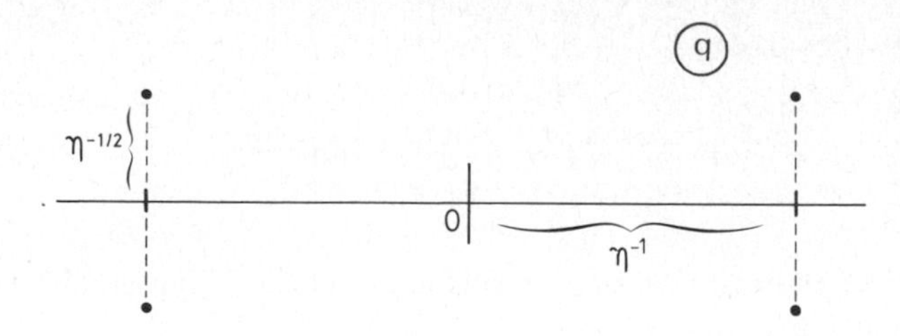

Fig. 2. The regularized spectrum when nonlinear corrections have been made.

and

$$(\partial_x - (\log f)')^{-1} g(x) = f(x) \int^x g(y)/f(y)\,\mathrm{d}y.$$

Thus we can rewrite (17) as

$$\binom{a}{b} - L^{-1}K\binom{a}{b} = L^{-1}R,$$

and now $1 - L^{-1}K$ is *essentially a second order differential operator* whose spectrum is as shown in fig. 2. Thus, the *spectrum has been regularized* through the ansatz for W_1. We can view this as a sort of WKB method. The inverse of $1 - L^{-1}K$ is now bounded by $\eta^{-4}\eta^{+1/2}\eta^{-1/2}\eta^{+1}\eta^{-1/2}$ (three distances and one imaginary part, as well as a factor η^{-4} for the leading coefficient). Thus we have a factor η^{-1}. The factor $L^{-1}R$ is of order η and thus we do not seem to find a small term as expected. However, L^{-1} is regularizing, and a more careful analysis of the eq. (15), which we do not reproduce here*, leads to an additional factor $\eta^{1/2}$, so that we are finally left with a *small perturbation problem*.

We have now described the main strategy for solving eq. (2). There is one more problem, which is typical for coinciding eigenvalues and which has to be solved. The function $l(x)$ at $+\infty$ does not behave like $\exp(-x/2)$ as one might expect, but can only be shown to behave as $(x+1)\exp(-x/2)$. This induces a *logarithmic divergence* in the inverse of L, similar to logarithmic corrections of marginal directions in the renormalization of field theories. Therefore, what we have said above is only valid in a domain $(-\infty, X]$ for some *finite* X. It is technically most convenient to choose $X = \mathcal{O}(|\log \eta|)$. Then we have to solve eq. (2) from X to $+\infty$. This is done with almost standard methods. For each n, one considers $a_{jn} = \partial_x^j W_n|_{x=X}$, $j = 0, 1, 2, 3$, and there is a two dimensional subset of values of a_{jn} (three dimensional when $n = 1$) for which the solution with initial

*Essentially an integration by parts.

conditions a_{jn} at X tends to zero at $+\infty$. This is a stable manifold theorem. The global solution is then obtained by matching, using the free parameters which arise from the restriction to $(-\infty, X]$ (these are the exponentially diverging solutions, and translation and phase invariance of the equation).

4. Conclusions

We have insisted on the basic ideas of the proof, because they are applicable to other equations of the same type, and suggest how to find an expansion parameter, and a converging expansion in such problems. Let us summarize the main steps:

0) Define fronts properly (eqs. (6)–(8)).

1) There is only one dangerous sector, the others are regularized by the Fourier frequency of the stationary solution.

2) The main sector ($n = 1$) leads to a *nonlinear amplitude* equation. This equation is a *singular perturbation* (derivatives of higher order with smaller coefficients) of a *second order nonlinear problem*.

3) The second order nonlinear amplitude equation can be solved by the "energy method" (not explained above; see, e.g., ref. 2) viewing x as "time" (here, the one dimensional character of the equation is used). This method will also work for the integro-differential equations encountered in the more "physical" models of solidification.

4) Having solved the second order equation, we *linearize* the main equation and the linearized second order problem can be inverted by explicit integration.

5) The singular perturbation is *regularized* by this procedure, as in the WKB method (where the singular term is $\mathrm{i}h\partial_x^2$ as $h \to 0$).

6) This leads to a contraction problem and hence to an algorithm which, in principle, converges to all orders in perturbation theory (in η).

7) The marginal nature of the second order problem leads to logarithmic divergences which can be handled by a *cutoff* in position space (at X). The large-x region is handled as in field theory by the large field method (in our case the smallness of U^3) by a stable manifold theory.

We hope that the above strategy clarifies the nature of the fronts, and allows a systematic and numerically stable study of the front problem for other equations.

However, many interesting questions remain open. While we are working on the "local stability" question, and are confident of solving the corresponding technical difficulties, the marginality conjecture remains open. However, we hope to have provided a tool for reliable studies of the question.

As a final question–conjecture we would like to known whether (or under which conditions) the *nodes* of the front move to their final position in a monotonic fashion (no "overshooting"). If this were true in the marginally stable case, we would have a sort of "maximum principle", and hence an analogy with the proofs of Aronson and Weinberger.

References

1) J.S. Langer and H. Müller-Krumbhaar, Mode selection in a dendrite-like nonlinear system, Phys. Rev. A **27** (1982) 499.
2) D.G. Aronson and H. Weinberger, Multidimensional nonlinear diffusions arising in population genetics, Adv. Math. **30** (1978) 30.
3) P. Collet and J.-P. Eckmann, The existence of dendritic fronts, Comm. Math. Phys., in press.
4) P. Collet and J.-P. Eckmann, in preparation.

Physica **140A** (1986) 105–113
North-Holland, Amsterdam

EQUILIBRIUM STRUCTURE AND RIGIDITY OF ALUMINA POLYMERS*

D.W. SCHAEFER, R.A. SHELLEMAN†, K.D. KEEFER and J.E. MARTIN
Sandia National Laboratories, Albuquerque, NM 87185-5800, USA

Alumina polymers are studied at various degrees of branching using *X*-ray and light scattering. A combination of static and dynamic data shows that lattice animal-like structures are formed which break apart and swell on dilution. We observe a threshold for large-scale polymerization when there are 2.5 bridging groups per Al. Below this value only polynuclear species occur whereas above this value the system gels.

1. Introduction

Branched inorganic polymers are typically prepared by the hydrolysis and condensation of soluble metal compounds. These polymers are particularly interesting because the degree of branching can be readily varied by changing the solution chemistry. In the case of silica polymers[1]), for example, a variety of structures have already been observed, including dense colloidal particles, fractally rough colloidal particles[2]), and conventional branched polymers[3]). Inorganic polymers provide a testing ground for modern theories of aggregation. These theories, in turn, can be used to predict new structures.

Inorganic condensates form an important class of precursors to ceramic materials. A typical structural ceramic, for example, is prepared from a concentrated slurry of colloidal particles which is compacted under pressure and sintered at high temperature to form the final product. By contrast, glass films formed by the sol–gel process generally involve polymeric precursors which are dried and sintered at relatively low temperatures. Depending on the application then, either dense colloidal particles or highly ramified polymeric structures may be desirable. The purpose of our study is to elucidate the factors which control precursor structure by understanding the polymerization process in solution. Here we concentrate on alumina polymerization.

*This work performed at Sandia National Laboratories, Albuquerque, NM and supported by the U.S. Department of Energy under Contract No. DE-AC-04-76DP00789.

†Present address: Department of Materials Science and Engineering, Pennsylvania State University, University Park, PA 16802, USA.

0378-4371/86/$03.50 © Elsevier Science Publishers B.V.
(North-Holland Physics Publishing Division)

In the present study, we believe we are observing *equilibrium* structures and suggest that the clusters are lattice animals. Although it is possible to grow lattice animals by a kinetic process, the present system is studied after aging and restructuring of the clusters. We find an interesting "critical point" in the growth behavior as a function of degree of hydrolysis and speculate that the size of the clusters, their swelling behavior, and their integrity reflects a "rigidity" threshold.

2. Fractals and scattering

Since most random growth processes produce fractal structures, it is convenient to characterize these structures using the concept of a fractal dimension, D. Basically D is the exponent which relates the mass, M, to the radius, R, of the cluster.

$$M \sim R^{D}. \tag{1}$$

If the object is compact then clearly $D = 3$, the dimension of space. The more open and ramified the structure is, the farther D departs from 3 ($1 < D < 3$). For dilute, quasi-monodisperse systems, one measures D directly from the power-law exponent relating the intensity, I, to the momentum transfer, K, (exponent $= -D$, provided D is not exactly 3) measured in a scattering experiment[4]). Random objects described by eq. (1) are called mass fractals. These structures have holes in all length scales and the surface area scales with the mass. If the slope of the scattering profile is < -3 then the system is not fractal in this sense. Rather such systems are uniform, possibly with rough surfaces. If one measures slopes between -3 and -4 the system is described as fractally rough[2,5]).

For fractally rough surfaces, the surface area, S, depends on the cluster radius as

$$S \sim R^{D_s}. \tag{2}$$

Note that for conventional smooth particles, $D_s = 2$, but for rough surfaces $2 < D_s < 3$. Bale and Schmidt[5]) have shown that the slope of the scattering profiles for uniform particles with rough surfaces is $D_s - 6$. Note that for smooth surfaces with $D_s = 2$, the scattered intensity, I, decays as K^{-4}.

The above discussion implies that scattering profiles can be used to determine the structure of clusters in solution. Clearly, compact colloidal particles can be distinguished from ramified chainlike species (mass fractals) by noting whether the slope is greater or less than -3. In addition, one might hope for a close correspondence between observed slopes and that predicted by kinetic or equilibrium models.

Further information on the structure can be obtained from dynamic light scattering. In this technique one measures the mean relaxation rate, $\langle \Gamma \rangle$, of fluctuations in the scattered light and infers the hydrodynamic behavior of the clusters. In addition to providing the hydrodynamic correlation range, information about the cluster size distribution as well as cluster flexibility is potentially available from the dependence of $\langle \Gamma \rangle$ on K. In this paper, dynamic as well as static measurements, are reported.

3. Experimental

We study the growth of alumina polymers from the condensation polymerization of hydrolyzed alumina. Polymerization occurs by elimination of a proton in the following reaction,

$$n\,\mathrm{Al(OH_2)_6^{3+}} \rightarrow [\mathrm{Al(OH_2)}_{6-r}]_n(\mathrm{OH})_{r(n-1)}^{[3n-r(n-1)]+} + (n-1)r\,\mathrm{H_3O^+}, \quad (3)$$

where n is the degree of polymerization and r is the number of bridging hydroxyls per Al. Eq. (3) is specific to octahedral coordination and does not represent all possible condensations. Our experiments were carried out with the nitrate salt of aluminum. The value of r is controlled by neutralizing the proton in eq. (3) with NaOH produced by the following reaction,

$$\frac{r}{2}\mathrm{Na_2CO_3} + r\,\mathrm{H_2O} \rightarrow \frac{r}{2}\mathrm{H_2CO_3} + r\,\mathrm{NaOH}. \quad (4)$$

Note that for large n eqs. (3) and (4) reactions produce r bridging hydroxyls per Al assuming complete reactions. Reactions are driven to completion because OH^- produced in eq. (4) is consumed in eq. (3) and because CO_2 is expelled.

Reactions were carried out at 0.5M Al in water solution. For $r \geqslant 2$, a white precipitate forms and/or the solution gels upon addition of Na_2CO_3. After aging at 70 °C, however, the precipitate dissolves. Depending on the value of r, gelation may reoccur at later times. For $r < 2.5 \pm 0.2$, the solution remains clear and does not gel after aging.

We studied the solutions after overnight aging at 70 °C. For $r = 2$, the solutions are clear and show very weak scattering. Guinier analysis of the small-angle X-ray scattering curves indicates very small structures with a radius of gyration of 3.2 Å. This result confirms previous observations by Rausch and Bale[6]), who noted that this value is consistent with the existence of $[Al_{13}O_4(OH)_{24}]^{7+}$, (Al_{13}), ions.

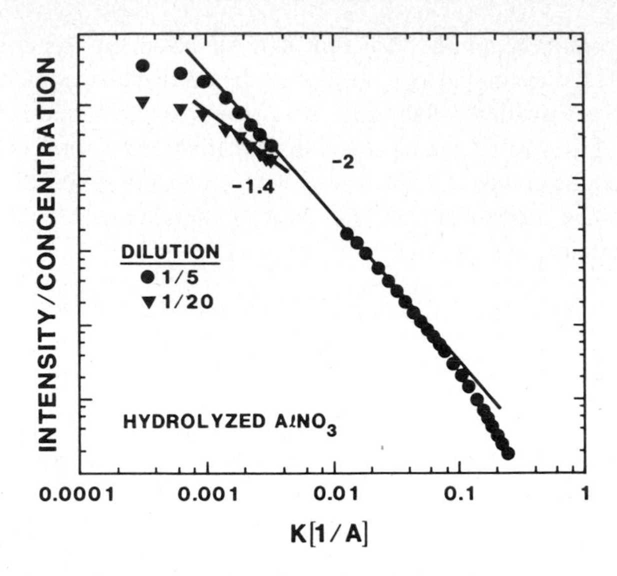

Fig. 1. Static scattering curves for alumina at $r = 2.5$ and two dilutions.

In this paper we concentrate on the $r = 2.5$ solution, which shows very strong scattering, but does not gel. The solution was studied at $\frac{1}{5}$ dilution (0.1M) by X-ray and light scattering and at $\frac{1}{20}$ dilution (0.025M) by light scattering. The combined static X-ray and light scattering data are shown in fig. 1. The 0.1M light scattering data were matched to the X-ray data by an arbitrary shift factor. The static data are consistent with the existence of very large aggregates (Guinier radius, $R_G = 1.4 \times 10^3$ Å) with a fractal dimension of 2.0 ± 0.1. The data in the X-ray region confirm previous observations of Rausch and Bale[6]).

Upon dilution, very interesting behavior is observed in the light scattering regime (lower curve of fig. 1). Although the R_G is unchanged, the slope in the intermediate regime increases from -2 to -1.4. In addition, the $K = 0$ intercept decreases on dilution, indicating a decrease by a factor of 3 in the molecular weight (assuming both systems are in the dilute limit). The concentration (c) dependence of the measured parameters is summarized in the table I. We take these results to mean that the clusters break up and swell on dilution.

Fig. 2 shows the results of the dynamic measurements. Consider first the power-law behavior in the large K limit. At least two explanations exist for the observed K dependence of $\langle \Gamma \rangle$. If the aggregates were totally flexible, scaling arguments show that $\langle \Gamma \rangle \sim K^3$ in the high-K limit. The fact that we observe $\langle \Gamma \rangle \sim K^{2.6}$ may simply be due to the fact that the data is not asymptotic. Since the observed behavior is very similar to that found in solution aggregates of silica[7]), however, we adopt Martin and Leyvraz's perspective and interpret this

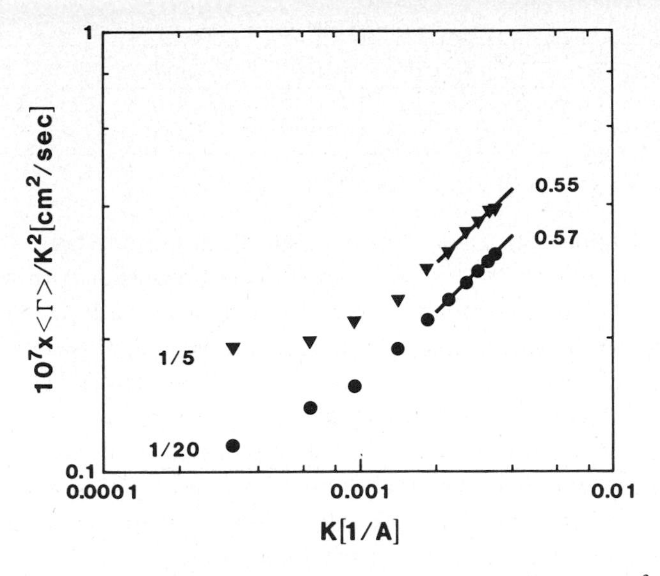

Fig. 2. Dynamic data corresponding to the light scattering data in fig. 1. $\langle \Gamma \rangle / K^2$ is the apparent diffusion constant.

slope in terms of a power-law distribution of cluster sizes[8]). That is, we assume the aggregates are rigid and that the deviation from K^2 behavior is entirely due to the distribution of sizes.

The power-law behavior of $\langle \Gamma \rangle$ in the high-K region suggests that they might be subject to scaling analysis. A plot of the scaled data is given in fig. 3. To scale

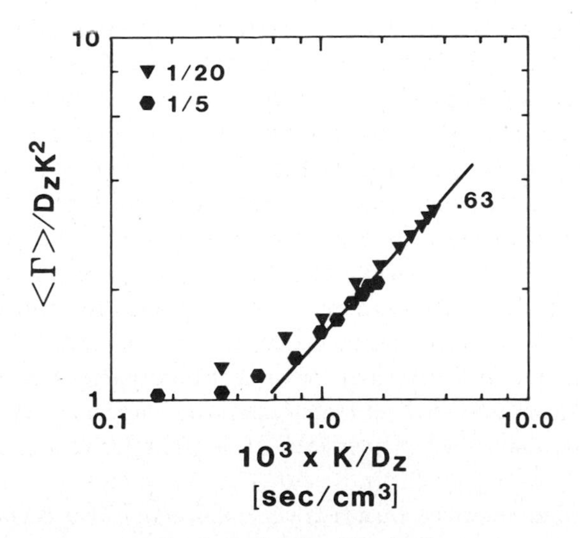

Fig. 3. Scaled data from fig. 2.

TABLE I

Al concentration	$I/C\ (K \to 0)$	R_G(Å)	R_H(Å)	D	α	τ
0.1M	3.3	1.4×10^3	1.35×10^3	2.0	0.55	1.8
0.025M	1	1.4×10^3	2.7×10^3	1.4	0.57	1.7

the data, we obtained the Z-averaged apparent diffusion constant, D_z, from the intercept of the plot of $\langle \Gamma \rangle$ vs. K^2. Since D_z is proportional to an apparent hydrodynamic radius, this radius may be used to scale both axes as in fig. 3. Upon scaling, the data are nearly superimposed in the high-K regime and show a limiting slope consistent with $\langle \Gamma \rangle \sim K^{2.6 \pm 0.15}$. This scaling behavior is also similar to that observed in solution aggregates of silica[7]).

We interpret the power-law exponent, α, which relate $\langle \Gamma \rangle$ and K, using the theory of polydisperse rigid fractals given by Martin et al.[8,9]): $\alpha = 3 - D(2 - \tau)$, where D is the fractal dimension of a single cluster, τ is the power-law exponent which characterizes the mass, M, distribution: $P(M) \sim M^{-\tau}$. Using this formula, in conjunction with D measured from the static light scattering data, we find that τ is essentially unchanged on dilution ($\tau = 1.74 \pm 0.20$). This result, of course, says nothing about the cut-offs. Clearly, the hydrodynamic correlation range obtained from the $K = 0$ limit of the dynamic light scattering data changes drastically on dilution.

4. Discussion

In order to interpret the above results we start with the most striking feature of alumina polymerization, namely, that below $r = 2.5 \pm 0.2$, the system is dominated by small species whereas above $r = 2.5$, the system gels. At $r = 2.5$ we have something like the critical point where large aggregated species exist in solution, but where the system does not actually gel. If we assume complete hydrolysis and condensation, then $r = 2.5$ corresponds to 2.5 crosslinks per aluminum. It is interesting to note that this value is the so called vector percolation threshold[10]) close to value of 2.4 and represents the minimum number of crosslinks of bonds required for a three-dimensional structure to attain structural rigidity. We speculate then that the absence of large aggregates below $r = 2.5$ is due to the breakup of under-coordinated species. For $r > 2.5$, large aggregates can grow and ultimately gel. The value $r = 2.5$ then represents the value below which the polymerization does not lead to structurally stable objects.

If the above idea is correct, then it is very reasonable that clusters prepared at $r = 2.5$ are structurally weak. Further, it is not surprising that, upon dilution, these structures both break apart and swell. Our static light scattering data contained in fig. 1 suggest that this is the case. Upon four-fold dilution, the molecular weight decreases by a factor of 3, the radius of gyration remains constant, and that the fractal dimension of the structures decreases from 2 to 1.4. We envision that the exceedingly tenuous polymers at $r = 2.5$ break apart and swell on dilution.

Interpretation of the effect of dilution on the dynamic data is less straight forward. From the scaling behavior in the intermediate regime, we conclude that the distribution of the aggregates is not changed. From the $K = 0$ limit of $\langle \Gamma \rangle / K^2$, however, we conclude that the apparent hydrodynamic radius, R_H, increases by a factor of 2 on dilution. The fact that the ratio R_H/R_G changes from < 1 at 0.1M to > 1 at 0.025M, suggests a radical change in the geometry of the clusters. Although the static data are consistent with swelling, we have no theory to interpret the change in the apparent hydrodynamic radius. In the absence of such a theory we simply conclude that the largest clusters break apart and that on dilution to 0.025M the entire ensemble swells.

The X-ray data on the 0.1M samples is consistent with the observations of Rausch and Bale[6]) on solutions at concentrations as high as 1M. We believe, therefore, that the observed fractal dimension, $D = 2$, represents the as-grown structure of the alumina polymers and does not reflect the effects of swelling. This value corresponds to the expected fractal dimension for lattice animals. Lattice animals are random structures in which all isomers of the same molecular weight are equally weighted[11]). Although lattice animals are found in percolation models[12]) far below the percolation threshold, we do not believe that alumina polymerization occurs by percolation. Basically, the polymers are mobile and the idea of static percolation seems unreasonable except, perhaps, very near the gel point. In the aged alumina system, we believe we have reached a state in which all structures of the same molecular weight are equally probable. Therefore, we observe the lattice-animal exponent.

Rausch and Bale[6]) speculated that the observed $D = 2$ may be due to sheet-like 2-dimensional structures. These structures are well known in alumina mineralogy and could possibly exist in solution. Our dilution results, however, suggest that this is not the case. Compact 2-dimensional structures would not swell upon dilution. Therefore, we opt in favor of the ramified lattice-animal structures for the interpretation of the observed slope of -2.

Axelos et al. observed the slope of -1.4 in their scattering studies of alumina polymers[13]). They speculated that this slope indicates 2-dimensional diffusion-limited aggregation. It should be noted that their scattering curves were taken within 100 sec of the initiation of aggregation and in no sense represent the same regime we are studying. Therefore, there is no correspondence between their observed D at early times and ours. We point out, nonetheless, that their interpretation in terms of simple diffusion-limited aggregation is unlikely even in early stage growth. If a 2-dimensional aggregate forms in a 3-dimensional space, it seems unlikely that growth could map onto a simple diffusion-limited process, since the sticking probability must depend on the orientation at the tips of the growing clusters. We suggest, therefore, that even in the early stages of growth, that 2-dimensional structures are not formed.

5. Conclusion

Our results on alumina polymers suggest the following ideas to explain the polymerization processes. Below $r = 2$, only small ions, possibly the Al_{13}, ion exist in solution at equilibrium. At $r = 2.5$, large clusters with a radius of gyration of 1,400 Å exist in solution and we interpret these to be lattice animals. At $r = 2.5$, the system crosses over from a sol to a gel. The structures which are observed at $r = 2.5$ are very tenuous and are easily disrupted on dilution. In addition, substantial swelling of the clusters is observed on dilution. We speculate that the $r = 2.5$ value corresponds to a rigidity threshold. At this value of r, the coordination of aluminum is just sufficient to give long range stability to the clusters.

Our ideas concerning lattice animals and rigidity are not necessarily inconsistent with the existence of the Al_{13} ion as the basic monomer unit. Since this ion meets the requirements for 2.5 bonds per aluminum, it is reasonable that clusters of these ions would also be very fragile, and would swell on dilution. At higher values of r, sufficient bridging oxygens would be available to more rigidly link the Al_{13} ions together and, therefore, lead to the effects we observe in this paper. In fact, the crossover at about 0.1 $Å^{-1}$ is consistent with more compact structures at small length scales[1]). We conclude, therefore, that the existence of a rigidity threshold would be observed if the clusters consisted of randomly linked Al_{13} monomers.

References

1) D.W. Schaefer and K.D. Keefer, Fractal aspects of ceramic synthesis in Better Ceramics Through Chemistry II, MRS Symp. Proc. C.J. Brinker, D.E. Clark and D.R. Ulrich, eds. (North-Holland, Amsterdam, 1986).
2) K.D. Keefer and D.W. Schaefer, Phys. Rev. Lett. **56** (1986) 2376.
3) D.W. Schaefer and K.D. Keefer, Phys. Rev. Lett. **53** (1984) 1383.
4) D.W. Schaefer and K.D. Keefer, Mat. Res. Soc. Symp. Proc. **32** (1984) 1.
5) H.D. Bale and P.W. Schmidt, Phys. Rev. Lett. **53** (1984) 596.
6) W.V. Rausch and H.D. Bale, J. Chem. Phys. **40** (1964) 3391.

7) J.E. Martin and D.W. Schaefer, Phys. Rev. Lett. **53** (1984) 2457.
8) J.E. Martin and F. Leyvraz, Phys. Rev. A **34** (1986).
9) J.E. Martin, D.W. Schaefer and A.J. Hurd, Phys. Rev. A **33** (1986) 3540.
10) M.F. Thorpe, J. Non-Cryst. Solids **57** (1983) 355.
11) H.J. Herrmann in On Growth and Form, H.E. Stanely and N. Ostrowsky, eds. (Boston, Martinus Nijhoff, 1986), p. 3.
12) D. Stauffer, Introduction to Percolation Theory (London, Taylor and Francis, 1985).
13) M. Axelos, D. Tchoubar, J.Y. Bottero and F. Fiessinger, J. Physique (Paris) **46** (1985) 1587.

Physica **140A** (1986) 114–123
North-Holland, Amsterdam

PATTERN GROWTH AND FLUID DISPLACEMENTS THROUGH POROUS MEDIA

Roland LENORMAND
Dowell Schlumberger, B.P. 90, 42003 Saint Etienne Cedex 1, France

A review is given of various mechanisms which occur during immiscible displacements in 2-dimensional networks of interconnected capillaries. We show how the physical laws governing meniscus equilibrium and flow conditions lead to different statistical models (mean field, percolation, DLA . . .).

1. Introduction

Many papers have been devoted to the use of Invasion Percolation[1–3]) (IP) and Diffusion Limited Aggregation[4,5]) (DLA) to describe the displacement of one fluid by another in a porous medium. These approaches are based on the randomness of the pore diameters of the porous solid associated with threshold effects for capillary forces (IP) or viscous instabilities (DLA).

In this paper I present some displacements in 2-dimensional porous media made of interconnected channels etched in a transparent plate (micromodels). First, I show how the various physical parameters describing the fluids and the geometry of the medium can explain the different mechanisms observed at microscopic scale. Second, I explain briefly how each mechanism can be described by a given statistical theory, not only the above-mentioned IP and DLA but also other forms of percolation, compact cluster growth, mean field approach, etc. The main results are collected in table I. Furthermore, these micromodels can be used to study the transition between two mechanisms. For instance, by tuning only one parameter such as the flow rate, the pattern of the injected fluid evolves continuously from IP to DLA (fig. 1).

2. Monophasic flow

The micromodels used in the experiments are made of transparent resin cast on a photographically etched mold[6]). The cross-section of each duct of the etched network is rectangular (fig. 2) with a constant depth ($x = 1$ mm) and a

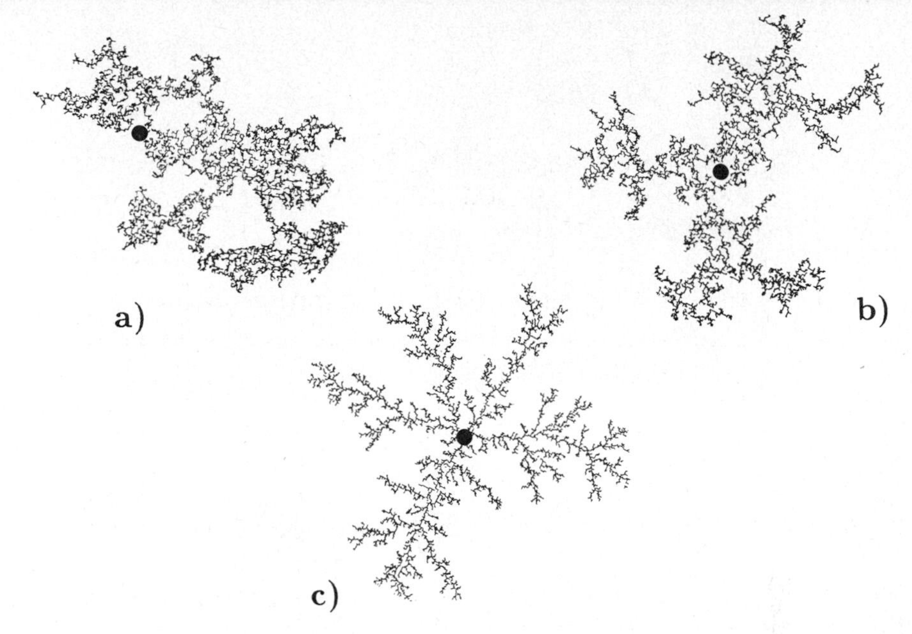

Fig. 1. Air displacing very viscous oil in a radial micromodel containing 250,000 ducts. a) Very low flow rate: viscous forces are negligible and the pattern corresponds to invasion percolation; b) intermediate flow rate: cross-over between percolation and DLA; c) high flow rate: the pattern is very similar to DLA

width d which varies from pore to pore with a given distribution and a random location. Various sizes of network are used for the experiments, the largest one containing more than 250,000 ducts. We will call *ducts* the cylindrical capillaries (bonds) and *pores* the volumes of the intersections (sites). Fig. 3 shows some close-ups of displacements.

The simplest property of a porous medium is the *porosity* ε, the ratio between the volumes of the voids and the bulk. Plotting ε as a function of the radius l of a

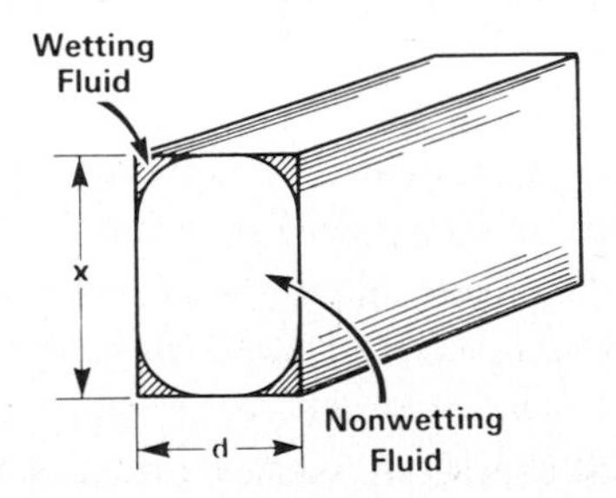

Fig. 2. Two fluid flow in a duct.

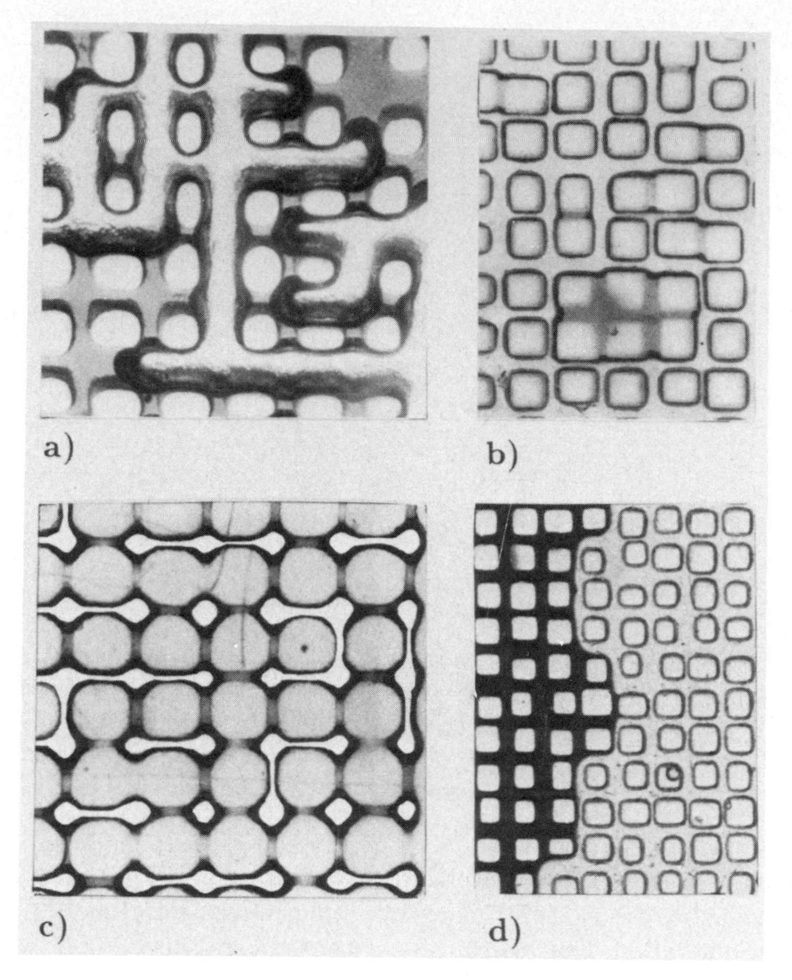

Fig. 3. Close-ups of various experiments in micromodels with air (in white) and oil (in grey). a) Drainage; b) imbibition at very low flow rate in "small pore" geometry: the wetting fluid (oil) flows as a film along the walls and collapses in the smallest ducts. Clusters are either single ducts or rectangles (compact cluster growth). c) Imbibition at very low flow rate in "large pores" geometry: the W fluid (grey) collapses in smallest ducts without filling pores (classical percolation). d) Imbibition without flow by film: the succession of collapses inside pores leads to a flat interface at large scale.

disc centered on a point M chosen at a random location leads to the curves shown in fig. 4. For values of l smaller than l_0, of the order of the grain size a, one can see the microscopic effects due to the pore randomness. For values larger than l_0 the porosity becomes constant and the medium can be studied as a *continuum*.

Another property which characterizes the porous matrix is the *permeability* k. When one fluid is flowing through a porous sample, the pressure drop ΔP

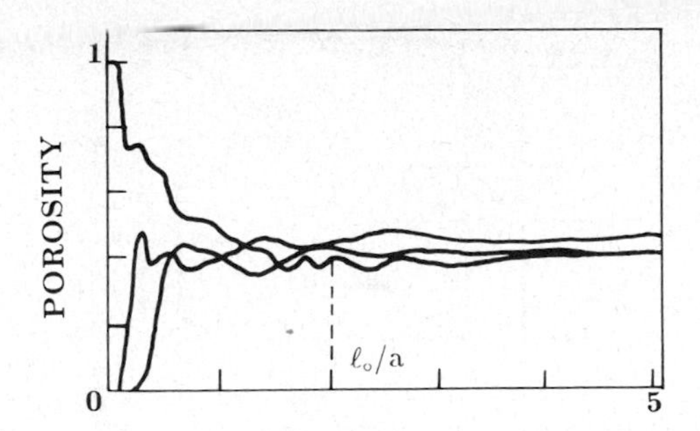

Fig. 4. Mean porosity inside a disc of radius l in the micromodel as a function of the dimensionless radius l/a, a being the mesh size of the square network.

between the two ends is linked to the length L, the cross-section S, the volumetric flow rate q and the viscosity μ of the fluid by Darcy's law: $\Delta P = \mu qL/kS$. From the geometrical properties, we can calculate the hydraulic conductance g of each channel of the network (analogous to a network of random resistors). The problem is to calculate the overall permeability k from the distribution law $f(g)$. This can be done by using an *effective medium* technique[7]). The principle is to look at only one channel (g) and replace all the other conductances by a mean value g_m. The local fluctuation due to the difference $g - g_m$ is calculated and we choose the value of g_m which leads to a zero average fluctuation. For the square network, this method leads to:

$$\int \frac{f(g)(g_m - g)\,dg}{g + g_m} = 0. \tag{1}$$

3. Capillary displacements

In this case viscous and gravity forces are assumed to be negligible and consequently, all the displacement mechanisms are linked to capillary forces and randomness due to the different sizes of pores in a porous medium.

Generally speaking, when one fluid (say oil) is slowly displacing another immiscible fluid (say water) in a capillary tube of diameter D_0, the fluid for which the contact angle θ (between the tube and the meniscus) is smaller than $\pi/2$ is called the wetting fluid (W); the other one is the nonwetting fluid (NW). The pressure in the NW fluids exceeds the pressure in the wetting fluid by a value P, called capillary pressure and linked to the interfacial tension γ by the Laplace law: $P = 4\gamma \cos\theta/D_0$.

A displacement where the NW fluid is pushing the wetting fluid is called *drainage* (D), the reverse is *imbibition* (I).

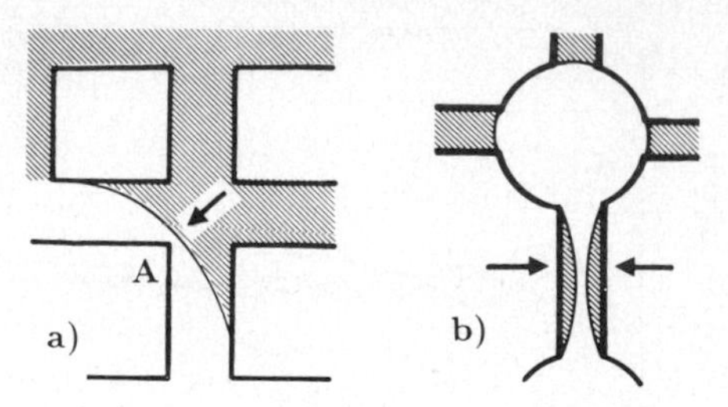

Fig. 5. Different mechanisms for the meniscus displacement during imbibition: a) small pore geometry: the meniscus collapses inside the intersection (pore) when it reaches point A of the walls; b) large pore geometry: the meniscus collapses inside a duct.

3.1. *Physical mechanisms*

Experiments show that both fluids can flow simultaneously in the same duct with different velocities, the wetting fluid remaining in the extreme corners of the cross-section and roughness of the walls (fig. 2). This effect explains the richness of mechanisms involved during displacements.

A displacement can be divided into three parts: i) flow of the injected fluid from the entrance towards the moving meniscus, ii) displacement of the meniscus, iii) flow of the displaced fluid towards the exit.

Meniscus displacements. This study has been presented in detail in previous papers[8,9]) and the main result is the effect of the geometry of the pores which can introduce a selection of imbibition mechanisms. We found two main cases: i) *large pores*, when the size of the pore is large compared with duct diameters; the meniscus is unstable inside a pore (fig. 5a). ii) *small pores*, in the opposite case, the meniscus collapses inside a duct (fig. 5b). We will see later the consequences of these two cases, the former leading to percolation type patterns, the latter to flat interfaces at large scales (table I).

Flow of the fluids. The nonwetting fluid can flow only in the bulk of the ducts. So the flow occurs only if a continuous path of ducts or pores filled with this phase exists either towards the entrance (during drainage) or the exit (during imbibition) of the network. Otherwise the NW fluid is trapped.

We observe 3 kinds of flow of the wetting fluid: i) flow in the bulk of the ducts, ii) along the corners when the NW fluid fills the central part of the duct, iii) by *film* along the roughness of the walls, but only in case of strong wettability and very low flow rate.

TABLE I
Various displacement mechanisms and related statistical models.

	Type of Flow	Large Pores	Small Pores
MONOPHASIC FLOW	Flow in the ducts	EFFECTIVE MEDIUM APPROACH	
DRAINAGE	Flow in the ducts	INVASION PERCOLATION	
IMBIBITION	No flow by film	INVASION PERCO. Dual network	COMPACT GROWTH
	Flow by film	BOND PERCOLATION	COMPACT CLUSTER GROWTH
VISCOUS FLOW	Stable	FLAT INTERFACE	
	Unstable	GRADIENT GOVERNED GROWTH or DIFFUSION LIMITED AGGREGATION	

3.2. *Drainage*

Capillary forces prevent the NW fluid from spontaneously entering a porous medium. It can only enter a duct (diameter D_0) when the pressure exceeds the capillary pressure. From a statistical point of view a duct with $D > D_0$ is an active or conductive bond and a duct with $D < D_0$ an inactive bond. The fraction p of active bonds can easily be deduced from the throat size distribution.

At a given pressure P, the injected fluid invades all the percolation clusters connected to the injection point (fig. 1a at large scale and fig. 2a at pore scale); this mechanism has been called invasion percolation[1–3,10]). During the displacement, the wetting phase is trapped in the network when the invading NW fluid breaks the continuous path toward the exit (in 2-D, the W fluid cannot escape by flowing via the corners).

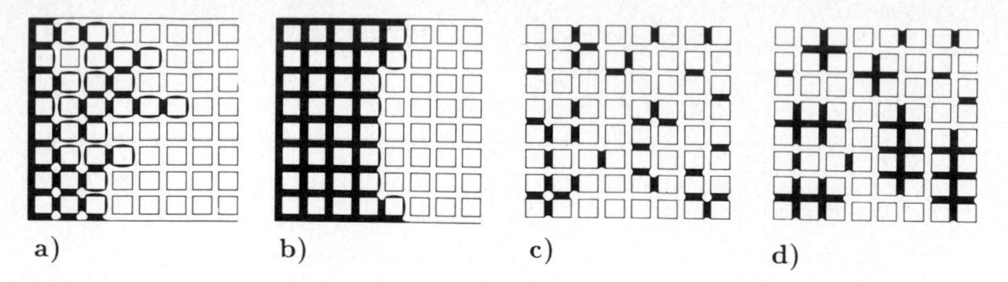

Fig. 6. Simulation of the different imbibition mechanisms. The injected wetting fluid is black. a) No flow by film, large pores (invasion percolation; b) no flow by film, small pores (flat interface); c) flow by film, large pores (classical percolation); d) flow by film, small pores (compact cluster growth).

3.3. *Imbibition*

The type of displacement depends upon two factors: the pore geometry (large or small pores) and the possibility of flow by film along the roughness. Simulations of the 4 types of displacement in small networks are shown in fig. 6.

No flow by film. For *large pores* (fig. 6a), the injected wetting fluid invades the network by a succession of collapses in the ducts (fig. 5b). Due to pressure effects, the smallest channel of the interface between the fluids is filled at each step and because of the randomness of the pore size in the network, this mechanism can be described by *invasion percolation*. However, the topology is not the same as in drainage, and we have shown[9,11]) that the process corresponds to *bond percolation* in the *dual* network (the sites of the dual network are the centers of the solid grains).

For small pores (fig. 6b), the meniscus instability inside a pore (fig. 5a) leads to filling the network line after line, without trapping. The result is a kind of *compact* crystal growth, with a shape related to the network's mesh (fig. 3d).

Flow by film. In this case, the wetting fluid can always flow in the bulk of the ducts or along the corners but it can also spread along the roughness of the walls and reach all the ducts in the network without continuity problems. The thickness of the film increases with time until a collapse occurs in the smallest duct and so forth. The resulting pattern depends on meniscus stability in the porc, rclatcd to the geometry:

For *large pores*, the meniscus remains stable in the pores, even if many ducts are filled with the wetting fluid (experiment fig. 3c and simulation fig. 6c). The number of filled ducts increases with time at random locations until the NW fluid becomes disconnected. The result is a *percolation* process in the dual network for the invading fluid or "classical" percolation for the displaced fluid.

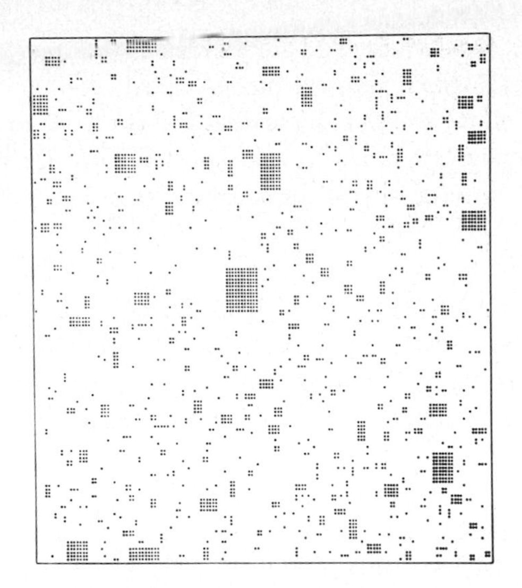

Fig. 7. Simulation of the compact cluster growth just below the threshold in a square network (150 × 150). The black dots represent the filled pores (sites), after[12]).

For *small pores*, the meniscus is unstable inside a pore when two adjacent ducts are filled and very quickly, the pore and the two other ducts are filled (fig. 6d). This leads to rectangle clusters which grow with time (experiments fig. 2b). For a critical fraction p^* of collapsed ducts, one of the clusters "swallows" the other and invades the entire network. We have shown[12]) that p^* does not have a finite value but tends to zero as the network size N increases ($p^* \propto 1/\log N$). Simulation, fig. 7, shows the situation just before the threshold. We called this mechanism *compact cluster growth*.

At a given stage of a capillary displacement, menisci in the pores do not "see" the exit because we are assuming a zero pressure drop in the fluids. Consequently, they are described by "local models", i.e., at each step, the interface between the two phases moves towards an adjacent site according to some local rules. The mechanism is different for viscous displacements.

4. Viscous displacements

In this case, capillary forces are negligible compared to viscous forces. However, we need a small amount of capillary effects, otherwise the fluids mix inside the ducts and the problem is more complex (miscible displacements).

Viscous displacements, either stable or unstable are governed by the pressure field between the entrance and the exit. Consequently, even in the case of a stable displacement, a local model based on some rules at the interface cannot be used for modeling viscous displacements. A model, called Gradient Governed Growth has been developed simultaneously by several authors[13-15]) to solve this problem, using both a continuum approach to calculate the pressure field and a discrete displacement of the interface which accounts for the granular structure of the porous medium.

Injecting the more viscous fluid. The displacement is stable and leads to a *flat interface* at large scale which corresponds to its stable solution of the potential flow (i.e. radial flow in the case of central injection). However, we expect fluctuations due to permeability variations at the microscopic scale l_0 previously defined.

Injecting the less viscous fluid. The displacement is unstable (analogy with Saffman–Taylor[16]) instability in a Hele Shaw cell). If we assume that the injected fluid has negligible viscosity, the growing pattern can also be represented by a model known as Diffusion Limited Aggregation[4,17]). A computation[18]) based on a network of *random* conductances, leads to similar results. The validity of this model is demonstrated by the similarity between computer simulations and experimental patterns. For instance, fig. 1c shows the displacement of a very viscous oil by air in a radial geometry.

5. Perspectives

Studies of the transition mechanisms are in progress[19]), especially the crossover between IP and DLA in drainage (fig. 1) and also between IP and "flat interface" which seems very similar to diffusion problems[20]).

References

1) R. Lenormand and S. Bories, C. R. Acad. Sc. Paris **291B** (1980) 279.
2) D. Wilkinson and J.F. Willemsen, J. Phys. A **16** (1983) 3365.
3) J.T. Chayes, L. Chayes and C.M. Newman, Commun. Math. Phys. **101** (1985) 383.
4) L. Paterson, Phys. Rev. Lett. **52** (1984) 1621.
5) K.J. Måløy, J. Feder and T. Jøssang, Phys. Rev. Lett. **55** (1985) 1885.
6) R. Lenormand and Zarcone, Proc. P.C.H. conf., Tel-Aviv (1984), Phys. Chem. Hydrodyn. **6** (1985) 497.
7) S. Kirkpatrick, Rev. Modern Phys. **45** (1973) 574.
8) R. Lenormand, C. Zarcone and A. Sarr, J. Fluid Mech. **135** (1983) 337.
9) R. Lenormand and C. Zarcone, Soc. Petrol. Eng. paper 13264 (1984).
10) R. Lenormand and C. Zarcone, Phys. Rev. Lett. **54** (1985) 2226.

11) R. Lenormand, C.R. Acad. Sc. Paris **291B** (1980) 279.
12) R. Lenormand and C. Zarcone in Kinetics of Aggregation and Gelation, F. Family and D.P. Landau, eds. (Elsevier, Amsterdam 1984).
13) A.J. DeGregoria, Phys. Fluids **28** (1985) 2933.
14) M. King and H. Sher, Soc. Petrol. Eng. paper 14366 (1985).
15) J.D. Sherwood and J. Nittmann, J. Physique **47** (1986) 15.
16) P.G. Saffman and G.I. Taylor, Proc. R. Soc. Lond. **A245** (1958) 311.
17) T.A. Witten and Sander, Phys. Rev. B **27** (1983) 5686.
18) J.D. Chen and D. Wilkinson, Phys. Rev. Lett. **55** (1985) 1892.
19) R. Lenormand, Soc. Petrol. Eng. paper 15390 (1986).
20) B. Sapoval, M. Rosso, J.F. Gouyet and J.F. Colonna, Solid State Ionics **18–19** (1986) 21.

Physica **140A** (1986) 124–133
North-Holland, Amsterdam

FRACTAL VISCOUS FINGERING: EXPERIMENTS AND MODELS

Johann NITTMANN

Dowell Schlumberger, 42003 Saint Etienne Cedex 1, France

Highly ramified fingering structures occur when water is forced under pressure to displace a high viscosity shear-thinning fluid in a Hele Shaw cell. We found that water fingers grow by splitting of the leading finger tips giving rise to arborescent structures which obey fractal scaling. Their Hausdorff dimension d_f is 1.70 for a wide range of experimental conditions. We show that diffusion limited growth models can accurately simulate the growth of these fractal patterns.

1. Introduction

The displacement of a high viscosity fluid by a low viscosity fluid in a Hele Shaw cell has recently attracted much interest from the scientific world. This is due to two reasons: (i) for the first time fractal structures have been found in laminar flow caused by a fluid dynamical instability[1]), (ii) there is a close relationship between the growth of fractal fingers and the formation of dendritic structures such as snow flakes[2,3]). A typical radial viscous fingering structure is shown in fig. 1 [4]) where water (black) displaces a thick water based polymer solution in a Hele Shaw cell. We observe that the leading finger tips are not stable but break up into two or three branches which are competing with each other for further growth. The winner very shortly experiences the same scenario. This gives the structure a tree-like appearance with a well defined finger width. What causes the creeping water fingers to grow in a self similar fashion and what physical parameters determine this characteristic finger thickness? Before answering these and other questions we would like to summarize the experimental results of fractal growth in a Hele Shaw cell.

2. Observing fractal fingering

1.1. *Stable Saffman–Taylor fingers*

The viscous finger instability was first studied by Saffman and Taylor[5]) using two immiscible fluids – water and oil in a linear Hele Shaw cell. For their range

0378-4371/86/$03.50

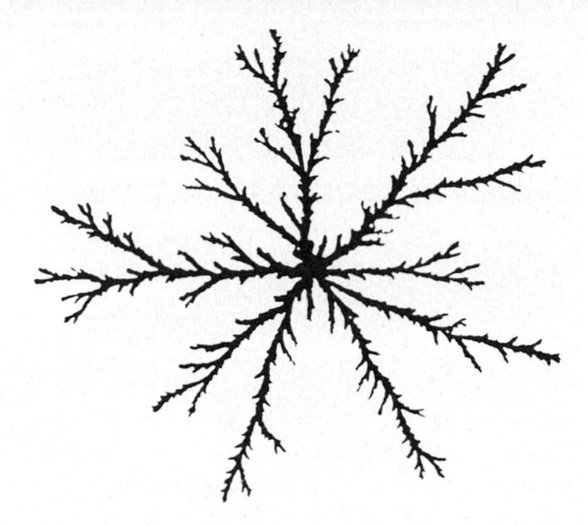

Fig. 1. Typical fractal viscous fingers with $d_f = 1.70$ created by pushing water under pressure into a radial Hele Shaw cell filled with a viscous polymer solution.

of experimental conditions single water fingers were observed which had stable tips (fig. 2a). The finger width of the linear finger varied as a function of the capillary number N_{Ca}

$$N_{Ca} = \frac{\mu u}{\sigma}, \tag{1}$$

with μ, u and σ being the fluid viscosity, flow velocity and interfacial tension, respectively. N_{Ca} represents the ratio of viscous forces to capillary forces. Chuoke et al.[6]) found, from a linear stability analysis, the following relation for the critical finger width λ_c (wavelength of a perturbation)

$$\lambda_c = 2\pi b(12 N_{Ca})^{-1/2}. \tag{2}$$

$\lambda_{max} = \sqrt{3}\,\lambda_c$ corresponds to the finger thickness we see during fingering of immiscible fluids. Perturbations with wavelength smaller than λ_c cannot grow.

1.2. *Tip splitting for high capillary numbers*

We found[1]) by repeating the Saffman–Taylor experiment, that an increase of the capillary number (achieved by increasing the flow rate) causes the break up of the single finger into two fingers of about half of the original finger width (fig. 2b). This indicates that the capillary forces (i.e. interfacial tension) play an important role in suppressing the splitting of the finger tip. To obtain tip splitting and ramified growth a small or vanishing interfacial tension is essential.

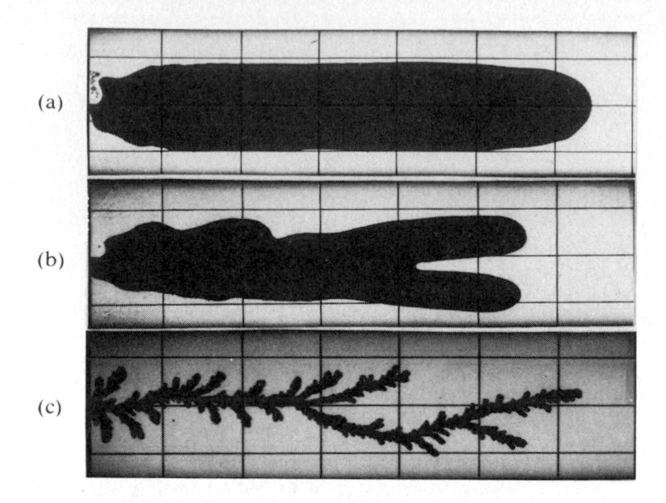

Fig. 2. Saffman–Taylor experiment in linear Hele Shaw cell (a) Water displacing oil, $N_{Ca} = 0.08$; (b) Water displacing oil, $N_{Ca} = 0.16$; (c) water displacing miscible polymer solution, $N_{Ca} \to \infty$.

To decrease the capillary forces further it is practical to use miscible fluids ($N_{Ca} \to \infty$). We used a water based polymer solution (scleroglucan), which gave rise to the structures in fig. 1 and fig. 2c. Here, a priori, zero interfacial tension exists because the fluids mix; however, on a time scale much larger than the experiments last[4]). The polymer solution is a shear-thinning non-Newtonian fluid. This means that the viscosity of the fluid is not constant (as in a Newtonian fluid) but a function of the shear rate (basically the velocity gradient) in the fluid. For a large shear rate the viscosity is small and vice versa. This enables an easy flow very close to the glass plates[7]) (where the velocity gradients are high) and prevents fingering in the direction normal to the glass plates as frequently observed for Newtonian fluids[8]). This ensures a strictly two-dimensional flow.

1.3. *The finger width*

Eq. (2) predicts a $\lambda_c \to 0$ for $N_{Ca} \to \infty$ (miscible fluids). We found[4]), however, that the finger width was finite. It is a linear function of the plate separation b and does not depend on any other parameters, e.g. the flow velocity as predicted from eq. (2). This is in agreement with experiments performed with Newtonian fluids by Paterson[8]). Paterson (viscous dissipation of energy) and recently De Gennes[9]) (viscoelastic relaxation) have proposed theories to explain finite finger thickness for Newtonian and visco-elastic fluids. Further below we propose a novel approach based on a tunable noise model where physical fluctuations give rise to a finite finger width.

Fig. 3. Water displacing fluidose (newtonian fluid, miscible) in radial Hele Shaw cell ($d_f = 1.85$).

1.4. *Quantitative description of a fluid dynamical instability: the fractal dimension*

Consecutive splitting of the leading finger tips gives rise to the ramified structures in figs. 1, 2c and 3 (water displacing fluidose). Further progress in analysing the patterns is only possible by computer image analysis. The first step consists of digitizing the experimental structures by mapping them on to a screen with typically 256 × 256 pixels. Each pixel belongs either to fluid 1 or 2 and its coordinates x and y are used in the data analysis. When calculating how a characteristic radius of the fingering pattern (e.g. the radius of gyration) changes as a function of the area occupied by the fingers, we found a linear relation on double logarithmic paper. The slope named Fractal Dimension[10]) (or Hausdorff Dimension) is an important quantity describing the structure. We have used several methods to calculate d_f, which are explained elsewhere[4,11]).

For fig. 2c we obtained the value[1]) $d_f = 1.39$, but it was clear that the lateral boundaries of the cell prevent the structure from fully developing. Therefore boundary free radial experiments[12]) were necessary. A typical result is shown in fig. 1. For this structure we found $d_f = 1.70 \pm 0.05$ [4]) for a large range of flow rates, plate separations and polymer concentrations. Similar fractal growth patterns have also been found for water displacing clay slurries[13]).

3. Governing equations for fingering growth

Which equations govern two-fluid flow in a Hele Shaw cell? If we neglect for the time being that one of our fluids is shear thinning (see further discussion in refs. 4 and 11) we can write for the flow velocity $\boldsymbol{u}$ of a single fluid in parallel plate flow[14]):

$$\boldsymbol{u} = -\frac{b^2}{12\mu}\nabla p. \tag{3}$$

This fundamental equation describing the global velocity field is the Darcy law. b is the constant plate separation, μ is the fluid viscosity and p is the fluid pressure. The same equation has to hold for two fluids (μ_1, μ_2) with the following boundary condition normal to the interface:

$$\frac{1}{\mu_1}\nabla p_1 = \frac{1}{\mu_2}\nabla p_2. \tag{4}$$

For incompressible fluids the conservation of mass reads:

$$\nabla \cdot \boldsymbol{u} = 0. \tag{5}$$

With eq. (3) and the assumption that the water has negligible viscosity compared to the displaced fluid one obtaines a Laplace equation[15]) for the fluid pressure within the displaced fluid:

$$\nabla^2 p = 0. \tag{6}$$

As boundary conditions for a circular domain with radius R we choose:

$$p_{(r=0)} = p_1, \tag{7}$$

$$p_{(r=R)} = 0. \tag{8}$$

p_1 is the constant injection pressure. Studies by Shraiman and Bensimon[16]) using a conformal mapping technique have shown that perturbing the initially stable flow causes cusp singularities to develop, since perturbations grow faster as they grow in amplitude. This is in agreement with the early conclusion we drew from extrapolating eq. (2). However cusp singularities have never been observed in experiments. Therefore, stabilizing forces (e.g. interfacial tension) are essential to inhibit this catastrophic instability. Interfacial tension introduces a pressure discontinuity Δp_{I} at the interface which for strictly two-dimensional flow is of the form:

$$\Delta p_{\mathrm{I}} = \frac{\sigma}{R_1} \tag{9}$$

with R_1 being the radius of curvature. Sharp tips have a zero radius of curvature and therefore an infinite interfacial tension which will prevent cusps from forming. The role of interfacial tension during finger growth has been studied by various authors[17]).

However the concept of interfacial tension as a stabilizing force for viscous fingering is not applicable if we consider miscible fluids. We believe that a different physical concept has to be used to explain finite viscous fingers for miscible fluids.

During an unstable displacement (the less viscous fluid pushing the more viscous fluid) any small perturbation can grow thereby changing the shape of the interface irreversibly. We have evidence from a computational model that a fingering structure such as fig. 1 is a consequence of microscopic fluctuations present in any physical system, particularly in the Hele Shaw cell. Fluctuations will occur in the form of imperfections in the glass plate surfaces which will lead to fluctuations in the plate separation distance b, injection pressure fluctuations or tiny variations in the polymer concentration. During unstable flow conditions an outward extension of the interface for example will be pushed by a slightly larger pressure gradient than that experienced by the neighbour elements and will therefore advance even further. With no other fluctuations intervening, it will become a cusp. However, owing to the presence of other fluctuations during the next infinitesimal small step it might slow down or move sidewards giving rise to a branching of the finger tip.

4. The tunable noise model

In order to model fluctuations as present in a Hele Shaw cell, we replace eq. (3) by a different statistical rule. Rather than allowing growth such that the flow velocities u_i are equal to the local pressure gradient, each interface element only has a certain probability $\Pr\{u_i\}$ of growing (of advancing) which we choose to be proportional to the local pressure gradient.

$$\Pr\{u_i\} = \frac{\nabla p_i}{\sum_k \nabla p_k}. \tag{10}$$

With this statistical rule, eqs. (6)–(8) are essentially the equations used in the

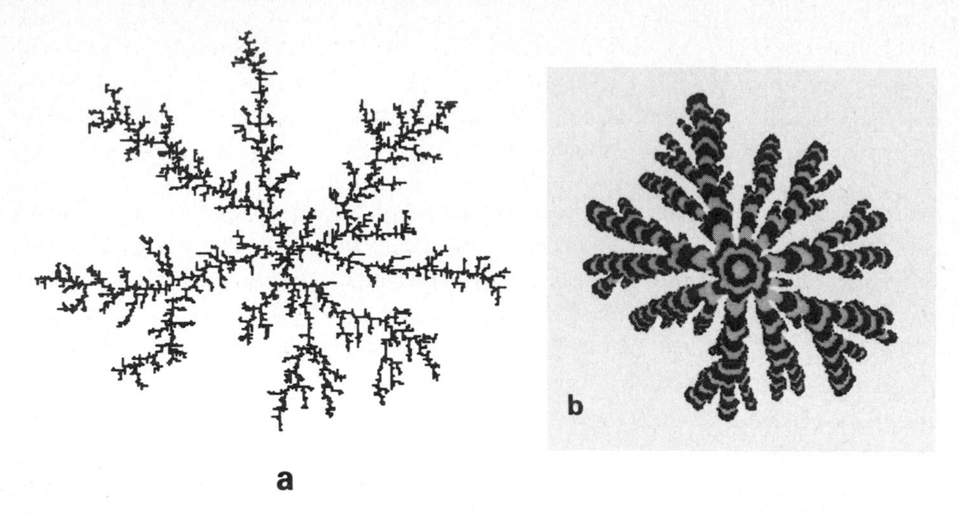

Fig. 4. (a) A typical DLA cluster made of 10000 sites; (b) viscous finger growth simulated by tunable noise model, $s = 50$, 15000 sites. Different shading is used to indicate new growth sites which have been added during a time interval.

dielectric breakdown model[18]) (DBM), which is closely related to the diffusion limited aggregation model[19]) (DLA). DBM and DLA structures have a fractal dimension equal (within statistical errors) to fractal viscous fingers. The appearance of a DLA cluster for example (fig. 4a) differs however from fig. 1 in that

(i) the spokes of the radial viscous fingers are straighter;

(ii) there is more foilage growing on DLA spokes;

(iii) the tip splitting angles, measured at the finger bifurcations have a narrower distribution (mean is $\sim 40°$) than for DLA.

These three points can be attributed to one phenomenon. The fluctuations which govern the growth of DLA clusters are larger than the fluctuations which cause viscous fingering in a Hele Shaw cell. Indeed as pointed out by Lenormand[20]), DLA patterns are indistinguishable from radial viscous fingering structures in two-dimensional porous media models with a large pore size distribution[21]). In porous media models each pore is connected in principle with three neighbouring pores. This is similar to a computer lattice. If the pores have different sizes then large (macroscopic) fluctuations exists. In a Hele Shaw cell, however, the gap between the glass plates corresponds to the pores. As the gap is constant, only microscopic fluctuations due to imperfections will exist in a Hele Shaw cell. We therefore need to reduce the noise inherited in DLA or DB models. To do so we base our model on DBM in which we introduce the following modification:

Eq. (8) implies that each interface element has a growth probability proportional to the pressure gradient between the element and its next neighbours across the interface. At each time step only one interface element is allowed to grow before eqs. (6)–(8) are solved again. The fact that the interface evolution depends on growth probabilities rather than on the exact local pressure gradients introduces noise and causes a deviation from eq. (2). To tune the amount of noise or deviation from eq. (2) we have developed the tunable noise model[3]). The noise is tuned by a parameter s which counts the number of times an interface element has to be chosen before it can actually grow. For DBM, $s = 1$. This corresponds to the maximum amount of noise possible. In the case of no noise ($s = \infty$, mean field limit) the interface grows in concentric circle. By tuning s we obtain a crossover between the two cases. We found that the fingerwidth λ_f varies logarithmically with s.

A typical radial viscous finger structure (15000 sites) with $s = 50$ is shown in fig. 4b. The structure now has straighter spokes and a narrower distribution of tip split angles. It has a fractal dimension slightly above 1.70, but will reach this value only for very large clusters. Please note that the structure is initially non-fractal. It first grows as a circle from which for a critical radius R_c (which is only a function of the parameter s) it breaks up into several radial branches.

The concept of noise reduction can also be applied to DLA as first pointed out by Tang[22]). Our parameter s now corresponds to the number of times a potential growth site has to be visited by random walkers before it becomes a cluster site. Using this concept Kertèsz and Vicsek[23]) found structures very different from fig. 5b. Dentritic fingers grew without tip splitting. It appears that a small difference in the boundary condition at the interface (Ball et al.[23])) is responsible for the difference.

5. Towards snowflakes and three-dimensional fractal fingering

Viscous fingers are characterized by their disordered growth. It seems therefore surprising that these disordered patterns are related to highly ordered structures whose formation has puzzled scientists for a long time – snowflakes[24]). However if one compares the equations of dendritic growth with eqs. (6)–(10) a striking similarity is present. In solidification the driving force is the temperature gradient between the solid and the liquid phase. The temperature gradient plays the role of our pressure gradient. Capillary forces introduce characteristic finger width in both cases.

Recent work by Ben Jacob et al.[2,25]) has indicated that anisotropy could be an important ingredient during snowflake formation. However theoretical models have not been able to produce convincing structures so far. We have been able to extend the tunable noise model to consider for anisotropy on a microscopic level.

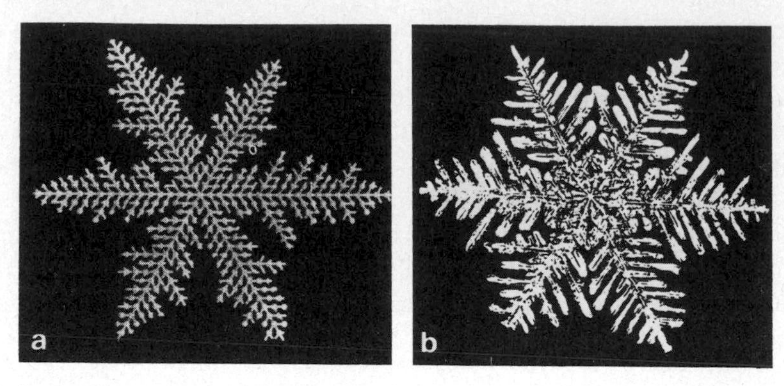

Fig. 5. Snow flakes patterns. (a) Tunable noise model with six-fold anisotropy; (b) a real snowflake (ref. 26).

Figs. 5a and b show a simulated and a real snowflake[26]) respectively. Although interfacial tension is absent one observes a striking resemblance. Further details are given in ref. 3.

So far our studies have focused on two-dimensional structures. Radial viscous fingers in a Hele Shaw cell are limited to two dimensions as the third dimension provides the necessary friction to build up viscous pressure. In porous media such as rock, fingering is three-dimensional as the narrow pores and throats provide the necessary friction forces. However, besides viscous fingering a new form of instability–acid fingering has recently been studied by Daccord and Lenormand[27]). They force acid under pressure into a soluble rock saturated with water. The acid etches the rock and leaves three-dimensional channels behind. These fingers can be made visible by molding the etched channels and by dissolving the rock after the mold has hardened. Here the viscous forces are not a function of fluid viscosities but of the different permeabilities between etched and unetched rock. This new exciting work is still in progress.

Fractal fingering started from a simple experiment of displacing fluids between parallel glass plates but its close relation to the physics of the evolution of solid–fluid and solid–acid interfaces gives it an important place in understanding the growth of physical, geological and biological structures.

Acknowledgement

Much of this work has been done in collaboration with G. Daccord and H.E. Stanley (Boston University) at the Dowell Schlumberger Research Laboratories. I would like to thank R. Lenormand and F. Rondelez for many helpful discussions. Fig. 3 was kindly provided by C. Portella.

References

1) J. Nittmann, G. Daccord and H.E. Stanley, Nature **314** (1985) 141.
2) E. Ben-Jacob, R. Godbey, N.D. Goldenfeld, J. Koplik, H. Levine, T. Mueller and L.M. Sander, Phys. Rev. Lett. **55** (1985) 1315.
3) J. Nittmann and H.E. Stanely, Nature **321** (1986) 663.
4) G. Daccord, J. Nittmann and H.E. Stanley, Phys. Rev. Lett. **56** (1986) 336.
5) P.G. Saffman and G.I. Taylor, Proc. R. Soc. **A245** (1958) 312. Much progress has been made recently in understanding the finger width selection of stable Saffman–Taylor fingers, see R. Combescot et al., Phys. Rev. Lett. **56** (1986) 2036 and further references therein.
6) R.L. Chuoke, P. Van Meurs and C.J. Van der Poel, J. Petrol Tech. **11** (1959) 64.
7) G. Daccord, J. Nittmann and H.E. Stanley, in On Growth and Form: Fractal and Non-Fractal Patterns in Physics,H.E. Stanley and N. Ostrowsky, eds. (Nijhoff, Dortrecht, 1985).
8) L. Paterson, Phys. Fluids **28** (1985) 26.
9) P.G. De Gennes (preprint).
10) B.B. Mandelbrot, The Fractal Geometry of Nature (Freeman, San Francisco, 1982).
11) J. Nittmann, G. Daccord and H.E. Stanley, in Fractals in Physics, L. Pietronero and E. Tosatti, eds. (North-Holland, Amsterdam, 1986).
12) L. Paterson, J. Fluid Mech. **113** (1981) 513.
13) H. Van Damme, F. Obrecht, P. Levitz, L. Gatineau and C. Laroche, Nature **320** (1986) 731.
14) H. Lamb, Hydrodynamics (Cambridge Univ. Press, London, 1932).
15) L. Paterson, Phys. Rev. Lett. **52** (1984) 1621. The case of a finite viscosity ratio has been studied by J.D. Sherwood and J. Nittmann, J. Phys. Paris **47** (1986) 15.
16) B.I. Shraiman and D. Bensimon, Phys. Rev, A **30** (1984) 2840.
17) L. Kadanoff, J. Stat. Phys. **39** (1985) 267. L. M. Sander, P. Ramanlal and E. Ben-Jacob, Phys. Rev. A **32** (1985) 3160. A. J. De Gregoria and L.W. Schwartz, J. Fluid Mech. **164** (1986) 383. D. Bensimon, Phys. Rev. A **33** (1986) 1302.
18) L. Niemeyer, L. Pietronero and H.J. Wiesmann, Phys. Rev. Lett. **52** (1984) 1033.
19) T.A. Witten and L.M. Sander, Phys. Rev. Lett. **47** (1981) 1499.
20) R. Lenormand, Physica **140A** (1986) 114, these proceedings.
21) J.D. Chen and D. Wilkinson, Phys. Rev. Lett. **55** (1985) 1892. K.N. Maloy, J. Feder and T. Jossang, Phys. Rev. Lett. **55** (1985) 2688.
22) C. Tang, Phys. Rev. A **31** (1985) 1977.
23) J. Kertèsz and T. Vicsek, J. Phys. A **19** (1986) L257; R.C. Ball, R.M. Brady, E.T. Samulski and B.R. Thompson, preprint.
24) J.S. Langer, Rev. Mod. Phys. **52** (1980) 1.
25) Similar studies on anisotropic fingering in etched Hele Shaw cells were performed simultaneously by R. Lenormand (private communication).
26) W.A. Bentley and W.J. Humphreys, Snow Crystals (Dover, New York, 1962).
27) G. Daccord and R. Lenormand, submitted to Nature.

Physica 140A (1986) 134–141
North-Holland, Amsterdam

TRANSITION OF GROWTH FORM FROM DENDRITE TO AGGREGATE

Yasuji SAWADA
Research Institute of Electrical Communication, Tohoku University, Sendai 980, Japan

Transition of growth form from dendrite to aggregate was experimentally studied in a solution growth system and in an electro-chemical decomposition system. The results indicated the following facts: The transition occurs through oscillation similar to the bifurcation scheme of the dynamical system. The growth orientation of a dendrite is not necessarily directed to one of the crystallographic axes. Irregular patterns can be formed without noise in an anisotropic system. A physical quantity related to this transition was measured for a metal aggregate system.

1. Introduction

The growth form is categorized into three universal types. Needle type, dendritic type and aggregate type. The first has generally a growth tip of ellipsoidal revolution of a definite curvature and grows with a definite velocity. The second type has generally speaking a growth tip of ellipsoidal revolution too. But it has side-branching activity. Side-branch is periodically produced in the vicinity of the growth tip and grows in a direction with some angle from the growth direction of the tip. The third type of the growth form is characterized by its irregularity. The tip irregularly splits from time to time as it grows, and the produced tips grow in random orientation until they are forced to stop growing by the effect of the other nearby branches. The existence of the surface tension and the crystal anisotropy of the real system makes breath-taking variety of the form found in nature, like snowflakes. But the three types of the growth form are considered most fundamental.

However none of them is understood in a satisfactory manner. For the needle crystal growth the selection of the growth velocity of the tip and the tip radius has rejected theoretical approach for a long time until only recently the hypothesis of "marginal stability" proposed by Langer[1]) was proved theoretically by his coworkers[2]) for small undercooling. The comparison of the theoretical result with experimental observation is under way.

The mechanism of side-branching activity was studied by several researchers using some simplified models[3–7]), not for a fully nonlocal system. The main

0378-4371/86/$03.50

questions here are in what condition of the parameters, such as undercooling or anisotropy strength, the side-branching activity appears, and in what condition the growth of the tip outruns the side-branching activity. In the geometric model the side-branching activity is stable only for a special value of parameters[4]), and in the boundary layer model the active location was found to depend on the velocity squared when a noise-driven side-branching mechanism is considered[6]). This may suggest that the dendrite and the needle type are qualtitatively indistinguishable, and the difference may be only quantitative.

The difference between the dendrite and the aggregate is more fundamental. The "diffusion limited aggregate" grows based on random walk of particles[8]). Thus the irregularity is based on randomness or stochasticity of the walking particles. Therefore some averaging in the growth algorithm may eliminate irregularities. This idea has, in fact, motivated some researchers to carry out simulation[9,10]). They found that the pattern thus formed was a needle type instead of random aggregate for cubic pixels. Prior to these numerical studies there were several works[11–13]) on hydrodynamical systems, the so called viscous fingering. Similarity of the governing mechanism of pattern formation between the crystal growth and viscous fingering has been noticed for some years, but it was only recent that experiments were compared between the two systems. Ben-Jacob et al.[13]) have shown that the pattern produced by a viscous fingering experiment looks like dendritic when the glass plate confining the fluid has some anisotropic modification. This result was considered to show that the anisotropy stabilizes the tip of the dendrite. Very recently Couder et al.[14]) have demonstrated that the dendritic pattern of a stabilized orientation can be formed without anisotropic modification on the glass plate. In Crystal Growth, Honjo et al.[15]) have recently shown that an irregular structure can be formed into an anisotropic system such as NH_4Cl, if the glass was irregularly scratched. However, a question still remains; is it possible that growth form can be changed from dendritic to irregular by changing an experimental parameter, but without giving any external perturbation in the system? This question may be equivalent to asking if the aggregate is deterministically irregular or stochastically irregular[16]). To answer the question partially an electro-chemical decomposition system was studied by two groups[17,18]) independently. The result showed that the transition from a dendrite to irregularity occurred for a critical value of the applied potential. But it was difficult to study in detail how the transition occurs, since the structure includes a wide range of length scales.

In this paper we present experimental results to answer the following questions:

1) Is the growth direction of a "free" dendrite always in a crystallographic axis?

2) Can an irregular pattern be formed from the aqueous solution of NH_4Cl, which is known to grow usually a fairly anisotropic dendrite?

3) Is there any physical quantity one can measure other than geometric shape which helps understanding of the transition from dendritic to irregular growth form?

2. Experimental

We have studied two systems: solution growth from an aqueous solution of NH_4Cl in a thin cell and electro-chemical decomposition of an aqueous solution of $ZnSO_4$ in a thin cell. An air gap of arbitrary thickness from a few microns to a hundred microns was produced with an accuracy to 1 μm between two optical flat glasses. 33.3 wt percent aqueous solution of NH_4Cl was introduced into the gap by the help of the capillary effect, and then shielded. The temperature was controlled to change the undercooling. The growth pattern was recorded and analyzed.

For the electro-chemical decomposition experiment we used a system already described before[17]). However, in the present work we measured current fluctuations simultaneously through the growing boundary for a fixed potential difference. The autocorrelation function was computed from $4k$ points during 20 sec. The concentration of $ZnSO_4$ was 1M for the present work.

3. Results and discussion

Fig. 1 shows examples of the growth tip for three undercoolings for a 10 μm cell. The relation between the measured tip velocity and the tip curvature is shown in fig. 2. The slope obtained from the least squares method is 2.00 ± 0.18, showing a perfect $v\rho^2$ = constant relation. The results suggest that for a quasi

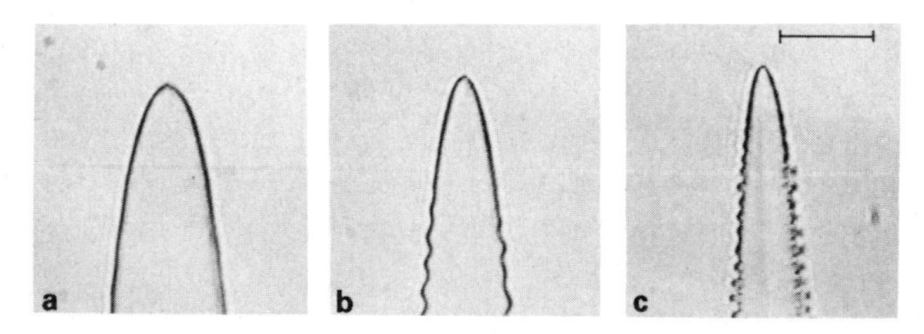

Fig. 1. Three examples of a parabolic tip with increasing undercooling from (a) to (c). The mark indicates 50 μm.

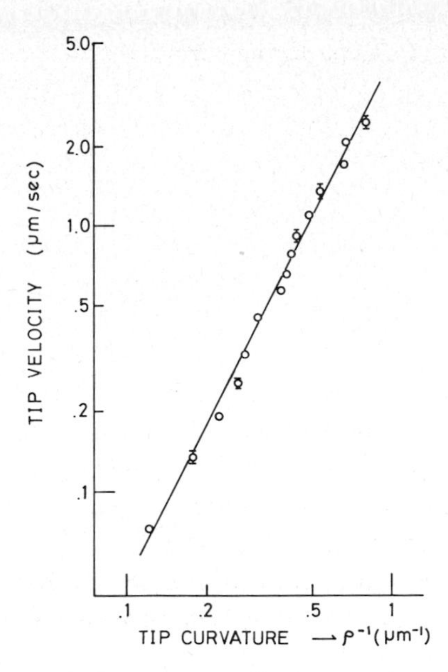

Fig. 2. The measured relation between the tip curvature and the tip velocity.

two-dimensional system where the system thickness is about one tenth of the diffusion length, the relation is the same with the one for a three-dimensional case[19]). However, when the thickness is below 5 μm, the growth form becomes faceted and the speed is reduced to a negligibly small value for a low undercooling condition[20]).

The growth direction of the dendrites for fig. 1 and fig. 2 is $\langle 100 \rangle$. This can be identified from a crystal shape left as grown for a while when the side branches approach that of equilibrium shape and the crystal facet becomes apparent (fig. 3). When the undercooling is reduced one observes a coexistence of dendritic growth of the parabolic tip and that of the oscillatory tip[20]). When the undercooling is high enough, the $\langle 100 \rangle$ direction is by far the most preferred orientation. As the undercooling is reduced, the $\langle 110 \rangle$ direction increases its relative strength of anisotropy (see fig. 4a). When the $\langle 110 \rangle$ direction is almost equally preferred with the $\langle 100 \rangle$ direction, the dendrite of oscillatory type appears. It seems that the oscillation of the tip is caused by the coupling between the tip growth in $\langle 110 \rangle / \langle 100 \rangle$ and the side-branching activity in $\langle 100 \rangle / \langle 110 \rangle$ through the diffusion field. The coupling is possible in this case as the two directions are only 45° from each other. Symmetric growth is observed only when the growth direction is either $\langle 100 \rangle$ or $\langle 110 \rangle$. It occurs sometimes, however, that an arbitrary direction is chosen initially as the growth orientation. Although side-branching activity is no more symmetric in this case, the growth direction can be stable for a fairly long

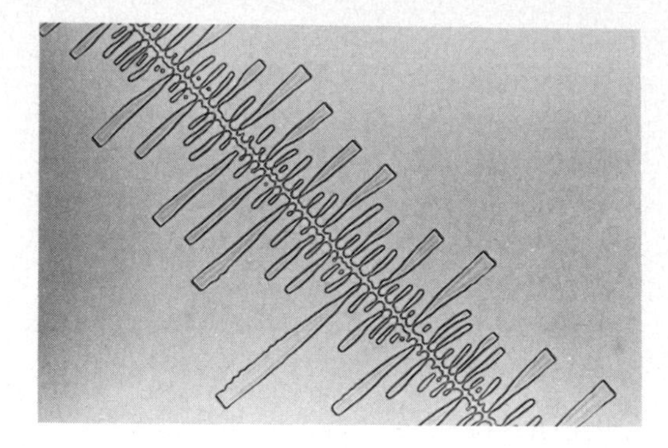

Fig. 3. A dendrite which grew in the ⟨100⟩ direction. The rectangular facet appears after several minutes after the growth.

time. In fig. 5 one can find three growth directions for which the side branches are symmetric (a), and the others which grow in none of the crystallographic directions are not symmetric, some being periodic (b) and some being erratic (c). Two observations are essential here. One is that the dendrite does not necessarily grow in a crystallographic direction. This is known for a directional growth system. But it was not known for a "free" dendrite system. This reminds us of the viscous fingering experiment by Couder[14]), where a bubble stabilizes an oscillatory dendrite. The other is that even in an anisotropic system an erratic pattern is formed. Fig. 6 is a photograph of a part of fig. 5 when it grew further. The pattern continues tip-splitting and screening. When undercooling is further reduced, the erratic pattern is more frequently observed. The two facts suggest strongly that the transition from a needle to an erratic pattern through dendritic growth is similar to a bifurcation scheme of a nonlinear dynamical system, i.e., fixed point, limit cycle, more frequencies and then erratic. Fig. 7 shows the power spectra of current fluctuation for zinc electro-chemical decomposition. In the

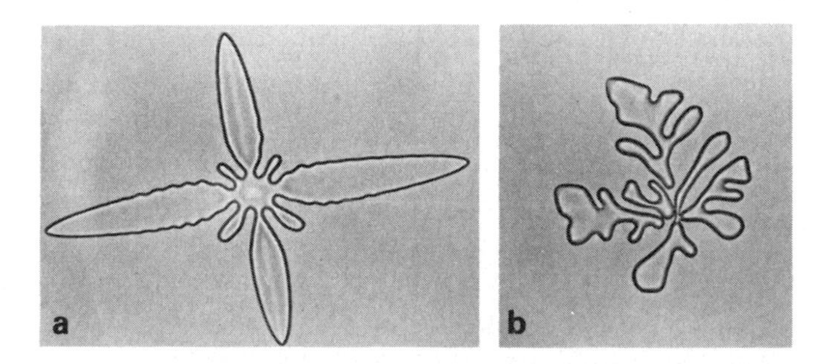

Fig. 4. (a) A dendrite showing cubic symmetry; (b) an irregular pattern.

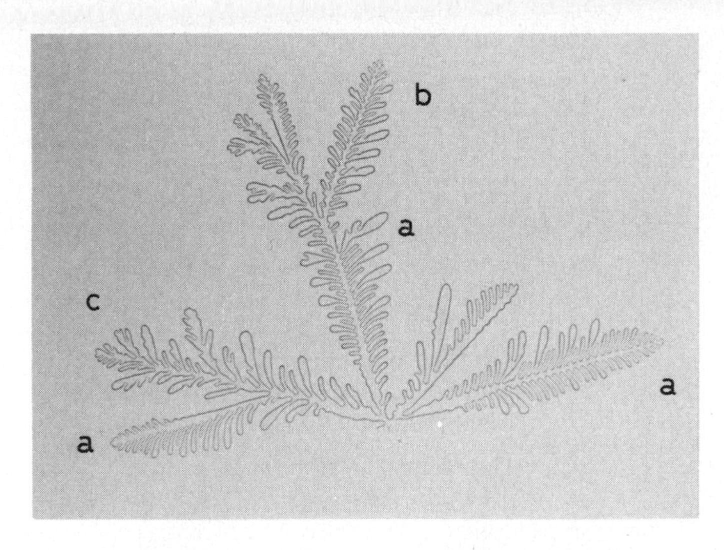

Fig. 5. An example of coexistence of (a) periodic and symmetric, (b) periodic but not symmetric, (c) not periodic nor symmetric branches, depending on the growth direction.

dendritic regime (fig. 7a) a single peak at ≈ 2 Hz is observed. The amplitude of the observation is $\approx 10\ \mu$A while the dc current is ≈ 1 mA. When the applied potential is reduced, one observes a period doubling (fig. 7b). When the applied potential is below 2.2 V, the spectrum is broad (fig. 7c). Finally when it is less than 1.8 V the spectrum shows a power law with an index equal to 2 (fig. 7d).

The frequency and the power of the peak were plotted in fig. 8. The frequency of the oscillation keeps increasing with increasing applied potential, but the power is highest near 3 V. When the potential is too high, side-branching activity is suppressed and the pattern becomes stringy[17]). And when the potential is smaller than ≈ 2 V, irregular aggregate is predominant.

The transition from aggregate to dendritic was observed previously[17,18]), but it was not noticed that the dendritic pattern is of oscillatory nature. Here we

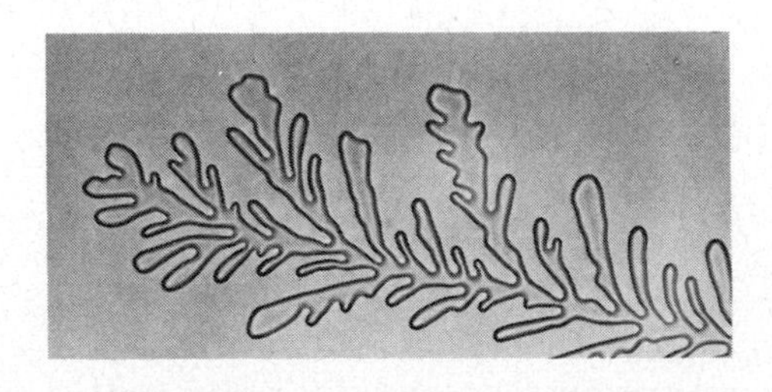

Fig. 6. A photograph of a part of fig. 5 when it grew further.

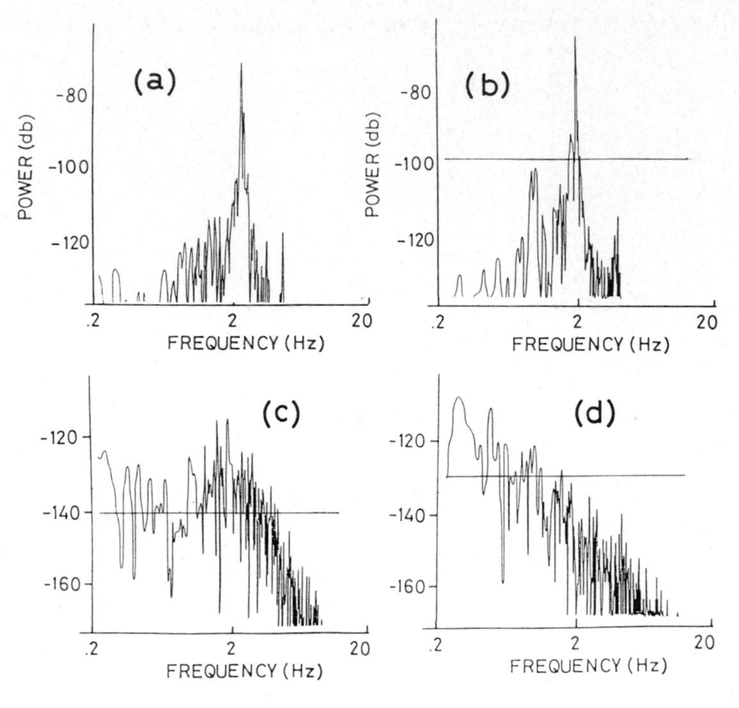

Fig. 7. Power spectra of current fluctuation during electro-chemical decomposition of zinc. (a) $V = 3.2$ V; (b) $V = 3.0$ V; (c) $V = 2.2$ V; (d) $V = 1.8$ V.

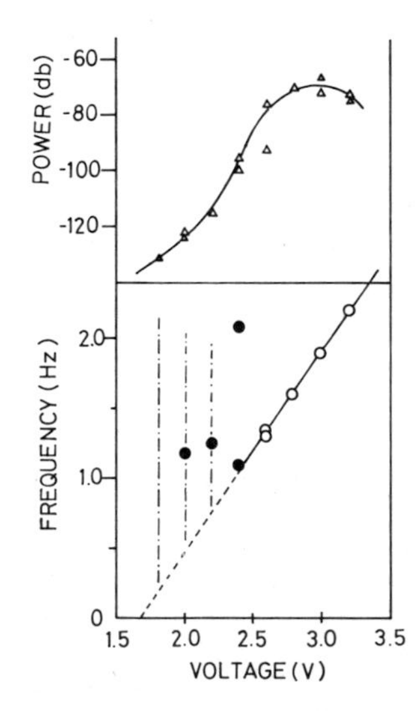

Fig. 8. The power and the frequency of the peak of the power spectra versus applied voltage. Open circle indicates isolated sharp peak; closed circle indicates a broad band.

showed the first measurement of a physical quantity rather than a geometrical quantity associated with the transition[21]).

References

1) J.S. Langer, Rev. Mod. Phys. **52** (1980) 1.
2) A. Barbieri, D.C. Hong and J.S. Langer, preprint.
3) R.C. Brower, D.A. Kessler, J. Kopliland and H. Levine, Phys. Rev. A **29** (1984) 1335.
4) D.A. Kessler, J. Koplik and H. Levine, Phys. Rev. A **30** (1984) 3161.
5) D.A. Kessler, J. Koplik and H. Levine, Phys. Rev. A **31** (1985) 1712.
6) R. Pieters and J.S. Langer, Phys. Rev. Lett. **56** (1986) 1948.
7) O. Martin and N. Goldenfeld, preprint.
8) T.A. Witten and L.M. Sander, Phys. Rev. Lett. **47** (1983) 1400.
9) J. Nittmann and H.E. Stanely, Nature **321** (1986) 663.
10) J. Kertesz and T. Vicsek, to appear in J. Phys. A.
11) L. Paterson, Phys. Rev. Lett. **52** (1984) 1621.
12) J. Nittmann, G. Daccord and H.E. Stanley, Nature **314** (1985) 141.
13) E. Ben-Jacob, R. Godbey, N. Goldenfeld, J. Koplik, H. Levine T. Muller and L.M. Sander, Phys. Rev. Lett. **55** (1985) 1315.
14) Y. Couder, O. Cardoso, D. Dupuy, P. Tavernier and W. Thom, to appear in Europhys. Lett.
15) H. Honjo, S. Ohta and M. Matsushita, to appear in J. Phys. Soc. Jpn.
16) L.M. Sander, Proc. 6th Int. Symp. on Fractal in Physics, L. Pietronero and E. Tosatti, eds. (North-Holland, Amsterdam, 1985), p. 241.
17) Y. Sawada, A. Dougherty and J.P. Gollub, Phys. Rev. Lett. **56** (1986) 1260.
18) D. Grier, E. Ben-Jacob, R. Clarke and L.M. Sander, Phys. Rev. Lett. **56** (1986) 1264.
19) B. Caroli, C. Caroli and B. Roulet, preprint and Y. Sawada and A. Tanaka in preparation.
20) H. Honjo, S. Ohta and Y. Sawada, Phys. Rev. Lett. **55** (1985) 841.
21) Y. Sawada and Y. Hayakawa, in preparation.

Physica **140A** (1986) 142–149
North-Holland, Amsterdam

KINETICS IN CHEMICALLY REACTIVE LIQUIDS

Frank H. STILLINGER
AT & T Bell Laboratories, Murray Hill, NJ 07974, USA

Atomic motions in condensed phases can usefully be described in terms of the geometry of the potential energy hypersurface in multidimensional configuration space. Discrete local minima in this hypersurface generate nonoverlapping basins which span the space. For systems which can undergo chemical reactions, the minima and their basins can be classified according to the numbers of reactant and product species present. Reaction rates are determined by passage over saddle-point regions between contiguous basins with different species populations. These principles are illustrated with molecular dynamics simulation results for molten sulfur, showing how the stable low-temperature medium of octameric rings (S_8) polymerizes and degrades at elevated temperature.

1. Introduction

Chemical reactions arguably constitute the most important class of time-dependent processes that occur in liquids. Understanding them in a general and comprehensive way still remains a major theoretical challenge. The present objective is to show how some recent developments in condensed-phase theory which focus on the differential geometry of potential-energy hypersurfaces[1]) can be adapted naturally and usefully to analyze chemical reactivity in liquids at the atomic level.

Consider a system of N identical atoms, interacting with potential energy $\Phi(\boldsymbol{r}_1, \ldots, \boldsymbol{r}_N)$. We suppose Φ is bounded below, and at least twice differentiable if all pair distances are positive. If wall forces are present to break translational and rotational symmetry, then we can expect Φ to exhibit a discrete set of local minima in the $3N$-dimensional configuration space. Quite general arguments suggest that for large N the number of distinguishable minima (mechanically stable atom arrangements) rises exponentially with N. It makes sense to classify minima on the basis of ϕ, their depth on a per-particle basis ($\phi = \Phi_{\min}/N$), and so we can denote the distribution of minima for large N by $\exp[N\sigma(\phi)]$. The combination $k_B\sigma(\phi)$ is the geometric packing entropy for those stable atomic arrangements with potential $N\phi$.

Each minimum is surrounded by a "basin," defined as the set of configurations $\boldsymbol{r}_1, \ldots, \boldsymbol{r}_N$ which are connected to the minimum by a steepest-descent trajectory

0378-4371/86/$03.50

(tangent to $-\nabla\Phi$) on the Φ hypersurface. Aside from a zero-measure set of boundary points, all configurations can be uniquely assigned in this way to basins. Consequently, essentially any atomic configuration can be resolved into an inherent packing part (the minimum configuration), and a vibrational deformation that displaces the configuration away from the minimum within the same basin.

This separation of inherent packing structure and vibration leads to an exact expression (in the large-N limit) for the classical canonical partition function as a simple quadrature over ϕ:

$$Z_N = \lambda^{-3N} \int \exp\left\{ N\left[\sigma(\phi) - \beta\phi - \beta f_v(\beta, \phi)\right]\right\} d\phi. \tag{1.1}$$

Here $\beta = 1/k_B T$ and λ is the thermal deBroglie wavelength. The quantity f_v is the vibrational free energy (per atom) for those minima whose depth is ϕ. Because N is large, the free energy can be obtained simply by evaluating the integrand in eq. (1.1) at its maximum with respect to ϕ. The value $\phi_m(\beta)$ which produces the maximum corresponds to the dominating set of packings for the given temperature. In the case of the thermodynamically stable liquid, these packings are amorphous.

If the system has not achieved equilibrium, eq. (1.1) is not applicable. Nevertheless the exhaustive division of configuration space into basins is a valuable tool for understanding kinetic processes that drive the system eventually to equilibrium.

2. Chemical bonds

For the simple case of liquified noble gases ϕ can be adequately represented by a sum of central pair potentials (such as the Lennard–Jones 12–6 function). However this is not possible in the more interesting cases that involve covalent chemical bond formation. As we shall see below, a combination of two and three-atom potentials qualitatively suffice to represent directional and saturable chemical bonds.

The presence of covalent bonds introduces a distinct separation between chemical and physical length and energy scales. Chemical bonds are much stronger and substantially shorter than those associated with Van der Waals forces. Table I contrasts the mean chemical bond lengths observed in crystals for several elements with the distances to nearest nonbonded neighbors[2]). On account of this separation, there is virtually never any ambiguity about which pairs of atoms are bonded and which are not; and this applies to all packings

TABLE I
Comparison of chemical bond lengths with distances to nearest nonbonded neighbors, in crystals of the elements[a, b]

Element	Mean bond length	Nearest nonbonded neighbor distance
B	1.76	> 2.50
C[c]	1.42	2.46
N	1.06	3.47
O	1.15	3.20
F	1.49	2.82
Si	2.29	3.73
P[d]	2.23	3.59
S	2.06	3.37
Cl	1.98	3.32
As	2.52	3.12
Se	2.37	3.44
Br	2.27	3.32

[a] From Donohue, ref. 2.
[b] All distances in ångstroms.
[c] Graphite.
[d] Black phosphorous.

whether crystalline or not. By introducing a cutoff distance r_c midway between the bonded and nonbonded distances (as those in table I), the criterion

$$r_{ij} \leqslant r_c \tag{2.1}$$

serves to identify all pairs ij which are chemically bonded.

Once pair connections (bonds) have been identified for any stable packing, the corresponding molecular species, the connected atom subsets, can also be identified. By convention we shall suppose that the molecular species identified at the potential minima are applicable throughout the entire respective basins, thereby eliminating ambiguities that might arise from large-amplitude vibrational distortions.

Chemical reaction rates can now be inferred from the manner in which a system, initially localized in a set of basins corresponding to a nonequilibrium population of molecules, spreads outward to equilibrate over the full set of basins.

3. Sulfur model

At low temperatures the stable form of elemental sulfur involves octameric ring molecules (S_8) with each atom chemically bonded to two nearest neighbors.

Without violating the divalency preference, alternative cyclic molecular species S_n can also form, with little change in bond lengths or angles[2]). At high temperatures, these cyclic species can break a bond to form diradical chains, and these chains can link to form high-molecular weight linear polymers. The polymerization process is reversible and cooperative, leading to a distinctive lambda transition in the liquid phase at 159°C[3]).

A combination of atom pair potentials v_2 and atom-triplet potentials v_3 has been identified[4]) which qualitatively reproduces the elaborate structural chemistry of sulfur, and which has permitted molecular dynamics simulations to be carried out for 1000 sulfur atoms. The function v_2 provides for chemical bonds, while v_3 limits them to divalency and enforces the correct bond angle.

Fig. 1 displays an atomic pair correlation function for a liquid at 126°C (melting point is 115°C), consisting entirely of S_8 rings. Vibrational deformation has been removed to enhance the details of short-range order. In other words a representative set of system configurations served as input to steepest descent

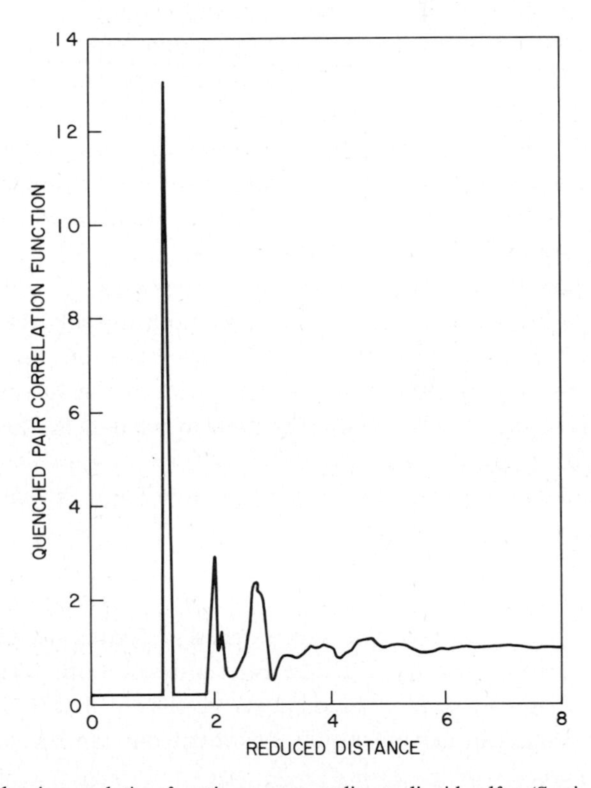

Fig. 1. Quenched pair correlation function corresponding to liquid sulfur (S_8 rings only), at 126°C. All contributing configurations are local potential minima produced by steepest-descent mapping on the potential hypersurface. Distances are measured in units equal to 1.69 Å.

mappings onto potential minima, and fig. 1 shows the pair correlation function averaged over those liquid-phase minima.

The most obvious feature illustrated by fig. 1 is the narrow and isolated peak at reduced distance 1.20. This comprises all bonded pairs (exactly 1000 in this instance). The continuous but structured distribution of other pairs beginning at about reduced distance 1.95 comprises nonbonded intramolecular and intermolecular pairs. Identification of bonds is quite unambiguous in this "quenched" pair correlation function, and the situation is equally unambiguous for quenched pair correlation functions evaluated when other sulfur molecular species (ring and chain) are present.

4. Chemical reactivity and reaction rates

Experimentally, cyclic S_8 molecules remain intact for several hours in the liquid at its melting point. Molecular dynamics simulation in real time proceeds about 10^{15} times more slowly, so no reactivity would be observed at this temperature in a simulation. However the reaction rates can be drastically increased by raising the temperature. In the high temperature regime it is quite feasible to monitor conversion of an initially pure medium of S_8 rings into other species. Following earlier remarks, this is done by extracting a series of system configurations from a high-temperature constant-energy simulation, subjecting each to a steepest-descent quench, and observing what molecular species are present at the corresponding series of potential minima.

Fig. 2 shows how the number of surviving S_8 rings declines with time for two samples suddenly heated from low temperature. Both initially contained 125 S_8 rings. One exhibited mean temperature 870 °C during the run, the other 1050 °C. Both show reactivity, but temperature dependence of the S_8-ring loss rate is obvious. In both of these runs the density was held fixed at the low-temperature liquid value, 1.805 g/cm^3.

The mean degree of polymerization, on a per-atom basis, is given by

$$d = \sum j^2 n_j / N, \tag{4.1}$$

where n_j is the number of molecular species present containing exactly j sulfur atoms. This quantity is exactly 8 at the beginning of both reaction runs, of course, and then rises as time proceeds. It is gratifying to see that the model agrees with experiment in demonstrating a spontaneous tendency to polymerize at high temperature.

Fig. 3 provides an histogram for the n_j distribution at the end of the 1050 °C run. The results are classified also by ring or chain character, the only molecular

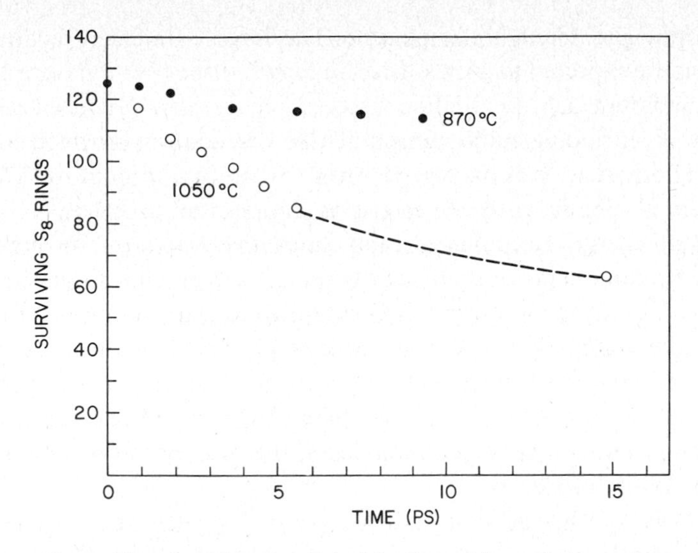

Fig. 2. Time-dependent survival of S_8 rings at elevated temperatures.

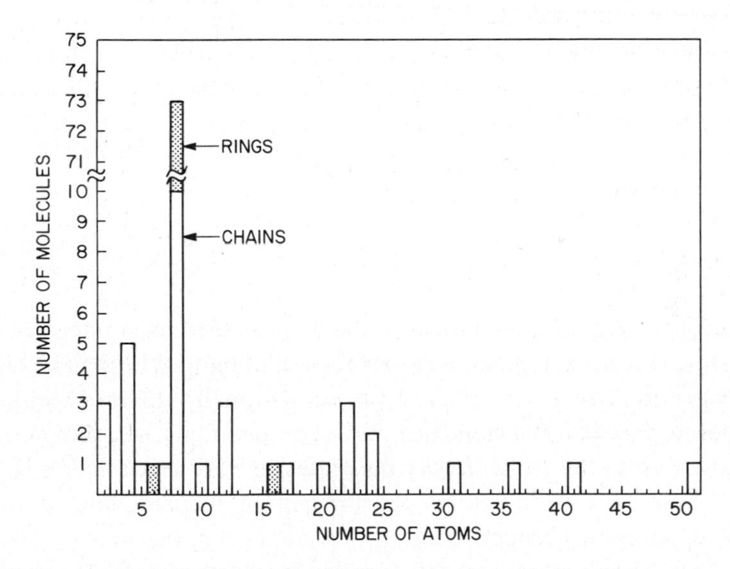

Fig. 3. Populations of ring (dark) and chain (light) species, containing j sulfur atoms, at the end of the 1050° C reaction run.

bonding topologies found to be present. The largest fragment, a diradical S_{51} chain, would be expected to grow further in length if the reaction were allowed to continue. Simultaneous degradation processes have also produced small fragments as well, including sulfur dimers. These degradation-polymerization reactions are endothermic. It is no surprise then to find that in both runs the depths of the potential minima encountered rise as the reaction proceeds.

Since total energy is conserved, and since the reactions are endothermic, kinetic temperature tends to drop slightly during a run. That would be expected to slow reaction rates, but there is a compensating feature that is an important dense medium effect. Diradical chain species produced in the early stages are particularly reactive, with chain ends attacking S_8 rings and breaking them open to form longer chains. A long diradical chain can even coil back on itself and in principal cause its own cleavage. In any case, the reaction becomes autocatalytic after its very early phase.

Initial stages of the reaction, at least for our model, show an unexpected feature. Specifically the n_j distribution is weighted toward even numbers j. For example, even at the end of the lower temperature (870 °C) run, no odd-j species were present, even though chains initially formed by ring opening had both polymerized and spontaneously fragmented. In the higher temperature run (1050 °C), the first appearance of an odd-j species occurred only after half of the run was completed. Indeed the histogram in fig. 3 for the end of the run shows a continuing bias in favor of even j's.

5. Conclusion

By joining conventional molecular dynamics calculations to steepest-descent mapping onto potential minima, a powerful simulation tool becomes available to study chemical reactions in condensed phases. Normally this will require use of more elaborate potential functions than those composed just of additive atom-pair interactions. But as the sulfur model illustrates, it may often suffice to supplement pair potentials with appropriate three-atom contributions to represent directed and saturable chemical bonds.

It has been shown that a cutoff-distance criterion, eq. (2.1), provides an unambiguous and nonarbitrary means for identifying chemical bonds and hence molecular species, for all potential minima and their respective surrounding basins. In the case of sulfur, it will be desirable in the future to search for the presence of unusual topologies for large ring molecules, namely interlinked rings ("catenane" analogs), and knotted structures (such as the elementary trefoil).

References

1) F.H. Stillinger and T.A. Weber, Science **225** (1985) 983.
2) J. Donohue, The Structures of the Elements (Krieger, Malabar, Florida, 1982).
3) B. Meyer, ed; Elemental Sulfur, Chemistry and Physics (Wiley-Interscience, New York, 1965).
4) F.H. Stillinger, T.A. Weber and R.A. LaViolette, J. Chem. Phys., to be published. F.H. Stillinger and T.A. Weber, J. Chem. Phys., to be published.

Physica **140A** (1986) 150–159
North-Holland, Amsterdam

THE TWO-DIMENSIONAL COULOMB GAS: A FIXED ION MODEL AND THE DYNAMIC CHARACTERIZATION OF THE DIELECTRIC–PLASMA TRANSITION

J.P. HANSEN

Laboratoire de Physique Théorique des Liquides, Université Pierre et Marie Curie, 75252 Paris Cedex 05 and Commissariat à l'Energie Atomique, Centre d'Etudes de Limeil-Valenton, B.P. No. 27, 94190 Villeneuve-Saint-Georges, France*

We show how the dielectric–plasma transition in the two-dimensional Coulomb gas, suitably mapped onto a delocalization transition by fixing one species of charges on a triangular lattice, can be characterized by a number of static and dynamic "diagnostics". The influence of the "ionization" equilibrium on the collective charge oscillation (plasmon) mode is investigated by molecular dynamics simulations.

1. The Coulomb gas: why two dimensions?

The solution of Poisson's equation,

$$\nabla^2\phi(r) = -C_d q\,\delta(\boldsymbol{r}) \tag{1}$$

(with $C_d = 2, 2\pi, 4\pi$ in $d = 1, 2, 3$ dimensions), yields a Coulomb potential $\phi(r)$ due to a charge q, which, while independent of d in Fourier space ($\hat{\phi}(k) \sim k^{-2}$), has different functional forms in $\boldsymbol{r}$-space ($\sim r, \log r, r^{-1}$, respectively). The resulting dielectric behaviour of the classical Coulomb gas of oppositely charged particles depends crucially on dimensionality. In one dimension the Coulomb gas is a dielectric at all temperatures: opposite charges are always bound in pairs[1]). In three dimensions the Coulomb gas is believed to be a conductor (or plasma) as long as it remains in a disordered (fluid) phase; in order to prevent the collapse of opposite charges the attractive r^{-1} singularity of the 3d Coulomb potential must be regularized at short distances (finite size rather than point charges). The two-dimensional case is the most interesting, since the corresponding Coulomb gas is a dielectric (as in 1d) below some density-dependent critical temperature, while it is a plasma (as in 3d) above that temperature. The existence of the low T

*Unité Associée au C.N.R.S.

0378-4371/86/\$03.50

insulator phase, where pairs of opposite charges form dipolar "molecules", can be ultimately traced back to the weakly binding character of the logarithmic potential at large r. The stability of the 2d Coulomb gas requires, just as in 3d, charges of finite size, since opposite point charges collapse below the reduced temperature $T^* = k_B T/q^2 = \frac{1}{2}$, when the Boltzmann factor $\exp\{-\log(r/L)/T^*\} \sim r^{-1/T^*}$ ceases to be integrable[2]).

The dielectric–plasma transition is the prototype of a Kosterlitz–Thouless (KT) transition[3,4]) in 2d systems which are characterized by a divergent response (susceptibility) to an external field; simultaneously the decay of spatial correlations changes from exponential in the high T phase to a power law below the critical T[5]). Other models which exhibit a KT transition are the XY model and the discrete gaussian model for the roughening transition[6]).

The KT transition in the 2d Coulomb gas may be characterized by *static* or *dynamic* "diagnostics". The former focus on the fluctuations of the total dipole moment of the sample

$$\boldsymbol{M} = \sum_{i=1}^{N} q_i \boldsymbol{r}_i. \tag{2}$$

The fluctuations of $\boldsymbol{M}$ determine the dielectric response function via the fluctuation–dissipation theorem[7]). An elementary calculation yields

$$\begin{aligned}\frac{\langle|\boldsymbol{M}|^2\rangle}{\Omega_d} &= -\frac{1}{2\Omega_d}\left\langle \sum_{i\neq j}\sum q_i q_j |\boldsymbol{r}_i - \boldsymbol{r}_j|^2 \right\rangle \\ &= -\frac{1}{2}\sum_\alpha \sum_\beta n_\alpha n_\beta q_\alpha q_\beta \int \left[g_{\alpha\beta}(r) - 1\right] r^2 \mathrm{d}^d r,\end{aligned} \tag{3}$$

where $g_{\alpha\beta}(r)$ is the pair distribution function between particles of species α and β, of densities $n_\alpha = N_\alpha/\Omega_d$, and Ω_d is the d-dimensional sample volume. In the plasma phase, the r.h.s. of eq. (3) can be calculated exactly from the perfect screening condition[8]) with the result

$$\frac{\langle|\boldsymbol{M}|^2\rangle}{\Omega_d} = \frac{d}{k_D^2}\sum_\alpha n_\alpha q_\alpha^2 = \frac{d}{2(d-1)\pi} k_B T, \tag{4}$$

where $k_D^2 = C_d \Sigma_\alpha n_\alpha q_\alpha^2 / k_B T$ is the square of the inverse Debye screening length. The result (4) leads to an infinite dielectric constant ε in the plasma.

In the insulator phase, the r.h.s. of (3) can be estimated by assuming that opposite charges form *independent* pairs, an approximation valid in the low density limit, with the result

$$\frac{\langle |\boldsymbol{M}|^2 \rangle}{\Omega_d} \underset{T\to 0}{\simeq} \frac{n}{2}\sigma^2 q^2, \tag{5}$$

where $n = n_+ + n_-$ and σ is the particle diameter; this leads to a finite value of ε. This *static* diagnostic has been used in a recent Monte Carlo study of the phase diagram of the 2d Coulomb gas of oppositely charged hard disks[9]).

In the work presented here, we have focussed on a *dynamic* characterization of the plasma–dielectric transition[10]), our main motivation being an investigation of the influence of an "ionization equilibrium" on collective behaviour in plasmas. The KT transition in 2d furnishes a recombination mechanism of opposite charges (positive "ions" and negative "electrons") in purely classical terms, without any reference to quantum mechanics. The most obvious dynamic diagnostic for a dielectric–plasma transition is the onset of electrical conduction, corresponding to the collective diffusion of charge which is characterized by an Einstein-type relation,

$$\lim_{t\to\infty} \frac{\langle |\boldsymbol{M}(t) - \boldsymbol{M}(0)|^2 \rangle}{S} = a + bt, \tag{6}$$

where $S \equiv \Omega_2$ is the area of the sample and $\boldsymbol{M}(t)$ the total dipole moment (2) at time t. In the dielectric phase ($T < T_1$) the total charge does not diffuse, so that

$$a = 2\frac{\langle |\boldsymbol{M}|^2 \rangle}{S}, \qquad b = 0,\ T < T_1, \tag{7a}$$

while in the plasma phase,

$$b = \frac{4\sigma}{k_B T}, \qquad T > T_1, \tag{7b}$$

where σ is the d.c. conductivity. A measurement of σ in a "molecular dynamics" (MD) simulation constitutes the key diagnostic for the dielectric–plasma transition.

2. The model: from recombination to localization

The usual symmetric 2d Coulomb gas is made up of N positive charges $+q$ and N negative charges $-q$ interacting via the pair potentials

$$v_{\alpha\beta}(r) = q^2\left[\left(\frac{\sigma_{\alpha\beta}}{r}\right)^{\nu} - Z_\alpha Z_\beta \log\left(\frac{r}{L}\right)\right], \tag{8}$$

where Z_α, $Z_\beta = \pm 1$ $(1 \leqslant \alpha,\ \beta \leqslant 2)$; the first term is a short-range repulsion $(\nu \gg 1)$, which need only act between opposite charges $(\sigma_{11} = \sigma_{22} = 0;\ \sigma_{12} \equiv \sigma > 0)$, while the second term is the Coulomb potential, with L and arbitrary scaling length. An equilibrium state of the system is characterized by a reduced density $\eta = \pi n \sigma^2$ and a reduced inverse temperature $\Gamma = q^2/k_B T = 1/T^*$.

We have simplified the model further by assigning an infinite mass to the positive "ions" and fixing them on the sites $\{\boldsymbol{R}_j\}$ of a hexagonal lattice, while the N interacting "electrons" (of mass m) move in the periodic field of the ions. In this way we transform the original two-component Coulomb gas into an inhomogeneous "one-component plasma" (OCP) of classical electrons in a periodic (rather than uniform) neutralizing ionic background. More importantly, we have mapped the KT plasma–dielectric transition onto a classical localization transition since the ion–electron recombination below the transition temperature T_1 now prevents electrons from diffusing throughout the ionic lattice. In this way we obtain an additional dynamic diagnostic of the KT transition, namely the vanishing of the self-diffusion coefficient D. If $\boldsymbol{r}_i(t)$ denotes the position of the ith electron at time t, the Einstein relation similar to (6) reads:

$$\lim_{t\to\infty} \left\langle |\boldsymbol{r}_i(t) - \boldsymbol{r}_i(0)|^2 \right\rangle = a' + b't, \tag{9}$$

$$a' = 2\left\langle |\Delta \boldsymbol{r}|^2 \right\rangle, \qquad b' = 0,\ T < T_1', \tag{10a}$$

$$b' = 4D, \qquad T > T_1', \tag{10b}$$

where $\langle |\Delta \boldsymbol{r}|^2 \rangle$ denotes the mean square displacement of an electron around its host ion in the dielectric phase, and the threshold temperature T_1' need not be identical with T_1 in eqs. (7), although the results of our simulations indicate that they must be very close. The advantage of the individual (D) over the collective (σ) diffusion diagnostic in an MD simulation is one of statistics: since D can be averaged over N independent estimates (one for each electron), the unavoidable statistical uncertainties are much smaller ($\sim N^{1/2}$ times!) for D than for σ.

In summary the model which we consider henceforth is characterized by the following potential energy for a given electron configuration $\{\boldsymbol{r}_i\}$:

$$V_N(\{\boldsymbol{r}_i\}) = C + \sum_i \Psi(\boldsymbol{r}_i) + \frac{1}{2} \sum_{i \neq j}\sum v(|\boldsymbol{r}_i - \boldsymbol{r}_j|), \tag{11}$$

where C is the Madelung energy of the ionic lattice, $\Psi(\boldsymbol{r}) = \Sigma_j v_{12}(|\boldsymbol{r} - \boldsymbol{R}_j|)$ is the periodic potential of the ions acting on the electrons and $v(r) \equiv v_{22}(r) = -q^2 \log(r/L)$.

3. Trajectories and averages

In an MD simulation the classical, coupled equations of motion of the N electrons, derived from the potential (11), are replaced by finite difference equations and integrated numerically with an adequate time step Δt[11]). We have carried out extensive simulations both in the constant energy (microcanonical) and in the constant kinetic energy (i.e., temperature) ensembles[10]). Most simulations were done for a sample of $N = 108$ electrons (and as many fixed ions) in a hexagonal cell, with periodic boundary conditions, along the isochores $\eta = 0.01$ and 0.0004, and over a range of couplings $20 \gtrsim \Gamma \gtrsim 1$. The "natural" time scale in the plasma phase is the inverse of the plasma frequency, $\omega_p = (2\pi n q^2/m)^{1/2}$.

A typical example of the electron trajectories generated in an MD run for $\eta = 0.01$ and $\Gamma = 4.3$ is shown in fig. 1. At this temperature "bound" and "free" (diffusing) electrons are seen to coexist ("ionization equilibrium"). Of course, if the simulation were carried on for a longer time, the bound electrons would eventually become delocalized, while the diffusing ones would be momentarily trapped. At this density, we observed complete localization (at least for times as long as $10^3\omega_p^{-1}$) for $\Gamma \gtrsim 5.7$, while the trajectories become more and more space-filling as Γ is decreased. Thus a "visual" estimate of the inverse transition temperature is $\Gamma_1 = 1/T_1^* \simeq 5.7$ for $\eta = 0.01$ and $\Gamma_1 \simeq 2.8$ for $\eta = 0.0004$: the transition temperature increases as the density is lowered, and we conjecture that $\lim_{\eta \to 0} \Gamma_1 = 2$, which is the "collapse" coupling for point particles[2]): Note that the fixed ion model leads to a low density transition temperature which is exactly twice that found for the symmetric Coulomb gas (mobile ions)[3,9]).

Statistical averages can be calculated along the phase space trajectory generated in the MD runs and lead to estimate of thermodynamic properties: static and time-dependent correlation functions, which allow a quantitative characterization of the KT transition. Some of the salient results are the following:

a) The constant area specific heat exhibits, along each isochore, a pronounced peak at a temperature $T_2^* > T_1^*$, which may be associated with the collective breakup of ion–electron pairs beyond the delocalization threshold; the peak sharpens and moves to higher temperatures as η decreases.

b) Delocalization is clearly signalled by the periodic one-particle density $\rho^{(1)}(\boldsymbol{r})$ of the electrons which spreads out as Γ is lowered, and takes on nonzero values at the Wigner–Seitz cell boundary below Γ_1, indicating that electrons may move between neighbouring ions.

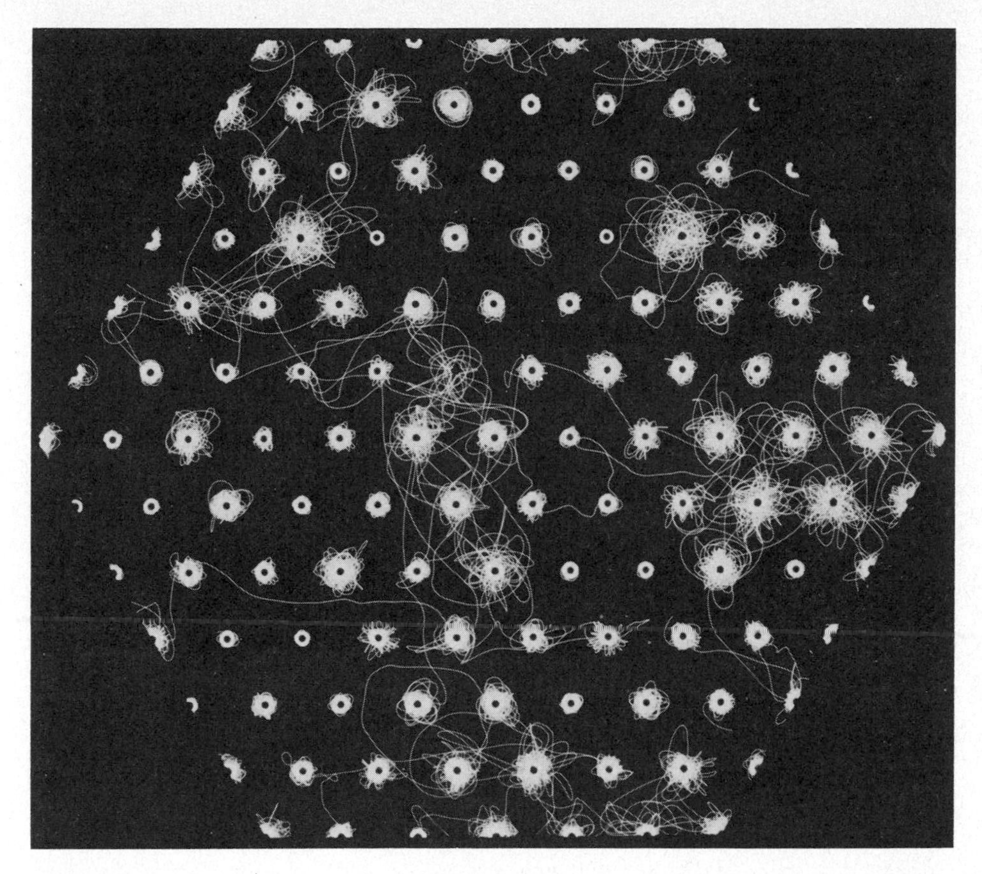

Fig. 1. Trajectories of 108 electrons in a hexagonal lattice of fixed ions at $\eta = 0.01$, $\Gamma = 4.3$, integrated over 400 ω_p^{-1}.

c) The static structure factor is the following density–density correlation function:

$$S(\boldsymbol{k}) = \frac{1}{N}\langle \rho_{\boldsymbol{k}} \rho_{-\boldsymbol{k}} \rangle, \tag{12a}$$

$$\rho_{\boldsymbol{k}} = \sum_{i=1}^{N} \mathrm{e}^{\mathrm{i}\boldsymbol{k}\cdot\boldsymbol{r}_i}. \tag{12b}$$

The long wavelength limit of $S(\boldsymbol{k})$ is not affected by the discrete (periodic) nature of the ionic background and should coincide with the uniform OCP limit obtained from fluctuation theory[12]):

$$S(\boldsymbol{k}) \underset{k\to 0}{=} \frac{k^2/k_\mathrm{D}^2}{1 + k^2/k_\mathrm{s}^2}, \tag{13}$$

where $k_s^2 = 1/\lambda_s^2 = k_D^2(\partial\beta P/\partial n)_T^{-1}$ is the inverse square screening length determined by the compressibility. The screening length λ_s increases as the temperature is lowered and appears to diverge at the transition, in agreement with intuition (the number of "free" electrons, capable of participating in collective screening, vanishes).

4. Individual versus collective dynamics

The KT transition, as the unrelated glass transition, may be most conveniently characterized by the self-diffusion threshold. D can be estimated from eq. (9), or equivalently by the time integral of the velocity autocorrelation function (ACF) $Z(t) = \langle \boldsymbol{v}_i(t) \cdot \boldsymbol{v}_i(0) > / \langle v_i^2 \rangle$. Two examples of $Z(t)$ and of its Fourier transform (spectrum) $\hat{Z}(\omega)$ are shown in fig. 2. The contrast between the high and low temperature results is striking. In the plasma phase, $Z(t)$ decays monotonically, and $D \sim \hat{Z}$ $(\omega = 0)$ is large; in the dielectric phase $Z(t)$ exhibits marked oscillations and the spectral analysis reveals a sharp resonance which can be associated with the rotational motion of individual electrons, and a broad high-frequency vibrational band; D is now strictly zero.

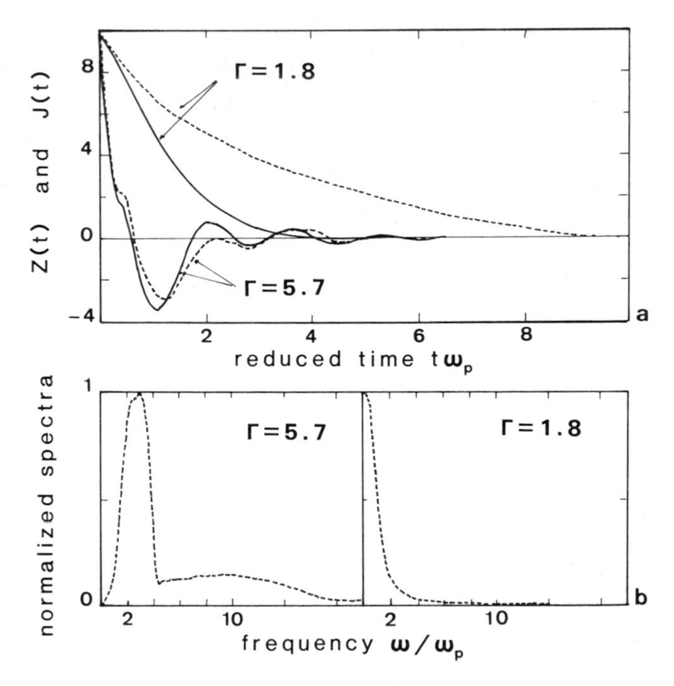

Fig. 2. (a) Normalized velocity and current ACF $Z(t)$ (full curves) and $J(t)$ (dotted curve) versus reduced time for $\eta = 0.01$, $\Gamma = 5.7$ and 1.8. (b) Normalized spectra $\hat{Z}(\omega)$ versus reduced frequency.

The electron conductivity σ, on the other hand, can be estimated from (6) or from the time integral of the electric current ACF:

$$J(t) = \frac{m}{2Nk_{\mathrm{B}}T}\langle \boldsymbol{j}(t)\cdot\boldsymbol{j}(0)\rangle, \tag{14}$$

where the microscopic electric current per unit charge $\boldsymbol{j}(t)$ is just the sum of electron velocities. In the dielectric phase, $J(t)$ turns out to be undistinguishable from $Z(t)$, within statistical uncertainties, showing that dynamics are dominated by individual particle motion (cross correlations between velocities of different electrons are practically negligible); $\sigma \sim \hat{J}$ $(\omega = 0)$ is zero just as D. In the plasma phase, on the other hand, collective behaviour becomes increasingly important as T increases, and $J(t)$ differs more and more significantly from $Z(t)$, decaying much more slowly; this feature can be attributed to the fact that electron–electron collisions conserve momentum, and hence the electric current, but change individual velocities.

The collective behaviour of the plasma is also characterized by the charge fluctuation spectrum $S(\boldsymbol{k}, \omega)$, which is the dynamic generalization of the static structure factor (12):

$$S(\boldsymbol{k}, \omega) = \frac{1}{2\pi}\int_{-\infty}^{+\infty} \mathrm{e}^{\mathrm{i}\omega t}\frac{1}{N}\langle \rho_{\boldsymbol{k}}(t)\rho_{-\boldsymbol{k}}(0)\rangle\,\mathrm{d}t. \tag{15}$$

We have computed $S(\boldsymbol{k}, \omega)$ for wave numbers $\boldsymbol{k}$ compatible with the periodic boundary conditions. As long as the wavelength $\lambda = 2\pi/k$ is significantly larger than the screening length λ_{s}, $S(\boldsymbol{k}, \omega)$ exhibits a sharp resonance at a frequency close to ω_{p}: this is a manifestation of a well-defined propagating plasma oscillation (plasmon) mode. For a fixed wave vector $\boldsymbol{k}$, the resonance broadens as the temperature is lowered, due to ion–electron recombinations. At a fixed temperature and density, the dispersion curve $\omega(k)$ of the plasmon is nonmonotonic, presumably due to coupling between the collective mode and individual particle motion in the ionic field.

The long wavelength limit of the charge fluctuation spectrum can be related to the complex a.c. conductivity:

$$\sigma(\omega) = \sigma'(\omega) + \mathrm{i}\sigma''(\omega) = \frac{\omega_{\mathrm{p}}^2}{2\pi}\int_0^{\infty} \mathrm{e}^{\mathrm{i}\omega t}J(t)\,\mathrm{d}t \tag{16}$$

by an elementary hydrodynamic calculation[13]),

$$s(\omega) = \lim_{k\to 0}\frac{S(\boldsymbol{k}, \omega)}{S(\boldsymbol{k})} = \frac{\sigma'(\omega)}{[\omega - 2\pi\sigma''(\omega)]^2 + [2\pi\sigma'(\omega)]^2}. \tag{17}$$

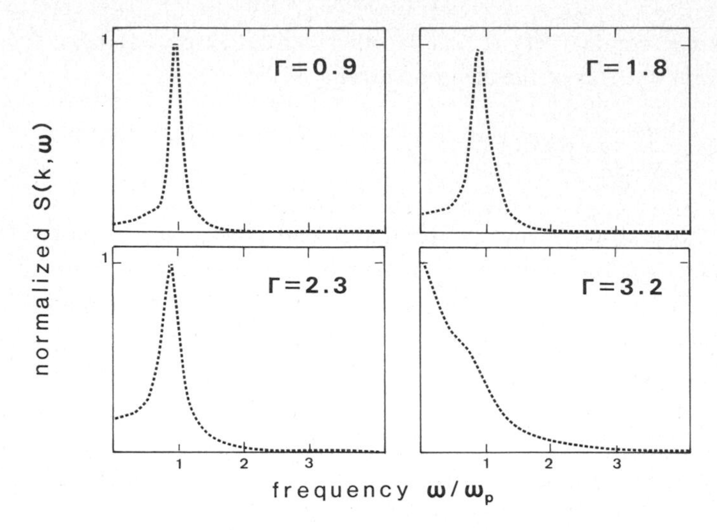

Fig. 3. Long wavelength charge fluctuation spectrum $s(\omega)$ for $\eta = 0.01$ and four couplings: $\Gamma = 0.9, 1.8, 2.3, 3.2$.

The results obtained along the isochore $\eta = 0.01$ for different couplings Γ in the plasma phase are shown in fig. 3. The infinite wavelength plasmon is seen to be increasingly damped as the temperature is lowered, due to electron localization. The resonance frequency drops with temperature, in qualitative agreement with the prediction of the collisionless Vlasov kinetic equation for the inhomogeneous electron gas in the periodic field of the ions[14]). Note, however, that mean field theory does not lead to a KT transition and predicts the 2d Coulomb gas to be a conductor at all temperatures.

5. Perspectives

We are presently extending our model in two directions. First we consider a Coulomb gas with multiply charged ions $(nq, -q)$; in that case we expect to find a cascade of n "recombinations"[15]), leading to as many KT transitions. The second natural extension of our MD simulations is to the mobile ion case; as indicated in this paper the KT transition temperature drops then by a factor of two in the low density limit, compared to the fixed ion situation.

Acknowledgements

I am indebted to Jean Clérouin, Angel Alastuey and Bernard Piller for their enthusiastic collaboration.

References

1) A. Lenard, J. Math. Phys. **2** (1961) 682.
2) E.H. Hauge and P.C. Hemmer, Phys. Norv. **5** (1971) 209; C. Deutsch and M. Lavaud, Phys. Rev. A **9** (1974) 2598.
3) M. Kosterlitz and D.J. Thouless, J. Phys. C **6** (1973) 1181.
4) J.M. Kosterlitz, J. Phys. C **7** (1974) 1046, **10** (1977) 3753.
5) J. Fröhlich and T. Spencer, Phys. Rev. Lett. **46** (1981) 1006.
6) S.T. Chiu and J.D. Weeks, Phys. Rev. B **14** (1976) 4978.
7) R. Kubo, Rep. Prog. Phys. **29** (1966) 255; P. Madden and D. Kivelson, Adv. Chem. Phys. **56** (1984) 567.
8) F. Stillinger and R. Lovett, J. Chem. Phys. **49** (1968) 1991; P.A. Martin and C. Gruber, J. Stat. Phys. **31** (1983) 691.
9) J.M. Caillol and D. Levesque, Phys. Rev. B **33** (1986) 499.
10) J. Clérouin and J.P. Hansen, Phys. Rev. Lett. **54** (1985) 2277.
11) See, e.g., J.P. Hansen and I.R. McDonald, Theory of Simple Liquids, 2d ed. (Academic, London, 1986).
12) P. Vieillefosse and J.P. Hansen, Phys. Rev. A **12** (1975) 1106.
13) See, e.g., J.P. Hansen and L. Sjögren, Phys. Fluids **25** (1982) 617.
14) A. Alastuey and J.P. Hansen, Europhys. Lett. **2** (1986) 97.
15) J.P. Hansen and P. Viot, J. Stat. Phys. **38** (1985) 823.

Physica 140A (1986) 160–168
North-Holland, Amsterdam

QUANTUM STATISTICS OF COULOMB SYSTEMS: THERMODYNAMIC FUNCTIONS AND PHASE TRANSITIONS*

W. EBELING
Sektion Physik der Humboldt-Universität, Berlin, DDR

W.D. KRAEFT
Sektion Physik / Elektronik der Ernst-Moritz-Arndt-Universität, Greifswald, DDR

and

D. KREMP AND G. RÖPKE
Sektion Physik der Wilhelm-Pieck-Universität, Rostock, DDR

Matter is considered as a quantum system of nonrelativistic point charges, the electrons and the nuclei (or other positive particles). After the discussion of several exact bounds and stability theorems the low and high density structures of the thermodynamic functions (corresponding to ideal gases) are given. The corner of correlations in the density–temperature plane is treated by means of thermodynamic Green functions. The realistic description of correlations requires partial summations corresponding to the effects of screening (chains), bound states (ladders) and chemical equilibria (quasi-particle summations). It is shown that Coulomb interactions may cause thermodynamic instabilities corresponding to the first order plasma phase transitions.

1. Introduction

Matter consists of charged particles, the negative electrons and the positive nuclei; this fundamental discovery due to Rutherford was presented 75 years ago to the Philosophical Society in Manchester. The law of interaction between point charges, Coulomb's law, was published 200 years ago. Between 1785 and 1791 Charles Augustin Coulomb presented seven papers (memoirs) on electricity and magnetism to the Academy of Science in Paris. Here we are concerned with the quantum statistics of macroscopic matter consisting of very many nonrelativistic electrons and nuclei. Since the properties of the materials we see and touch are almost entirely determined by Coulomb forces, as they manifest themselves in the collective behaviour of interacting electrons and nuclei, this is a basic problem of

*Correspondence address: W.D. Kraeft, Sektion Physik/Elektronik der Ernst–Moritz–Arndt–Universität, DDR-2200 Greifswald, Domstr. 10*a*, DDR.

0378-4371/86/$03.50 © Elsevier Science Publishers B.V.
(North-Holland Physics Publishing Division)

statistical mechanics. We would like to understand how the underlying Coulomb forces give rise to the effective interactions and quasi-particles, which appear in the usual statistical mechanics, and we would like to have a theory which is based on first principles only. Another reason to deal with Coulomb systems is more practical. There are many systems, e.g., dense plasmas, metals and semiconductors, where bare Coulomb interactions appear explicitly and determine all properties[1]). The location of various real plasmas can be represented in the density–temperature plane[2]). We may imagine that the thermodynamic functions and other physical quantities are like landscapes over this plane.

Nonideality $\Gamma > 1$ is restricted to a small part in the n–T plane which is called "corner of correlations"[1,3]).

2. Stability theorems and other exact results

In 1967 Dyson and Lenard proved in a seminal paper[4]) the existence of a lower bound for a neutral Coulombic system consisting of N_e electrons and positive point charges,

$$E \geqslant -CN_e. \tag{1}$$

In the thermodynamic limit ($N_e \to \infty, V \to \infty, n_e = N_e/V =$ finite) the Fermi character of the electrons is absolutely necessary to prevent the collapse of the system. The existence of a lower bound proportional to N_e is called H-stability of the system. Charged bosons do not show H-stability. Essentially the increase of the kinetic energy of the electron subsystem with N_e and the corresponding increase of pressure with n_e prevents the collapse of the system. Without the Fermi character of the electrons, matter in bulk would collapse into a condensed high-density phase. The proof of eq. (1) given by Dyson and Lenard is a mathematical "tour de force"; a simple elegant derivation of eq. (1) with a greatly improved constant ($C \approx 23$ Ry) was obtained by Lieb and Thirring[5]) by using a theorem stating that the Thomas–Fermi approximation yields a lower bound to the energy

$$E \geqslant E_{TF}. \tag{2}$$

H-stability is not sufficient to guarantee the basic thermodynamic conditions, extensivity and stability of the thermodynamic functions. Lebowitz and Lieb could prove however that for Coulombic systems the free energy density exists in the thermodynamic limit and satisfies the stability conditions[5])

$$\frac{1}{k} = \rho^2 \frac{\partial^2}{\partial \rho^2} f \geqslant 0, \qquad c_V = -T \frac{\partial^2}{\partial T^2} f \geqslant 0;$$

$$\rho = \sum_a n_a m_a, \qquad n_a = N_a / V. \tag{3}$$

We may obtain informations about the low density behaviour of the free energy from the Lebowitz–Peña theorem[6])

$$\beta f \to \beta f_{\rm id}^{\rm B} = \sum_a n_a \ln\left(n_a \lambda_a^3 / e\right) \quad \text{if } n_a \lambda_a^3 \to 0. \tag{4}$$

We may conclude from this that in the limit of low density the Coulomb system behaves like an ideal Boltzmann gas consisting of the elementary particles (the electrons and the nuclei). Especially we may conclude from eq. (4) that matter is always ionized in the limit of sufficiently low density. The low density structure of the free energy is known since the work of Mayer (1950) and Friedman (1962); it reads[2])

$$\begin{aligned}\beta\left(f_{\rm id}^{\rm B} - f\right) = {} & A_0(T)\rho^{3/2} + A_1(T)\rho^2 \ln \rho \\ & + A_2(T)\rho^2 + A_3(T)\rho^{5/2} \ln \rho + A_4(T)\rho^{5/2} \\ & + \mathcal{O}\left(\rho^3 \ln \rho\right) + \mathcal{O}\left(\rho^3\right).\end{aligned} \tag{5}$$

The first two corrections to the Boltzmann term are of classical nature (limiting laws). Beginning from $\mathcal{O}(\rho^2)$ the quantum character of the charges plays an essential role. The coefficients $A_0(T) \cdots A_4(T)$ are exactly known without any approximations with respect to h^2 and e^2.

3. The method of Green functions

Most quantum-statistical methods describe – as a rule – only a certain part of the density–temperature plane. The great advantage of the technique of Green functions is that it covers, at least in principle, the entire "corner of correlations". Let us define as usual the n-order Green functions (GF) by

$$G_n(1 \cdots n 1' \cdots n') = \frac{1}{(\mathrm{i}\hbar)^n} \langle T\{ a(1) \cdots a(n) a^\dagger(n') \cdots a^\dagger(1') \} \rangle. \tag{6}$$

The single-particle GF and the two-particle GF can be used to derive the thermodynamic functions as well as other macroscopic properties as, e.g., transport coefficients and optical properties; cf. ref. 2. To evaluate the GF we may use the theorem that the n-particle GF is given as the sum of the contributions corresponding to all topologically nonequivalent connected Feynman diagrams starting from the open ends $1, 2, \ldots, n$ and leading to the open ends $1', 2', \ldots, n'$. For practical purposes one has to restrict the calculation to certain subclasses of the diagrams. In order to get realistic descriptions of Coulombic systems, infinite classes of diagrams, which correspond to the following effects, have to be summed:

(i) screening of the interactions (chain summations);

(ii) formation of bound states of groups of elementary particles such as atoms and molecules (summation of ladder diagrams);

(iii) chemical equilibria between free and bound states (quasi-particle summations).

In general the summation of certain classes of diagrams is equivalent to the solution of linear integral equations[1]).

In order to calculate the thermodynamic functions we shall use the formulae

$$n(\beta, \mu) = \frac{1}{V} \sum_{p} \int \frac{d\omega}{\pi} f_1(\omega) \operatorname{Im} G_1(p\omega), \tag{7}$$

$$p(\beta, \mu) = \int_{-\infty}^{\mu} d\mu' \, n(\beta, \mu'). \tag{8}$$

With the approximations for G_1 described above, we arrive at expressions for the equation of state which contain not only single-particle but also n-particle contributions. For example one obtains[7,8,9])

$$n(\beta, \mu) = \sum_{p} f_1(\mathscr{E}_1) + \sum_{nP} g_2(E_{nP}) + \sum_{P} \int d\omega \, g_2(\omega) \frac{d}{d\omega} \sin \delta(\omega), \tag{9}$$

where $\delta(\omega)$ is a generalized density-dependent scattering phase. The energy levels include first-order shifts

$$\mathscr{E}_1 = \frac{p^2}{2m} + \Delta_p^{\text{free}} + \Delta_p^{\text{bound}}, \tag{10}$$

$$E_{nP} = E_{nP}^0 + \Delta_{nP}^{\text{free}} + \Delta_{nP}^{\text{bound}}. \tag{11}$$

By integration of eq. (7) according to eq. (8) one obtains

$$p = p_{\mathrm{id}}^{(1)}(\beta,\mu) + p_{\mathrm{id}}^{(2)}(\beta,\mu) + \ldots + p_{\mathrm{int}}(\beta,\mu). \tag{12}$$

Here the first term corresponds to free one-particle states, the second one to two-particle bound states and the last one to the interactions of quasi-particles.

So far the theory has been worked out in the grand canonical ensemble. This representation is appropriate for the description of chemical bonds; however, additive forces are not described well. Further phase transitions of first order are difficult to recognize in the grand canonical representation. Therefore, the transition to a mixed representation, the chemical picture[1,3]), is advantageous. The recipe for the performance of this transition is given by the principle of equivalence: bound states are to be treated on the same footing as free particles[3,8]). Following this idea we introduce chemical potentials of quasi-particles by

$$\begin{aligned} \mu_1^{(\nu)} &= \mu + \Delta_\nu(0), \\ \mu_2^{(n)} &= \mu(1) + \mu(1') + E_n^0 + \Delta_n(0) \end{aligned} \tag{13}$$

where $\Delta_\nu(0)$ and $\Delta_n(0)$ are shifts taken at the bottom of the energy bands. The band index ν has been introduced for the case that $\mathscr{E}_1$ shows an energy band structure as, e.g., in crystalline states. In the new representation we may easily go to the canonical ensemble ending with the expression[13])

$$f(\beta, n_1^{(\nu)}, n_2^{(n)}\ldots) = \sum_\nu f_{\mathrm{id}}^{(\nu)}(\beta, \mu_{1,\mathrm{id}}^{(\nu)}) + \sum_n f_{\mathrm{id}}^{(n)}(\beta, \mu_{2,\mathrm{id}}^{(n)}) - p_{\mathrm{int}}(\beta, \mu_{1,\mathrm{id}}^{(\nu)}). \tag{14}$$

The closing relation, which yields the necessary relation between densities and chemical potentials, reads

$$\begin{aligned} n_1^{(\nu)} &= \sum_p f_1(\mathscr{E}_1^0, \beta, \mu_{1,\mathrm{id}}^{(\nu)}), \\ n_2^{(n)} &= \sum_P g_2(E_{nP}^0, \beta, \mu_{2,\mathrm{id}}^{(n)}). \end{aligned} \tag{15}$$

Finally we have to require that the free energy be minimized with respect to the division into the different chemical species,

$$f(\beta, n_1^{(\nu)}, n_2^{(n)}, \ldots) = \min, \tag{16}$$

$$n = \sum_\nu n_1^{(\nu)} + \sum_n n_2^{(n)} + \ldots. \tag{17}$$

We still note that the border between bound and free states may also be considered as a free variational parameter. The relations (13)–(17) altogether represent the "chemical quasi-particle picture" which describes many effects which are typical for Coulombic systems like ionization, formation of atoms and molecules, band structure of electronic states, etc.

4. Thermodynamics and phase transitions

Let us first consider a fully symmetrical plasma consisting of positive and negative charges with equal masses $m = m_+ = m_-$. Following the prescriptions given above we may write the thermodynamic functions per electron in the following form:

$$\eta(r_s, \tau) = \eta_{id} + \eta_{xc}. \tag{18}$$

Here η is any of the thermodynamic functions

$$\eta\{G_{xc}/N_e;\ F_{xc}/N_e;\ U_{xc}/N_e\} \tag{19}$$

or

$$p_{xc}V = F_{xc} - G_{xc}, \qquad TS_{xc} = U_{xc} - F_{xc},$$

measured in the energy unit

$$\mathrm{Ry}^* = \frac{m}{2m_e\mathscr{E}_r^2}\,\mathrm{Ry} \tag{20}$$

($\mathscr{E}_r$ is the relative dielectric constant). Instead of density and temperature we use

$$r_s^* = \frac{d}{a_B^*}, \qquad \tau^* = \frac{T}{\mathrm{Ry}^*}, \qquad a_B^* = 2\mathscr{E}_r\frac{m_e}{m}. \tag{21}$$

For definiteness we shall now use unshifted energy levels and the Brillouin–Planck–Larkin convention for the sum over states of the atom[1-3]). Then the interaction contribution includes only the Hartree–Fock exchange term and the Montroll–Ward correlation term. We note that for symmetrical plasmas the contributions of the Feynman diagrams with an odd number of lines to the thermodynamic functions cancel due to electroneutrality and equality of masses. Further the exchange-correlation contribution is related by a simple scaling relation[10]) to the properties of a one-component electron plasma (OCP)

$$\eta_{xc}(r_s^*, \tau^*)/\mathrm{Ry}^* = 8\eta_{xc}^{\mathrm{OCP}}(r_s, \tau)/\mathrm{Ry}. \tag{22}$$

For the OCP the limiting properties are well known as, e.g.,

$$\eta_{xc} = \begin{cases} \sqrt{3}\, d_0 \tau \Gamma^{3/2}\left(1 - d_1 \sqrt{3}\, \tau^{1/2} \Gamma^{3/2}\right) - 3 d_H \tau^2 \Gamma^3 + \ldots, & \Gamma \ll 1; \\ a_H r_s^{-1} + a_W r_s^{-1}, \quad r_s \gg 1,\ \tau \ll 1; & \\ a_H r_s^{-1} + b_0 \ln r_s + b_1, \quad r_s \ll 1,\ \tau \ll 1. & \end{cases} \tag{23}$$

The meaning of the coefficients which are explicitly given in table I are: d_0 is the Debye contribution, d_1 is the quantum correction, d_H is the high-temperature Hartree–Fock term, a_H is the low-temperature Hartree–Fock term, a_W is the Wigner lattice term and b_0, b_1 are the Gell-Man–Brueckner terms. Further we have

$$\Gamma = \frac{e^2}{kTd} = \frac{2}{\tau r_s}. \tag{24}$$

The limiting structure (23) suggests the construction of Padé approximations[2,10,11]. One possibility is

$$\eta_{xc}^{OCP} = -\frac{C_1\left(a_H + a_W(r_s)\right) + 2\sqrt{6}\, d_0 r_s^{11/2} \tau^{5/2} + 24 d_H r_s^4 \tau^2}{C_1 r_s + C_2 r_s^4 \tau^2 + 2\sqrt{6}\, d_1 r_s^{11/2} \tau^2 + r_s^7 \tau^3}, \tag{25}$$

$$a_W(r_s) = 2 b_0 r_s \ln\left\{1 + \left[r_s^{1/2} \exp\left(-\frac{b_1}{2 b_0}\right) + 2 b_0 a_W^{-1} r_s\right]^{-1}\right\}. \tag{26}$$

The free constants are fitted to numerical results[12] in the following way:

$$C_1 = 50 + r_s^3, \qquad C_2 = 2.3.$$

Table I

Limiting law coefficients which appear in the Padé approximation.

	G_{xc}/N_e	F_{xc}/N_e	U_{xc}/N_e
d_0	0.50000	0.33333	0.50000
d_1	0.5305	0.3979	0.9947
d_H	0.1250	0.0625	0.21875
a_H	1.2218	0.91633	0.91633
a_W	1.1674	0.87553	0.87553
b_0	0.06218	0.06218	0.06218
b_1	0.1140	0.0933	0.0933

Using these results and the condition of thermodynamic stability

$$\frac{\partial}{\partial n_e}\left(\frac{G}{N_e}\right) \geqslant 0, \tag{27}$$

we obtain the critical parameters

$$\tau_{cr} \approx 0.08, \quad r_s^{cr} = 3.4. \tag{28}$$

For positronium plasmas this gives $T_{cr} \approx 6000$ K and for electron–hole plasmas in Ge $T_{cr} \approx 8$ K. The latter value is in reasonable agreement with experimental findings which yield $T_{cr} = 6$ K. More realistic theories of *e–h* plasmas include details of the band structure and the small asymmetry of the masses[10,13]). We underline that the phase transition observed in *e–h* plasmas is nearly entirely determined by Coulombic interactions which means it is a so-called plasma phase transition. The theory of Coulomb-influenced phase transitions in metallic plasmas as in Cs follows in principle the same lines as given above but requires a more careful account of different types of composite particles[8]). The present theory already approaches the experimental critical point

$$T_{cr} = 1924 \text{ K}, \qquad \rho_{cr} = 0.379 \text{ g/cm}^3, \qquad p_{cr} = 9.25 \text{ MPa},$$

but it is not yet appropriate to explain all details. The interplay between the ordinary liquid–gas phase transition and the transition from the dielectric to the metallic state seems to be responsible for the anomalous form of the phase transition border line in the reduced density and temperature representation for Cs and Hg. In contrast to systems which are metallic at normal pressures and show one critical point above 10^3 K at pressures in the MPa region, the theory of systems which are atomic or molecular fluids at normal pressures, predicts two phase transitions. The first phase transition, which is determined by Van der Waals forces, is entirely located in the region of neutral fluids with

$$T_c^{(1)} \sim 10^1\text{–}10^3 \text{ K}, \qquad p_c^{(1)} \sim 10^5\text{–}10^7 \text{ Pa}.$$

The second phase transition, which is mainly determined by Coulombic interactions, is located in the plasma region at

$$T_c^{(2)} \sim 10^3 \sim 10^5 \text{ K}, \qquad p_c^{(2)} \sim 10^8\text{–}10^{10} \text{ Pa}.$$

The required Mbar pressures are now at the border of the experimentally accessible region; they are given, however, in many natural systems as, e.g., in the interior of stars and the giant planets. A careful experimental study of Coulombic phase transitions is still in its very beginning[14,15]). Theoretically predicted phase diagrams for H and Xe-systems are presented in ref. 16.

References

1) W. Ebeling, W.D. Kraeft and D. Kremp, Theory of Bound States and Ionization Equilibrium in Plasmas and Solids (Akademie-Verlag, Berlin, 1976), revised Russian transl. (Mir, Moscow, 1979).
2) W.D. Kraeft, D. Kremp, W. Ebeling and G. Röpke, Quantum Statistics of Charged Particle Systems (Akademie-Verlag, Berlin, 1986 and Plenum, New York, 1986).
3) W. Ebeling, Physica **43** (1969) 293, **73** (1974) 573, **130A** (1985) 587.
4) F.J. Dyson and A. Lenard, J. Math. Phys. **8** (1967) 423, **9** (1968) 698.
5) E.H. Lieb, Rev. Mod. Phys. **48** (1976) 553.
6) J.L. Lebowitz and R.E. Peña, J. Chem. Phys. **59** (1973) 1362.
7) W. Ebeling, W.D. Kraeft and D. Kremp, Beitr. Plasmaphysik **10** (1970) 237; Physica **51** (1971) 146.
8) R. Redmer and G. Röpke, Physica **130A** (1985) 523; G. Röpke, L. Münchow and H. Schulz, Nucl. Phys. A **379** (1982) 536; Phys. Lett. B **110** (1982) 21.
9) H. Stolz and R. Zimmermann, Phys. Stat. Sol. (b) **94** (1979) 133; **131** (1985) 151; D. Kremp, W.D. Kraeft and A. Lambert, Physica **127A** (1984) 72.
10) W. Ebeling, W. Richert, W.D. Kraeft and W. Stolzmann, Phys. Stat. Sol. (b) **104** (1981) 193.
11) W. Ebeling and W. Richert, Phys. Lett. A **108** (1985) 80; Phys. Stat. Sol. (b) **128** (1985) 467.
12) S.H. Vosko, L. Wilk and M. Nusair, Canad. J. Phys. **58** (1980) 1200; F. Perrot and M. Dharma-Wardana, Phys. Rev. A **30** (1984) 2619; W.D. Kraeft and W. Stolzmann, Physica **79A** (1979) 306; W. Stolzmann and W.D. Kraeft, Ann. Physik **36** (1979) 338.
13) M. Rösler, R. Zimmermann and W. Richert, Phys. Stat. Sol. (b) **121** (1984) 609.
14) T.M. Rice, J.C. Hensel, T.C. Phillips and G.A. Thomas, Solid State Physics, vol. 32 (Academic, New York, 1977).
15) F. Hensel, Thermophysical properties of metallic fluids in the sub- and supercritical region, in Proc. 8th Symp. Thermophys. Prop., June 15–18, Gaithersburg MD, 1981, Angew. Chem. Int. Ed.; V.E. Fortov and I.T. Yakubov, Physics of Nonideal Plasmas (Chernogolovka, 1984), in Russian.
16) W.D. Kraeft, W. Ebeling, D. Kremp and G. Röpke, Ann. Physik, submitted.

Physica **140A** (1986) 169–174
North-Holland, Amsterdam

UNIVERSAL CONDUCTANCE FLUCTUATIONS IN DISORDERED METALS

Patrick A. LEE

Department of Physics, Massachusetts Institute of Technology, Cambridge, MA 02139, USA

Recent results on the conductance fluctuations in disordered metals are reviewed and a heuristic argument for the universal conductance fluctuation is given.

Recently it has been found both experimentally[1–3]) and theoretically[4–8]) that the electrical conductance in disordered metals exhibits fluctuations as the magnetic field B or the chemical potential μ is varied. The RMS magnitude of these fluctuations for a given sample is the same as that from sample to sample and is of the order $\delta G \approx e^2/h$, independent of the degree of disorder, the sample size and spatial dimensions, provided the temperature is low enough that the inelastic scattering length L_{in}, is larger than any of the sample dimensions[5,7]). $L_{\text{in}} = (D\tau_{\text{in}})^{1/2}$ is the distance an electron diffuses during the inelastic scattering time τ_{in} if D is the diffusion constant. It is to be emphasized that such fluctuations are not time-dependent noise. Instead, the conductance $G(\mu, B)$ is a deterministic, albeit fluctuating, function of its arguments for a given realization of the impurity configuration.

The basic physics of the conductance fluctuations has to do with the fact that on a length scale shorter than L_{in}, the electron wavefunction is coherent and the conductance must be treated quantum mechanically. Instead of the classical Boltzmann equation, it is more fruitful to consider conductance as being proportional to the total transmission probability of an electron through the disordered region. Indeed, it can be shown that the generalized Landauer formula[9,10]),

$$G = (e^2/h)\sum_{\alpha\beta}|t_{\alpha\beta}|^2, \tag{1}$$

adequately describes the conductance G on the disordered region. Here $t_{\alpha\beta}$ is the transmission amplitude from incoming channel α on the left to outgoing channel β to the right, where α, β may label the transverse momentum of the electron outside the disordered sample. The number of channels N is approximately given

0378-4371/86/$03.50

by $(Lk_F)^{d-1}$ where L is the transverse dimension. Universal conductance fluctuation is the statement that

$$(\Delta G)^2 \equiv \langle G^2 \rangle - \langle G \rangle^2$$
$$= C(e^2/h)^2 \tag{2}$$

where C is the number of order unity which is weakly dependent on dimensionality, for $d \leqslant 4$. Here $\langle\,\rangle$ denotes averaging over impurity configurations and G is given by eq. (1) for each impurity configuration. The original derivation of eq. (2) used Feynman diagrams. Here we offer a heuristic argument which may provide new insights on the problem.

Let us first consider the fluctuation of each transmission probability $|t_{\alpha\beta}|^2$. Quantum mechanically the transmission amplitude can be written as

$$t_{\alpha\beta} = \sum_{i=1}^{M} A_{\alpha,\beta}(i), \tag{3}$$

where $A_{\alpha\beta}(i)$ represents the probability amplitude due to the ith Feynman path that connects channel α to β. Semiclassically, we can think of the Feynman path i as classical random walks from the left or the sample to the right. Note that unlike the more familiar example of ballistic transport, where the important Feynman paths are restricted to a narrow tube of radius k_F^{-1} connecting the end points, here the presence of disorder means that the electron motion is diffusive, and the important Feynman paths are the random walks which cover much of the sample and are numerous. It is not a simple problem to determine M or $A_{\alpha\beta}(i)$, but fortunately that will not be necessary for our argument, as long as M is very large. We assume that $A_{\alpha\beta}(i)$ are independent complex random variables. This assumption is not as restrictive as it appears, because if a set of Feynman paths $\{i\}$ are correlated, we can simply denote $\Sigma_{\{i\}} A_{\alpha\beta}(i)$ by a new variable which will be uncorrelated. With this assumption, we can calculate the fluctuation in $|t_{\alpha\beta}|^2$, defined by $\Delta|t_{\alpha\beta}|^2 \equiv [\langle |t_{\alpha\beta}|^4\rangle - \langle |t_{\alpha\beta}|^2\rangle]^{1/2}$. Ignoring terms of order unity compared with M,

$$\langle |t_{\alpha\beta}|^4 \rangle = \sum_{ijkl} \langle A^*_{\alpha\beta}(i) A_{\alpha\beta}(j) A^*_{\alpha\beta}(k) A_{\alpha\beta}(l) \rangle$$
$$= 2\left\langle \sum_i |A_{\alpha\beta}(i)|^2 \right\rangle^2$$
$$= 2\langle |t_{\alpha\beta}|^2 \rangle^2. \tag{4}$$

From eq. (4) we immediately conclude that

$$\frac{\Delta\left\langle |t_{\alpha\beta}|^2 \right\rangle}{\left\langle |t_{\alpha\beta}|^2 \right\rangle} = 1. \tag{5}$$

Eq. (5) states that the relative fluctuation of each transmission probability is of order unity. We note that $|t_{\alpha\beta}|^2$ is directly measurable in the analogous problem of the propagation of classical waves through a random medium. For example, it is the intensity of the transmitted light in a given direction β if a laser beam is incident in a direction α. Large fluctuations in the reflection intensity were reported recently[11]). Eq. (5) is a well-known result in the literature on speckle patterns[12]) and was first discussed in the context of electrical conductance by Buttiker et al.[13]).

With eqs. (5) and (1) we can attempt to estimate the fluctuation in G. To do this we need to know something about the correlation between $|t_{\alpha\beta}|^2$ and $|t_{\alpha'\beta'}|^2$. The simplest assumption is that different channels are uncorrelated. This assumption was made by Buttiker et al.[13]), and as we shall see, leads to a lower bound on the fluctuations that are smaller than eq. (2). To see this we note that according to eq. (1), we have to add an N^2 object, each of which has fluctuations of the order $|t_{\alpha\beta}|^2$, according to eq. (5). One is led to the conclusion that

$$\Delta G' \approx (e^2/h)(N^2)^{1/2}\left\langle |t_{\alpha\beta}|^2 \right\rangle. \tag{6}$$

Using Ohm's law $G = \sigma L^{d-2}$, $\sigma = (e^2/h)k_{\mathrm{F}}^{d-1}\ell$ where ℓ is the mean free path and $N = (Lk_{\mathrm{F}})^{d-1}$, we conclude that

$$\left\langle |t_{\alpha\beta}|^2 \right\rangle \approx \frac{\ell}{L}\frac{1}{N} \tag{7}$$

so that

$$\Delta G' \approx \frac{e^2}{h}\frac{\ell}{L}, \tag{8}$$

which is smaller than the correct result given by eq. (2). This discrepancy has led Imry[14]) to introduce the interesting idea of effective channel number.

Now we shall show that the correct answer can be obtained if we consider the reflection amplitude $r_{\alpha\beta}$ instead of $t_{\alpha\beta}$. Let us introduce

$$R = (e^2/h)\sum_{\alpha\beta}|r_{\alpha\beta}|^2. \tag{9}$$

By unitarity, $G + R = (e^2/h)N$, so that the fluctuation in R is the same as the fluctuation in G. Using similar reasoning that led to eq. (5) we obtain

$$\frac{\Delta\left\langle |r_{\alpha\beta}|^2 \right\rangle}{\left\langle |r_{\alpha\beta}|^2 \right\rangle} = 1. \tag{10}$$

If we assume that $|r_{\alpha\beta}|^2$ are uncorrelated from channel to channel, we obtain instead of eq. (6),

$$\Delta R \approx (e^2/h) N \left\langle |r_{\alpha\beta}|^2 \right\rangle. \tag{11}$$

Using eq. (7) and unitarity, we can estimate that

$$\left\langle |r_{\alpha\beta}|^2 \right\rangle \approx \frac{1}{N}\left(1 - \frac{\ell}{L}\right). \tag{12}$$

This simply expresses the fact as long as $\ell \ll L$, most of the incoming beam is reflected into N reflecting channels. Combining eqs. (12) and (11) we obtain the universal conductance result

$$\Delta G = \Delta R \approx (e^2/h). \tag{13}$$

Upon comparing eqs. (13) and (8) we see that the innocent assumption of uncorrelated $|r_{\alpha\beta}|^2$ imposes correlations among $|t_{\alpha\beta}|^2$ in order to reproduce the same conductance fluctuations. We note that the reflection amplitude is probably dominated by a few scattering events, whereas $t_{\alpha\beta}$ must involve multiple scattering in order to traverse the sample, so that it is not unreasonable to expect more correlations among channels in the transmission. The direct measurements of correlations among $|t_{\alpha\beta}|^2$ in light scattering experiments will be very interesting.

The above argument is only intended to be heuristic, and not a replacement of the more rigorous diagrammatic analysis. In particular, this argument appears to be independent of dimensionality, whereas eq. (2) is expected to break down for $d > 4$. In addition, in the case when the width of the sample W exceeds the length L, diagrammatic analysis indicates that eq. (13) should be enhanced by the geometrical factor W/L, whereas the heuristic argument does not seem to yield this factor. Nevertheless, this argument emphasizes the need to introduce correlations among $|t_{\alpha\beta}|^2$ and should shed new light on our understanding of the problem.

We conclude by noting that the physical picture of the conductance being given by a sum over many Feynman paths leads naturally to the conclusion that the conductance of a given sample is extremely sensitive to the motion of a single impurity. Again the calculation was done using Feynman diagrams[15,16]), but the following physical argument gives the correct answer. Let us consider each Feynman path as a random walk consisting of segments of tubes with length ℓ and cross-section area $k_F^{-(d-1)}$, i.e., the electron propagates ballistically between elastic collisions. The volume occupied by each Feynman path is then $k_F^{-(d-1)}\ell(L/\ell)^2$. We conclude that each Feynman path covers a fraction f of the total volume L^d,

$$f = \frac{1}{(k_F\ell)^{d-1}}\left(\frac{L}{\ell}\right)^{2-d}. \tag{14}$$

Conversely, a given impurity is visited by the same fraction f of all the Feynman paths. The motion of a single impurity will change the phase of the amplitude A_i corresponding to the fraction f of all the paths, so that its effect on $(\Delta G)^2$ is the same as that of changing the entire impurity configuration, except reduced by f. We therefore conclude that moving a single impurity changes the conductance by

$$\Delta G_1 \approx (e^2/h) f^{1/2}. \tag{15}$$

In particular, note that ΔG_1 is independent of sample size in two dimensions. This unusual result should be amenable to direct experimental test. It also allows us to understand the magnitude of $1/f$ noise in disordered metals due to defect migration[15]).

Acknowledgements

I am thankful to S.C. Feng and A.D. Stone for collaborations and discussions on the work reported here. I also acknowledge the support of the NSF grant DMR84-41030.

References

1) C.P. Umbach, S. Washburn, R.B. Laibowitz and R.A. Webb, Phys. Rev. B **30** (1984) 4048.
2) G. Blonder, Bull. Am. Phys. Soc. **29** (1984) 535.
3) J. Licini, D. Bishop, M. Kastner and J. Melngailis, Phys. Rev. Lett. **55** (1985) 1987.
4) A.D. Stone, Phys. Rev. Lett. **54** (1985) 2692.
5) P.A. Lee and A.D. Stone, Phys. Rev. Lett. **55** (1985) 1622.
6) P.A. Lee, A.D. Stone and H. Fukuyama, preprint.
7) B.L. Altshuler, Pis'ma Zh. Eksp. Teor. Fiz. **41** (1985) 530, transl.: JETP Lett. **41** (1985) 648.
8) B.L. Altshuler and D.E. Khmelnitskii, Pis'ma Zh. Eksp. Teor. Fiz. **42** (1985) 291, transl.: JETP Lett. **42** (1986) 359.

9) R. Landauer, Phil. Mag. **21** (1970) 863.
10) D.S. Fisher and P.A. Lee, Phys. Rev. B **23** (1981) 6851.
11) S. Etemad, R. Thompson and M.J. Andrejco, Phys. Rev. Lett. **57** (1986) 575.
12) J.W. Goodman, J. Opt. Soc. Am. **66** (1976) 1145.
13) M. Buttiker, Y. Imry, R. Landauer and S. Pinhas, Phys. Rev. B **31** (1985) 6207.
14) Y. Imry, Europhysics Lett. **1** (1986) 249.
15) S.C. Feng, P.A. Lee and A.D. Stone, Phys. Rev. Lett. **56** (1986) 1960; Erratum **56** (1986) 2772.
16) B.L. Altshuler and B. Spivak, JETP Lett. **42** (1986) 447.

Physica **140A** (1986) 175–182
North-Holland, Amsterdam

THE AHARONOV–BOHM EFFECT IN NORMAL METALS – NON-ENSEMBLE AVERAGED QUANTUM TRANSPORT

R.A. WEBB, S. WASHBURN, C.P. UMBACH, F.P. MILLIKEN, R.B. LAIBOWITZ
and A.D. BENOIT*

IBM T.J. Watson Research Center, P.O. Box 218, Yorktown Hts., NY 10598, USA

The Aharonov–Bohm effect in normal metal rings together with the associated conductance fluctuations in wires are clear signatures that experiments are now being performed in a quantum regime where conventional transport concepts do not apply. Ensemble averaging and taking the thermodynamic limit should not be used if comparisons to experiments are to be made. It is shown that by varying both the size of the sample as well as the temperature, we can study how ensemble averaging and energy averaging transforms a coherent quantum system into a classical one.

The ideas of averaging and taking the thermodynamic limit are standard concepts used in statistical physics to calculate the bulk properties of condensed matter systems. The electrical resistance of a wire is an example of an average property that can be successfully computed using these techniques. The basic idea is to assume that, on average, the electrons in the sample move without scattering for some distance l (the mean free path), scatter off of impurities or boundaries (specularly, elastically or inelastically), and repeat this process many times. By averaging over all possible scattering sites and taking the limit as the volume V of the system approaches infinity with the density of electrons N/V a constant, the electrical resistivity can be accurately computed for a bulk system. Of course if we could measure the resistance of a very small part of the sample containing only 30 atoms or so we would not be surprised to discover that this value is different from the bulk average. At this microscopic size scale both the quantum mechanical properties of the electron and the details of the individual scattering centers are needed to compute the resistance. Generally we need not worry about these microscopic details because once the size of the system is much larger than some typical correlation length the average properties are thought to be a good description of the system.

The surprising result of recent theoretical[1–5]) and experimental[6–8]) work on the subject of transport in small systems is that this correlation length can be

*Permanent address: CNRS, Grenoble, France.

0378-4371/86/$03.50 © Elsevier Science Publishers B.V.
(North-Holland Physics Publishing Division)

very large at low temperatures, many millimeters in some cases. In this case the transport properties can only be properly determined using the quantum mechanical description first given by Landauer[9]). Theoretical results clearly show that once the inelastic scattering length (the distance the electron travels before losing phase memory in its wave function) is longer than the size of the system, the measured transport properties depend crucially upon the individual details of the scattering, and the phases of all the electrons in the sample can become highly correlated. In addition, dissipation arises not from the scattering of electrons from impurity sites but rather from contact to thermal reservoirs.

If a wire is formed into a ring and a magnetic field is applied to the enclosed area, the electrical resistance is predicted and found experimentally to oscillate periodically as the flux is changed. The period of oscillation is h/e, the normal metal flux quantum. The amplitude of these conductance oscillations is on the order of e^2/h at low temperatures. This of course is a clear manifestation of the Aharonov–Bohm effect[10]). The important point here is that it is occurring in a highly disordered conductor exhibiting a large electrical resistance. If the magnetic field is applied to the entire ring, including the wires forming the ring, the periodic conductance oscillations become superimposed upon a randomly fluctuating background resistance. The rms value of these fluctuations is also found to be approximately equal to e^2/h. This effect (known as the Universal Conductance Fluctuations[4])) is also an Aharonov–Bohm effect but occurring within the wires forming the ring. These fluctuations are present in any wire. The condition for their existence at low temperatures is simply that the phase coherence length of the electrons be comparable to the sample.

Advances in sample fabrication techniques now permit structures to be built where all dimensions of a device can be much smaller than 1 μm. Thus at low temperatures we can study samples which are in this new quantum transport regime and, by varying the sample size or raising the temperature, we can study the transition to the more familiar classical regime. In what follows, we will primarily discuss how this transition occurs.

The basic idea behind the Aharonov–Bohm effect in normal metals is that electron wave packets encircling a magnetic flux will exhibit a phase shift due to the magnetic vector potential A. In a metallic ring, small enough so that the electron states are not randomized by inelastic (or magnetic) scattering during the traversal of the arm of the ring, an interference pattern should be present in the magnetoresistance of the device. Fig. 1 displays some recent results confirming the existence of the Aharonov–Bohm effect in disordered normal metal rings. The data are from a gold ring, 39 nm thick, 39 nm wide and 0.82 μm in diameter and is shown in the inset to fig. 1b. The periodic h/e oscillations are clearly visible and are superimposed upon a randomly fluctuating background magnetoconductance. The amplitude of the h/e oscillations is modulated by another

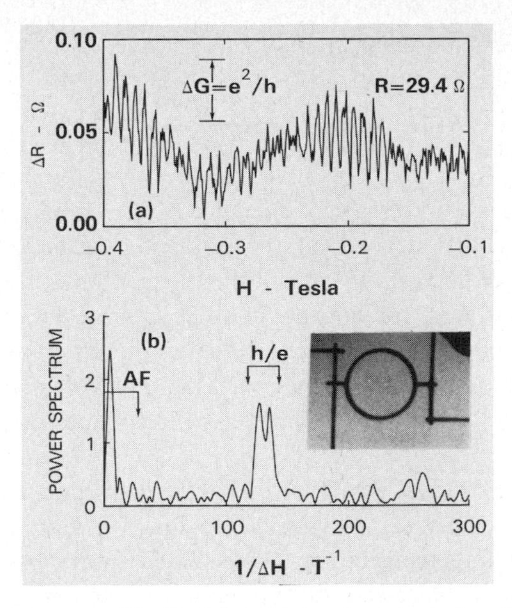

Fig. 1. (a) Magneto-resistance at 50 mK of the gold ring shown in the inset illustrating the modulated Aharonov–Bohm oscillations and the aperiodic conductance fluctuations. (b) The Fourier transform of the data over a ± 0.5 T field range.

aperiodic function of field. The Fourier transform of these data is shown in fig. 1b. The expected location of the h/e oscillations is indicated by the arrows and is based upon the measured inside and outside area of the ring. The structure in the Fourier transform at small $1/\Delta H$ arises from the aperiodic conductance fluctuations. The characteristic field scale for one aperiodic fluctuation is approximately two flux quanta in the area of the wire normal to the field and is indicated by an arrow. It has been shown experimentally that both the h/e and aperiodic conductance fluctuations are unattenuated by the magnetic field up to fields as large as 16 T. Recent theoretical work has shown that in the metallic conduction regime these fluctuations are universal in character. That is, they do not depend on the magnitude of the electrical resistance or the material from which the sample is made. The rms value of the conductance fluctuations at low temperatures is always on the order of e^2/h so long as the phase coherence length is longer than the sample. The same is true of the h/e oscillations. Typical low temperature values for the magnitude of the fluctuations are 0.2–0.5 and 0.5–1.0 e^2/h for the h/e oscillations and aperiodic fluctuations respectively.

The conductance of a real ring with many parallel conduction channels has been calculated quantum mechanically by Büttiker, Imry, Landauer, and Pinhas[1]). They show that a generalization of the Landauer conductance formula[9]) can be used to predict many of the observed properties of rings with finite width. One way of summarizing these results is

$$G = A(H) + \frac{e^2}{h}\left[B(H) + C(H)\cos\left(\frac{2\pi\Phi}{\Phi_0} - \beta\right)\right], \tag{1}$$

where the first term is the average conductance of the sample, and B and C are numbers of order 1 that depend on the scattering details and represent the universal conductance fluctuations[4,5]) and the modulation of the h/e oscillations respectively. The third term contains the periodic h/e oscillations where Φ is the flux enclosed by the loop. The parameter β is a random offset which shifts the phase of the oscillations at zero magnetic field and can also be field dependent. The theory predicts[11]) that β is sample dependent and can be any number between 0 and 2π.

Several types of averaging are involved in the transition from the quantum coherent transport regime to the classical regime. At first thought, the most significant is that of finite temperature. There should always exist some characteristic energy scale such that electrons with energies differing by more than this amount will exhibit different interference patterns. A first guess to this energy scale would naturally be the spacing between electron energy levels in the sample. For the gold ring displayed in fig. 1, this energy is about 0.1 mK and is inversely proportional to the volume of the system. Fortunately this estimate is not correct. Stone and Imry[4]) have shown that the correct estimate is based upon the work by Thouless. The basic idea for a wire is that the relevant energy scale is given by the sensitivity of the electron energy levels to the boundaries. This scale is given by $\Delta E = \hbar/t$ where t is the time it takes to diffuse from one end of the sample to the other. Since the starting assumption is that the phase coherence length is larger than the size of the system, this time is set by L^2/D, where L is the sample length and D is the diffusion constant. If $k_B T$ is smaller than this correlation energy, E_C, there will be no effect of temperature on the quantum transport properties. If the thermal energy is larger than E_C, we expect the quantum conductance fluctuations to average to zero stochastically as

$$\Delta G = \frac{e^2}{h}\left(\frac{E_C}{k_B T}\right)^{1/2}. \tag{2}$$

Fig. 2 shows data for both the magnitude of the universal conductance fluctuations and the Aharonov–Bohm effect measured in a single gold ring over a temperature range where the phase coherence length is larger than the size of the sample[6]). The solid line is the expected theoretical behavior if only energy averaging is important with E_C computed to be 35 mK. The agreement with energy averaging theories is clearly excellent.

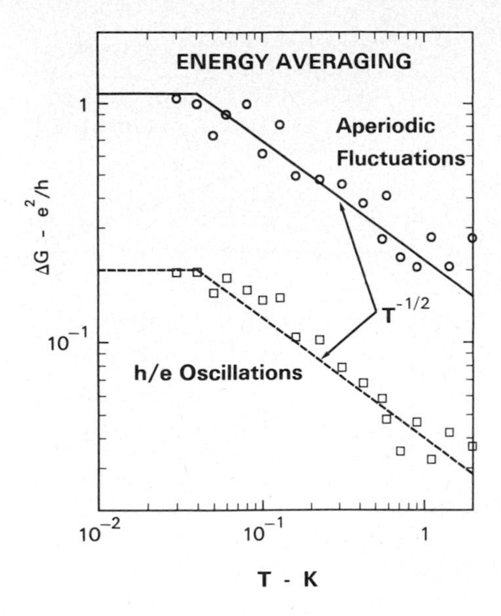

Fig. 2. Temperature dependence of the aperiodic conductance fluctuations and the h/e oscillations for the gold ring shown in fig. 1. Straight lines are theoretical fits.

A second type of averaging that is important in large samples occurs when the phase coherence length is shorter than the sample length. In the absence of energy averaging, we can break up the sample into small sections, each L_ϕ long, and ask how the quantum conductance fluctuations are averaged to zero. Within each section, the conductance fluctuations will be $\Delta G = \Delta R_s/R_s^2 = e^2/h$ where R_s is the resistance of a section L_ϕ long. The number of sections in the entire sample is $N = L/L_\phi$, and the resistance fluctuation ΔR measured over the entire sample should be $\Delta R = N^{1/2}\Delta R_s$. Thus the measured conductance fluctuation $\Delta R/R^2$ will average to zero as $N^{-3/2}$ or,

$$\Delta G = \frac{e^2}{h}\left(\frac{L_\phi}{L}\right)^{3/2}. \tag{3}$$

This type of averaging has been observed for the Aharonov–Bohm effect in recent experiments on series arrays of silver rings[12]). Fig. 3a shows the results of measurements on four array samples containing 1, 3, 10, and 30 square rings, 75 nm wide, 20 nm thick and 0.94 μm on a side. The inset to fig. 3a shows one of the samples used in these measurements. The phase coherence length was longer than the size of an individual ring but shorter than the distance between rings. The measured rms amplitude of the h/e oscillations are displayed as a function of the number of series rings. A correction for the lead resistance connecting the rings has been made so as to only consider the effects of averaging segments that

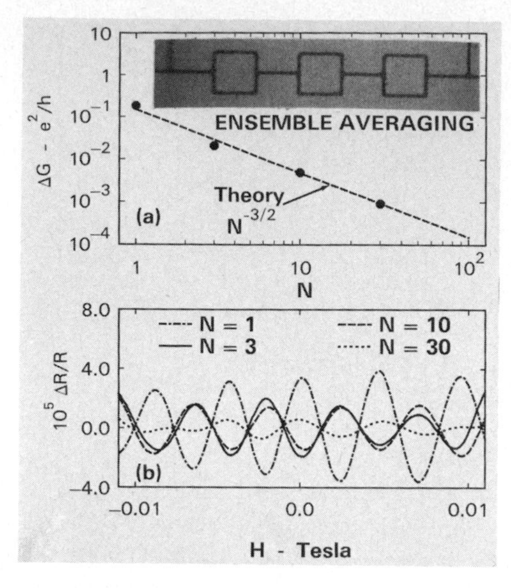

Fig. 3. (a) Amplitude of the h/e oscillations as a function of the number N of uncorrelated rings. Inset is a photo of the series three ring sample. (b) Magnetic field dependence of the measured h/e amplitude for 1, 3, 10 and 30 ring samples demonstrating the random phase for each sample.

contribute to the Aharonov–Bohm effect. The expected $N^{-3/2}$ behavior with a 0.2 e^2/h prefactor is confirmed by these data. The amplitude of the h/e oscillations near zero magnetic field are shown in fig. 3b for all four samples. Note that two of the samples have a maximum resistance at $H = 0$ while the other two have a minimum resistance. This clearly demonstrates the sample specific nature of the phase of the h/e oscillations as predicted by Gefen[2]).

A third type of averaging occurs for the Aharonov–Bohm oscillations when the phase coherence length becomes smaller than half the circumference of a single ring. In this case the quantum interference effects will average to zero exponentially with increasing sample size. The reason for the exponential averaging rather than the more benign power law averaging is that quantum interference is much more sensitive to phase memory over the entire circumference of the ring than ensemble averaging of uncorrelated fluctuations. The length scale for h/e oscillations is fixed by the diameter of the ring and does not change as the phase coherence length shrinks. Energy averaging, on the other hand, actually becomes less important when the phase coherence length is short. In general the phase coherence length can be written as $L_\phi = (L_{in}^{-2} + 2L_s^{-2})^{-1/2}$ where L_s is the temperature independent spin flip scattering process, and L_{in} is the temperature dependent inelastic scattering length. In one dimension, L_{in} is found to have a simple power law dependence of the form $L_{in} = A/T^p$ with A a constant and p

varying from $\frac{1}{4}$ to 1 depending on the nature of the scattering process. In this case the amplitude of the conductance fluctuations for a ring takes the form

$$\Delta G = \frac{e^2}{h}\left[\frac{\pi^2 \hbar D}{L_\phi^2 k_B T}\right]^{1/2} \exp\left(-\pi r/L_\phi\right). \tag{4}$$

If L_ϕ is short and D not too small, E_c can be quite large.

This exponential averaging of the h/e oscillations has been observed in recent experiments on antimony rings[13]). Fig. 4 displays the magnitude of both the universal conductance fluctuations and the h/e oscillations as a function of temperature for a Sb ring 0.85 μm in diameter, 0.08 μm thick, and 0.036 μm wide. The solid lines are the theoretically predicted behavior (from eqs. (3) and (4)) for the data with $E_c > 2$ K, $L_{in} = 0.26\ \mu\text{m}/\text{T}^{1/2}$, and $L_s = 0.76\ \mu$m. These values were in good agreement with the independently determined L_ϕ using fits to the weak localization behavior occurring near zero field. The only adjustable parameter was the magnitude of the fluctuations which was found to be 0.2 e^2/h for the Aharonov–Bohm oscillations and 0.56 e^2/h for the aperiodic conductance fluctuations. Both of these are in excellent agreement with values determined from previous experiments on Au and Ag rings and lines. The electrical resistance of this sample was nearly 1 kΩ indicating that the mean free path was 7 nm or nearly 1000 times smaller than L_ϕ.

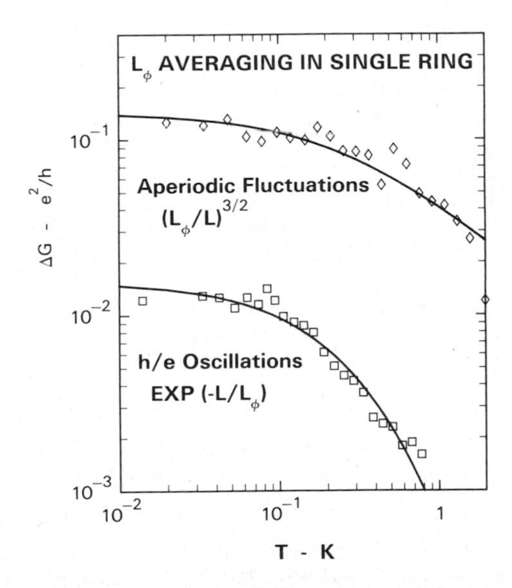

Fig. 4. Temperature dependence of the aperiodic fluctuations and the h/e oscillations for a single Sb ring in the limit where the phase coherence length is shorter than the sample size. Solid lines are theoretical fits with no energy averaging.

The quantum transport observed in disordered normal metal samples is a clear indication that conventional approaches to calculating electrical resistances using ensemble averaging and thermodynamic limit concepts will not accurately predict the properties of even rather large samples at low temperatures. Although the experimental observations of these new quantum coherent effects is rather recent, much has already been learned about transport in the diffusive limit and in the near future we believe that measurements extending into the ballistic regime will be possible. Thus we should be able to study the complete evolution of a quantum system into a classical one and quantitatively compare with theoretical predictions.

Acknowledgements

We wish to thank our colleagues M. Azbel, M. Büttiker, Y. Gefen, Y. Imry, R. Landauer, P. Lee and D. Stone for important contributions enhancing our understanding of the role of averaging discussed in this work.

References

1) M. Büttiker, Y. Imry, R. Landauer and S. Pinhas, Phys. Rev. B **31** (1985) 6207 and references cited therein.
2) M. Büttiker, Y. Imry and R. Landauer, Phys. Lett. A **96** (1983) 365; Y. Gefen, Y. Imry and M.Ya. Azbel, Surf. Sci. **142** (1984) 203 and Phys. Rev. Lett. **52** (1984) 129; Y. Gefen, private communication.
3) Y. Imry, Europhys. Lett. **1** (1986) 249; For a recent review see Y. Imry in Directions in Condensed Matter Physics, G. Grinstein and E. Mazenko, eds. (World Scientific Publ., Singapore, 1986).
4) A.D. Stone, Phys. Rev. Lett. **54** (1985) 2692; P.A. Lee and A.D. Stone, Phys. Rev. Lett. **55** (1985) 1622; A.D. Stone and Y. Imry, Phys. Rev. Lett. **56** (1986) 189.
5) B.L. Al'tshuler, JETP Lett. **41** (1985) 648; B.L. Al'tshuler and D.E. Khmel'nitskii, JETP Lett. **42** (1985) 359; B.L. Al'tshuler and B.Z. Spivak, JETP Lett. **42** (1985) 447; B.L. Al'tshuler and B.I. Shklovskii, preprint.
6) R.A. Webb, S. Washburn, C.P. Umbach and R.B. Laibowitz, Phys. Rev. Lett. **54** (1985) 2696; S. Washburn, C.P. Umbach, R.B. Laibowitz and R.A. Webb, Phys. Rev. B **32** (1985) 4789;
7) R.A. Webb, S. Washburn, C.P. Umbach and R.B. Laibowitz, in Localization, Interaction, and Transport Phenomena in Impure Metals, G. Bergmann, Y. Bruynseraede and B. Kramer, eds., (Springer, Heidelberg, 1985); C.P. Umbach, S. Washburn, R.B. Laibowitz and R.A. Webb, Phys. Rev. B **30** (1984) 4048.
8) V. Chandrasekhar, M.J. Rooks, S. Wind and D.E. Prober, Phys. Rev. Lett. **55** (1985) 1610; S. Datta et al. Phys. Rev. Lett. **55** (1985) 2344.
9) R. Landauer, Phil. Mag. **21** (1970) 863.
10) Y. Aharonov and D. Bohm, Phys. Rev. **115** (1959) 485.
11) M. Büttiker and Y. Imry, J. Phys. C **18** (1985) L467; M. Büttiker, to be published.
12) C.P. Umbach, C. van Haesendonck, R.B. Laibowitz, S. Washburn and R.A. Webb, Phys. Rev. Lett. **56** (1986) 386.
13) R.A. Webb, F.P. Milliken, S. Washburn, C.P. Umbach and R.B. Laibowitz, to be published.

Physica **140A** (1986) 183–190
North-Holland, Amsterdam

LOCALIZATION OF LIGHT: THE QUEST FOR THE WHITE HOLE

Ad LAGENDIJK, Meint P. VAN ALBADA and Martin B. VAN DER MARK
Natuurkundig Laboratorium, Universiteit van Amsterdam, Valckenierstraat 65, 1018 XE Amsterdam, The Netherlands

The experimental observation of weak localization of light is discussed and some general aspects of Anderson localization of light are treated.

1. Introduction

Localization of light is a fascinating new field in which modern developments of condensed matter physics are applied to the well-established field of linear optics. Localization as introduced by Anderson[1]) refers to a dramatic change in the propagation of an electron when it is subject to a spatially random potential[2]).

Several approaches can be used to describe the phenomenon of localization in a disordered medium. One method is to look at the transport properties of the electron. In this picture localization is concerned with the vanishing of the diffusion coefficient. This view of localization has gotten a thorough foundation due to the important original work of Götze[3]), and later workers[4]), who used transport equations of the mode-coupling type. At localization the interferences of scattered waves cannot be neglected anymore, but indeed their influence becomes essential ("there is life after a mean-free path").

In three dimensions a certain critical degree of disorder is needed before the localization will set in. This can be simply understood on the basis of the heuristic Yoffe–Regel criterium:

$$\lambda_{\mathrm{mf}} \lesssim \lambda/(2\pi),$$

in which λ_{mf} is the mean-free path of the electron, and λ is its wavelength. If this condition is met the wave character of the wave (function) is completely lost and localization is established. Although the vast majority of theories of localization have been developed for the Schrödinger wave equation, it is quite clear that the concept is much broader. In principle for almost any wave equation localized solutions can be obtained when solved for a random medium. This becomes

0378-4371/86/$03.50

particularly clear when one realizes that the Yoffe–Regel criterium can be applied to all wave phenomena. Several workers have recently suggested experiments in which localization of waves could be observed[5–8]). Localization of electromagnetic waves is for many reasons of fundamental importance. One deals with localization of *vector* waves described by a very important, well-established and well-studied set of equations basically different from the Schrödinger wave equation.

Before rushing into experiments on multiple scattering of light, one should realize that there is a huge body of literature on single[9]) and multiple-scattering of light[10,11]). In the majority of these treatments interference is neglected, however. In the next section we will discuss our experimental observation of interference effects in strongly scattering media. In the last section we will speculate on how localization of light could be measured.

2. Weak localization

If the mean-free path is not short enough yet to observe Anderson localization, one might still be able to observe interference effects. There is a particular class of interference effects which is known as weak localization, and it should be considered to be a precursor of Anderson localization.

Suppose we consider the propagation of a wave in a random medium from point A to B (see fig. 1). One should consider all possible paths from A to B, and in fig. 1 two of them, I and II, are depicted. If the amplitudes along these paths are a_{I} and a_{II}, respectively, the probability for arriving at B is $(a_{\mathrm{I}} + a_{\mathrm{II}})^2 = a_{\mathrm{I}}^2 + a_{\mathrm{II}}^2 + 2a_{\mathrm{I}}a_{\mathrm{II}}$, in which the incoherent contribution embraces the sum of the squares, whereas the crossproduct refers to the coherent or interference contribution. When one considers all possible paths all interference terms will have different signs and magnitudes, and very likely cancel. This random phase approximation is obviously very wrong for a medium close to the localization transition, but one can already point out much weaker conditions under which this approximation breaks down. Consider the transport of a wave in a random

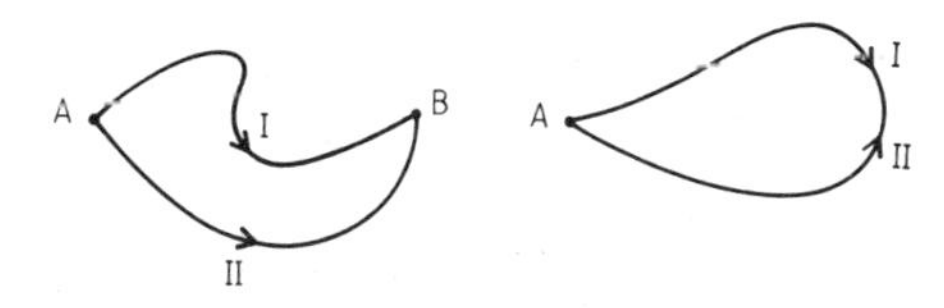

Fig. 1. Left: two possible paths (I and II) for wave propagation from A to B in a random medium. Right: a possible path (I) and its time-reversed counterpart (II) contributing to the returning probability for wave propagation from A.

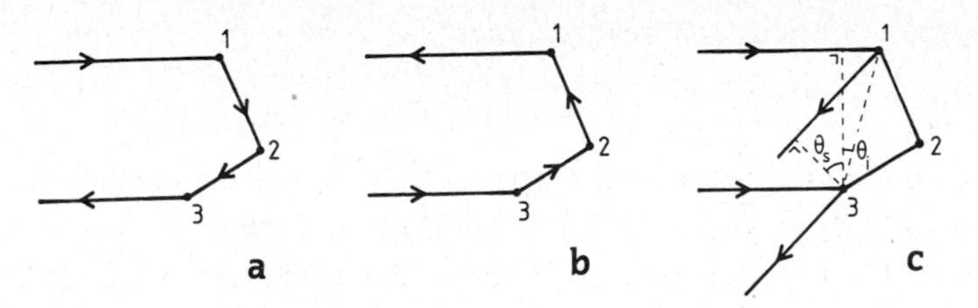

Fig. 2. Two third-order backscattering amplitude diagrams connected by time-reversal symmetry (a and b). Two interfering third-order scattering amplitude diagrams, slightly away from 180° backscattering (c) (angles exaggerated here).

medium from point A onto itself (see fig. 1). Let us focus on arbitrary path I, and path II, which is the same as path I but traversed in opposite sense. The contributions of path I and II always add up coherently, because they have the same phase. If one would add them up incoherently one would underestimate the returning probability by a factor of two.

The argument given so far refers to a real-space picture. Now we would like to translate this view into the wave-vector domain which is more appropriate for scattering. The scattering equivalent of the returning probability is backscattering. In fig. 2 we have depicted three-particle scattering events under backscattering conditions. In that figure two events are depicted which are related by time inversion (reverse off all momenta) (a and b). These two amplitude diagrams always add up coherently and the backscattering is a factor of two larger than expected on the basis of incoherent scattering. In fig. 3 we give the corresponding scattering diagrams for the intensity. The ladder diagram represents the incoherent contribution (the diagrams (a) and (b) of fig. 2 squared), and the crossed diagram represents the interference term (cross product of the diagrams (a) and (b) in fig. 2). At exactly 180° backscattering the crossed-diagram contribution is equal to the contribution of the ladder diagram[12,13]). Moving away a little from the backscattering direction (diagram (c) of fig. 2) the interference contribution will quickly diminish. A feeling can be obtained from the simple construction in fig. 2. The two interfering paths are constructed and the phase difference between

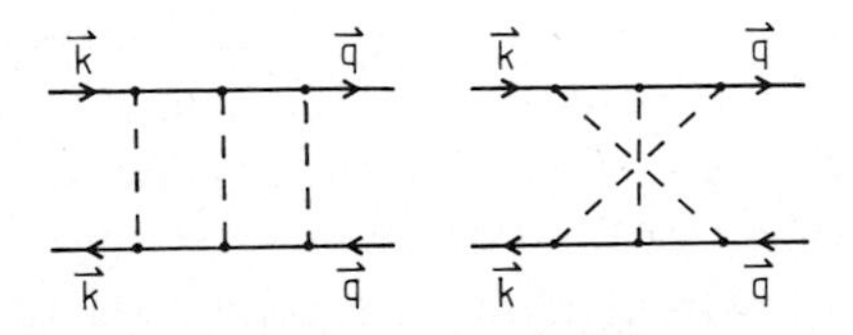

Fig. 3. Two types of diagrams contributing to the intensity. Ladder: incoherent contribution; crossed: interference contribution.

the two equals

$$d\{\sin(\theta_s + \theta_i) - \sin\theta_i\} = 2d\cos\{\tfrac{1}{2}(2\theta_i + \theta_s)\}\sin\{\tfrac{1}{2}(\theta_s)\},$$

where d is the distance between the first and last scattering center. The angle θ_s describes the deviation from the 180° backscatter direction. For a Nth-order scattering event the distance d will be of the order $N^{1/2}\lambda_{mf}$. If the phase difference is of order one (d of order $\lambda/2\pi$) there will be no constructive interference anymore. This corresponds to an increased backscattering within a cone having a full apex angle of order $\lambda/2\pi\lambda_{mf}$ (due to the $N^{1/2}$ factor higher-order events have narrower cones associated with them). This phenomenon is referred to as weak localization, and its observation, together with its connection to localization has recently been reported[14,15]).

Before we will discuss this increased backscattering in more detail we would like to point out the connection between weak localization and Anderson localization. The diagrams responsible for the increased backscattering (the most-crossed diagrams) are the same diagrams expected to be responsible for Anderson localization. Indeed an increased backscattering gives rise to a hampered propagation. The monitoring of the critical backscattering gives precise (angularly resolved) information about the most-crossed diagrams (that is interference between time-reversed events). The phenomenon of weak localization has been pointed out first for electrons[16,12]) and has been demonstrated experimentally.

We have to deal now with an essential and crucial complication when considering interference of electromagnetic waves. Up to now we have implicitly assumed to deal with scalar waves. However, electromagnetic waves are vector waves, and this constitutes a cardinal extension. In practice it means that the polarization of the waves becomes important. Without going into details we point out that by using time reversal for the Maxwell equations one winds up with the following conclusion: for backscattering polarized parallel to the incoming beam an enhancement factor of two is expected (if single scattering is excluded), and for the perpendicular polarization an enhancement factor close to one should be observed. The enhancement factor for the perpendicular component is due to lower-order scattering and for long light paths there is no perpendicularly polarized interference scattering anymore[14,17]).

The increased backscattering cone can easily be observed for concentrated suspensions of polystyrene spheres in water, and enhancement factors are between 1.5 and 2.0 for the parallel and 1.0 and 1.3 for the perpendicular polarization. Lower-order scattering shows many complicating and nonuniversal features. For many reasons one would like to separate out these lower order contributions. In the first place localization is concerned with the long light paths (many scattering centers). Furthermore for longer light paths scattering can very likely be well described by a simple and universal model of isotropic scattering.

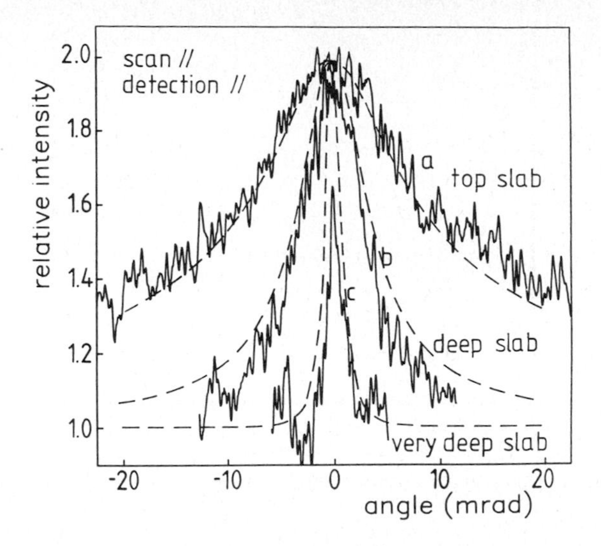

Fig. 4. Experimental cone shapes from a 9.6 volume % 0.483 μm particle suspension of polystyrene in water. Curve (a) corresponds to a slab thickness of 15 μm, curve (b) corresponds to the 45–15 μm and curve c corresponds to the 1500–70 μm difference scattering. Curve (c) is resolution limited. The FWHM of the absolute scan (not shown here) is 3.7 mrad, and the enhancement factor is 1.8. All results refer to parallelly polarized backscatter (detected light has the same polarization as the incoming laser beam). Dashed lines: theoretical curves.

We have developed a method to probe the contribution of exlcusively higher-order scattering (long light paths). Subtracting the backscattering patterns of slabs of thickness L_1 and L_2 ($L_2 > L_1$), the low-order scattering contributions coming from the front layer of the sample ($L < L_1$) cancel out, and what remains is the contribution of light that has "seen" the deeper part of the slab ($L_1 < L < L_2$).

One expects that the longer light paths get out of phase easier than the short light paths (remember the $N^{1/2}$-factor). Consequently the long light paths should yield narrower contributions to the cone of enhanced backscattering. This is splendidly demonstrated in the experimental results depicted in fig. 4. This figure shows the shapes of contributions to the backscattering cone from a 1500 μm slab of a 9.6% suspension of 0.482 μm particles, originating from one "full" slab (corrected for single scattering) and two "difference" slabs. From the figure it can be clearly seen that after elimination of the single scattering contribution the enhancement factor for the parallel light component is very nearly equal to 2 (curves (a) and (b)) until the limit of experimental resolution is reached (curve (c)).

From the theoretical side we have solved rigorously the summation of most-crossed diagrams for isotropic point scatterers for a *finite* slab[18]). All our

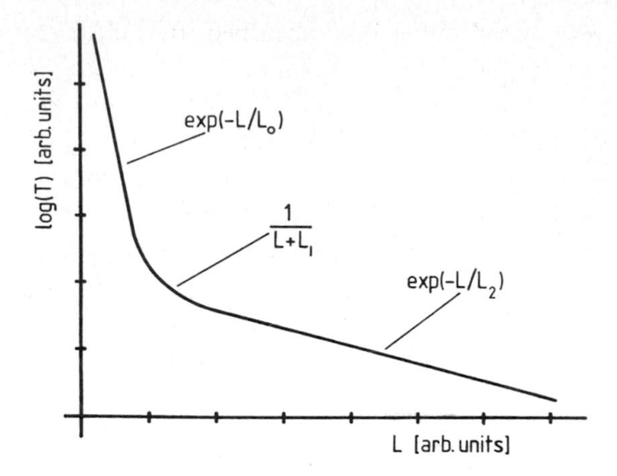

Fig. 5. Schematic presentation of the transmission as a function of sample thickness.

experimental results from our differential technique are compared with this new theory. In fig. 4 we have also presented the theoretical results. The agreement is very good, and this gives confidence in the theory. The advantage of having a good theory of weak localization is that discrepancies with this theory for very short mean-free path samples can be attributed then (hopefully) to critical strong localization effects.

3. Anderson localization

Ideally a medium in which light is localized acts as a trap for light in it, and as a perfect reflector for light outside. The best name to describe this phenomenon is a "white hole". There are several ways to observe Anderson localization, and we will discuss them very briefly.

One way is to look at the transmission as a function of slab thickness L. One should first realize what a "normal" strongly scattering medium should do. In fig. 5 we have plotted schematically the expected behavior. First one expects a strong decrease of the transmission because of the scattering out of the coherent beam. This gives rise to an exponential decay with as decay length the scattering mean-free path. Then one should arrive at the diffusive regime, and the dependence should be roughly as $(1/L)$. The third regime arises because eventually the diffuse light, which traverses very long distances in the sample, gets absorbed. The characteristic length L_2 associated with this absorption is given by $(L_2)^2 = \frac{1}{3}\lambda_{\mathrm{mf,s}}\lambda_{\mathrm{mf,t}}W/(1 - W)$, in which transport and scattering mean-free path have been used, and where W, the albedo, gives the probability in a single-scattering event of being scattered rather than absorbed. It is clear that albedo's very close to one (typical values for (1 − W are 10^{-5}) can still give rather short characteristic inelastic decay lengths. For our suspensions the general picture of fig. 5 has routinely been observed, with the observation of the diffusion regime over more than a decade. If Anderson localization would occur the diffusive regime should be absent, or at least modified dramatically. Anderson has speculated recently on the scaling behavior of transmission and backscattering[19]).

Another way of monitoring the Anderson localization transition would be the observation of the backscattering. The width of the critical backscattering cone is a good probe. However, in the backscattering there is always a substantial fraction caused by lower-order scattering. This has experimentally very clearly been demonstrated with our difference technique. This light is scattered in the front layer of the cell, and light is certainly not in the localized regime there. So in backscattering one really wants to probe the long light paths. One way is to use our difference technique, another way is to monitor the central part of the backscatter cone, since the long light paths contribute only there.

Localization is concerned with the slowing down of diffusion, and all types of exotic time-dependent measurements can be thought of to directly probe the light transport. Furthermore one might expect that the localization transition might be associated with various fluctuation phenomena which could be observed.

4. Conclusion

The study of localization of light or more generally of electromagnetic waves is a brand-new area of physics which has opened up due to the observation of weak localization of light. In the near future one can expect many new exciting observations to be made. One can think of many interesting parts of the electromagnetic spectrum where interesting observations can be made. Many different kinds of random media can be envisaged. Whether or not Anderson localization of electromagnetic waves can or will be observed remains a challenging question.

Acknowledgement

This work is part of the research program of the "Stichting voor Fundamenteel Onderzoek der Materie (FOM)", which is financially supported by the "Nederlandse Organisatie voor Zuiver-Wetenschappelijk Onderzoek (ZWO)".

References

1) P. W. Anderson, Phys. Rev. **109** (1958) 1492.
2) For a recent collection of papers on Anderson localization, see: Anderson Localization, Y. Nagaoka and H. Fukuyama, eds. (Springer, Berlin, 1982).
3) W. Götze, J. Phys. C **12** (1979) 1279; W. Götze, Phil. Mag. B **43** (1981) 219.
4) D. Vollhardt and P. Wölfle, Phys. Rev. B **22** (1980) 4666.
5) S. John and M.J. Stephen, Phys. Rev. B **28** (1983) 6358.
6) S. John, Phys. Rev. Lett. **53** (1984) 2169.
7) T.R. Kirkpatrick, Phys. Rev. B **31** (1985) 5746.
8) S. DasSarma, A. Kobayashi and R.E. Prange, Phys. Rev. Lett. **56** (1986) 1280.
9) For excellent introductions to single-particle scattering: H.C. van de Hulst, Light scattering by Small Particles (Dover, New York, 1981); C.F. Bohren and D.R. Huffman, Absorption and Scattering of Light by Small Particles (Wiley, New York, 1983).
10) A. Ishimaru, Wave Propagation and Scattering in Random Media, Vols. I and II (Academic, New York, 1978).
11) H.C. van de Hulst, Multiple Light Scattering, Vols. I and II, (Academic, New York, 1980).
12) G. Bergmann, Physica **126B** (1984) 229; Kmel'nitskii, Physica **126B** (1984) 235.
13) E. Akkermans and R. Maynard, J. Physique Lett. **46** (1985) L1045.
14) M.P. van Albada and A. Lagendijk, Phys. Rev. Lett. **55** (1985) 2692.
15) P.E. Wolf and G. Maret, Phys. Rev. Lett. **55**, (1985) 2696.
16) E. Abrahams, P.W. Anderson, D.C. Licciardello and T.V. Ramakrishnan, Phys. Rev. Lett. **42** (1979) 673.
17) E. Akkermans, P.E. Wolf and R. Maynard, Phys. Rev. Lett. **56** (1986) 1471.
18) M.P. van Albada, M.B. van de Mark and A. Lagendijk, to be published.
19) P.W. Anderson, Phil. Mag. B **52** (1985) 505.

Physica **140A** (1986) 191–197
North-Holland, Amsterdam

CRYSTALLIZATION OF AMORPHOUS METAL–INSULATOR MIXTURES

G. DEUTSCHER and Y. LAREAH

Department of Physics and Astronomy, Tel Aviv University, Ramat Aviv, Tel Aviv, Israel

A review is given of the various crystallization morphologies in metal–insulator mixtures. Granular structures have high threshold values while random ones obey the rule of Sher and Zallen. Annealing of amorphous mixtures leads to a number of new, diffusion limited, morphologies, including a "dense branching morphology" colony and a metal, single crystal, DLA like structure growing in the amorphous matrix.

The study of percolation on lattices has shown that the value of the threshold varies with the type of Bravais lattice and with the dimensionality. On the contrary, values of the critical indices are found to depend only on the dimensionality of the lattice. However, Sher and Zallen[1]) have pointed out to the existence of an underlying universality in the values of p_c. Using values of p_c known either exactly (as for the 2D triangular lattice) or from numerical work (Monte Carlo or series expansion), they remarked that the volume fraction X_c, occupied at $p = p_c$, is independent of the lattice type in 2 and 3 dimensions.

Actually, the value of X_c in 2D should be equal to $\frac{1}{2}$ [1]), and in 3D the accepted figure is $X_{c3} = 0.16$. One must emphasize that these values apply only to a random continuum in which both constituents play symmetric roles. In 2D, the width of the occupied and empty channels should be equal. In a 3D mixture, the "metallic" and "insulating" grains should fill the space at random, and be of similar size.

Historically, the first systems that were studied were sputtered films (such as $Al–Al_2O_3$, $W–Al_2O_3$) that had threshold values around 50% [2]). This seemed to contradict the prediction of Sher and Zallen, unless one would consider these films as two dimensional. But such an interpretation presented two difficulties: first, the films were rather thick in terms of the number of layers of grains (10 or more); second, the accurate determination of the conductivity index gave the value 1.9, in good agreement with 3D values obtained from simulations and in clear disagreement with 2D values. This apparent paradox is due to the structure of these "granular metals". It has been suggested[3]) that this structure is the result of the nucleation and growth of the metallic grains, the insulating constituent

0378-4371/86/$03.50

being rejected at the periphery of the grain. Growth of the metallic grain stops when a continuous insulating coating is formed, presenting a barrier to the further diffusion of the metal. By this process, the initially amorphous mixture is transformed into a granular metal, whose grain size is governed by the insulator concentration. If R is the average grain's size and a the thickness of the insulator coating, in the limit of large grain sizes the insulator volume fraction X_I is of the order of $aR^2/R^3 \propto R^{-1}$. This is what is observed experimentally at small insulator concentrations[4]).

However, the surface tension between the grain and its surrounding sets a lower limit to R, below which the grain would not be stable. When the insulator concentration is increased beyond the value that corresponds to that size, the thickness of the insulating barriers starts to build up, at random. Since the electrical resistance of a tunnelling barrier increases exponentially with its thickness, this build up quickly leads to the progressive, random, disconnection of grains. The conductivity is then governed by a bond percolation transition – hence the critical behavior observed by Abeles et al. and others. This case provides a nice illustration of universality in the presence of short range interactions.

Random systems

The granular structure does not represent the stable form of metal–insulator mixtures, since the insulator is amorphous. Indeed, crystallization of the insulator results in completely different structures. In fact, a variety of structures may be obtained, depending on the exact conditions of preparation. If the metal and the insulator are co-deposited by evaporation in a vacuum system onto a substrate held at sufficiently high temperature to ensure the crystallization of the insulator, a random structure is obtained.

The deposition temperature that is necessary for the crystallization of the insulator depends on the particular eutectic system studied. In In–Ge and Pb–Ge, crystallization of the Ge occurs already when the films are deposited onto a substrate held at room temperature[3]). In Al–Ge, a substrate temperature of at least 160°C is necessary[5]).

In all known cases where both constituents crystallize during deposition, a random structure is obtained. It appears that the metal and insulator grains crystallize at random, are often of similar sizes and play symmetric geometrical roles in the structure. An examination of the structure reveals black inside white, white inside black, on a variety of scales. Below a certain scale, the structure actually looks self-similar.

The threshold for metallic conductivity is close to 15% in all known random films composed of metallic and insulator grains, if their thickness is large enough, and definitely confirms that the threshold of about 50% observed in the films

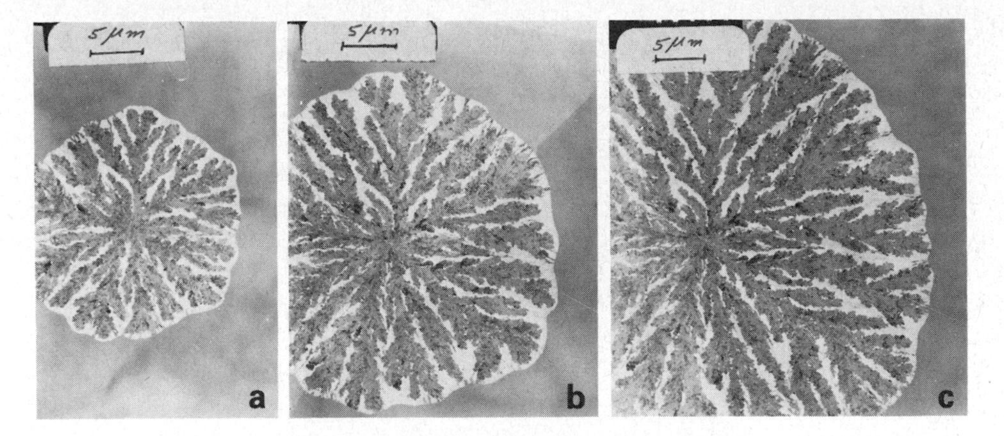

Fig. 1.

where the metal is crystalline and the insulator amorphous, results from their granular structure.

Other morphologies

Figs. 1 to 7 display the different morphologies that can be obtained by the *annealing* of amorphous Al–Ge mixtures. This annealing is done in the electron microscope, the amorphous film having been deposited on a copper grid whose windows are covered by a thin film of a material such as carbon that is transparent to the electron beam and sufficiently strong to provide a substrate for

Fig. 2.

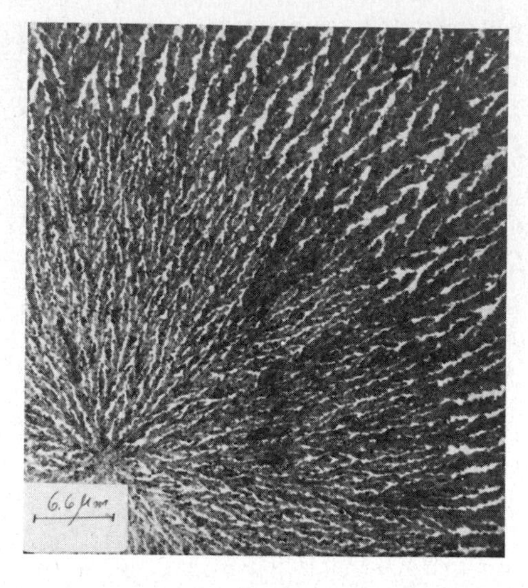

Fig. 3.

the amorphous films. The latter are about 500 Å thick. Annealing can be done either by heating the copper grid with a small furnace, or by concentrating the electron beam on a specific location.

Figs. 1a–c show the growth of a colony at a fixed grid temperature (furnace heating). The grey background is the amorphous matrix. The colony shows a radial structure limited by an envelope that shows hexagonal symmetry at the beginning (1a) but tends to become circular at long times (1c). The white envelope is mostly composed of Al, the black regions in the interior of the colony are composed of Ge grains.

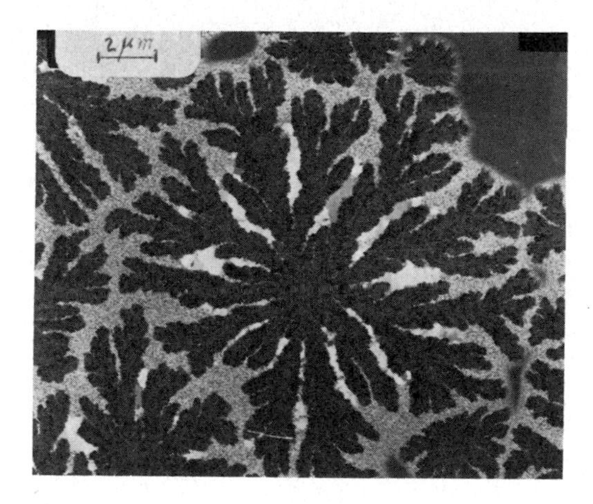

Fig. 4.

Fig. 5.

Fig. 2 is an enlargement showing more clearly the structure of the Al envelope. Fig. 3 shows the colony structure obtained when heating first at 296°C (inner part) and then at 270°C (outer region). The scale of the structure becomes larger at lower temperature, at the same time the growth speed is reduced.

Fig. 4 shows a colony grown from a granular film. Tip splittings of the colony are particularly clear.

Dark field electron microscopy gives interesting clues on the nucleation and growth process of the colony. Ge grains have random orientation and nucleate behind the Al envelope (fig. 5). On the other hand, the Al grains are quite close to having all the same orientation, coherent over long distances (fig. 6). This suggests that the nucleation of a single Al grain has initiated the nucleation and growth of the entire colony. The Al envelope, from which all the Al grains are generated, keeps this original orientation.

The structure of these colonies has been called the "dense branching morphology"[6]). It is similar to one of the morphologies that can be obtained when a fluid is invading another one. On the other hand, the structure of the Al front resembles sometimes that of DLA (fig. 7).

These DLA like structures are obtained by pulse heating of the matrix with the electron beam. They grow very quickly. They resemble the morphology obtained when a fluid under pressure invades a viscoelastic medium, such as water injected in clay[7]).

It is clear that much remains to be done to control and understand the various morphologies that can be obtained from the crystallization of amorphous alloys.

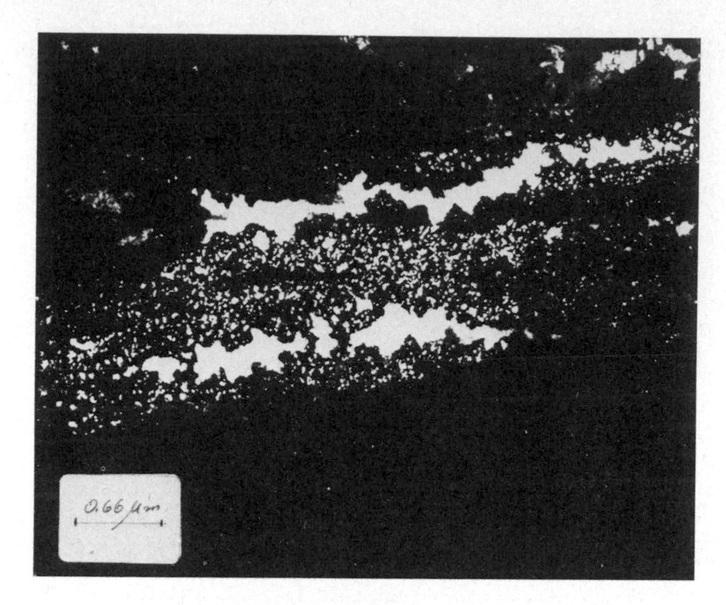

Fig. 6.

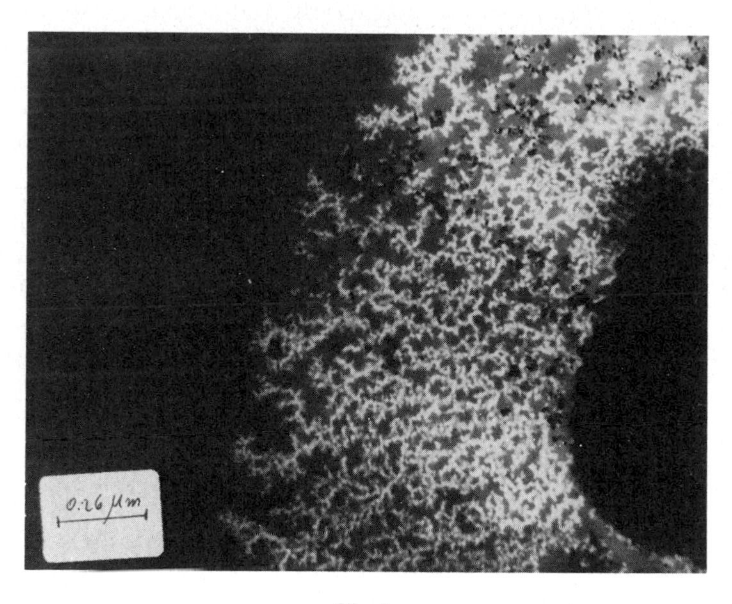

Fig. 7.

In particular, the role of surface tension and of noise is probably important. Granular, random, DBM and DLA have been identified, indicating the richness of these out of equilibrium growth processes.

Acknowledgements

This work was supported in part by the Oren Family Chair for Experimental Solid State Physics. One of us (G.D.) is indebted to Les Houches Summer School of Theoretical Physics.

References

1) H. Sher and R. Zallen, J. Chem. Phys. **53** (1970) 3759.
2) B. Abeles, Adv. Phys. **24** (1975) 407.
3) G. Deutscher, M. Rappaport and Z. Ovadyahy, Solid State Commun. **28** (1978) 593.
4) Y. Shapira and G. Deutscher, Thin Solid Films **87** (1982) 29.
5) A. Kapitulnik, M. Rappaport and G. Deutscher, J. Physique Lett. **42** (1981) L-541.
6) E. Ben-Jacob, G. Deutscher, P. Garik, Nigel Goldenfeld and Y. Lareah, submitted to Phys. Rev. Lett.
7) Vandamme and Laroche, unpublished (1986).

Physica **140A** (1986) 198–203
North-Holland, Amsterdam

EXPERIMENTAL STUDIES OF THE DYNAMICS OF FLUID PHASE TRANSITIONS IN TWO DIMENSIONS

Charles M. KNOBLER

Department of Chemistry and Biochemistry, University of California, Los Angeles, CA 90024, USA

Two types of experiments are described by which the dynamics of phase transitions in two-dimensional fluid systems can be studied. Phase transitions in monolayers of amphiphiles adsorbed on water can be followed by fluorescence microscopy. Light scattering and direct microscopic observation of breath figures, which are patterns formed when fluids are condensed on solid surfaces, provide information about the growth and coalescence of droplets.

1. Introduction

The present keen interest in the dynamics of first-order phase transitions is motivated both by the challenge of understanding the complex nonlinear processes involved and by the important practical applications in physics, metallurgy and polymer science. Significant progress has been made toward a theoretical description of the evolution of a system when it is quenched from a one-phase region into a two-phase region of a phase diagram, but much of our understanding of the phenomena comes from computer simulations and from experiments[1]). Although simulations of three-dimensional systems have been carried out, many of the computations have been restricted to two dimensions. It is of interest, therefore, to be able to compare the simulations with experiment, both as a step toward understanding three-dimensional problems and in order to investigate phenomena that are associated with the lower dimensionality.

Such experiments are just beginning to be performed; I will describe two of them and present some of the preliminary results. One of the experiments, which involves monolayers of fatty acids adsorbed on water, is an almost exact two-dimensional analog of studies of the dynamics of phase separation in fluid mixtures that we have carried out in three dimensions[2]). The other, which is concerned with droplet growth on a solid surface, is not entirely a two-dimensional problem. The essential growth phenomena are found to be restricted to the surface, however, hence there is a close connection to strictly two-dimensional systems.

2. Phase transitions in Langmuir monolayers

The archetypical two-dimensional systems are monolayers adsorbed on liquid surfaces. Direct observation of the physical state of such systems, i.e. of the nature of the monolayer phases, can now be carried out by fluorescence microscopy[3,4]) and we have begun[5,6]) to employ this technique to study monolayers of pure fatty acids and their mixtures. Monolayers in which a small fraction of the molecules have been labeled with a fluorescent marker are prepared on a clean water surface, as in the classical Langmuir trough procedure[7]). The texture of the monolayer is observed with a fluorescence microscope as a function of the thermodynamic state, which is determined by the temperature, surface pressure and surface concentration. The images are most conveniently viewed with a television camera and the evolution of the system after a change in state can be recorded on video tape for later analysis. Features as small as 2 μm can be distinguished.

Surface pressure-area isotherms of fatty acids fall into two main types[7]), which are shown schematically in fig. 1; characteristic regions of the isotherms are traditionally associated with specific phases. In the simplest isotherm, which is found, for example, in stearic acid at low temperature, there is only a single horizontal portion, which is associated with a first-order transition between a gaseous and a liquid phase. Other isotherms, such as those for oleic acid, have an additional feature at small areas, which has been identified with a transition between so-called liquid–expanded and liquid–condensed states. The order of this transition had been in doubt because the isotherms did not appear to be

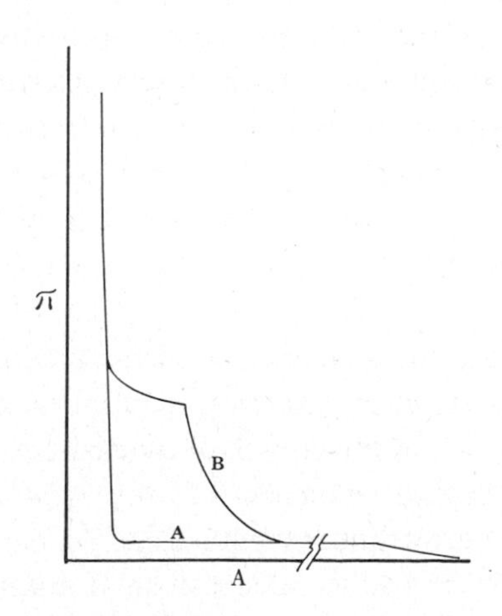

Fig. 1. Typical surface pressure-area isotherms for monolayers of fatty acids. A) Single fluid phase; B) Liquid–expanded and liquid–condensed phases.

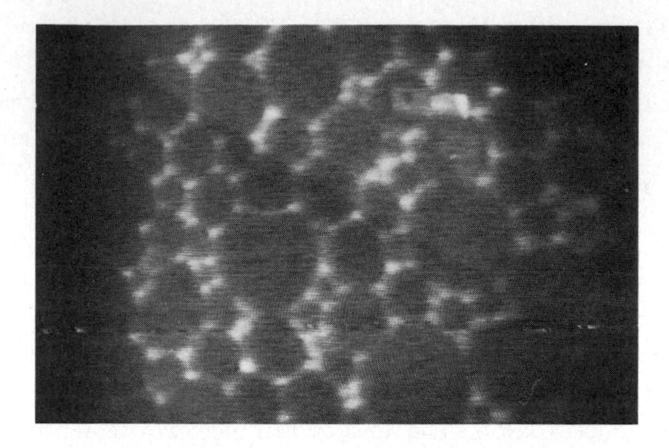

Fig. 2. Fluorescence microscope photo of the foam structure in stearic acid monolayers in the liquid–gas coexistence region. The largest cells are 100 μm in size.

horizontal. Recent studies[8]), however, give strong evidence that the transition is first order.

Two groups have already used the fluorescence microscope to study the phase behavior of phospholipid monolayers[3,4,9,10]). Their investigations have focussed primarily on the formation of solid phases and on the growth of chiral crystallites. Our studies have so far been concentrated on the gas–liquid phase transition in stearic acid[5]) and on both the gas–liquid and the expanded–condensed liquid transition in pentadecanoic acid[6]).

Quench measurements on stearic acid in the gas–liquid coexistence region, in which the area is rapidly increased, reveal the existence of a two-dimensional foam structure, fig. 2, whose evolution can be followed with time. (Foam structures are ubiquitous; they have been observed[6]) in pentadecanoic acid as well, with several chromophores, and with concentrations of labeled molecules ranging from 0.1 to 100%.) Such random two-dimensional cellular structures can be described[11]) in terms of the distribution of the number of edges surrounding a cell, $p(n)$, and the average cell size, expressible in terms of the area or a linear dimension, d.

In two dimensions the average number of edges surrounding a cell must be six in the limit of a large system, provided that the number of edges that join at each vertex is three. An analysis[5]) of the growth of a quasi-two-dimensional soap foam has shown that the second moment of the cell-side distribution $\mu_2 = \Sigma(n - \langle n \rangle)^2 p(n)$ and d grow linearly with time. At early times $p(n)$ is nearly Gaussian, but it becomes skewed as cells with small numbers of sides disappear.

Photographs of the foam at various times after a quench have so far been analyzed for several experiments; the number of cells in each photo varies from

15 to 120. Within the experimental precision the $\langle n \rangle = 6$. In none of the photos is $p(n)$ Gaussian; it is always markedly skewed toward larger sizes. There is a large scatter in the second moments, but it is evident that they increase with time. Our failure to find a Gaussian distribution may therefore be the result of our inability to resolve the cell structure at early times. The average linear dimension grows as $t^{(0.55 \pm 0.05)}$, rather than the value of unity found in the soap foams and more consistent with the $t^{1/2}$-law found in two-dimensional grain growth[12]).

3. Breath figures

Breath figures are the patterns formed when a vapor condenses as a liquid on a cold surface. If the liquid wets the surface, a uniform film is formed, which appears dark. An assembly of droplets, which scatters light and appears white, forms when there is only partial wetting.

The formation of breath figures involves heterogeneous nucleation and subsequent growth. Although there is considerable literature on heterogeneous nucleation, there is surprisingly little known about the growth of fluid phases on surfaces. The problem is by itself intriguing and there are connections as well to general questions of the kinetics of phase transitions and of wetting. For this reason we[13]) have recently begun a detailed experimental study of breath figures.

The experiment consists of streaming a gas saturated with water across a glass plate that has been cooled. The resulting patterns can be studied by microscopic observation and by light scattering. The parameters that can be varied are the nature of the substrate, its temperature, the temperature of the gas, the gas flow rate and its degree of saturation. Treatment of the substrate surface, by silanization, for example, allows the contact angle to be altered.

Measurements for a surface on which the contact angle is 90° have provided the most interesting results. Three different growth regimes are observed, depending on the time. The characteristics of the droplet patterns in each regime can be described in terms of the average droplet radius $\langle R \rangle$, the polydispersity of the droplet size $g = \Delta R / \langle R \rangle$, the mean distance between droplets $\langle a \rangle$ and the surface coverage $\varepsilon^2 = (2\langle R \rangle / \langle a \rangle)^2$, quantities that determine the form of the scattering factor.

At early times, typically $t \leqslant 1$ s after initiation of the flow, the surface is covered by a homogeneous pattern of small droplets with both $\langle R \rangle$ and $\langle a \rangle$ smaller than 2 μm. One can observe furious activity caused by coalescence of droplets.

In the intermediate time regime, typically $1 < t \leqslant 300$ s and $2 < \langle R \rangle \leqslant 300$ μm, there is well-defined order, as demonstrated by the structure factor, which exhibits a characteristic ring, fig. 3. The droplets grow at essentially constant

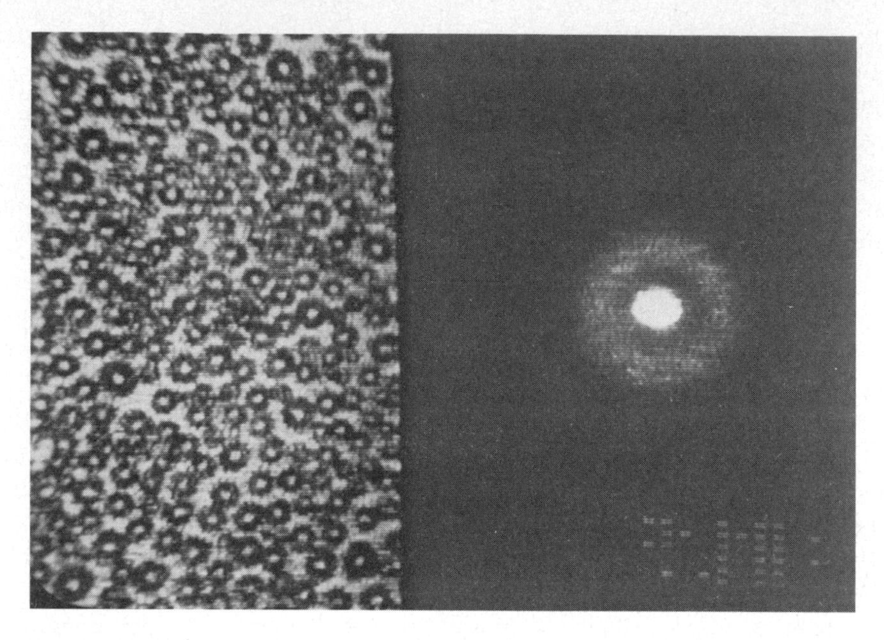

Fig. 3. Breath figures. The left-hand portion is the direct-space image and the right-hand portion is the optical transform, i.e. the small-angle scattering. The average droplet size is 50 μm.

polydispersity ($g \approx 0.2$) and surface coverage ($\varepsilon^2 \approx 0.5$). Under these conditions there is a single peak in the structure factor whose position q_m corresponds to $\langle a \rangle^{-1}$. If one normalizes the scale by q_m, the pattern remains remarkably constant with time.

Throughout this time period, there are two distinct mechanisms of droplet growth. The radii of isolated droplets grow as t^n, with $n = 0.23 \pm 0.05$. When neighboring droplets become sufficiently large to touch, however, this growth is interrupted by coalescence. Two or more droplets merge very rapidly to produce a step change in the radius, but the power-law growth remains unchanged between these breaks. The average growth rate, which is a combination of the two mechanisms, can be assessed from the time dependence of q_m^{-1}, which because of the constancy of ε is proportional to $\langle R \rangle$. We find $q_m^{-1} \sim t^m$ with $m = 0.75 \pm 0.05$, independent of flow rate and supersaturation.

At very long times, typically $t > 300$ s, small droplets form between the large droplets and the distribution of sizes becomes bimodal. The pattern of growth of the large droplets remains unchanged, however, until gravity causes them to become anisotropic and to begin to flow (the surface is vertical).

If a droplet grew directly from the vapor, the rate of increase in its volume would be proportional to its surface area, hence R would increase linearly with time. The very much slower growth that is observed is therefore evidence that another mechanism is dominant. The dependence on supersaturation is also inconsistent with the direct growth from the vapor[13]). A mechanism in which embryos of critical radius condense on the glass surface and then diffuse to growing droplets is roughly consistent with our results, but it gives $R \sim t^{1/3}$ for an isolated droplet. The mechanism can be patched up to give slower growth, but not in a very convincing manner.

We have so far been unable to develop a model of the growth that includes coalescence. If, however, one assumes (in accord with experiment) that the droplet pattern remains self-similar, it can be shown[14]) that the relation between the growth exponents $m = 3n$ must hold. This is indeed observed.

Acknowledgements

This work was supported by the U.S. National Science Foundation.

References

1) For a general review see J.D. Gunton, M. San Miguel and P.S. Sahni, in Phase Transitions and Critical Phenomena, vol. 8, C. Domb and J. Lebowitz, eds. (Academic Press, London, 1983).
2) C.M. Knobler, in Decomposition of Alloys: The Early Stages, Proc. 2nd Acta-Scripta Met. Conf., P. Haasen, V. Gerold, R. Wagner and M.F. Ashby, eds. (Pergamon, New York, 1984) p. 55; N.-C. Wong and C.M. Knobler, J. Chem. Phys. **69** (1978) 725.
3) V. van Tscharner and H.M. McConnell, Biophys. J. **36** (1981) 409.
4) M. Lösche, E. Sackmann and H. Möhwald, Ber. Bunsenges. Phys. Chem. **87** (1983) 848; M. Lösche and H. Möhwald, Rev. Sci. Instr. **55** (1984) 1968.
5) B. Moore, C.M. Knobler, D. Broseta and F. Rondelez, Faraday Symp. Chem. Soc. No. 20 (in press).
6) J.F. Baret, N. Suresh, F. Rondelez and C.M. Knobler, to be published.
7) G.L. Gaines, Jr., Insoluble Monolayers at Liquid–Gas Interfaces (Interscience, New York, 1966).
8) N.R. Pallas and B.A. Pethica, Langmuir **1** (1985) 509.
9) V.T. Moy, D.J. Keller, H.E. Gaub and H.M. McConnell, J. Phys. Chem. **90** (1986) 3198.
10) M. Lösche and H. Möhwald, Europ. Biophys. J. **11** (1984) 35.
11) D. Weaire and N. Rivier, Contemp. Phys. **25** (1984) 59.
12) P.S. Sahni, D.J. Srolovitz, G.S. Grest, M.P. Anderson and S.A. Safran, Phys. Rev. B **28** (1983) 2705.
13) D. Beysens and C.M. Knobler, Phys. Rev. Lett. **57** (1986) 1433.
14) J.-L. Viovy, personal communication.

Physica **140A** (1986) 204–209
North-Holland, Amsterdam

SHEAR FLOW PROBLEMS IN CRITICAL BINARY MIXTURES

Akira ONUKI

Research Institute for Fundamental Physics, Kyoto University, Kyoto 606, Japan

Systems near critical points can be brought into a state far from equilibrium by a relatively small disturbance. A brief review is presented on such phenomena in critical fluid mixtures under shear or stirring.

1. Introduction

Because of limited space I give mostly physical pictures for the phenomena in critical fluids. First we introduce experiments closely related to my review. Beysens et al. measured light scattering from a critical binary mixture streaming through a capillary tube[1-3]). The structure of the critical fluctuations was found to be drastically changed for $S\tau_\xi > 1$, where S is the shear rate and τ_ξ is the average life time of the fluctuations. Goldburg et al. initiated light scattering experiments from turbulent critical binary mixtures[4]). The main issue is how the two phases emerge and are mixed in the unstable or metastable regions. Very recently Hashimoto et al. have started to study entangled polymer mixtures under shear. Effects are even more drastic than in the usual low molecular weight case[5]).

Also I point out abundance of shear flow problems in polymeric systems. Most of them remain to be investigated in the future despite their importance.

2. Critical fluctuations under strong shear

We are interested in the strong shear case $S\tau_\xi > 1$, where most fluctuations are deformed before being dissipated thermally[6,7]). Here $\tau_\xi = 6\pi\eta\xi^3/k_BT$, where η is the viscosity and $\xi = \xi_0(T/T_c - 1)^{-\nu}$ is the correlation length. This condition can be rewritten in the one phase region at the critical composition as

$$T/T_c - 1 < \bar{\tau}_s \equiv \left(6\pi\eta\xi_0^3 S/k_BT\right)^{1/3\nu}. \tag{1}$$

Then we introduce a wave number k_c by

$$(k_B T/6\pi\eta)k_c^3 = S \quad \text{or} \quad k_c\xi = (S\tau_\xi)^{1/3}. \tag{2}$$

Hereafter the fluctuations with wave numbers greater than k_c are denoted by SWF and those with numbers smaller than k_c by LWF. The LWF are strongly elongated along the flow and their lifetime and the correlation length are no longer given by τ_ξ and ξ. On the other hand, SWF are little affected by shear.

A renormalization group theory[6]) showed that a mean field theory can be used once SWF have been coarse-grained in the theory[8]). The LWF are suppressed below the equilibrium level such that the nonlinear coupling among LWF can be taken into account by a normal perturbation scheme in three dimensions.

To explain this simplifying result let us start with a mean field (Langevin) equation for the order parameter $\psi(\boldsymbol{r}, t)$,

$$\frac{\partial}{\partial t}\psi = -Sy\frac{\partial}{\partial x}\psi + \lambda_0\nabla^2\left[r_0 - \nabla^2 + \frac{1}{6}u_0\psi^2\right]\psi + \theta. \tag{3}$$

Here λ_0, r_0, and u_0 are bare (unrenormalized) coefficients and θ is the random source related to λ_0 via the fluctuation–dissipation relation. In (3) ψ is convected by the mean flow, but is decoupled from the velocity field fluctuations. This is allowable only for $d > 4$, d being the spatial dimensionality.

Our finding is that (3) can be used for $d > 2.4$ to describe the dynamics of LWF if the coefficients are renormalized. In particular, in the long wavelength limit, $k \ll k_c$, the equation can be linearized in the form

$$\frac{\partial}{\partial t}\psi_{\boldsymbol{k}} = Sk_x\frac{\partial}{\partial k_y}\psi_{\boldsymbol{k}} - \lambda_\infty(\hat{\boldsymbol{k}})k^2\left[\overline{r_\infty}(\hat{\boldsymbol{k}}) + k^2\right]\psi_{\boldsymbol{k}} + \theta_{\boldsymbol{k}}, \tag{4}$$

where $\psi_{\boldsymbol{k}}(t)$ is the Fourier component of $\psi(\boldsymbol{r}, t)$, and $\theta_{\boldsymbol{k}}(t)$ is a random force. To first order in $\varepsilon = 4 - d$ we obtain[7])

$$\lambda_\infty(\hat{\boldsymbol{k}}) \sim k_c^{-18\varepsilon/19}\left[1 + \varepsilon A_\lambda(\hat{\boldsymbol{k}})\right], \tag{5}$$

$$\overline{r_\infty}(\hat{\boldsymbol{k}}) = \xi_0^{-2}(T/T_c - 1)(k_c\xi_0)^{\varepsilon/3}[1 + \varepsilon A_T(\hat{\boldsymbol{k}})] + \varepsilon k_c^2 A_s(\hat{\boldsymbol{k}}) + \tfrac{1}{2}u^* k_c^{\varepsilon} m^2. \tag{6}$$

Here ξ_0 is the microscopic length appearing in $\xi = \xi_0(T/T_c - 1)^{-\nu}$, $m = \langle\psi\rangle$ is the average order parameter measured from the critical value, and $u^* = (16\pi^2/3)\varepsilon + \dots$. The lower cut-off of the renormalization group contribution is k_c for $S\tau_\xi > 1$, while it is ξ^{-1} for $S\tau_\xi < 1$. This results in fractional powers of the form k_c^ζ with ζ of order ε in (5) and (6). The corrections $A_\lambda(\hat{\boldsymbol{k}})$, $A_T(\hat{\boldsymbol{k}})$, and $A_s(\hat{\boldsymbol{k}})$ arise from the fluctuations with wave numbers of order k_c and weakly depend on the direction $\hat{\boldsymbol{k}} = \boldsymbol{k}/k$.

Let us further lower T at the critical composition $m = 0$. Then an instability is first triggered for $\boldsymbol{k}$ parallel to the z axis. The second term on the right hand side of (6) serves to lower the critical temperature,

$$\Delta T = T_c(S) - T_c(0) = -\left[0.0832\varepsilon + \mathcal{O}(\varepsilon^2)\right]\bar{\tau}_s T_c(0), \tag{7}$$

where T_c has been rewritten as $T_c(0)$. The shift would be about one-tenth of the cross-over value $\bar{\tau}_s T_c$ in three dimensions. There is no shift of the critical composition.

The steady state variance $I_k = \langle|\psi_k|^2\rangle$ can be obtained from (4). It is simply $[\bar{r}_\infty(\hat{\boldsymbol{k}}) + k^2]^{-1}$ for $k_x = 0$, and is generally expressed in an integral form for $k_x \neq 0$. The overall behavior for $k < k_c$ is roughly described by

$$I_k \sim \left[\bar{r}_\infty(\hat{\boldsymbol{k}}) + ck_c^{8/5}|k_x|^{2/5} + k^2\right]^{-1}, \tag{8}$$

where $c \sim 1$. If $k \lesssim k_c$ and $0 < T - T_c(S) \lesssim T_c\bar{\tau}_s$, $I_k \propto |k_x|^{-2/5}$ for most directions of $\boldsymbol{k}$. The mean field behavior $I_k \propto (T - T_c(S))^{-1}$ is observable only when

$$\bar{\tau}_s \gtrsim (T - T_c(S))/T_c \gtrsim \bar{\tau}_s|k_x/k_c|^{2/5}. \tag{9}$$

A sizable temperature region exists only for $|k_x/k_c|^{2/5} \ll 1$. Surely, I_k grows as $T \to T_c(S)$, but the growth is much weakened.

Beysens et al. measured I_k as functions of $\boldsymbol{k}$, T, and S and determined the critical temperature shift ΔT by observing abrupt turbidity increase for $T < T_c(S)$. Their results are in general accord with the theory. Afterwards predictions were also made on another effect in critical fluid, shear-induced anisotropy in the dielectric tensor (birefringence and dichroism)[9]). Experiments agreed fairly with the theory[10]), but there is still a discrepancy[11]).

Oxtoby predicted a logarithmic shear dependence of the viscosity for $S\tau_\xi > 1$ from a mode coupling theory[12]), while we obtained $\eta \propto k_c^{-\varepsilon/19}$ from the ε expansion scheme[13]). The two predictions are virtually the same and seem to explain some data[14]). A weak normal stress effect should also exist[13]), but such experiments are nonexistent.

3. Spinodal decomposition under shear

Second we consider a spinodal decomposition process under shear, which occurs for $T < T_c(S)$ at the critical composition[15]). Beysens and Perrot realized a periodic spinodal decomposition of a critical binary mixture by periodically tilting the capillary tube through which the fluid passed[16]). A spinodal ring could be observed when the tube was horizontal and the shear was small. Then it changed into a sharp, bright streak as the tilting angle was increased. This arose from elongation of domains.

Hashimoto's group has found that unstable polymer mixtures tend to a stationary homogeneous state under stationary shear[5]). The scattering is characterized by a sharp streak at high shear. These finds should also be applicable to usual binary mixtures.

For $S\tau_\xi < 1$ the anisotropy of domains is not strong and their size a is of order $\sigma/\eta S$ for most parameter regions near T_c, σ being the surface tension[15]). For $S\tau_\xi > 1$ domains are rapidly elongated and become strongly anisotropic.

Furthermore a considerable increase of the macroscopic viscosity is expected. This is because domains are continuously deformed and broken, producing small-scale velocity fields. The excess energy dissipation will be of order $\sigma a^2 S$ within each domain for $S\tau_\xi < 1$. Equating this with $(\Delta\eta)S^2a^3$ we find the excess viscosity as $\Delta\eta \sim \sigma/aS \sim \eta$. For $S\tau_\xi > 1$, $\Delta\eta$ will decrease as $\Delta\eta/\eta \sim (S\tau_\xi)^{-p}$ with p of order 1. A large normal stress effect also exists for $T < T_c(S)$.

It is thus very intriguing to investigate the domain structure induced by shear. Such studies are of great technological importance for the polymer case.

4. Nucleation under weak shear

Let us supercool a fluid mixture at the off-critical condition in the presence of low shear. Droplets must be torn when the capillary pressure ($\sim \sigma/a$) is much less than the shear stress ηS [17,18]). Here σ is the surface tension and a is the droplet size. This results in a upper limit of the droplet size, $a^* \sim 0.1\sigma/\eta S$. Noticeable droplets can emerge only when the critical radius $r_c(\sim \xi/\phi)$ is smaller than a^*, ϕ being the volume fraction much less than 1. This condition is rewritten as $S\tau_\xi < \phi(\ll 1)$.

Here we have assumed a sharp interface and the resultant surface tension. However, even if $S\tau_\xi \ll \phi$, there is still a possibility that shear can affect the birth process of the critical nucleus sufficiently near the criticality. Localized fluctuations might be greatly deformed by shear before they could develop a well-defined interface and grow up to the critical size. At present we cannot give definite predictions because details of the birth process remain unknown[19]).

Another interesting aspect is that droplets coagulate in shear flow. Let us assume an initial distribution of an appreciable number of droplets greater than r_c. Then the average size will increase exponentially with a rate of order $S\phi$ up to a size of order a^* [18]). The growth will thus saturate in a time of order $1/S\phi$, whereas it undergoes a critical slowing down in the absence of shear[19]).

5. Turbulent critical binary mixtures

Ruiz and Nelson first suggested a stirring experiment near the criticality[20]). They were interested in a transient mixing process. But their conclusions have not yet been confirmed by experiments[21]). Here we consider another aspect, the fluctuations in a stationary state which is realized after a mixing time. If not far from the criticality, the characteristic size of the fluctuations is much smaller than the inverse of the Kolmogorov cut-off wave number k_d. Then the fluctuations should be deformed by spatially homogeneous, random shear, because the size of the smallest eddies is of order $1/k_d$ [22]). The magnitude of the shear is typically of order $S_d = (\eta/\rho)k_d^2$. Its inverse is the turn-over time of the smallest eddies and may be identified with the correlation time of the shear.

Let the concentration be at the critical value. If $S_d\tau_\xi > 1$, the fluctuations should be suppressed strongly as $I_k \sim 1/k_c^2$ for $k \lesssim k_c$ [23]). Here k_c is defined by (2) where S should be replaced by S_d. If $S_d\tau_\xi \ll 1$ and $T < T_c$, the fluid will be composed of finely divided domains scattering light strongly. In this case I concluded that the spinodal decomposition should be stopped effectively at a particular time of order $1/S_d$ resulting in a stationary state.

Satten and Ronis have presented a model with a special assumption for the velocity field, predicting the existence of a critical point[24,25]). On the other hand, I expect no particular critical temperature in contrast to the laminar shear case.

6. Shear flow problems in other systems and summary

Many other systems are known to be strongly affected by shear. For example, entangled polymers exhibit a strong non-Newtonian effect as $\eta \propto S^{-p}$ with $p \sim 0.8$ for $S > 1/\tau_R$, where τ_R is a very long relaxation time characterizing the viscoelastic properties[26,27]). Undoubtedly there should be some drastic structural change of entanglements in such strong shear[28]). Scattering experiments from streaming polymers seem to be most promising to study the nonlinear rheological behavior on the molecular level.

Another challenging subject is the rheology of polymeric liquid crystals[27,29]). Mention should also be made of a variety of shear-induced melting phenomena in colloidal suspensions[30]). A considerable body of simulations and kinetic theories have been reported on (noncritical) particle systems under shear[31]), where shear must be enormously large, however. Surprisingly, Evans and Hanley claimed the existence of thermodynamics in the presence of shear[32]), but the relationship between their conclusions and ours remains unclear.

In real systems significant shear effects can emerge only when some slow relaxation process is affected by shear. Most conspicuous should be effects on critical fluctuations, entanglements in polymers, and various ordered structures.

References

1) D. Beysens, M. Gbadamassi and L. Boyer, Phys. Rev. Lett. **43** (1979) 1253.
2) D. Beysens and M. Gbadamassi, J. de Physique **40** (1979) L-565; Phys. Rev. A **22** (1980) 2250.
3) D. Beysens, M. Gbadamassi and B. Moncef-Bouonz, Phys. Rev. A **28** (1983) 2491.
4) D.J. Pine, N. Eswar, J.V. Maher and W.I. Goldburg, Phys. Rev. A **29** (1984) 308.
5) T. Hashimoto and T. Takebe, in preparation.
6) A. Onuki and K. Kawasaki, Prog. Theor. Phys. Supp. **64** (1978) 436; Ann. Phys. **121** (1979) 456.
7) A. Onuki, K. Yamazaki and K. Kawasaki, Ann. Phys. **131** (1981) 217.
8) B.I. Halperin and P.C. Hohenberg, Rev. Mod. Phys. **49** (1977) 435.
9) A. Onuki and K. Kawasaki, Physica **111A** (1982) 607.
10) Y.C. Chou and W.I. Goldburg, Phys. Rev. Lett. **47** (1981) 1155; D. Beysens and M. Gbadamassi, Phys. Rev. Lett. **47** (1981) 846.
11) D. Beysens, R. Gastaud and F. Decruppe, Phys. Rev. A **30** (1984) 1145.
12) D.W. Oxtoby, J. Chem. Phys. **62** (1975) 1463.
13) A. Onuki and K. Kawasaki, Phys. Lett. A **75** (1980) 485.
14) K. Hamano, T. Nomura, T. Kawazura and N. Kuwahara, Phys. Rev. A **26** (1982) 1153 and references quoted therein.
15) T. Imaeda, A. Onuki and K. Kawasaki, Prog. Theor. Phys. **71** (1984) 16; A. Onuki, Phys. Rev. A **34** (1986) 3528.
16) D. Beysens and F. Perrot, J. de Physique **45** (1984) L-31.
17) G.I. Taylor, Proc. Roy. Soc. A **146** (1934) 501; J.M. Rallison, Ann. Rev. Fluid Mech. **16** (1984) 45.
18) A. Onuki and S. Takesue, Phys. Lett. A **114** (1986) 133.
19) W.I. Goldburg, in: Light Scattering Near the Phase Transition, H.Z. Cummins and A.P. Levanyuk, eds. (North-Holland, Amsterdam), p. 531.
20) R. Ruiz and D.R. Nelson, Phys. Rev. A **23** (1981) 3224; Phys. Rev. A **24** (1981) 2727.
21) N. Eswar, J.V. Maher, D.J. Pine and W.I. Goldburg, Phys. Rev. Lett. **51** (1983) 1272; C.K. Chan, J.V. Maher and W.I. Goldburg, Phys. Rev. A **32** (1985) 3117.
22) R.H. Kraichnan, Phys. Fluids **11** (1968) 945.
23) A. Onuki, Phys. Lett. A **101** (1984) 286.
24) G. Satten and D. Ronis, Phys. Rev. Lett. **55** (1985) 91; Phys. Rev. A **33** (1986) 3415.
25) J. Aronovitz and D.R. Nelson, Phys. Rev. A **29** (1984) 2012.
26) W.W. Graessley, Faraday Symp. Chem. Soc. **18** (1983) 7.
27) M. Doi and S.F. Edwards, The Theory of Polymer Dynamics, to be published.
28) M. Sekiya and M. Doi, J. Phys. Soc. Jpn. **51** (1982) 3672; A. Onuki, J. Phys. Soc. Jpn. **55** (1986) 1041.
29) Papers in Faraday Discussions of the Chemical Society No. 79, 1985.
30) N. Clark and B. Ackerson, Physica **118A** (1983) 221.
31) Papers in Physica **118A** (1983).
32) D.J. Evans and H.J.M. Hanley, Phys. Lett. A **79** (1980) 178; Phys. Lett. A **80** (1980) 175.

Physica **140A** (1986) 210–211
North-Holland, Amsterdam

SOME EXACT RESULTS FOR POLYMER MODELS

Deepak DHAR

Theoretical Physics Group, Tata Institute of Fundamental Research, Homi Bhabha Road, Bombay 400005, India

Some recent exact results concerning lattice models of directed and undirected branched polymers are briefly reviewed.

I would like to describe here very briefly some of the recent work on directed and undirected animals.

The directed animals problem is a simple variant of the conventional (undirected) animals problem, the latter being a natural lattice model of branched polymers. Introducing the directionality constraint makes the problem much more tractable. In particular, the generating function of the directed site animals on a d-dimensional lattice can be shown to be equal to the average density of particles in a related crystal growth model on the same lattice[1]). In special cases, the latter can be obtained as the equilibrium density of a $(d-1)$ dimensional lattice gas with extended hard core interactions, the density being evaluated at negative values of chemical activity by analytic continuation.

The generating function of directed animals on a square lattice can be related to the density of a one-dimensional lattice gas with nearest neighbour exclusion. This is easily solved, and gives the result[1-3]) that for large n, the number of square lattice directed animals with n sites increases as $3^n n^{-1/2}$, and that their average width varies as $n^{1/2}$.

These results have been obtained by several other techniques as well. Nadal et al.[4]) and Hakim and Nadal[5]) obtained them by the transfer matrix method, finding the largest eigenvalue and eigenvector of the transfer matrix. Gouyou Beauchamps and Viennot[6]) obtained them by establishing a one-to-one correspondence between the animals and a special class of one-dimensional random walks. They can also be obtained by establishing equivalence to Baxter's hard square lattice gas model with next nearest neighbor interactions[2]). None of these methods have succeeded in determining the value of the dynamic exponent z exactly.

0378-4371/86/$03.50

For the three dimensional problem, we have used its equivalence to the exactly solved hard hexagon model[7]), to obtain the exponents

$$\theta_{\text{directed}}(d=3) = \tfrac{5}{6} = 2\nu_{\perp}(d=3).$$

The density profile in the scaling limit has been recently studied by Cardy[8]).

The directed animal exponents in d-dimensions are related to the exponents of undirected animals in $(d+1)$ dimensions, and to the Lee–Yang edge singularity in $(d-1)$ dimensions[9–12]). These relations, in particular, imply that for the four dimensional undirected animals $\theta = \frac{11}{6}$, $\nu = \frac{5}{12}$. For the Lee–Yang edge singularity in two dimensions, the exponent $\sigma = -\frac{1}{6}$.

For details, the original papers may be consulted.

References

1) D. Dhar, Phys. Rev. Lett. **51** (1983) 853.
2) D. Dhar, Phys. Rev. Lett. **49** (1982) 959.
3) D. Dhar, M.K. Phani and M. Barma, J. Phys. **A15** (1982) L279.
4) J.P. Nadal, B. Derrida and J. Vannimenus, J. Phys. (Paris) **43** (1982) 1561.
5) V. Hakim and J.P. Nadal, J. Phys. **A16** (1983) L213.
6) D. Gouyou Beauchamps and G. Viennot, Adv. in Appl. Math. (in press); see also G. Viennot, in: Seminaire N. Bourbaki, Asterisque **121–122** (Soc. Math. France, Paris, 1985) 225.
7) R.J. Baxter, Exactly Solved Models in Statistical Mechanics (Academic, New York, 1982).
8) J.L. Cardy, J. Phys. **A18** (1985) 2771.
9) H.E. Stanley, S. Redner and Z.R. Yang, J. Phys. **A15** (1982) L569.
10) S. Redner and Z.R. Yang, J. Phys. **A15** (1982) 177.
11) J.L. Cardy, J. Phys. **A15** (1982) 593.
12) N. Breur and H.K. Jansen, Z. Phys. **B48** (1982) 347.

Physica **140A** (1986) 212–218
North-Holland, Amsterdam

LEVY WALKS WITH APPLICATIONS TO TURBULENCE AND CHAOS

Michael F. SHLESINGER
Physics Division, Office of Naval Research, 800 North Quincy Street, Arlington, VA 22217, USA

Joseph KLAFTER
Corporate Research Science Laboratory, EXXON Research and Engineering Co., Annandale, NJ 08801, USA

and

Bruce J. WEST
La Jolla Institute, Division of Applied Nonlinear Problems, 3252 Holiday Court, Suite 208, La Jolla, CA 92037, USA

Diffusion on fractal structures has been a popular topic of research in the last few years with much emphasis on the sublinear behavior in time of the mean square displacement of a random walker. Another type of diffusion is encountered in turbulent flows with the mean square displacement being superlinear in time. We introduce a novel stochastic process, called a Levy walk which generalizes fractal Brownian motion, to provide a statistical theory for motion in the fractal media which exists in a turbulent flow. The Levy walk describes random (but still correlated) motion in space and time in a scaling fashion and is able to account for the motion of particles in a hierarchy of coherent structures. We apply our model to the description of fluctuating fluid flow. When Kolmogorov's $-\frac{5}{3}$ law for homogeneous turbulence is used to determine the memory of the Levy walk then Richardson's $\frac{4}{3}$ law of turbulent diffusion follows in the Mandelbrot absolute curdling limit. If, as suggested by Mandelbrot, that turbulence is isotropic, but fractal, then intermittency corrections to the $-\frac{5}{3}$ law follow in a natural fashion. The same process, with a different space–time scaling provides a description of chaos in a Josephson junction.

1. Laws of diffusion

We are all familiar with the Brownian motion of a pollen mote introduced into physics by Einstein and its connection to the diffusion equation. Other laws of diffusion slower than Brownian motion have since been investigated including random walks on random walk paths, random walks on fractals, and motion in energetically disordered solids. Fractal Brownian motion[6]) has the mean square displacement

$$\langle R^2(t)\rangle \sim t^{2H}, \qquad 0 < H < 1. \tag{1}$$

0378-4371/86/$03.50

For $H = \frac{1}{2}$ Brownian motion is recovered. For $H > \frac{1}{2}$ the motion has persistence, and antipersistence for $H < \frac{1}{2}$. The remainder of this paper will be devoted to developing a generalization of fractal Brownian motion, utilizing space–time scaling, to allow us to go beyond the $\langle R^2(t)\rangle \sim t^2$ limit. The motivation for this research is to develop a statistical theory of diffusion in a fully developed turbulent flow.

It may surprise some readers to learn that more rapid diffusion than allowed by eq. (1) was found by Richardson[1]) in 1926 in his study of smoke and balloon diffusion in the turbulent atmosphere. He found that the *relative* mean square displacement of two objects was given by

$$\langle R^2(t)\rangle \sim t^3. \tag{2}$$

He wrote that this is "due to eddies of many sizes acting together", and rationalized this empirical law by assuming that the relative diffusion constant of points separated by a distance R varied as $D(R) \sim R^{4/3}$. Let $P(R, t)$ be the probability that two points which begin together at $t = 0$ are separated by a distance R at time t. Then substituting the diffusion current $j(R, t) = D(R)\,\partial P(R, t)/\partial R$ into the continuity equation yields Richardson's equation

$$\partial P(R, t)/\partial t = \partial/\partial R\left[D(R)\,\partial P(R, t)/\partial R\right], \tag{3}$$

which together with $D(R) = R^{4/3}$ and the closure rule $\langle R^{4/3}\rangle = \langle R^2\rangle^{2/3}$ recovers eq. (2).

Batchelor[2]) arrived at eq. (2) by choosing $D = D(t) = t^2$, and Hentschel and Procaccia[3]) further generalized to $D = D(R, t) = R^a t^b$ with a and b chosen in regard to the study of the fractal shape of clouds. We wish to present a more microscopic derivation of enhanced diffusion results such as $\langle R^2(t)\rangle \sim t^3$, and a variant which allows for spatial intermittency. Our main ingredients include Kolmogorov's famous $-\frac{5}{3}$ law and the use of Levy distributions. While Kolmogorov's work on turbulence is known to many physicists, his 1954 mathematics book with Gnedenko[4]), Limit Distributions for Sum of Independent Random Variables, which focuses on Levy distributions is not. This article is one step toward fulfilling their phophecy about Levy distributions that "it is probable that the scope of applied problems in which they play an essential role will become in due course rather wide".

2. Levy flights[5,6])

When the vogue in probability theory was to prove the central limit theorem under the weakest possible conditions, Levy[5]) at École Polytechnique in the

1920's and 1930's searched for exceptions to it. He began by asking a scaling question about a random walk process. Let $X_1, \ldots, X_N$, be identically distributed random variables with zero mean. Levy asked, when can

$$Y_N = \frac{1}{N^{1/\beta}} \sum_{i=1}^{N} X_i \tag{4}$$

with probability density $F(y)$ have the same probability density $p(x)$ as the X_i? This is essentially the question of when does a part have the same nature as the whole, which is of course the paradigm of fractals, the field invented and developed by Levy's student Mandelbrot[6]).

X_i represents the ith step in a random walk, and $N^{1/\beta} Y_N$ is the random variable whose outcome is the position of the random walk after N steps. When $\langle X_i^2 \rangle$ is finite, then a scale exists and with $\beta = 2$ the solution to Levy's problem is the gaussian distribution. An infinity of other solutions exists parametrized by β. If for large $R = |\boldsymbol{R}|$,

$$p(R) \sim R^{-1-\beta} \qquad (0 < \beta < 2), \tag{5}$$

then the Fourier transform of $p(R)$, $\tilde{p}(k) \sim 1 - \text{const.}\ |k|^{\beta} \sim \exp(-\text{const.}\ |k|^{\beta})$, and by the convolution theorem

$$\tilde{F}(k) \sim \exp(-\text{const.}\ N|k|^{\beta}). \tag{6}$$

The mean square displacement for any $N \geqslant 1$ is

$$\langle Y_N^2 \rangle = -\frac{\partial^2}{\partial k^2} \tilde{F}(k=0) = \infty \qquad (0 < \beta < 2) \tag{7}$$

so there is no characteristic size in this process, even for a single step when $N = 1$. Steps of all sizes occur and one can show that a self-similar set of points of fractal dimension β is visited as $N \to \infty$.

The main concern is that the $\langle X_i^2 \rangle = \infty$ and this infinity has delayed the use of Levy's work in physical applications[7]).

3. Levy walks[8])

3.1. *Formalism*

Levy flights jump between successively visited sites, however distant (see fig. 1). We now explore a stochastic process which visits the same sites as the Levy flight, plus all the intermediate sites on the straight line segment connecting the successive Levy flight points (see fig. 2). We call this a Levy walk as it walks rather than flies over the Levy trajectory.

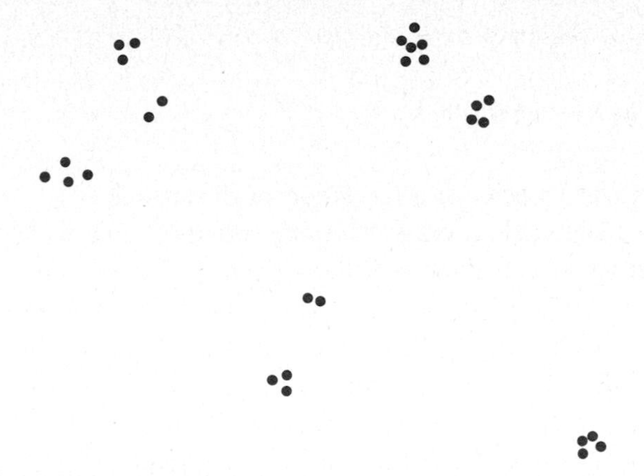

Fig. 1. A portion of the self-similar set of points visited by a Levy flight. The concept of a velocity does not enter into a Levy flight process.

For the flight we need only specify $p(\boldsymbol{R})$, the probability that a jump of $\boldsymbol{R}$ occurs. The continuous motion of the walk is generated by $\Psi(\boldsymbol{R}, t)$ the probability density that the vector displacement $\boldsymbol{R}$ is traversed in a time t. We write $\Psi(\boldsymbol{R}, t)$ as

$$\Psi(\boldsymbol{R}, t) = \psi(t|\boldsymbol{R})\, p(\boldsymbol{R}), \tag{8}$$

where $p(\boldsymbol{R})$ is defined as for the flight and $\psi(t|\boldsymbol{R})$ is the conditional probability density that this transition takes a time t given that it was a displacement of $\boldsymbol{R}$.

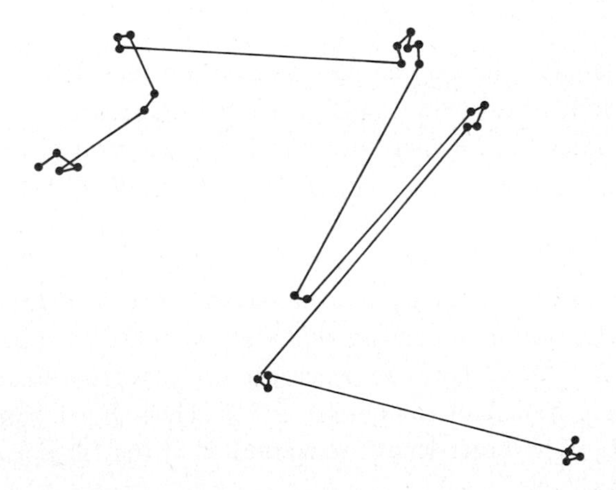

Fig. 2. For the equivalent Levy walk the Levy flight sites become turning points in the Levy walk trajectory. Different persistence lengths are traversed with different velocities according to Kolmogorov's scaling for fully developed turbulence.

We choose

$$\psi(t|\boldsymbol{R}) = \delta(R - V(R)t) \tag{9}$$

with $V(R)$ being the velocity of a transition of distance R.

We call the Levy flight points "turning points", and let $Q(\boldsymbol{R}, t)$ be the probability to reach one of these at time t. Then

$$Q(\boldsymbol{R}, t) = \sum_{\boldsymbol{R}'} \int_0^t Q(\boldsymbol{R} - \boldsymbol{R}', t - \tau)\Psi(\boldsymbol{R}', \tau)\,\mathrm{d}\tau + \delta(t)\delta_{\boldsymbol{R},0}, \tag{10}$$

i.e., one reaches point $\boldsymbol{R} - \boldsymbol{R}'$ at time $t - \tau$ and then covers the remaining displacement $\boldsymbol{R}'$ in a straight line walk segment in a time τ. The probability $P(\boldsymbol{R}, t)$ to be at any site $\boldsymbol{R}$ at time t is given by

$$P(\boldsymbol{R}, t) = \sum_{\boldsymbol{R}'} \int_0^t Q(\boldsymbol{R} - \boldsymbol{R}', t - \tau)\left[\int_1^\infty \Psi(\lambda\boldsymbol{R}', \lambda\tau)\,\mathrm{d}\lambda\right]\mathrm{d}\tau. \tag{11}$$

The Q term brings the walker to $\boldsymbol{R} - \boldsymbol{R}'$ at time $t - \tau$ and the term in brackets allows for jumps of $\lambda\boldsymbol{R}'$ to be completed in times $\lambda\tau$ for all $\lambda \geqslant 1$. All of these walk segments have velocity R'/τ and thus pass through point $\boldsymbol{R}$ at time t, but except for the $\lambda = 1$ case these walk segments pass through and do not end at point $\boldsymbol{R}$.

3.2. *Turbulence*

We can now finally make our connection to turbulence. We consider a random walker in a random vortex field. Larger vortices induce larger persistence lengths for the walker's motion. The distribution of these lengths is given by $p(R)$ in eq. (5). Also, the larger the vortex, the larger will be the walker's velocity since vortex size scales with vortex energy. We use eq. (9) for $\psi(t|\boldsymbol{R})$ and calculate $V(|\boldsymbol{R}|)$ using Kolmogorov's scaling[9] arguments on inertial range turbulent dissipation. Let the average kinetic energy E_R associated with a scale R be $E_R \sim V_R^2$. If the rate of energy dissipation ε_R across this scale is constant then $\varepsilon_R \sim E_R/t_R \sim V_R^3/R$. Thus $V(R) \sim R^{1/3}$. One can generalize this argument to only allow points R which lie on a fractal of dimension d_f[10]. Then $E_R \sim V_R^2 q_R$, where $q_R = (R/R_0)^{E-d_f}$, R_0 is an outer length scale and E is the Euclidean dimension[11]. This leads to

$$V(R) \sim R^{1/3+(E-d_f)/6} \equiv R^{1/3+\mu/6}. \tag{12}$$

A calculation of the mean square displacement yields

$$\langle R^2(t)\rangle \sim \begin{cases} t^{3+3\mu/(4-\mu)}, & \beta \leqslant \frac{1}{3}(1-\mu), \\ t^{2+6(1-\beta)/(4-\mu)}, & \frac{1}{3}(1-\mu) \leqslant \beta \leqslant (10-\mu)/6, \\ t, & \beta \geqslant (10-\mu)/6. \end{cases} \tag{13}$$

The first case in eq. (13) recovers Richardson's law (with an intermittency correction) and corresponds to the mean time $\langle t\rangle$ of a jump being infinite, i.e., no characteristic jump time exists. This result was previously obtained by Hentschel and Procaccia[11]), using a scaling argument. The calculations for $\beta \geqslant \frac{1}{3}(1-\mu)$ are novel. The final case is asymptotically equivalent to Brownian motion and occurs when the mean square time $\langle t^2\rangle$ to complete a jump is finite. This shows that Kolmogorov's scaling does not necessarily imply Richardson's law.

3.3. *Chaos*

In one dimension, if $V(R) = V$ independent of R then, it can be shown that[8])

$$\langle R^2(t)\rangle \sim \begin{cases} t^2, & 0 < \beta \leqslant 1, \langle t\rangle = \infty, \\ t^{2-\beta}, & 1 < \beta < 2, \langle t\rangle < \infty, \langle t^2\rangle = \infty, \\ t \ln t, & \beta = 2, \langle t^2\rangle \text{ log divergent}, \\ t, & \beta > 2, \langle t^2\rangle < \infty. \end{cases} \tag{14}$$

The results have been useful for describing chaos in a Josephson junction where R is a parameter rather then a distance. In this latter case, $R > 0$ represents the number of times the voltage phase has rotated in a CW direction, and $R < 0$ the number of rotations in a CCW direction. The phase rotation rate depends on voltage and not the number of rotations in a given direction. Thus the time to complete R rotations is proportional to R so a coupled space–time memory is needed. The t^2 case corresponds to chaos, or broad band voltage noise, and also to $\langle t\rangle = \infty$. These results partially cover the range of behavior of fractal Brownian motion.

It was our generalization of scaling persistance lengths with velocities which enabled us to generate faster transport and go beyond the fractal Brownian motion results.

Geisel et al.[12]) have previously studied this chaos problem using a nonlinear mapping which involved in a universality exponent z. The choice $\beta = 1/(z-1)$ brings the Levy statistical description into the same form as the mapping treatment. No map has yet been proposed which can recover the turbulence results of eq. (13).

References

1) L.F. Richardson, Proc. Roy. Soc. London, Ser. A **110** (1926) 709.
2) G.K. Batchelor, Proc. Cambridge Philos. Soc. **48** (1952) 345.
3) H.G.E. Hentshel and I. Procaccia, Phys. Rev. A **29** (1984) 1046.
4) B.V. Gnedenko and A.N. Kolmogorov, Limit Distributions for Sums of Independent Random Variables (Addison-Wesley, Reading, MA, 1968).
5) P. Levy, Theorie de l'Addition des Variables Aleatoires (Gauthier-Villars, Paris, 1937).
6) B.B. Mandelbrot, The Fractal Geometry of Nature (Freeman, New York, 1983).
7) A.S. Monin and A. M. Yaglom, Statistical Fluid Mechanics (MIT Press, Cambridge, MA, 1975), vol. 2.
8) M.F. Shlesinger and J. Klafter, Phys. Rev. Lett. **54** (1985) 2551; in: On Growth and Form, H. E. Stanley and N. Ostrowski, eds. (Martinus Nijhoff, Amsterdam, 1985).
9) A.N. Kolmogorov, C.R. (Dokl.) Acad. Sci. USSR **30** (1941) 301.
10) B.B. Mandelbrot, J. Fluid. Mech. **62** (1974) 331; in: Turbulence and the Navier–Stokes Equation, R. Temann, ed. (Springer, Berlin, 1976).
11) H.G.E. Hentschel and I. Procaccia, Phys. Rev. A **27** (1983) 1266.
12) T. Geisel, J. Nierwetberg and A. Zacherl, Phys. Rev. Lett. **54** (1985) 616.

Physica 140A (1986) 219–224
North-Holland, Amsterdam

CONFORMAL INVARIANCE AND CRITICAL BEHAVIOR

John L. CARDY
Department of Physics, University of California, Santa Barbara, CA 93106, USA

A review is given of recent work on the application of conformal invariance to systems at critical points, with emphasis on the classification of such systems in two dimensions using the constraints of positivity and modular invariance on a torus.

That conformal invariance should be a symmetry of a large class of critical systems was realized a long time ago by Polyakov[1]). However, it is only in the last few years, following seminal work by Belavin, Polyakov and Zamolodchikov[2]) which uses many ideas originally developed in the context of string theory, that its consequences have been fully explored. Like any other symmetry in physics, it is realized in different physical systems in different ways. The conformal group in two dimensions consists of all possible analytic functions, and is therefore much bigger and more powerful than in higher dimensions where it has only a finite number of generators. For this reason we shall limit this review to two-dimensional behavior. Some of the mathematics behind this subject is fairly technical, so we shall omit all such details here, and refer the reader to a forthcoming review article[3]) and the original papers. We shall, however, mention some more recent developments which were not included in that review.

The idea that a critical system should be scale invariant is familiar. It finds a simple explanation within the renormalization group: on large distance scales the behavior is determined by a fixed point hamiltonian, which is invariant under renormalization group transformations corresponding to a scale change $r \to r' = b^{-1}r$. In addition there exist *scaling operators* which transform simply under such a rescaling: $\phi(r) \to b^{-x}\phi(r')$, where x is the *scaling dimension* of ϕ, related to its renormalization group eigenvalue $y = 2 - x$. If the system is rotationally invariant at large distances (even anisotropic systems can often be made rotationally invariant by a suitable anisotropic rescaling) then we can also assign a *spin* s to ϕ. In that case, the correlation function has the form

$$\langle\phi(z,\bar{z})\phi(z',\bar{z}')\rangle = (z - z')^{-2h_\phi}(\bar{z} - \bar{z}')^{-2\bar{h}_\phi}, \tag{1}$$

where $x = h_\phi + \bar{h}_\phi$, $s = h_\phi - \bar{h}_\phi$ and we are using $(z, \bar{z})$ as complex coordinates.

0378-4371/86/$03.50

A *conformal* transformation corresponds to a renormalization group transformation with a *nonuniform* $b(r)$. Because the fixed point hamiltonian is assumed to be short-ranged, it should be invariant under this larger group, and the scaling operators should transform as above, with b replaced by its local value $b(r)$. Not all coordinate transformations $r \to r'$ are allowed; only those which correspond locally to a scale change plus a rotation. This means that they have no shear component. In two dimensions, they correspond to mappings by analytic functions $z' = f(z)$. To summarize, systems which are scale invariant, rotationally invariant and have short-range interactions are expected to be also conformally invariant. The transformation law for the general correlation function is

$$\langle \phi(z_1, \bar{z}_1)\phi_2(z_2, \bar{z}_2)\cdots\rangle$$

$$= \prod_j f'(z_j)^{h_j} \overline{f'(z_j)}^{\bar{h}_j} \langle \phi_1(z_1', \bar{z}_1')\phi_2(z_2', \bar{z}_2')\cdots\rangle. \tag{2}$$

In writing (2) we must understand that while both sides are evaluated with the fixed point hamiltonian, they will in general correspond to different geometries. One of the elementary but fruitful applications of conformal invariance has exploited this to obtain information about finite-size scaling[4]). If, however, we wish to preserve the geometry (for example working in the infinite plane) by considering an infinitesimal transformation $z' = z + \alpha(z)$ with $\alpha(z) \to 0$ as $z \to \infty$, then $\alpha(z)$ necessarily has singularities at which the transformation is no longer conformal. Under a general infinitesimal *non*-conformal transformation $r_\mu' = r_\mu + \alpha_\mu$ the hamiltonian no longer remains at the fixed point, but will acquire an additional term which, on general grounds, must be of the form $-(1/2\pi)\int(\partial\alpha^\mu/\partial r_\nu)T_{\mu\nu}\,\mathrm{d}^2r$. This defines the *stress tensor* $T_{\mu\nu}$. The fact that the hamiltonian should be scale, rotation and translation invariant implies that $T_{\mu\nu}$ is traceless, symmetric and divergenceless, so that it has two independent components, which may be taken as $T(z)$, depending on z but not $\bar{z}$, and likewise $\overline{T}(\bar{z})$. If we now consider a transformation with $\alpha(z) \propto 1/(z - z_0)$, the hamiltonian on the right hand side of (2) will change by an amount proportional to $T(z_0)$. Expanding this out of the Boltzmann factor, we get the *conformal Ward identity*

$$\langle T(z_0)\phi_1(z_1)\phi_2(z_2)\cdots\rangle = \sum_j \left[\frac{h_j}{(z_0 - z_j)^2} + \frac{1}{z_0 - z_j}\frac{\partial}{\partial z_j}\right]$$

$$\times \langle \phi_1(z_1)\phi_2(z_2)\cdots\rangle, \tag{3}$$

where the $\bar{z}$ dependence has been suppressed. By choosing $\alpha(z)$ to have more singularities, we obtain further identities which relate the connected correlation functions with an arbitrary number of T's to those with none. In what follows we shall also need the full correlation functions including the disconnected parts. These will involve

$$\langle T(z)T(z')\rangle = (c/2)(z-z')^{-4}. \tag{4}$$

The z dependence of this follows from the fact that T has $x=2$, $s=2$, but, since the normalization of T is fixed by its definition, we are not free to choose the coefficient. c is called the *conformal anomaly number*, and its value is a universal property of the particular system.

The next important ingredient is the operator product expansion of T with an arbitrary operator, which has the form

$$T(z)\phi(z_1) = \sum_{n=0}^{\infty} (z-z_1)^{n-2} L_{-n}\phi(z_1). \tag{5}$$

The meaning of this is that the correlation functions of either side with an arbitrary operator are equal as $z \to z_1$. This defines the operators $L_{-n}\phi$ with dimensions $(h_\phi + n, \bar{h}_\phi)$. Further operators may be generated by repeated short distance expansions with T and $\bar{T}$. They all belong to the *conformal tower* of the *primary* operator ϕ. By substituting the expansion (5) in the Ward identity (3) and its generalizations one may calculate the correlation functions of any operator $\mathcal{O}$ in the conformal tower, whose z-dependence will be of the form $\langle \mathcal{O}(z)\mathcal{O}(0)\rangle = Az^{-2h_{\mathcal{O}}}$. In a theory with positive Boltzmann weights, the coefficients A must all be nonnegative. Friedan, Qiu and Shenker[5]) showed that, for theories with $c<1$, this requires the value of c to be quantized according to $c = 1 - 6/m(m+1)$, where $m = 3, 4, \ldots,$ and that the allowed values of the scaling dimensions are $(h_{pq}, h_{\bar{p}\bar{q}})$, with $1 \leqslant q \leqslant p \leqslant m-1$, $1 \leqslant \bar{q} \leqslant \bar{p} \leqslant m-1$, where

$$h_{pq} = \frac{(p(m+1)-qm)^2 - 1}{4m(m+1)}. \tag{6}$$

This is the Kac formula[6]). This result explains the long-standing puzzle as to why so many critical exponents in two dimensions are rational numbers. Belavin et al.[2]) showed that the correlation functions of operators whose scaling dimensions are given by the Kac formula satisfy linear differential equations, which in some simple cases, can be solved. This program was pursued by Dotsenko and Fateev[7]), who used the fact that any value of $c \leqslant 1$ can be realized by taking a gaussian model (theory of a free field ϕ), and computing correlation functions with a "charge" $e^{i\alpha\phi}$ at infinity, where $c = 1 - 24\alpha^2$. In this way they were able to calculate the 4-point functions of arbitrary operators.

Although the Kac formula gives the allowed values of the scaling dimensions, it does not determine which actually appear in a given universality class. This can be answered by imposing the condition of modular invariance on the torus, which we now describe. One of the most useful conformal mappings is given by $z' = (l/2\pi)\ln z$, which maps the whole z-plane onto a cylinder, an infinitely long strip of width l with periodic boundary conditions. If we apply the transformation law (2) we find[8]) that the correlation function $\langle\phi(u,v)\phi(0,0)\rangle$ (where $z' = u + \mathrm{i}v$) behaves for $u \to \infty$ like $\exp(-2\pi/l)(x_\phi u + \mathrm{i}s_\phi v)$. This means that if we write the transfer matrix for the strip as $\mathrm{e}^{-\hat{H}}$, and the momentum operator across the strip as $\hat{k}$, then to each scaling operator of dimensions $(h, \bar{h})$ corresponds an eigenstate of $\hat{H}$ and $\hat{k}$ with eigenvalues $E_0 + (2\pi/l)(h + \bar{h})$ and $(2\pi/l)(h - \bar{h})$, respectively. This explains the previously observed connection between finite-size scaling amplitudes in strips and critical exponents[9]). It implies that by numerically diagonalizing $\hat{H}$ one may find the scaling dimensions of all the operators, whether relevant or irrelevant. This has been applied to various models by Gehlen and Rittenberg[10]) and Saleur[11]). It has also been shown[12]) that E_0, which is the free energy per length of the strip, is related to the conformal anomaly number by $E_0 = fl - (\pi c/6l) + \dots$, where f is the nonuniversal bulk free energy per unit area. This result gives a way one may "measure" c for a given model.

Now consider the system defined on the torus formed by identifying opposite edges of a parallelogram whose vertices in the complex plane are $(0, 1, \tau, 1 + \tau)$. The partition function for the torus is given in terms of the operators $\hat{H}$, $\hat{k}$ of the cylinder by $Z(\tau) = \mathrm{Tr}\exp(-\hat{H}\,\mathrm{Im}\,\tau + \mathrm{i}\hat{k}\,\mathrm{Re}\,\tau)$. Because of the periodic boundary conditions, it follows that $Z(\tau + 1) = Z(\tau)$. Also, we may rotate the parallelogram so that its vertices are at $|\tau|(0, 1, \tau^{-1}, 1 + \tau^{-1})$. Since Z depends only on the shape of the parallelogram and not on its overall dimensions (disregarding the bulk term) we conclude that $Z(\tau^{-1}) = Z(\tau)$. These two relations imply that Z is invariant under the modular group generated by these two operations. On the other hand, for the models in the classification of Friedan et al.[5]) we know almost everything about the spectra of $\hat{H}$ and $\hat{k}$, except for the number of primary operators with given dimensions allowed by the Kac formula. It is quite remarkable that these spectra turn out to be just of the form to allow the modular invariance of Z, with only a finite number of constraints on the unknowns. These take the form of linear Diophantine equations[13]). The idea now is that different universality classes correspond to different solutions of these equations, for a given value of c. For each value of m in (6) there is always the solution corresponding to taking all allowed $s = 0$ operators just once. The scaling dimensions of the relevant operators equal those found on one of the solution manifolds of a sequence of generalized SOS models solved by Andrews,

Baxter and Forrester[14]). Huse[15]) identified these with the universality classes of critical, tricritical, etc., Ising models. In addition, Itzykson and Zuber[16]) and Gepner[17]) have found other solutions which involve operators with $s \neq 0$. The first two of these correspond to the critical and tricritical 3-state Potts models, but, so far, microscopic models corresponding to the other solutions have not been identified. The number of solutions of the modular invariance constraints appears to grow with m, indicating the existence of many unexplored universality classes. By relaxing the invariance condition to that of a subgroup of the modular group, one allows different kinds of boundary conditions which allow the disorder and parafermion operators of these models to appear[18]). Recently Friedan and Shenker[19]) have shown how correlation functions on the torus can in principle be extracted from the knowledge of Z on surfaces of higher genus.

Many models of interest in statistical mechanics do not satisfy the positivity requirement. In that case the quantization conditions do not apply, and there is no a priori reason to use the Kac formula. Nevertheless, it has been shown[7,20]) that the scaling dimensions of standard models do appear to fit. Examples are the continuous Q-state Potts model (with $\sqrt{Q} = 2\cos(\pi/m + 1)$), and the $\mathcal{O}(N)$ model (with $N = 2\cos(\pi/m)$). In both cases, certain values of p and q in (6) give exponents known by other means. However, one does not know what limits the allowed values in these cases. It is possible[21]) to construct field theories with $c > 1$ which are like Heisenberg models with an extra term in the hamiltonian, the Wess–Zumino term. This has no direct interpretation in terms of classical statistical mechanics. However Affleck[22]) has argued that they describe the zero-temperature critical behavior of quantum antiferromagnetic chains. Fateev and Zamolodchikov[23]) have shown that these theories may be viewed as a direct sum of a gaussian model and a theory with a Z_N symmetry. This latter has $c = 2(N-1)/(N+2)$. The exponents found by these authors agree with those found by Andrews et al.[14]) on their second solution manifold, identified by Huse[15]) as a Z_N critical point.

There have been many other interesting developments in this field, including the incorporation of supersymmetry into conformal invariance[24]) and applications to systems with boundaries[25]). At this stage it appears that we understand how to fit most known two-dimensional universality classes into the conformal classification scheme. What is less well understood is how the internal symmetries of these models mesh with their conformal properties.

Acknowledgement

This work was supported by NSF Grant No. PHY83-13324.

References

1) A.M. Polyakov, Sov. Phy. Zh. Eksp. Teor. Fiz. Pis. Red. **12** (1970) 271 [JETP Lett. **12** (1970) 381].
2) A.A. Belavin, A.M. Polyakov and A.B. Zamolodchikov, J. Stat. Phys. **34** (1984) 763; Nucl. Phys. **B241** (1984) 333.
3) J.L. Cardy, in Phase Transitions and Critical Phenomena, C. Domb and J. Lebowitz, eds. (Academic, New York, 1986), vol. 11.
4) See, for example, P. Kleban, G. Akinci, R. Hentschke and K.R. Brownstein, J. Phys. A **19** (1986) 437; N.C. Bartelt and T.L. Einstein, J. Phys. A **19** (1986), to appear.
5) D. Friedan, Z. Qiu and S. Shenker, Phys. Rev. Lett. **52** (1984) 1757; Vertex Operators in Mathematics and Physics, J. Lepowsky, ed. (Springer, New York, 1984) p. 419.
6) V.G. Kac, in Group Theoretical Methods in Physics, W. Beiglbock and A. Bohm, eds., Lecture Notes in Physics **94** (Springer-Verlag, New York, 1979), p. 441.
7) Vl.S. Dotsenko and V.A. Fateev, Nucl. Phys. **B240** (1984) 312, **B251** (1985) 691.
8) J.L. Cardy, J. Phys. A **17** (1984) L385.
9) See, for example, M.P. Nightingale and H.W.J. Blöte, J. Phys. A **16** (1983) L657.
10) G.V. Gehlen and V. Rittenberg, J. Phys. A, to be published.
11) H. Saleur, J. Phys. A, to be published.
12) H.W.J. Blöte, J.L. Cardy and M.P. Nightingale, Phys. Rev. Lett. **56** (1986) 742; I. Affleck, Phys. Rev. Lett. **56** (1986) 746.
13) J.L. Cardy, Nucl. Phys. **B270** (1986) 186.
14) G.E. Andrews, R.J. Baxter and P.J. Forrester, J. Stat. Phys. **35** (1984) 193.
15) D.A. Huse, Phys. Rev. B **30** (1984) 3908.
16) C. Itzykson and J.-B. Zuber, Nucl. Phys. **B275** [FS 17] (1986) 580.
17) D. Gepner, Princeton preprint.
18) J.L. Cardy, Nucl. Phys. **B275** [FS 17] (1986) 200; J.-B. Zuber, Phys. Lett. B **176** (1986) 127.
19) D. Friedan and S. Shenker, preprint EFI 86-18A.
20) V. Singh and B.S. Shastry, Pramāna-J. Phys. **25** (1985) 519.
21) E. Witten, Comm. Math. Phys. **92** (1984) 455; V. Kniznik and A.B. Zamolodchikov, Nucl. Phys. **B247** (1984) 83.
22) I. Affleck, Phys. Rev. Lett. **56** (1986) 746; Nucl. Phys. **B265** (1986) 409.
23) V.A. Fateev and A.B. Zamolodchikov, Sov. Phys. JETP **62** (1985) 215.
24) M.A. Bershadsky, V.G. Kniznik and M.G. Teitelman, Phys. Lett. B **151** (1985) 31; D. Friedan, Z. Qiu and S. Shenker, Phys. Lett. B **151** (1985) 37; W. Boucher, D. Friedan and A. Kent, Phys. Lett. B **172** (1986) 316.
25) J.L. Cardy, Nucl. Phys. **B240** (1984) 514.

Physica **140A** (1986) 225–231
North-Holland, Amsterdam

RIGOROUS STUDIES OF CRITICAL BEHAVIOR

Michael AIZENMAN*

Departments of Mathematics and Physics, Rutgers University, New Brunswick, NJ 08903, USA

Presented here are some basic results on the phase structure and the critical behavior in general classes of translation invariant ferromagnetic systems, and in independent percolation models. Some of these results were derived by means of nonlinear partial differential inequalities involving the order parameter. While the inequalities are satisfied in the discrete lattice systems, their structure contains hints of relations with suitable field theories ("ϕ^4" for ferromagnetic systems, and "ϕ^3" in case of percolation).

1. Introduction

The basic questions about the global phase structure and the critical behavior in statistical mechanics concern regimes which are beyond the direct reach of perturbative methods. Remarkable achievements of this field are the renormalization group and field theoretic techniques, which yield results of both qualitative and quantitative nature–without however establishing that these apply to the specific models which are explicitly written down. My goal here is to address the central issues – concerning the critical behavior – *within the models*, i.e., from the first principles of statistical mechanics.

In this brief note I present some basic results, which demonstrate that rigorous analysis is possible even in the absence of exact solutions, and make some more general comments on the methods which are illustrated by these examples. First, however, it should be emphasized that this neither is, nor is intended to be, a review of rigorous results on critical behavior. Such a review should include many other contributions which have recently been made, some of which are in fact quite topical. Somewhat more expanded presentations of recent fundamental results on percolation models may be found in ref. 1, although the above caveat applies also to these articles.

*Research supported in part by the NSF grant PHY-8605164 A02.

2. Some basic nonperturbative issues

Let me demonstrate the issues which one would like to be able to analyze in the context of the familiar ferromagnetic Ising model. (This choice is only partially guided by pedagogical considerations. It also happens to be the case that the results for this model are more complete than for others.) To fix the notation, we write the Hamiltonian as

$$H = -\sum_{\{x,y\}} \sigma_x \sigma_y J_{x,y} - h/\beta \sum_x \sigma_x, \tag{1}$$

and denote

$$M(\beta, h) = \langle \sigma_x \rangle_{\beta,h}, \tag{2}$$

where $\langle \ \rangle_{\beta,h}$ represents the thermal average, with the weights $e^{-\beta H(\sigma)}$, corresponding to the equilibrium state at the inverse-temperature β, and an external field (βh). The σ-variables are one component Ising spins, with $\sigma = \pm 1$ and symmetric a priori weights. However, all of our discussion generalizes also to other distributions, e.g., the continuous φ^4 variables, derivable from Ising spins by the Griffiths–Simon method[2]).

We shall now focus on the cases where the interaction is ferromagnetic and translation invariant, i.e.,

$$J_{x,y} \equiv J_{0,y-x} \geqslant 0. \tag{3}$$

In fact, some of the most interesting current discussions concern situations where these conditions are not satisfied. I shall not discuss here the preliminary results which have been obtained for such models, which exhibit effects of randomness and frustration, and focus instead on some even more fundamental issues.

As is generally known, the models outlined above exhibit phase transitions, which, however, can occur only at $h = 0$ (by the celebrated Lee and Young theory[3]). At high temperatures, i.e., low enough values of β, the equilibrium states are weak perturbations of the independent product state. Various high-temperature methods apply, and show that there is a unique Gibbs (equilibrium) state, and that spin–spin and other correlations fall off rapidly. (Rapid falloff means, for $\langle \sigma_x, \sigma_y \rangle$, exponential decay if the interactions are of finite range or decay exponentially, and a similar power law to that of J if $J_{x,y} \approx c/|x-y|^s$.) On the other extreme, for low temperatures the quantity $M(\beta, h)$ is discontinuous at $h = 0$, where there are two different Gibbs states related by a symmetry operation (global spin flip). Low-temperature expansions (which extend the original argument of Peierls[4]) converge, and provide answers to various questions, as long as β is large enough.

Typically, many of the more interesting questions about such models concern regimes which are not directly covered by the expansions mentioned above. That is the case with questions concerning the *phase structure* (in the intermediate-temperature regime where neither expansion converges), the *numerical computation* of the critical temperature, the possible *continuity of the phase transition* at $(\beta, h) = (\beta_c, 0)$, properties of the *critical exponents*, *scaling limits*, and the effects of *long range interactions*.

Similar questions arise in the study of a variety of other models. In particular, the rigorous study of the critical behavior in ferromagnetic systems was intertwined with the study of *percolation*, in such a way that the analysis of each benefited, in turn, from the progress made in the other.

To emphasize the analogy, let us consider the following mixed system of independent site and bond variables. The bonds – which are just pairs of lattice sites – are occupied (= connecting) independently with the probabilities:

$$\text{Prob}\,(\text{the bond } \{x, y\} \text{ is occupied}) = 1 - e^{-\beta J_{x,y}} \qquad (\approx \beta J_{x,y} \text{ if } \beta J \ll 1), \tag{4}$$

with $J_{x,y}$ satisfying (3), and the sites are independently colored green, with

$$\text{Prob}\,(x \text{ is green}) = 1 - e^{-h} \qquad (\approx h \text{ if } h \ll 1). \tag{5}$$

The *percolation analog* of the magnetization $(\langle \sigma_0 \rangle_{\beta, h})$ is the following quantity:

$$M(\beta, h) = \text{Prob}\,(\text{the origin is connected to a green site}). \tag{6}$$

M is readily seen to bear the following relation with the cluster size distribution:

$$\begin{aligned} M(\beta, h) &= 1 - \sum_{n<\infty} P_n e^{-hn} \\ &\equiv \sum_{1 \leqslant n \leqslant \infty} P_n (1 - e^{-hn}), \quad \text{for } h > 0, \end{aligned} \tag{7}$$

where P_n = Prob (the connected cluster of the origin has exactly n sites). In particular, $M(\beta, h) \to P_\infty(\beta)$ in the limit $h \to 0$. As an analog of the (untruncated) spin–spin correlation we have the two point function $\tau(x, y)$ = Prob (x is connected to y).

For the sake of concreteness let us define the critical point β_c as the threshold for the spontaneous symmetry breaking, or percolation, in our second example. In either case the definition is:

$$\beta_c = \inf\left\{0 \leqslant \beta \leqslant \infty \mid \lim_{h \to 0+} M(\beta, h) \neq 0\right\}. \tag{8}$$

3. Some general results

The following theorem (which includes the results of a number of authors) addresses some of the questions mentioned above.

Theorem. i) In any translation invariant Ising model, with ferromagnetic pair interactions on $\mathbb{Z}^d$ (in any dimension), the high-temperature and low-temperature phases (along the $h = 0$ line) extend up to a common critical point, in the sense that for any $\beta < \beta_c$ ($\equiv \beta_c^{(\text{sp.mg.})}$) the spin–spin correlation functions (as well as the higher correlations) decay rapidly–in the terminology introduced above (meaning exponentially fast in the finite range case). In particular, the susceptibility $\chi \equiv \partial M(\beta, h)/\partial h$ is finite up to β_c (which is the heart of the matter). Furthermore:

ii) With some finite constants C,

$$\begin{aligned} &M(\beta,0) \geqslant C(\beta - \beta_c)^{1/2}, && \text{for } \beta \geqslant \beta_c \ (h = 0),\\ &M(\beta_c, h) \geqslant Ch^{1/3}, && \text{for } h \geqslant 0 \ (\beta = \beta_c), \\ &\chi(\beta,0) \geqslant C(\beta - \beta_c)^1, && \text{for } \beta \leqslant \beta_c \ (h = 0). \end{aligned} \tag{9}$$

(These bounds show that the critical exponents defined by taking the suitable limits in $M(\beta,0) \approx C(\beta - \beta_c)^{\hat{\beta}}$, $M(\beta_c, h) \approx Ch^{1/\delta}$ and $\chi(\beta,0) \approx C(\beta - \beta_c)^{\gamma}$ satisfy

$$\hat{\beta} \leqslant \tfrac{1}{2}, \qquad \gamma \geqslant 1, \qquad \delta \geqslant 3, \tag{10}$$

where each inequality is saturated in the mean-field approximation. Furthermore, the divergent lower bound on χ implies that the correlation length diverges as $\beta \to \beta_c - 0$, i.e., the phase transition is "continuous".) Finally:

iii) For the nearest neighbor (nonvanishing) interactions, and long range interactions which are "reflection positive"[5]) and satisfy $J_{x,y} \geqslant C/|x - y|^s$, if either $d \geqslant 4$ or $s \geqslant d/2$, then

$$\hat{\beta} = \tfrac{1}{2}, \qquad \gamma = 1, \qquad \delta = 3, \tag{11}$$

where each equality means that the logarithmic ratio yielding the given exponent converges to the specified value. (In fact, m and χ are proven to satisfy converse bounds to those in ii), which in the interior of the above region of values of (d, s) differ only by the constants C, and at the boundary of that regime allow at most "logarithmic corrections".)

A closely analogous theorem holds for independent percolation models, for which in part i) one should only make the substitutions explained in section 2, and in part ii) the inequalities should be replaced by bounds which yield

$$\hat{\beta} \leqslant 1, \qquad \gamma \geqslant 1, \qquad \delta \geqslant 2. \tag{12}$$

However, instead of a full analog of part iii) we have so far only a certain "dynamical" diagrammatic condition (which does contain a hint of $d = 6$ being the upper critical dimension). The criteria for mean-field values of the critical exponents of Ising models are just as predicted by Wilson and Fisher[6]) and Fisher, Ma and Nickel[7]).

A key ingredient for the derivation of the above statements are few *nonlinear differential inequalities*, which are obtained for some key quantities of the models. Among the most important such relations (which it is convenient to first derive in finite volumes) are

$$\frac{\partial \chi}{\partial \beta} \leqslant |J| \chi^2, \quad \text{at } h = 0, \text{ with } |J| = \sum_x J_{0,x}, \tag{13}$$

$$\frac{\partial M(\beta, h)}{\partial \beta} \leqslant |J| M(\beta, h) \frac{\partial M(\beta, h)}{\partial h}, \tag{14}$$

satisfied both in Ising[8]) and percolation models[9]), and the following revealingly different pair[10]):

$$M(\beta, h) \leqslant h \frac{\partial M(\beta, h)}{\partial h} + M(\beta, h) \beta \frac{\partial M(\beta, h)}{\partial \beta} + M(\beta, h)^2 \quad \text{for percolation,} \tag{15}$$

$$M(\beta, h) \leqslant h \frac{\partial M(\beta, h)}{\partial h} + M(\beta, h)^2 \beta \frac{\partial M(\beta, h)}{\partial \beta} + M(\beta, h)^3 \quad \text{for Ising models.} \tag{16}$$

All the above inequalities have very suggestive explanations in the context of natural geometric representations – which for percolation need almost no introduction and for Ising models may be found in ref. 11. (Alas, the proofs are not entirely free from a certain amount of technicalities.) It may be interesting to know that these representations contain diagrammatic hints of relations of Ising model with a φ^4-field theory and of percolation with a φ^3-field theory, which are particularly evident in the last two inequalities! As is well known, such relations have been postulated, and exploited, in earlier nonrigorous – and some rigorous – works. Furthermore, there are some opposite bounds, which show that the factors which are discarded in some of these inequalities are dominated by terms associated with "bubble diagrams", in case of Ising models (and φ^4 fields), and "triangle diagrams" for percolation.

The theorems cited above include some very recent as well as some much older results. I leave it as an exercise to the reader to find how the inequalities (13)–(16) (augmented by $\partial M(\beta, h)/\partial h \leqslant M(\beta, h)/h$, which is valid for both models) imply the essential statement in part i) of the theorems cited above (which was derived in collaborations with D. Basky and R. Fernandez[10]). Of the critical exponent bounds in part ii), the one on γ is the oldest, having been derived by Glimm and Jaffe[12]) from (13). The simplest route to the other critical exponents in parts ii) passes through inequalities like $M(\beta, h) \leqslant h\partial M(\beta, h)/\partial h + |J|M(\beta, h)^3/h + M(\beta, h)^2$ (Ising) and $M(\beta, 0+) \equiv M^* \leqslant M^*\beta\partial M^*/\partial\beta + M^{*2}$ (for percolation, where $M^* = P_\infty$), which are obtained after the elimination of one of the derivatives in (15) and (16). These particular inequalities were first derived directly[13]) by Fröhlich and Sokal, and, correspondingly, Chayes and Chayes, who applied them for the general bound on δ in Ising models and $\hat{\beta}$ in percolation. Another route to the bounds on $\hat{\beta}$ and $\delta[a, n]$ also provides some interesting information on the critical temperature and its finite volume approximants[14]). The last mentioned work was based on one of the opposite bounds to the inequalities listed above. Such bounds, and other related applications can be found in refs. 10, 15 and 16.

4. Remarks on the method

Although results like those presented above apply mostly to models with some special properties, this collection includes a number of the principal examples studied in statistical mechanics and field theory. Furthermore, some generally useful principles have emerged, of which I would like to mention here the following.

–Differential inequalities involving few relevant parameters can be quite revealing, even though we are dealing with systems characterized by infinitely many degrees of freedom. The reason for that is perhaps best explained by Wilson's renormalization group picture.

–An effective way of using at least some expansions (which are generically viewed as sums with finite radii of convergence) is as tools for the extraction of relations between physical quantities, which persist into nonperturbative regimes.

–Finite volume, and other cutoffs (like the external field), may permit application of relations based on a "high-temperature" expansion also to the low-temperature regime.

–The strength of a method may be related to its weakness. Methods which give less information (like the inequalities discussed above) have a chance of providing nontrivial results about regions which offer resistance to some more robust approaches.

5. Related results

Directly related to the results mentioned in section 3 are the proof of a sharp transition in the three dimensional $\mathbb{Z}(2)$ pure gauge model, from a confining phase where the expectation value of the Wilson loop variable – describing external charges – decays by an "area law", to a phase characterized by a finite self-energy, i.e. a "perimeter law" (a joint work with A. Messager). The percolation results apply as well to oriented percolation, and thus yield also some information on the dynamical "contact process"[10]).

We have recently seen more results for which percolation related techniques were applied to ferromagnetic models. Such methods turned out to be very useful for the study of randomly dilute ferromagnets, and for the analysis of the special effects seen in the borderline case of one dimensional $1/r^2$ interactions[17]). It may be interesting to note that the results of some of the works quoted in ref. 17 were derived by means of rigorous renormalization type arguments (which are also facilitated by some geometric representations).

References

1) C.M. Newman, Percolation theory: a selective survey of rigorous results, in: Proc. SIAM Workshop on Multiphase Flow, Washington D.C. 1986, G. Papaniclaou, ed., to appear; M. Aizenman, General results in percolation theory, in: Proc. Taniguchi Symposium on Probabilistic Methods in Mathematical Physics, Katata and Kyoto, 1985, N. Ikeda, ed.
2) R. Griffiths, J. Math. Phys. **10** (1969) 1559; B. Simon and R. Griffiths, Commun. Math. Phys. **33** (1973) 145.
3) T.D. Lee and C.N. Young, Phys. Rev. **83** (1952) 410.
4) R. Peierls, Proc. Cambridge Phil. Soc. **32** (1936) 477.
5) J. Fröhlich, R. Israel, E.H. Lieb and B. Simon, Commun. Math. Phys. **62** (1978) 1.
6) K. Wilson and M.E. Fisher, Phys. Rev. Lett. **28** (1972) 240.
7) M.E. Fisher, S. Ma and B.G. Nickel, Phys. Rev. Lett. **29** (1972) 917.
8) R. Griffiths, C. Hurst and S. Sherman, J. Math. Phys. **11** (1970) 790; J.L. Lebowitz, **35** (1974) 87.
9) M. Aizenman and C.M. Newman, J. Stat. Phys. **36** (1984) 107.
10) M. Aizenman and D.J. Barsky, Sharpness of the phase transition in percolation models, submitted to Commun. Math. Phys.; M. Aizenman, D.J. Barsky and R. Fernandez, Remark on the sharpness of the phase transition in Ising type models, to be submitted to J. Stat. Phys.
11) M. Aizenman, Commun. Math. Phys. **86** (1982) 1.
12) J. Glimm and A. Jaffe, Phys. Rev. D **10** (1974) 536.
13) J. Fröhlich and A.D. Sokal, in preparation; J.T. Chayes and L. Chayes, Phys. Rev. Lett. **35** (1975) 327.
14) M. Aizenman, Phys. Rev. Lett. (1985).
15) J. Fröhlich, Nucl. Phys. B **200** (1982) 281.
16) M. Aizenman and R. Graham, Nucl. Phys. B **225** (1983) 261; D.C. Brydges, J. Fröhlich and A.D. Sokal, Commun. Math. Phys. **91** (1983) 141; M. Aizenman and R. Fernandez, J. Stat. Phys. **44** (1986) 393; R. Fernandez, J. Fröhlich and A.D. Sokal, in preparation.
17) M. Aizenman, J.T. Chayes, L. Chayes and C.M. Newman, Discontinuity of the order parameter in one dimensional $1/|x-y|^2$ Ising and Potts models, in preparation.

Physica **140A** (1986) 232–239
North-Holland, Amsterdam

MICROSCOPIC ORIGIN OF HYDRODYNAMIC EQUATIONS: DERIVATION AND CONSEQUENCES*

J.L. LEBOWITZ

Departments of Mathematics and Physics, Rutgers University,
New Brunswick, NJ 08903, USA

We describe some recent progress in deriving autonomous hydrodynamic type equations for macroscopic variables from model stochastic microscopic dynamics of particles on a lattice. The derivations also yield the microscopic fluctuations about the deterministic macroscopic evolution. These grow, with time, to become infinite when the deterministic solution is unstable. A form of microscopic pattern selection is also found.

1. Introduction

It is no secret that there does not exist at present anything resembling a rigorous derivation of the hydrodynamic equations governing the time evolution of macroscopic variables from the laws governing the dynamics of their microscopic constituents. These are generally autonomous equations of the form

$$\frac{\partial \boldsymbol{M}(\boldsymbol{r}, t)}{\partial t} = \boldsymbol{F}(\boldsymbol{M}, \nabla \boldsymbol{M}, \ldots), \tag{1.1}$$

where $\boldsymbol{M}(\boldsymbol{r}, t)$ denotes a "full set" of macroscopic variables such as the particle, momentum and energy, density of a fluid. $\boldsymbol{F}$ depends on the type of problems considered (and contains parameters specific to the system considered). Examples include the Euler and Navier–Stokes equations for fluids and diffusion-reaction type equations for chemically reacting mixtures[1]).

The essential mathematical element involved in the transition from microscopic to macroscopic evolution equations involves the suitable rescaling of space and time. By such rescalings one can take account in a precise way of the central fact that there are a very large number of atoms in a macroscopic drop of fluid each undergoing a very large number of "changes" (e.g., collisions) in a macroscopic instant "dt". The macroscopic view is therefore a "blurred" one: summing over a large number of elementary events. This brings in the "law of large

*Supported by NSF Grant DMR81-14726-02.

0378-4371/86/$03.50

numbers" which is crucial for obtaining deterministic autonomous macroscopic equations, like (1.1), *not just for averages*, but for the *almost sure value* of quantities which fluctuate on the microscopic scale. Controlling the fluctuations is clearly important if the deterministic equations are to describe what is actually observed in an experiment.

The question of fluctuations is unfortunately often overlooked in discussions of this subject. This may lead then to the erroneous conclusion that the origin of the macroscopic hydrodynamical equations is directly attributable to the highly irregular behavior of trajectories of nonlinear dynamical systems. This behavior, which occurs already in systems with a *few* degree of a freedom, does not however have any intrinsic mechanism for suppression of fluctuations. The time evolution of such a nonlinear system consisting of a few particles will therefore not exhibit the kind of deterministic behavior associated with macroscopic variables discussed here; cf. van Kampen's book[2]).

The mathematical difficulties encountered in carrying out the derivations from realistic microscopic dynamics are enormous – possibly insurmountable. It is only in very special cases, when the dynamics is essentially trivial (hard rods in one dimension, long range forces, harmonic crystals), that hydrodynamical type equations can be derived for nonequilibrium systems evolving via Hamiltonian dynamics. Things are a little better when we consider quasi-equilibrium situations, such as the time evolution of the relative concentrations in a fluid mixture whose components are mechanically identical, but differ in their "color", e.g., some chemical isomers or spin up and down ^{3}He atoms: the system, ignoring color, is in true thermal equilibrium. Unfortunately, even for these intrinsically linear nonequilibrium phenomena, the problem is too difficult to solve completely. All we can do in general is reduce the problem to that of the behavior of "test particles" in an equilibrium fluid[3]).

2. Lattice systems with stochastic dynamics

The main progress which has been made in recent years concerns systems whose microscopic dynamics involve some stochastic elements, e.g., particles (or spins) on a lattice with stochastic dynamics. What I shall describe now is joint work with Anna de Masi and Pablo Ferrari[4]) and some very recent extensions of it[5,6]).

In this work we study interacting particle (spin) systems on a lattice under the combined influence of general Glauber (spin flip) and simple exclusion (spin exchange) dynamics. We prove that when the conserving exchanges occur on a microscopic time scale, with rate ε^{-2}, $\varepsilon \ll 1$, while the Glauber rate is of order unity the magnetization m looked at on a macroscopic spatial scale of ε^{-1}, evolves according to an autonomous nonlinear diffusion-reaction equation,

$$\frac{\partial m(\boldsymbol{r}, t)}{\partial t} = \nabla^2 m + F(m), \qquad \boldsymbol{r} \in \mathbb{R}^d, \tag{2.1}$$

where $F(m)$ is a polynomial which can be "adjusted" by choosing suitable Glauber rates. Note that while m is no longer conserved on the macroscopic scale, it is conserved by the "fast" exchanges and so still serves as a good macroscopic variable.

The proof involves showing that the microscopic state is in fact close to "local equilibrium". The deviations from local equilibrium produce fluctuations about the deterministic solutions of (2.1). These fluctuations form a Gaussian field with an amplitude $\varepsilon^{d/2}$ around stable solutions of (2.1), but grow (exponentially or like some power of t) around unstable solutions.

We believe that despite their very special (and arbitrary) microscopic dynamics these models capture the essential features of the mechanism leading to hydrodynamical laws in real systems. They may also describe, at least qualitatively, the microscopic influence on fluctuations and on pattern selection, cf. examples in section 4.

3. Microscopic formulation

Our system is a simple cubic lattice in d-dimensions, at each site of which there is a spin $\sigma(\boldsymbol{x}) = \pm 1$. The configuration of the lattice $\boldsymbol{\sigma} = \{\sigma(\boldsymbol{x}), \boldsymbol{x} \in \mathbb{Z}^d\}$, changes with time via two mechanisms; a Glauber dynamics in which a spin flips at a site $\boldsymbol{x}$, with a rate $c(\boldsymbol{x}; \boldsymbol{\sigma})$ and a Kawasaki dynamics in which unequal spins at neighboring sites exchange, with a rate ε^{-2}. In particle language, $\sigma(x) = \pm 1$ denotes empty and occupied sites. Glauber dynamics then corresponds to creation or annihilation of particles while exchanges correspond to particles jumping to empty sites.

There are very few restrictions on $c(\boldsymbol{x}; \boldsymbol{\sigma})$. In particular Glauber dynamics can, but need not, be chosen to satisfy detailed balance for the equilibrium state of an Ising model with specified interactions at some reciprocal temperature β. The exchange process on the other hand, being independent of the spin configuration on neighboring sites, acts *as if* the system were at an infinite temperature, $\beta = 0$. It is very fast compared to the flip rate taking place on a macroscopic time scale.

We now define the macroscopic magnetization density by rescaling space by ε^{-1}. Let $\Lambda_{\boldsymbol{r}}^{\varepsilon}$ be a cubical box with sides of length ε^{-1} centered on $\boldsymbol{r} \in \mathbb{R}^d$. The magnetization density on this scale is

$$m^{\varepsilon}(\boldsymbol{r}, t; \boldsymbol{\sigma}) = \varepsilon^{d} \sum_{x \in \Lambda_{r}^{\varepsilon}} \sigma(x; t), \tag{3.1}$$

where $\sigma(\boldsymbol{x}; t)$ is the value of the (random) spin variable at site $\boldsymbol{x}$ at time t. $m^{\varepsilon}(\boldsymbol{r}, t; \cdot)$ is thus a random variable whose probability distribution depends on the initial state of the system. We shall assume the latter to have good cluster properties and that as $\varepsilon \to 0$,

$$\varepsilon^{-d}|\langle\sigma(\boldsymbol{x})\rangle_{\varepsilon} - m_{0}(\varepsilon \boldsymbol{x})| \to 0, \qquad \boldsymbol{x} \in \mathbb{Z}^{d},$$

where $m_0(\boldsymbol{r})$ is a smooth function of $\boldsymbol{r}$, $|m_0(\boldsymbol{r})| \leqslant 1$, $\boldsymbol{r} \in \mathbb{R}^{d}$.

Theorem. In the limit $\varepsilon \to 0$, $m^{\varepsilon}(\boldsymbol{r}, t; \boldsymbol{\sigma}) \to \int_{\Lambda_r} m(r', t)\,\mathrm{d}\boldsymbol{r}'$, where the integration is over the unit cube centered at $\boldsymbol{r}$ and $m(\boldsymbol{r}, t)$ is a *deterministic* (non-fluctuating) function of $\boldsymbol{r}$ and t satisfying eq. (2.1) with initial condition $m(\boldsymbol{r}, 0) = m_0(\boldsymbol{r})$ and $F(m) = -2\langle\sigma(\boldsymbol{x})c(\boldsymbol{x}; \boldsymbol{\sigma})\rangle_m$ – the average being taken with respect to the Bernoulli product measure, in which $\langle\sigma(\boldsymbol{x})\rangle = m$ for all $\boldsymbol{x}$. This measure is the same as the local equilibrium state for a system at infinite temperature with macroscopic magnetization $m(\boldsymbol{r}, t)$.

To see the microscopic fluctuations in the magnetization – corresponding to the deviations of the probability distribution at time t from a product measure – we must magnify them in an appropriate way.

Let

$$\phi^{\varepsilon}(\boldsymbol{r}, t; \boldsymbol{\sigma}) = \varepsilon^{-d/2}\left[m^{\varepsilon}(\boldsymbol{r}, t; \boldsymbol{\sigma}) - \int_{\Lambda_r} m(\boldsymbol{r}', t)\,\mathrm{d}\boldsymbol{r}'\right],$$

then $\phi^{\varepsilon}(\boldsymbol{r}, t; \cdot) = \int_{\Lambda_r}\phi(\boldsymbol{r}', t)\,\mathrm{d}\boldsymbol{r}'$; $\phi(\boldsymbol{r}, t)$ a random Gaussian field satisfying an Ornstein–Uhlenbeck type stochastic equation. The equal time correlations of the fluctuation field ϕ, $c(\boldsymbol{r}, \boldsymbol{r}'; t)$, satisfy a linear inhomogeneous equation whose coefficients depend on the solution of (2.1).

4. Examples

We shall illustrate the above analysis with two one dimensional examples, i.e., choices of Glauber rates $c(x; \boldsymbol{\sigma})$, x a site on the 1-d lattice. The first example will exhibit the effect of microscopic fluctuations about an unstable solution while the second will show a microscopic pattern selection. We expect this behavior to be typical also for higher dimensional cases:

Case I.

$$c(x;\boldsymbol{\sigma}) = 1 - \gamma\sigma(x)[\sigma(x+1)+\sigma(x-1)] + \gamma^2\sigma(x+1)\sigma(x-1). \tag{4.1}$$

For γ between -1 and 1 this Glauber rate satisfies detailed balance for a one dimensional Ising model with n.n. interaction J, $\gamma = \tanh \beta J$. Calling q the coordinate along the line, eq. (2.1) now takes the form

$$\frac{\partial m(q,t)}{\partial t} = \frac{\partial^2 m}{\partial q^2} - \frac{\mathrm{d}}{\mathrm{d}m}\left[\frac{\gamma^2}{2}m^4 - (2\gamma-1)m^2\right]. \tag{4.2}$$

Eq. (4.2) has a unique solution for any initial magnetization $m_0(q)$ in the interval $[-1,1]$. In particular it has the time independent spatially homogeneous solution $m(q,t) = m_0(q) = 0$. This solution is stable for $\gamma < \frac{1}{2}$, i.e., when J is antiferromagnetic ($J < 0$) or when $0 \leqslant \gamma \leqslant \gamma_c = \frac{1}{2}$ (a "mean field" critical temperature). It becomes unstable at $\gamma = \gamma_c$ and for $\gamma > \gamma_c$ there exist two stable uniform solutions with spontaneous magnetizations $\pm\gamma^{-1}(1-\gamma/\gamma_c)^{1/2}$. The equal time covariance of the fluctuation field $c(q,q';t)$ for the $m_0(q) = 0$ solution has the form[4])

$$\begin{aligned} c(q,q';t) &- \left[1 - m^2(q,t)\right]\delta(q-q') \\ &= 8\gamma\int_0^t \mathrm{d}s\,\frac{\exp\left\{\left[-(q-q')^2/8s\right] - 4(1-\gamma/\gamma_c)\,s\right\}}{\sqrt{8\pi s}} \\ &\xrightarrow[\substack{t\to\infty\\ \gamma<\gamma_c}]{} \frac{\gamma}{(\gamma_c-\gamma)^{1/2}}\exp\left[-2(\gamma_c-\gamma)|q-q'|\right]. \end{aligned} \tag{4.3}$$

For $\gamma > \gamma_c$ the fluctuations grow exponentially with time around this unstable solution while for $\gamma = \gamma_c$ they grow like $\sqrt{t}$. To understand the long time behavior of this systems for $\gamma \geqslant \gamma_c$ one has to go beyond linear Gaussian fluctuations. This has been done by De Masi, Presutti and Vares[5]). They prove that at long times, before taking the hydrodynamical limit, the evolution departs from that predicted by (4.2), i.e., $m(q,t) = 0$. Instead the microscopic state becomes a nontrivial mixture of states with different magnetizations.

Case II.

$$c(x;\sigma) = \tfrac{1}{8}[1-\sigma(x)][2+\sigma(x-1)+\sigma(x+1)]. \tag{4.4}$$

This corresponds in particle language to a particle creating a new particle (giving birth) on an adjacent empty site with a rate $\frac{1}{2}$ (compared to the rate of jumping to such a site ε^{-2}). It can also be interpreted in terms of a one dimensional cross section of a flame front with convection[7]); $\sigma(x) = 1$ or -1 represents a burned or unburned cell (or region). Equation (1.1) expressed in terms of the macroscopic density $p(q,t) = \frac{1}{2}[1+m(q,t)] \in [0,1]$ has the form

$$\frac{\partial p(q,t)}{\partial t} = \frac{\partial^2 p}{\partial q^2} + p(1-p). \tag{4.5}$$

This is a classical diffusion-reaction (D.R.) equation studied by Kolmogarov, Petrovsky and Piscounov (KPP) and others[1]).

An important feature of this type of D.R. equation is that it admits travelling front solutions: $p(q,t) = u_c(q-ct)$, where $u_c(q)$ satisfies the equation

$$\begin{gathered} u_c''(q) + cu_c'(q) + u_c(1-u_c) = 0, \\ u_c(q) \to 1 \quad \text{as } q \to -\infty, \qquad u_c(q) \to 0 \quad \text{as } q \to \infty, \quad \text{for } c > 0, \end{gathered} \tag{4.6}$$

which has solutions for all speeds $|c| \geqslant c^*$. The marginal speed c^* which equals 2 for our case is singled out in the sense that for all initial conditions such that $\rho_0(q) \to 1$ as $q \to -\infty$, $\rho_0(q) = 0$ as $q \to \infty$ converges as $t \to \infty$ to $u_{c^*}(q - \lambda(t))$ for appropriate $\lambda(t)$ where $\lambda(t)/t \to c^*$ [1]).

This "selection principle" was investigated from a physical point of view by Langer and coworkers[8]). Their interest in this problem stems from a desire to understand pattern selection principles for equations describing dendritic growth of a solid front moving into a melt.

We show in ref. 6 that as seen from the rightmost particle, our microscopic model has precisely one invariant distribution and that the average velocity, $V(\varepsilon)$, of this particle satisfies $\varepsilon V(\varepsilon) \to 2$, as $\varepsilon \to 0$.

The significance of this result for other pattern selection problems is not clear to us at the present time. We are currently studying the fluctuations about these travelling solutions.

5. Discussion

Let me discuss now briefly the case where the dynamics is exclusively of the Kawasaki type. When the jump rates are spatially symmetric the resulting hydrodynamical equation for the macroscopic density $\rho(q,t)$ is just (2.1) with $F = 0$, i.e., a linear diffusion equation. When the rates are asymmetric, then in one dimension with $p > \frac{1}{2}$ the probability of right jumps, one obtains in the now appropriate scaling limit, in which both time and space scale like ε^{-1}, the inviscid Burger's equation[9])

$$\frac{\partial \rho(q,t)}{\partial t} = (1-2p)\frac{\partial}{\partial q}[\rho(1-\rho)]. \tag{5.1}$$

This is the analogue of the Euler equations for fluids. To obtain also the viscosity term $\sim \partial^2\rho/\partial q^2$ on the right side of (5.1) (analogous to the Navier–Stokes equations) one needs to make the asymmetry $p - \frac{1}{2}$ small, of order ε and go to the scaling of time like ε^{-2} appropriate for dissipation[9]).

This difference in scalings needed for the nondissipative and dissipative cases occurs also in the real fluid equations. Without the asymmetry parameter $p - \frac{1}{2}$ to play with we can only expect to get the Euler equations as scaling limits. The Navier–Stokes equations should then be seen as the leading correction in some asymptotic series. This situation arises also in recent work using cellular automata dynamics (which can be thought of as a deterministic limit of stochastic particle dynamics, or vice versa) to solve the Navier–Stokes equations[10]). It also occurs if one starts with the Boltzmann equation rather than with a microscopic particle system. The Boltzmann equation is itself of form (1.1) describing the autonomous evolution of quasi-macroscopic variables; instead of $\boldsymbol{M}(\boldsymbol{r},t)$ we have $f(\boldsymbol{r},\boldsymbol{v},t)$, $\boldsymbol{v}$ a velocity variable, as our basic set. Once the Boltzmann equation is accepted the derivation from it of hydrodynamical equations for the appropriate set of $\boldsymbol{M}(\boldsymbol{r},t)$ proceeds, with some measure of success, via Chapman–Enskog type expansion methods. These yield the Euler equations in the leading order, which is in fact the scaling limit mentioned earlier. The Navier–Stokes, Burnett and other higher order equations are then expected to form some kind of asymptotic series in the gradients.

Acknowledgements

I would like to take this opportunity to thank Errico Presutti and Herbert Spohn for their generous advice and technical help in the course of the development of these ideas.

References

1) J. Smoller, Shock Waves and Reaction-Diffusion Equations (Springer, New York, 1983) and references therein; D. Aronson and H. Weinberger, in: Nonlinear Diffusion in Population Genetics, Combustion, and Nerve Propagation, Springer Lecture Notes in Math. **446** (1975) 5–49.

2) N.G. van Kampen, Stochastic Processes in Physics and Chemistry (North-Holland, Amsterdam, 1981).
3) J.L. Lebowitz and H. Spohn, Microscopic basis for Fick's law for self-diffusion, J. Stat. Phys. **28** (1982) 539.
4) A. De Masi, P.A. Ferrari and J.L. Lebowitz, Rigorous derivation of reaction–diffusion equations with fluctuations, Phys. Rev. Lett. **55** (1985) 1947; Reaction–diffusion equations for interacting particle systems, J. Stat. Phys. **44** (1986) 589.
5) A. De Masi, E. Presutti and M.E. Vares, Escape from the unstable equilibrium in a random process with infinitely many interacting particles, J. Stat. Phys. **44** (1986) 645.
6) M. Bramson, P. Calderoni, A. De Masi, P.A. Ferrari, J. L. Lebowitz and R.H. Schonmann, Microscopic selection principle for a diffusion–reaction equation, J. Stat. Phys. (1986) Dec.
7) A.R. Kerstein, Computational study of propagating fronts in a lattice gas model, J. Stat. Phys. (1986) Dec.
8) E. Ben Jacob, H. Brand, G. Dee, L. Kramer and J.S. Langer, Pattern propagation in nonlinear dissipative systems, Physica **14D** (1985) 348.
9) A. De Masi, E. Presutti and E. Scacciatelli, The weakly asymmetric simple exclusion process, preprint, Rome University (1986).
10) U. Frisch, B. Hasslacher and Y. Pomeau, Lattice gas automata for the Navier–Stokes equation, Phys. Rev. Lett. **56** (1986) 1505; S. Wolfram, Cellular automaton fluids 1: basic theory, J. Stat. Phys. (1986) Nov.

Physica **140A** (1986) 240–250
North-Holland, Amsterdam

A MODEL FOR CRYSTALLIZATION: A VARIATION ON THE HUBBARD MODEL

Elliott H. LIEB

Departments of Mathematics and Physics, Princeton University, Jadwin Hall, P.O. Box 708, Princeton, NJ 08544, USA

A quantum mechanical lattice model of fermionic electrons interacting with infinitely massive nuclei is considered. (It can be viewed as a modified Hubbard model in which the spin-up electrons are not allowed to hop.) The electron–nucleus potential is "on-site" only. Neither this potential alone nor the kinetic energy alone can produce long range order. Thus, if long range order exists in this model, it must come from an exchange mechanism. N, the electron plus nucleus number, is taken to be less than or equal to the number of lattice sites. We prove the following: (i) For all dimensions, d, the ground state has long range order; in fact it is a perfect crystal with spacing $\sqrt{2}$ times the lattice spacing. A gap in the ground state energy always exists at the half-filled band point (N = number of lattices sites). (ii) For small, positive temperature, T, the ordering persists when $d \geq 2$. If T is large there is no long range order and there is exponential clustering of all correlation functions.

1. Introduction

This lecture concerns joint work with Kennedy[4]), first announced in ref. 10. It also contains some new work on the Peierls instability, not previously reported.

Much attention has been paid to the question of proving the existence of long range order in model statistical mechanical systems in which the basic atomic constituents interact with short range forces. An important example is a lattice spin system in which the spin at each site represents the localized spin of an atom located at that site and where the short range, pairwise interaction (Ising or Heisenberg) reputedly comes from an interatomic exchange energy. Another problem – so far unsolved – is the existence of periodic crystals which are supposed to come from short range (e.g., Lennard–Jones) interatomic potentials.

In the real world, however, these interactions are not given a priori; it is ultimately itinerant electrons and their correlations that give rise to the long range ordering. In other words, a deep unsolved problem is to derive magnetism or crystallization from the Schrödinger equation – or some caricature of it. The construction of a simple model based on itinerant electrons, and the rigorous derivation of ordering from it, is a challenge for mathematical physicists.

A lattice model of itinerant electrons that is believed to display ferro and antiferromagnetism – if it could be solved – is the Gutzwiller–Hubbard–

Kanamori model[1-3]). We are also unable to solve it, but we have succeeded in proving that a simplified version of it does display crystallization. It is a toy model but it is, to our knowledge, the first example of this genre. Roughly, it has the same relation to the Hubbard model as the Ising model has to the quantum Heisenberg model. Here we shall give a brief report of our results, the full details of which will appear elsewhere[4]).

The Hubbard model, which is the motivation for our model, is defined by the second-quantized Hamiltonian

$$H^{\mathrm{H}} = \sum_{\sigma} \sum_{x,\, y \in \Lambda} t_{xy} c^{\dagger}_{x\sigma} c_{y\sigma} + 2U \sum_{x \in \Lambda} n_{x\uparrow} n_{x\downarrow}, \tag{1}$$

with the following notation: $\sigma = \pm 1$ denotes the 2 spin states of the electrons; Λ is a finite lattice; $c_{x\sigma}$ is a fermion annihilation operator for a spin σ electron at $x \in \Lambda$; $n_{x\sigma} = c^{\dagger}_{x\sigma} c_{x\sigma}$ is the number operator for spin σ at x. Electrons interact only at the same site with an energy $2U$, and $t_{xy} = t_{yx}$ is the hopping amplitude from x to y.

The crucial assumption will be made that Λ is the union of two sublattices $A \cup B$ such that $t_{xy} = 0$ unless $x \in A$, $y \in B$ or $x \in B$, $y \in A$. The number of sites in Λ, A and B are denoted by $|\Lambda|$, $|A|$ and $|B|$. The two sublattices need not be isomorphic. Thus, for example, a face-centered cubic lattice is allowed with A = face centers and B = cube corners. Λ is said to be *connected* if every $x, y \in \Lambda$ can be joined by a "path" through nonzero t's.

In our model we assume that one kind of electron (say $\sigma = -1$) does not hop. One can say that these electrons are infinitely massive. The Hamiltonian is then

$$H = \sum_{x,\, y \in \Lambda} t_{xy} c^{\dagger}_{x} c_{y} + 2U \sum_{x \in \Lambda} n_x W(x) \tag{2}$$

with $n_x = c^{\dagger}_x c_x$ (the subscript σ is omitted since the dynamic electrons have $\sigma = +1$) and with $W(x) = +1$ if a fixed electron ($\sigma = -1$) is at x and $W(x) = 0$ otherwise. It will be recognized immediately that (2) is just a fancy way to say that the moveable electrons are independent, with a single particle Hamiltonian

$$\tilde{h} = T + V, \tag{3}$$

with T being the $|\Lambda|$-square matrix t_{xy} and $V_{xy} = 2UW(x)\delta_{xy}$. It is convenient to write $\tilde{h} = h + U$ with

$$h = T + US \tag{4}$$

with $S_{xy} = s_x \delta_{xy}$, and $s_x = 1$ (resp. -1) if x is occupied (resp. unoccupied). The $\{s_x\}$ are like Ising spins.

We shall henceforth call the moveable particles "electrons" and the fixed particles "nuclei". This terminology is most appropriate if $U < 0$, for then H does represent a lattice system of electrons and nuclei in which all Coulomb interactions except the on-site electron–nucleus attraction and the on-site infinite nuclear repulsion are regarded as "screened out". This conforms with the spirit of the original Hubbard model. The electron number, the nuclear number and the total particle number, all of which commute with H, are, respectively,

$$N_{\mathrm{e}} = \sum_{x \in \Lambda} n_x, \qquad N_{\mathrm{n}} = \tfrac{1}{2} \sum_{x \in \Lambda} [s_x + 1] = \sum_{x \in \Lambda} W(x), \qquad \mathcal{N} = N_{\mathrm{e}} + N_{\mathrm{n}}. \quad (5)$$

It is to be emphasized that W does *not* represent a disordered potential. We take the "annealed", not the "quenched" system. The ground state for fixed N_{e} and N_{n} is defined by taking the ground state of H (with respect to the electrons) for each W and then minimizing the result with respect to the location of the nuclei. The ground state energy will be denoted by $E(N_{\mathrm{e}}, N_{\mathrm{n}})$. Likewise, for positive temperature, we take $\mathrm{Tr}\, \mathrm{e}^{-\beta \mathrm{H}}$ with respect to *both* the electron variables and the nuclear locations.

Since $t_{xy} = 0$ for x, y on the same sublattice, the spectra of H and H^{H} are invariant under $t_{xy} \to -t_{xy}$ (all x, y). There is also a hole–particle symmetry. If $c_x^\dagger \to c_x$, $c_x \to c_x^\dagger$, $n_x \to 1 - n_x$ and $t_{xy} \to -t_{xy}$, then $H(U) \to H(-U) + 2UN_{\mathrm{n}}$. If $W(x) \to 1 - W(x)$, then $H(U) \to H(-U) + 2UN_{\mathrm{e}}$. A similar symmetry holds for H^{H}. Thus, the $U > 0$ and $U < 0$ cases are similar – from the mathematical point of view.

Our results are of two kinds. The first concerns the ground state which we prove always has perfect crystalline ordering and an energy gap (defined later). The second concerns the positive temperature ($1/kT = \beta < \infty$) grand canonical state. For large β and dimension $d \geq 2$, the long range order persists. For small β it disappears and there is exponential clustering of the nuclear correlation functions.

2. The ground state

Theorem 1. (*a*) *Let $U < 0$. Under the condition $\mathcal{N} \equiv N_{\mathrm{e}} + N_{\mathrm{n}} \leq 2|A|$, the ground state (i.e., we minimize $E(N_{\mathrm{e}}, N_{\mathrm{n}})$ over the set $N_{\mathrm{e}} + N_{\mathrm{n}} \leq 2|A|$) occurs for $N_{\mathrm{e}} = N_{\mathrm{n}} = |A|$ and a minimizing nuclear configuration is $W(x) = W_A(x) \equiv 1 (x \in A), 0 (x \notin A)$. Under the condition $\mathcal{N} \leq 2|B|$, the ground state is $N_{\mathrm{e}} = N_{\mathrm{n}} = |B|$ and the B sublattice is occupied ($W = W_B$). If Λ is connected, these are the only ground states, i.e., if $|A| > |B|$ the ground state is unique; if $|A| = |B|$ it is* *doubly degenerate. No assumption is made about the sign or magnitude or periodicity of the* t_{xy} *other than* $t_{xy} = 0$ *for* $x, y \in A$ *or* $x, y \in B$.

(*b*) *Let* $U > 0$. *Under the condition* $\mathcal{N} \geq |A| + |B|$, *there are two ground states*: $N_e = |A|$, $N_n = |B|$, $W = W_B$ *and* $N_e = |B|$, $N_n = |A|$, $W = W_A$. *If* Λ *is connected, these are the only ground states.*

The condition $\mathcal{N} = |\Lambda|$ is called the *half-filled* band. If $|A| = |B| = |\Lambda|/2$, the crystal occurs at the half-filled band. If Λ is a cubic lattice, for example, this means that the ground state is a cubic lattice of period $\sqrt{2}$ oriented at 45° with respect to Λ.

Theorem 1 relies heavily on the fact that the electrons are fermions. The ground state would be completely different if they were bosons. For bosons and for Λ a cubic lattice, the nuclei would all be clumped together in the ground state instead of being spread out into a crystal. By using rearrangement inequalities it is possible to describe this clumping quantitatively.

Next, we define the *energy gap*. Actually two different definitions are of interest. First, let

$$E(\mathcal{N}) \equiv \min\{E(N_e, N_n) | N_e + N_n = \mathcal{N}\}. \tag{6}$$

The *chemical potential* is defined by

$$\mu(\mathcal{N}) \equiv E(\mathcal{N}+1) - E(\mathcal{N}). \tag{7}$$

We say there is a *gap of the first kind* at $\mathcal{N}$ if

$$\mu(\mathcal{N}) - \mu(\mathcal{N}-1) \geq \varepsilon_1 > 0, \tag{8}$$

with ε_1 being independent of the size of the system. We say there is a *gap of the second kind* at N_e, N_n if

$$E(N_e + 1, N_n) + E(N_e - 1, N_n) - 2E(N_e, N_n) \geq \varepsilon_2 > 0. \tag{9}$$

In other words, the nuclear number is fixed in the second definition.

A gap is one indication that the system is an insulator, for it implies that it costs more energy to put a particle into the system than is gained by removing one. The first kind of gap is relevant if one views our model as an approximation to the Hubbard model; the second is relevant from the "electrons and nuclei" point of view.

Theorem 2. Assume that Λ is not only connected but that every $x, y \in \Lambda$ can be connected by a chain with $|t_{ab}| \geq \delta$ for every a, b on the chain. Also, assume that $\|T\| \leq \tau$. The following energy gaps exist with $\varepsilon_2 \geq \varepsilon_1 > 0$ and depending only on δ, τ and U:

$U < 0$: First kind at $\mathcal{N} = 2|A|$ and at $\mathcal{N} = 2|B|$. Second kind at $N_e = |A|$, $N_n = |A|$ and at $N_e = |B|$, $N_n = |B|$.

$U > 0$: First kind at $\mathcal{N} = |A| + |B|$. Second kind at $N_e = |A|$, $N_n = |B|$ and at $N_e = |B|$, $N_n = |A|$.

In order to give the flavor of our methods, the proof of Theorem 1 will be given here. The proof of Theorem 2 is more complicated.

Proof of theorem 1. Let $\lambda_1 \leq \lambda_2 \leq \cdots$ be the eigenvalues of h in (4). They depend on the nuclei. For N_e electrons the ground state energy, E, of H satisfies

$$E - UN_e = \sum_{j=1}^{N_e} \lambda_j \geq \sum_{\lambda_j < 0} \lambda_j = \tfrac{1}{2}[\operatorname{Tr} h - \operatorname{Tr}|h|]. \tag{10}$$

But $\operatorname{Tr} h = U\Sigma s_x = 2UN_n - U|\Lambda|$ and $|h| = \{T^2 + U^2 + UJ\}^{1/2}$ with $J_{xy} = t_{xy}[s_x + s_y]$. Since the function $0 < x \to x^{1/2}$ is concave, $f(y) = \operatorname{Tr}\{T^2 + U^2 + yUJ\}^{1/2}$ is concave in $y \in [-1, 1]$. But $f(-1) = f(1)$ (since $\operatorname{spec}(h)$ is invariant under $T \to -T$), so $f(1) \leq f(0)$, with equality if and only if $J \equiv 0$. Thus

$$E \geq U\mathcal{N} - \tfrac{1}{2}U|\Lambda| - \tfrac{1}{2}\operatorname{Tr}(T^2 + U^2)^{1/2}. \tag{11}$$

If Λ is connected, the only ways to have $J \equiv 0$ are either $W = W_A$ or W_B. Consider $U < 0$ and $\mathcal{N} \leq 2|A|$, whence, from (11),

$$E \geq U\left(2|A| - \tfrac{1}{2}|\Lambda|\right) - \tfrac{1}{2}\operatorname{Tr}(T^2 + U^2)^{1/2}. \tag{12}$$

If $W = W_A$ then, as is easily seen, h has precisely $|A|$ negative and $|B|$ positive eigenvalues. Thus, if $W = W_A$ and $N_e = |A|$, then (12) is an equality. The other cases are similar.

3. Grand canonical ensemble

First, we define the partition function Ξ. A nuclear configuration is denoted by $S = \{s_x\}$, $s_x = \pm 1$, and the λ_j are the eigenvalues of h in (4). If μ_n, μ_e are the nuclear and electronic chemical potentials,

$$\Xi = \sum_S \exp\left[\tfrac{1}{2}\beta\mu_n\left(\sum_x s_x + |\Lambda|\right)\right] \prod_{j=1}^{|\Lambda|} \left\{1 + \exp\left[-\beta\left(\lambda_j + U - \mu_e\right)\right]\right\}. \tag{13}$$

The product in (13) is just the well known Fermi–Dirac grand canonical partition function for the electrons. We want to choose μ_e, μ_n so that $\langle N_e \rangle = \frac{1}{2}|\Lambda|$ and $\langle N_n \rangle = \frac{1}{2}|\Lambda|$, or $\langle \Sigma s_x \rangle = 0$. From the fact that if $T \to -T$, $\operatorname{spec}(h) \to \operatorname{spec}(h)$, one has that when $S \to -S$, $\operatorname{spec}(h) \to -\operatorname{spec}(h)$. It is then easy to see that the desired chemical potentials are $\mu_e = \mu_n = U$. Since $\Sigma\lambda_j = U\Sigma s_x$, (13) becomes in this case (after dropping an irrelevant factor $2^{|\Lambda|}/e^{\beta U|\Lambda|/2}$)

$$\Xi = \sum_S \exp[-\beta F(S)] \tag{14}$$

with

$$-\beta F(S) = \sum_{j=1}^{|\Lambda|} \ln\cosh\left(\tfrac{1}{2}\beta\lambda_j\right) = \operatorname{Tr}\ln\cosh\left[\tfrac{1}{2}\beta h\right]$$

$$= \operatorname{Tr}\ln\cosh\left[\tfrac{1}{2}\beta\left(T^2 + U^2 + UJ\right)^{1/2}\right]. \tag{15}$$

Thus, (14) is like an Ising model partition function but with a complicated, temperature dependent "spin–spin" interaction, $F(S)$, given by (15) in terms of the eigenvalues of h. With respect to this "spin measure" we can talk about the presence or absence of long range nuclear order in the thermodynamic limit. In order to discuss this limit we henceforth restrict ourselves to a translation invariant nearest neighbor hopping on a cubic lattice in d dimensions.

What we are able to prove is summarized in the schematic fig. 1 and in

Theorem 3. For all U and sufficiently large β there is long range order for $d \geq 2$ (the same kind as in the ground state). For all U and sufficiently small β there is none; indeed there is exponential decay of all nuclear correlation functions.

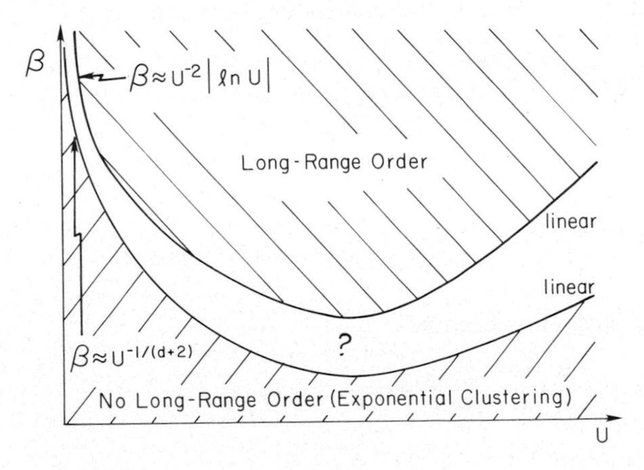

Fig. 1.

Presumably there is no intermediate phase, but we cannot prove this. For large U, β_c is clearly linear in U. For small U, we have the bound on the lower $\beta_c \sim U^{-1/(2+d)}$ and the bound on the upper $\beta_c \sim |\ln U| U^{-2}$. Our guess is that the true state of affairs is $\beta_c \sim U^{-2}$.

For large β we use a Peierls argument; for small β we use Dobrushin's uniqueness theorem. A sketch of our proof – omitting many important details – is the following. For simplicity, we here consider only large $U > 0$.

Define, for $x > 0$,

$$P(x) = \ln \cosh x^{1/2}. \tag{16}$$

[We note in passing that P is concave, and we see from the last expression in (15), using the proof in Theorem 1, that $F(S)$ has its minima at precisely the same values of W (or S) as in Theorem 1.] P is a Pick (or Herglotz) function with the representation

$$P'(x) = \tfrac{1}{2}x^{-1/2} \tanh x^{1/2} = \sum_{k=0}^{\infty} \left[\left(k + \tfrac{1}{2}\right)^2 \pi^2 + x\right]^{-1}. \tag{17}$$

We are interested in $x = \frac{1}{4}\beta^2(T^2 + U^2 + UJ)$ with $J_{xy} = t_{xy}(s_x + s_y)$.

Long range order (large β)

Choose S, and then define antiferromagnetic contours in the usual way. If γ is a connected contour component and $\Gamma + \gamma$ is the whole contour, we want to prove that $\Delta \equiv F(\Gamma + \gamma) - F(\Gamma) \geq C|\gamma|$ for a suitable constant $C = C(U)$.

Obviously, $J_{\Gamma+\gamma} = J_\Gamma + J_\gamma$. To remove γ, we change s_x to $-s_x$ inside γ. For $0 \leq t \leq 1$, define $J(t) = J_\Gamma + tJ_\gamma$. Then, assuming for simplicity that Γ lies entirely outside γ, we have, by differentiating (15) and using (17), that

$$-\beta\Delta = \tfrac{1}{4}\beta^2 U \sum_{k=0}^{\infty} \int_0^1 \mathrm{d}t \, \mathrm{Tr}\,(G_k J_\gamma), \tag{18}$$

with

$$G_k = \left[\left(k + \tfrac{1}{2}\right)^2 \pi^2 + \tfrac{1}{4}\beta^2\left(T^2 + U^2 + UJ_\Gamma + tUJ_\gamma\right)\right]^{-1}. \tag{19}$$

Integrating by parts, this becomes

$$-\beta\Delta = -(\beta^4 U^2/16) \sum_k \int_0^1 \mathrm{d}t \, (1-t) \, \mathrm{Tr}\, G_k J_\gamma G_k J_\gamma + \cdots . \tag{20}$$

[The $\cdots$ terms in (20) come from $t = 0$ in the partial integration. They are small and easily bounded for large U, but they have to be treated more judiciously when U is small.] If A, B, D are matrices with $A \geq B \geq 0$, then $\operatorname{Tr} AD^{\dagger}AD \geq \operatorname{Tr} BD^{\dagger}BD$. Also, $A \geq B > 0 \Rightarrow 0 < A^{-1} \leq B^{-1}$. Moreover, $T^2 \leq (2d)^2$ and $UJ \leq T^2 + U^2$. Using this, we have the matrix inequality

$$G_k \geq \left[\left(k + \tfrac{1}{2}\right)^2 \pi^2 + \tfrac{1}{2}\beta^2\left((2d)^2 + U^2\right)\right]^{-1}. \tag{21}$$

Summing on k,

$$\Delta \geq (\text{const.}) U^2 \left((2d)^2 + U^2\right)^{-3/2} \operatorname{Tr}\left(J_\gamma\right)^2.$$

Clearly $\operatorname{Tr}(J_\gamma)^2 = (\text{const.})|\gamma|$. Thus, for U large, $C(U) \geq (\text{const.})U^{-1}$, and thus long range order exists in $d \geq 2$ if β/U is large enough.

Absence of long range order (small β)

Dobrushin's uniqueness theorem[5-7]), together with the modification in[8]), gives the following criterion for exponential clustering. We have to bound the change in F when we change the spins at x and y (taking the worst case with respect to the other spins). Call this f_{xy}. The requirement is that for some $m > 0$ and all x,

$$\beta \sum_y f_{xy} \exp\left[m|x - y|\right] < 1. \tag{22}$$

By an argument similar to the preceding (large U)

$$\begin{aligned} f_{xy} &\cong \left(\beta^3 U^2/16\right) \sum_k \operatorname{Tr} G_k J_x G_k J_y \\ &\cong \left(\beta^3 U^2/16\right) \sum_k |G_k(x, y)|^2. \end{aligned} \tag{23}$$

Here, $J_x = \delta_x T + T\delta_x$ and $G_k(x, y)$ is the x, y matrix element of (19).

To implement (22) we now require an *upper* bound on $G_k(x, y)$ that has exponential decay. For this purpose the Combes–Thomas[9]) argument is ideal. Let Q be the matrix with elements $Q_{xy} = \delta_{xy} e^{n \cdot x}$ and with $|n| = 1$. Then

$$Q G_k^{-1} Q^{-1} = G_k^{-1} + R_k \equiv L_k. \tag{24}$$

The "remainder" R_k can be bounded for large U: $\|R_k\| \leq C\beta^2 U$ for some constant C. Similar to (21), for large U

$$G_k^{-1} \geq \left[\left(k + \tfrac{1}{2}\right)^2 \pi^2 + \tfrac{1}{8}\beta^2 U^2\right] \equiv \alpha_k. \tag{25}$$

Thus,

$$\|L_k^{-1}\| \leq \left[\alpha_k - C\beta^2 U\right]^{-1}. \tag{26}$$

Since $|(L_k^{-1})_{xy}| \leq \|L_k^{-1}\|$, we have from (24), $|(QG_kQ^{-1})_{xy}| \leq \|L_k^{-1}\|$, and thus

$$|(G_k)_{xy}| \leq \exp\left[n \cdot (y - x)\right]\left\{\alpha_k - C\beta^2 U\right\}^{-1}. \tag{27}$$

This holds for all n, no summing on k,

$$f_{xy} \leq (\text{const.})U^{-1}\mathrm{e}^{-2|x-y|}.$$

Hence (22) holds with $m = 1$ if β/U is small enough.

4. The Peierls instability

Another itinerant electron model in which translation invariance is partially broken was introduced by Peierls[11]); see also ref. 14. Consider a linear chain of $2N$ atoms (arranged in a ring) and let w_i $(i = 1, \ldots, 2N)$ denote the distance from atom i to atom $i + 1$ (with $2N + 1 \equiv 1$). If $d > 0$ is the equilibrium distance, the distortion energy associated with the w_i is taken to be

$$U(\{w\}) = \kappa \sum_{i=1}^{2N} (w_i - d)^2. \tag{28}$$

Here $\kappa > 0$ is a measure of stiffness.

Now suppose there are $2N$ (spin $\frac{1}{2}$) electrons ($\frac{1}{2}$ filled band) which can hop between nearest neighbor atoms, with an amplitude $t_i = t(w_i)$ to hop from atom i to $i + 1$ and from $i + 1$ to i. The function $t(w)$ will be explained shortly. The ground state energy of these $2N$ electrons is

$$K(\{w\}) = 2 \sum_{j=1}^{N} \lambda_j(T), \tag{29}$$

where $\lambda_1 \leq \lambda_2 \leq \cdots \leq \lambda_{2N}$ are the eigenvalues of the $2N \times 2N$ matrix T having matrix elements $T_{i,i+1} = T_{i+1,i} = t_i = t(w_i)$ and $T_{ij} = 0$ otherwise. The total Hamiltonian is

$$H(\{w\}) = K(\{w\}) + U(\{w\}) \tag{30}$$

and the problem is to determine the ground state energy

$$E = \min_{\{w\}} H(\{w\}) \tag{31}$$

and the configuration(s) $\{w\}$ that attains it.

The function $t(w)$ should be something like $a\,\mathrm{e}^{-cw}$, but here we shall adopt the choice of Su, Schrieffer and Heeger[12]):

$$t(w) = b - c(w - d) \tag{32}$$

with $b, c > 0$.

One guess for a minimizing $\{w\}$ is $w_i = W = \text{const}$. However, Peierls observed that if we take $w_i = W + (-1)^i\delta$ (a *dimerized configuration*) the energy will be lowered. The reason is that when $\{w\}$ has period two, a gap in the spectrum of T opens up at $k = \pm\pi/2$; since there are just enough electrons to fill the lower band, the K energy will be lowered by an amount (const.) $\delta^2 \ln|\delta|$ for small δ. (This comes from the fact that the perturbation theory is singular at the band edges.) On the other hand, U depends on δ like $\kappa\delta^2$, so it is favorable to have $\delta \neq 0$. This breaking of translation invariance is called the Peierls instability. The interesting question, which was open for a long time, is this: Is the translation invariance merely broken from period one to period two, or does something more complicated happen? Conceivably the minimizing $\{w\}$ could have a longer period – or perhaps no periodicity at all.

Our answer to this question is that period two always holds.

Theorem 4. For all b, c, d, κ, N there are precisely two minima for E,

$$w_i = W + (-1)^i\delta \quad \text{or} \quad w_i = W - (-1)^i\delta, \tag{33}$$

with W and δ being constants depending on κ, b, c, N. Moreover, there is a gap, i.e., a jump in the chemical potential:

$$\Delta\mu \equiv [E(2N+1) - E(2N)] - [E(2N) - E(2N-1)] > \varepsilon, \tag{34}$$

with ε depending on κ, b, c but independent of N, d. (Here $E(2N+1)$ is defined as in (30) and (31) but with K replaced by $2\Sigma_{j=1}^{N}\lambda_j(T) + \lambda_{N+1}(T)$ and, for $E(2N-1)$, K is replaced by $2\Sigma_{j=1}^{N}\lambda(T) - \lambda_N(T)$.)

As in the proof of Theorem 1, the proof of Theorem 4 uses the concavity of $\operatorname{Tr}\sqrt{X}$ with respect to X, and with $X = T^2$. The crucial observation is that for the dimerized configuration (33), T^2 is translation invariant.

It turns out that if the U in (28) is modified, an "integrable system" is obtained. In terms of t, using (32), we have

$$(w_i - d)^2 = \alpha + \beta t_i + \gamma t_i^2, \tag{35}$$

with α, β, γ constants. The modification is

$$U(\{w\}) = \alpha + \sum_{j=1}^{2N} \left[-\beta \ln t_i + \gamma t_i^2 \right], \tag{36}$$

with α, β, γ arbitrary constants (but (32) is unchanged). This system is presumably not very different from ours (but there is no proof). It was investigated in ref. 13. However, the solution in ref. 13 covers all electron numbers *except* the half-filled band ($2N$ electrons), which is the most interesting case since it is only here that one expects to see the instability manifest itself in the form proposed by Peierls – namely the partial breaking of translation invariance into a period two invariance.

Acknowledgement

The support of the U.S. National Science Foundation, grant PHY-8515288, is gratefully acknowledged.

References

1) M.C. Gutzwiller, Phys. Rev. Lett. **10** (1963) 159–162; Phys. Rev. **134** (1964) A923–941, **137** (1965) A1726–1735.
2) J. Hubbard, Proc. Roy. Soc. (London), Ser. A **276** (1963) 238–257, **277** (1964) 237–259.
3) J. Kanamori, Prog. Theor. Phys. **30** (1963) 275–289.
4) T. Kennedy and E.H. Lieb, An itinerant electron model with crystalline or magnetic long range order, Physica **138A** (1986) 320.
5) R.L. Dobrushin, Theory Probab. Appl. **13** (1968) 197–224.
6) L. Gross, Commun. Math. Phys. **68** (1979) 9–27.
7) H. Föllmer, J. Funct. Anal. **46** (1982) 387–395.
8) B. Simon, Commun. Math. Phys. **68** (1979) 183–185.
9) J.M. Combes and L. Thomas, Commun. Math. Phys. **34** (1973) 251–270.
10) Int. Conf. Mathematical Aspects of Statistical Mechanics and Field Theory, Groningen, The Netherlands, 1985, N.M. Hugenholtz and M. Winnink, eds., Springer Lecture Notes in Physics, Vol. 257 (Springer, New York, 1986).
11) R.E. Peierls, Quantum Theory of Solids (Clarendon, Oxford 1955), p. 108.
12) W.P. Su, J.R. Schrieffer and A.J. Heeger, Phys. Rev. Lett. **42** (1979) 1698–1701.
13) S.A. Brazovskii, N.E. Dzyaloshinskii and I.M. Krichever, Sov. Phys. JETP **56** (1982) 212–225.
14) H. Fröhlich, Proc. Roy. Soc. A **223** (1954) 296–305.

Physica **140A** (1986) 251–260
North-Holland, Amsterdam

SOME OPEN QUESTIONS IN THE ELASTICITY OF POLYMER NETWORKS

J. BASTIDE

Institut Charles Sadron (C.R.M.), 6, Rue Boussingault, 67083 Strasbourg Cedex and Laboratoire Léon Brillouin, CEA, Saclay, 91191 Gif-sur-Yvette Cedex, France

and

F. BOUE

Laboratoire Léon Brillouin, CEA , Saclay, 91191 Gif-sur-Yvette Cedex, France

We briefly present a description of a rubber, of the classical theories for rubber mechanical properties and of our small angle neutron scattering experiments, as a representative sample of observations at the molecular scale. We then comment about the comparison of our data with calculations from the classical models ("phantom network" and "junction affine"). Discrepancies appear, which are an opportunity to discuss the questions of defects, pendant chains and trapped entanglements and to make some remarks on the complicated connectivity of polymer networks even without chemical defects.

1. Introduction

The current position of problems in rubber elasticity can be seen in two points. First, the macroscopic properties (essentially mechanical properties) are grossly explained by a corpus of classical models[1]), which when observed in detail are sometimes difficult to test and sometimes in disagreement with these properties. Second, several microscopic techniques are in progress: again the results agree grossly with the classical theories but also show new discrepancies, even more striking. All over the world you will be told that their origin is "defects" in the network; some of these explanations are classical; some are recent[2,3]). Recent also are some different theories that are quite promising, such as those considering a network as the result of gelation process, for which some of the gelation theories still have consequences on the network structure far away from the gel point[4–6]). To be brief and not too awkward, we had to *choose* among all these points: we will only recall the spirit of the classical models and some of the macroscopic discrepancies, describe our experiment as a microscopic technique, among several others of deep interest[7]), use a few representative data to show some discrepancies, discuss only a part of the "defect" point of view and present

0378-4371/86/$03.50 © Elsevier Science Publishers B.V.
(North-Holland Physics Publishing Division)

selfishly some of our own questions, quoting only some others, maybe more valuable, as references.

2. What is a rubber?

A rubber is essentially a *liquid* of very large chains, mainly linear flexible molecules connected together in the form of a tridimensional net. In practice it is made by synthesizing some polymer chains, and linking them by their extremities with polyfunctional molecules, or bridging them by some side groups (generally several, even a lot) or creating some branching during the polymerization itself. There must be at least one infinite cluster constituting the net (to which one associates generally an average "mesh"). The chains are in the molten state and very flexible. Because of that, three quantities scale differently when going toward the macroscopic size[5]):

–the bulk modulus is the same as for a classical liquid (mainly from Van der Waals interactions);

–the enthalpic part of the deformation energy is here due mainly to the variation of angles between bonds and vanishes;

–the variation of distances between points changes the entropy of the chains and this leads to the main part of the shear modulus, which is noticeably smaller than the bulk modulus.

On the chemical point of view the main difficulty is to control the distribution and the efficiency of the major actors: the crosslinks.

Close materials are gels (rubbers swollen by a diluent) and entangled melts. In this last case the interlinking of the chains is not chemical: polymer melts of long chains exhibit rubber behaviour on time range depending on the chains' length: they can at this stage be visualized as networks of physical entanglements, and can be observed experimentally parallel to real networks.

The duality (solid/liquid) of rubbers has many applications, not only their very high deformability, which makes their importance in technology surely greater than any of us imagine.

3. A few words about classical predictions compared with mechanical data

The point here is to express the elastic force, the origin of which is entropic as evoked above. One has then to calculate the entropy for a strained net. Among many theories presented in refs. 1 and 2, we use that of Graessley[8]) for a brief presentation of their spirit. One represents a rubber as a set of crosslinks connected by Gaussian springs (representing the chains). It will be obvious to the modern reader that the crosslinks should move. However, the complete freedom for all of them is difficult to treat. A representative trick, the one of ref. 8, is to describe the rubber as an ensemble of "micronetworks" containing "mobile" junctions (○) surrounded by "fixed" junctions (×), whose positions are affinely transformed in the macroscopic strain when one deforms the network. Micronetworks of different orders look like

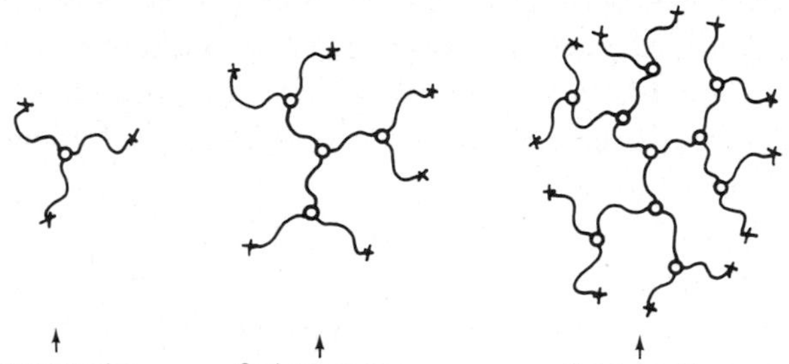

1st generation 2nd generation 3rd generation

The deformation entropy of a micronetwork, corresponding to a displacement of the fixed points, is a generalization of that of a single chain connected to two fixed points. The free energy of a net is obtained by summing over the ensemble of fixed points, and its derivative with respect to the displacement gives the elastic force f.

Trees of Gaussian chains obey linear equations as do electrical circuits. Because of this linearity, all the mean positions of the mobile junctions are also affinely transformed in the macroscopic strain. f is the sum of two terms: one proportional to the number of semimobile chains (×—○), the other to that of mobile chains (○—○). A lower limit for f – known as the "phantom network" limit and found also by the Deam–Edwards model – is obtained by reducing to nought the number of semimobile chains. Then in case of uniaxial deformation (λ ratio of final to initial length of the sample, assumed to be incompressible)

$$f/A_0 = ((\phi - 2)/\phi)\nu kT(\lambda - 1/\lambda^2), \quad (1)$$

where f is the force, A_0 the area of the unstrained sample, ν the number of chains per unit volume and ϕ the functionality of the crosslinks. An average fluctuation of the mobile points around their mean positions corresponds to this limit, fluctuation which is independent of the strain. Then, one instantaneous end-to-end vector is the convolution of its affinely transformed average by the so-called "phantom network" fluctuation, depending on ϕ and N, the polymerization index of the mesh.

A common feeling is that this description gives too much freedom to the chains; real chains are not "phantom": they have their own volume and cannot cross each other. An easy remedy for that is to transfer all the lack of freedom on the poor crosslinks by reducing their fluctuations, ultimately down to zero. Then, the prefactor $\phi - 2/\phi$ becomes unity and each instantaneous end-to-end vector is now completely affinely transformed. This is believed to form the upper limit for the elastic force. These two models provide, therefore, limits for mechanical data as well as for microscopic observations (both models allow simple calculations of all the quantities involved). Regarding the mechanical data, it must be stressed that these models give a "window" for the force but that the ratio $f/\lambda - 1/\lambda^2$ undergoes in practice a clear unpredicted softening when λ increases.

4. Study of the deformation at a molecular scale: our SANS experiment disagrees with the theoretical form factors

It is now well known that mixing deuterated polymer chains with normal ones (nondeuterated) produces some fluctuations in space of the neutron equivalent of the index for light, the diffusion length, which is different for deuterium and normal hydrogen. This scatters a neutron beam, and gives the Fourier transform of correlations in position of the monomers of the chains:

$$S(\boldsymbol{q}) \sim \left\langle \sum_{\substack{i,j \\ \text{(one chain)}}} \exp\left(\mathrm{i}\boldsymbol{q} \cdot \boldsymbol{r}_{ij}\right) \right\rangle,$$

where $\boldsymbol{q}$ is the scattering vector, $\boldsymbol{r}_{ij}$ the vector connecting two diffractors of one of the tagged particles and $\langle\ \rangle$ average over all objects and time, equivalent to a thermal average.

In our experiment, the tagged object is not a single mesh, but long chains in a melt crosslinked afterwards, forming long "labeled paths" inside the network, through many meshes. By doing so, a linking deficiency will only produce punctually a longer mesh, not a pendant chain which could behave completely differently from a two ends-linked one. Also the larger scale under access is the size of the path, much larger than in the case of the labeled mesh.

Crosslinking has been done by gamma irradiation of a melt slightly swollen (volume fraction of polymer 80%) with cyclopentane[9]), which produces active radicals. The gelated samples are then dried. This does not create any detectable change of the chain conformation, as checked by neutron observation.

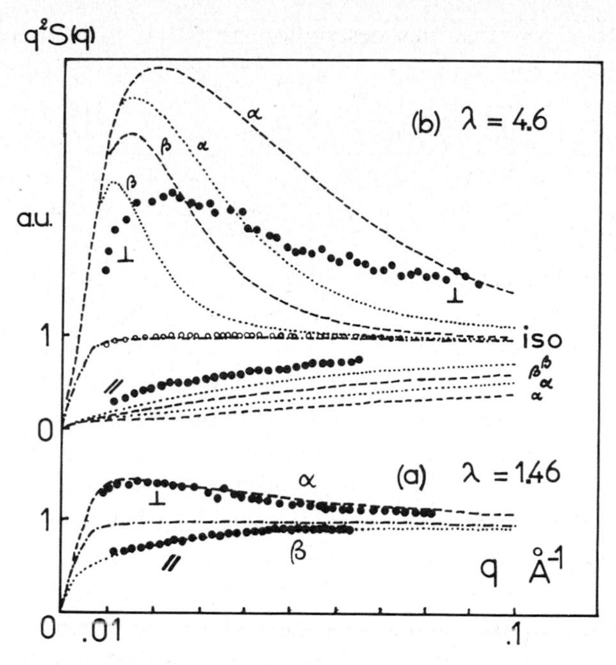

Fig. 1. Form factor in Kratky–Porod representation ($q^2S(q)$ vs q) for a labeled path (molecular weight $> 2 \times 10^6$) in a strained rubber for two elongation ratios, $\lambda = 1.46$, and $\lambda = 4.6$ in directions of observation perpendicular (●, ⊥) and parallel (●, ∥) to the stretching. Duration of relaxation is $t = 30$ min, at 150 °C. ○ are data for the unstrained material. The -·-· curve is the calculated form factor for a Gaussian chain. ····· is for "phantom network" calculations[18,9]): (a) $M = 50000$; (b) $M = 10000(\alpha)$ and $50000(\beta)$. ----- is for "junction affine" calculations[17,9]): (a) $M = 10000$; (b) $M = 10000(\alpha)$ and $50000(\beta)$. M is the molecular weight of the "mesh", i.e. the average chain segment between two consecutive crosslinks on the labeled path

Polystyrene is a glass at room temperature. We have heated the samples at temperature larger than $T_g = 100$ °C, the glass transition temperature. Then we have stretched them at an elongation ratio λ, and let the stress relax at constant λ during a time t. After that time we have stopped the relaxation by quenching back down to the glassy state. Several values have been used for λ and t. From the whole set[10]) we extract two sets of curves (fig. 1), both corresponding to a "large" time, 30 min at 150 °C, at which the deformed rubber is expected to be close to equilibrium under strain. The measured form factors should be comparable with the theoretical ones. The latter depend on the molecular weight of the mesh (i.e., chain segment between two consecutive links on the labeled chain), which we can only estimate rather grossly as $M = 20000$ [10]). Therefore we take $M = 10000, 20000, 50000$, and calculate for two limit models. For small λ, 1.46, the experimental points fit approximately some calculated curves, but not with

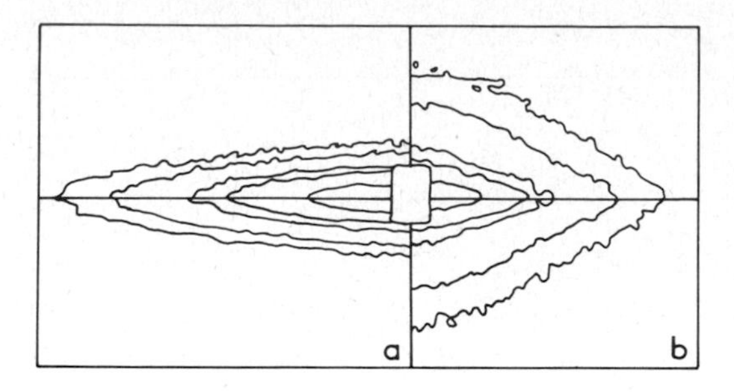

Fig. 2. Isointensity curves for $\lambda = 4.6$. (a) left hand side: small duration of relaxation t, 30 s, 110 ° C. (b) right hand side: same λ, large duration of relaxation t, 30 min, 150 ° C.

the same parameters in direction perpendicular to the elongation (junction affine model, $M = 10000$) and in parallel (phantom network, $M = 50000$, a large value). One could say that the behavior is intermediate between the two models, but *not in the predicted way* for which the same capability of rearrangement should show up simultaneously for all directions (i.e. the fitting curves in the two directions should correspond to the same model, with the same mesh size).

For large λ (4.6) any reasonable agreement seems to be lost: in parallel the chains appear less deformed than for a phantom network with $M = 50000$; in perpendicular the chains appear much more deformed at small distances (q large), and less deformed than any model prediction at large distance (q small).

4.1. *Isointensity curves*

We gain some additional information in considering those curves which join all cells of the same counting level on the bidimensional detector (fig. 2). Calculations based on the classical assumptions lead to curves quite similar to ellipses, for any λ and t. This is what we obtain experimentally for small λ, whatever t, and for large λ, but at small t only (fig. 2a). However, for large t, the shape changes noticeably toward that of a *losange* (fig. 2b). We have no current neat explanation of this. But note how to make it more spectacular: take the sample of fig. 2b (150 ° C, 30 min); let it come back to a shorter elongation ratio quite fast (return to $\lambda = 3$ at 110 ° C, for example), and cool it down as soon as this λ is reached: the losangic shape is noticeably enhanced.

5. Can "classical" defects in the network explain the discrepancies?

5.1. *Pendant chains*

Pendant denotes the portions of chains connected to the network by one extremity only. At least, the extremities of our labeled chains are pendant (maybe also parts in the middle if the chain is accidentally cut). We believe that the fraction of pendant chains is low from correlations with other experiments on gamma crosslinked networks (study of gelation, and of gels compared to model gels), but we cannot prove it. It is true that the presence of large amounts of such defects would move the experimental points towards the form factor of undeformed chains, which could be compatible with part of the discrepancies observed at $\lambda = 4.6$. But among other counterarguments, let us recall the behaviour for $\lambda = 1.46$; in the perpendicular direction the chains appear slightly too deformed, a little beyond the upper limit produced by the models. This makes it difficult to support the idea of a large fraction of pendant chains. Moreover, classical models computed with pendant chains do not produce losanges for isointensity curves.

5.2. *Trapped entanglements*

In crosslinked systems entanglements may exist as well as in melt, and some of them may not be destroyed by the brownian motion because they are trapped between some real crosslinks. Their presence would influence the elastic modulus, by increasing it, and thus they behave as crosslinks. Similarly, for any estimate of M, from any method, we cannot tell whether the elementary chains connect real crosslinks only or trapped entanglements. Some theories consider the latter as not directly effective, but just able to shift the material closer to the "junction affine" description; for others, trapped entanglements behave as real crosslinks. Following one or the other of these two arguments, the form factor should stay between the two classical models. However, one might object that the chains can slip through the entanglements to reduce the stress*. This slipping could lead to the different behaviours that we observe, especially when also present for melt samples, as the losange shapes for example.

6. Complexity of a network with no "classical" defects

On very general grounds, one observes that a polymer network is very unlikely to be mappable on a crystallographic lattice. It is more comparable to a glass. If one builds a network of rigid rods with flexible links, one observes that there are

*It is possible to include in the Deam–Edwards model the trapped entanglements modeled as slip links, i.e., the length of the part of chain between two of them, or between one and a real crosslink, varies with time. The main effect can be summarized as the fact that, under deformation, slip links have (after averaging over the three directions of space) more chemical length available between real crosslinks and thus more freedom. See ref. 3.

some "holes" in it, and some very long loops. Is the effect of holes and long loops negligible in rubber elasticity?

6.1. *Holes in a self-similar connectivity*[13])

Let us build a polymer net with the topology of a Sierpinsky gasket[11,12]), for example

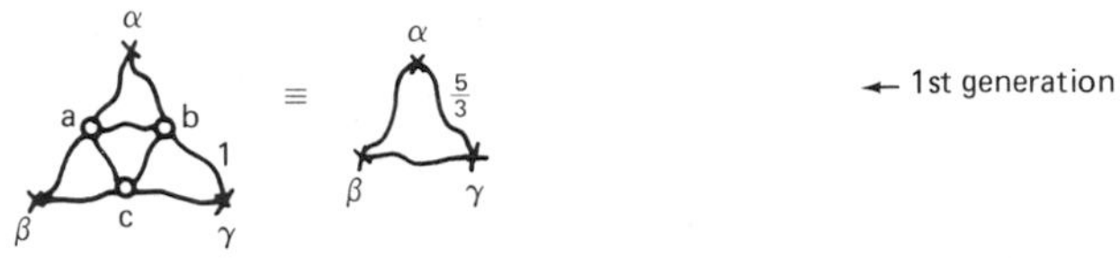

α a b 1 β c γ ≡ α $\frac{5}{3}$ β γ ← 1st generation

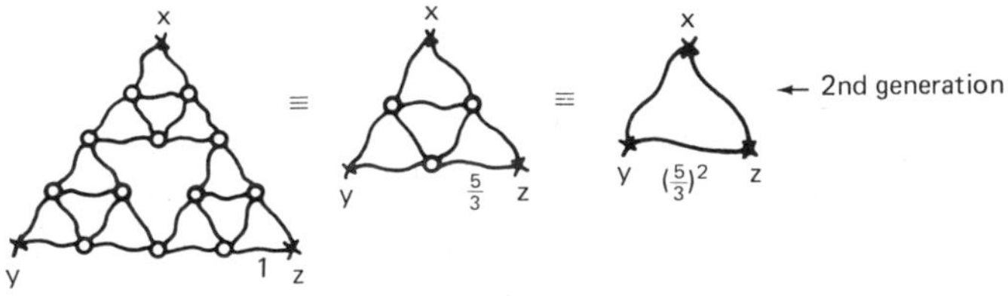

x y 1 z ≡ x y $\frac{5}{3}$ z ≡ x y $(\frac{5}{3})^2$ z ← 2nd generation

Each segment is in practice a polymer chain. At a certain nth order, we can stop the self-similarity and connect the elements in a more classical tridimensional network. Other possibilities exist, a tridimensional self-similar gasket with tetrahedra, for example, again up to a certain scale. The elementary chains can be monodisperse; the functional is always four, and nevertheless the system will have holes. It is then very interesting to calculate the modulus of this object by using the micronetwork approach: take a first order element, affinely displace the three external points α, β, γ, and free the three internal points a, b, c. This object is now equivalent to a triangle α, β, γ (with a larger edge, i.e., a chain with $N' = \frac{5}{3}N$). We can repeat that at the next order, keeping the three new external points fixed as in the first stage of ref. 8. After a certain number of generations, the elastic modulus becomes smaller than predicted by the micronetwork treatment (i.e., the "phantom network" limit). Also, the fluctuations of the chains junctions become very large; the deformation of an elementary chain segment could become smaller than in the phantom network prediction. So, holes may exist even without chemical defects, increasing the chain freedom and decreasing the modulus.

6.2. *Connectivity fluctuations*

Another question, perhaps related to the former one, is the way a couple of junctions is connected by the network. In a system at rest, crosslinked *and* observed in bulk, the chains appear nearly gaussian, as checked by SANS for example. Thus their end-to-end mean square distance is $R \sim N^{1/2}$. If N is large enough a sphere of radius R, containing say four couples of crosslinks directly connected by one mesh, contains equally a lot of crosslinks indirectly connected. Both are as closed, one calls the first kind "chemical" neighbours, the second kind "spatial" neighbours[1]). One sees in particular that the chemical length of the shortest path connecting a couple of junctions is submitted to fluctuations. Our guess is that the local deformation would reflect the local connectivity[10,14,15]). It will be an increasing function of the length of the shortest path. Then the average distance between two chemical neighbour crosslinks could be less deformed than the affine value, the couples of spatial neighbours accounting for a larger part of the deformation*. This is opposite to the classical models which make no distinction between couples of links, and some effect could be implicated in the discrepancies observed by SANS.

7. Conclusion

All together the physics of rubber is still an open problem. At the theoretical level, the old classical models are useful, but have some counterexamples. There is no real reason why the junction affine model should be the upper limit for the force (the model restricts the freedom of the crosslinks, but that of the chains could also be restricted). Neither is there a reason why the phantom network should be the lower limit (see section 6.1). At a microscopic scale, although the models produce a gross description, discrepancies are systematically observed that are difficult to explain only with defects of pendant chain type. Apart from the role of the trapped entanglements, we can imagine two origins for these discrepancies. The first would be an inadequate estimate of the junction fluctuations, keeping the idea of affine displacement of the mean positions of the junctions. This is for example what happens in our treatment of the Sierpinski network. The second origin would be a lack of systematic affine displacement of all mean positions, as in section 6.2. How this second origin would be related to the "nonphantomness" of the chains remains to be clarified. Another point of view is that the network will keep a memory of its gelation stage, even far beyond

*An extreme example is the network of rigid rods connected by flexible links ("universal joints")[14]): it can be deformed, or even swollen with no change of the rod length, i.e., the distance between chemical neighbours, and has a quasi-gaussian elasticity[16]).

the gel point[4,5]): this may also lead to disagreements with classical pictures, but we do not see currently if this is connected to the first or the second point evoked above.

Acknowledgements

The work presented is the result of interactions, discussions and collaborations with H. Benoit, S.F. Edwards, S.J. Candau, A. Lapp, L. Leibler, M. Daoud, J. Herz and C. Picot. We would like to thank them very much. Institut Charles Sadron and Laboratoire Leon Brillouin are laboratoires of the CNRS; SANS experiments have also been performed at Institut von Laue Langevin, Grenoble, France.

References

1) The major original models (see text) are all reviewed in P.J. Flory, Proc. Roy. Soc. **351** (1976) 351.
2) A more recent model is presented in R.T. Deam and S.F. Edwards, Phil. Tran. R. Soc. London, Ser. A **280** (1976) 1296. It is a basis for ref. 3.
3) R.C. Ball, M. Warner, M. Doi and S.F. Edwards, Polymer **22** (1981) 1010.
4) P.G. deGennes, J. Phys. Lett. **38** (1977) L355.
5) S. Alexander, in Physics of Finely Divided Matter, Springer Proc. in Phys. 5 (Springer, New York, 1985).
6) M. Daoud, E. Bouchaud and G. Jannink, to be published in Macromolecules.
7) See, for example, for other SANS experiments; M. Beltzung, C. Picot and J. Herz, Macromolecules **17** (1984) 663; R.W. Richards and A. Davidson, Macromolecules, to be published in 1986.
8) W.W. Graessley, Macromolecules **8** (1975) 186.
9) See more details in J. Bastide, J. Herz and F. Boue, J. Phys. **46** (1985) 1967.
10) F. Boue, J. Bastide and J. Herz, to be submitted to Macromolecules.
11) B. Mandelbrot, Fractals: Form, Chance, and Dimensions (Freeman, San Francisco, 1977).
12) R. Rammal and G. Toulouse, J. Phys. Lett. **44** (1983) L13.
13) J. Bastide and F. Boue, to be submitted.
14) J. Bastide, C. Picot and S.J. Candau, J. Macromol. Sci. Phys. B **19** (1981) 13.
15) S.J. Candau, J. Bastide and M. Delsanti, Adv. Pol. Sci. **44** (1982) 27.
16) F. Boue and S.F. Edwards, to be published.
17) R. Ullmann, Macromolecules **15** (1982) 1395.
18) M. Warner and S.F. Edwards, J. Phys. A Math. Gen. **11** (1978) 1649.

Physica **140A** (1986) 261–268
North-Holland, Amsterdam

KINETICS OF PHASE TRANSITION IN POLYMER GELS

Toyoichi TANAKA

Department of Physics and Center for Materials Science and Engineering, Massachusetts Institute of Technology, Cambridge, MA 02139, USA

Remarkable progress has been made recently in the understanding of the kinetics of the phase transition in polymer gels. This rich, scientific problem with substantial technological significance will be reviewed.

1. Introduction

Polymer gels, consisting of cross-linked polymer networks immersed in liquid, are known to undergo volume–phase transition; when external conditions such as temperature, solvent composition, or osmotic pressure change, a gel reversibly swells or shrinks, but does so discontinuously[1–5]). The volume change at the transition depends on both the degree of ionization and the stiffness of constituent polymer chains and can be as large as one thousand times[2]). The phenomena are interpreted in a similar way to the gas–liquid phase transition, and appear to be universal to any gels. Indeed, the phase transition has been observed in various synthetic polyvinyl gels[6]), and in gels made of natural polymers such as DNA, polysaccharide and polypeptide[7]).

2. Dynamical properties

There have been some studies on the dynamical properties of the phase transition, the most important of which may be the demonstration of the critical divergence and critical slowing-down in the thermal fluctuations of polymer network as revealed in the experiments of dynamic light scattering[8–10]), and ultrasonic absorption[11]). In contrast, the kinetics of the gel phase transition has not yet been fully explored.

3. Kinetics of linear response of gel

The important role of the collective diffusion of polymer networks was pointed out in the kinetics of volume change of gels[12]). It was shown, both experimentally

0378-4371/86/$03.50

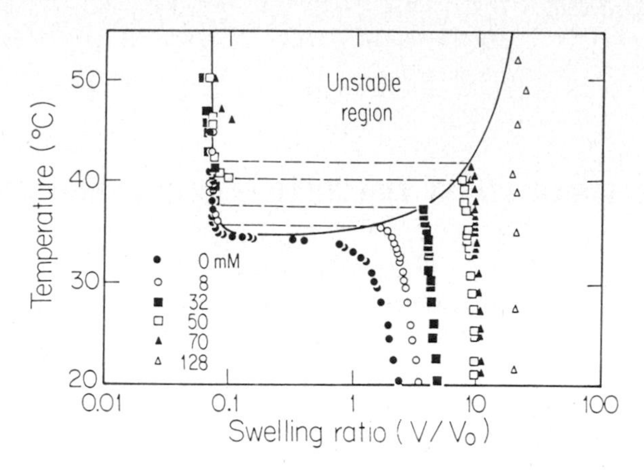

Fig. 1. Equilibrium volume of ionized N-isopropylacrylmide (NIPA) gels in water show discontinuous volume–phase transition in response to temperature. The values shown in the diagram indicate the amount of ionizable group (sodium acrylate) incorporated in 700 mM NIPA.

and theoretically, that the motion of a gel polymer network during the process of swelling or shrinking, is described by a collective diffusion equation with a diffusion coefficient defined as the ratio of the osmotic compressibility K of the network and the frictional coefficient f between the network and the liquid: $D = K/f$. Since it is a diffusion process, the time for a gel to swell or shrink is proportional to the square of the characteristic length L, such as the gel radius,

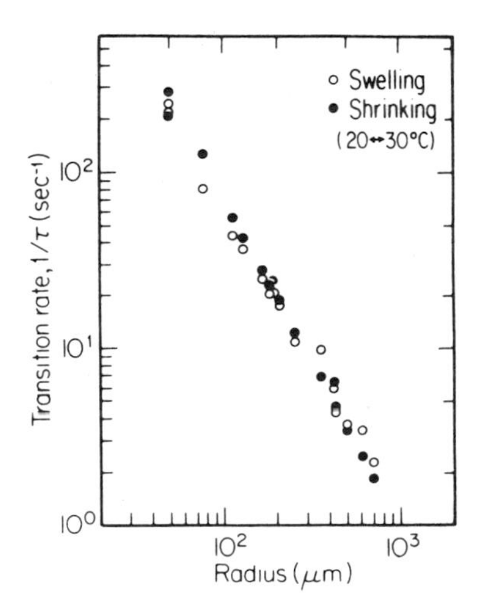

Fig. 2. Relaxation time of the volume change of a gel is proportional to the square of its equilibrium radius.

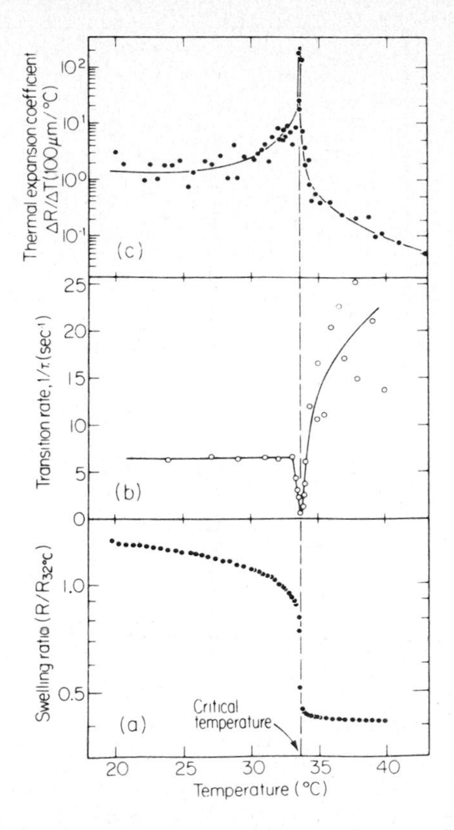

Fig. 3. (a) Equilibrium radius of an isopropylacrylamide gel sphere as a function of temperature. At lower temperatures the gel is swollen and at higher temperatures it is shrunken. At about 34°C the swelling curve becomes infinitely sharp, which corresponds to the critical point. (b) Relaxation time of gel volume change in response to a temperature jump, as a function of temperature. The relaxation becomes infinitely slow at the critical point. (c) Thermal expansion coefficient, the relative radius change per temperature increment, also diverges at the critical point.

or the wavelength of fluctuations. Its proportional coefficient is given by the collective diffusion coefficient D: $T = L^2/D$. This equation indicates that the characteristic time of volume change is determined by its size and diffusion coefficient. The square dependence on the size has been experimentally confirmed both in macroscopic kinetics and microscopic dynamics[12-14]). Then, how is the collective diffusion coefficient (D) determined?

4. Critical kinetics

Recently, experiments have been performed on the kinetics of volume changes of gels near the second order transition, where the volume change is continuous,

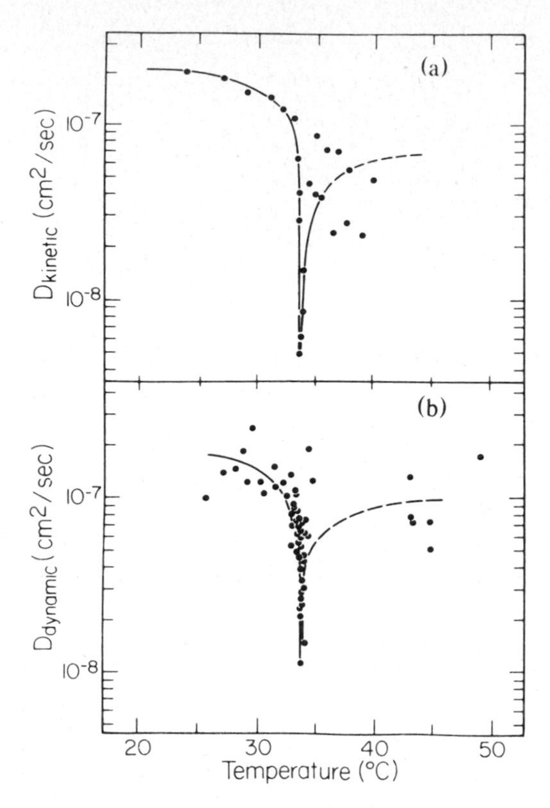

Fig. 4. (a) Collective diffusion coefficient D of a N-isopropylacrylmide gel as determined by the kinetics of volume change, as a function of temperature. It diminishes at the critical point. (b) Collective diffusion coefficient as determined from the density fluctuations by use of photon-correlation spectroscopy. The agreement between the results obtained from dynamics of microscopic fluctuations and from kinetics of macroscopic volume change is excellent considering the difficulty in the dynamic experiments.

but has an infinite slope with respect to the temperature. This state corresponds to the critical point. It was observed that the volume change at the critical point becomes infinitely slow, indicating that the diffusion coefficient diminishes at that point[15]). These experiments have clearly demonstrated the importance of the relative position of the gel state within its phase diagram in determining its kinetics of volume transition.

5. Pattern formation during swelling phase transition

In continuous volume changes, a gel retains a smooth surface during the process. In the discontinuous volume transitions, polymorphological patterns

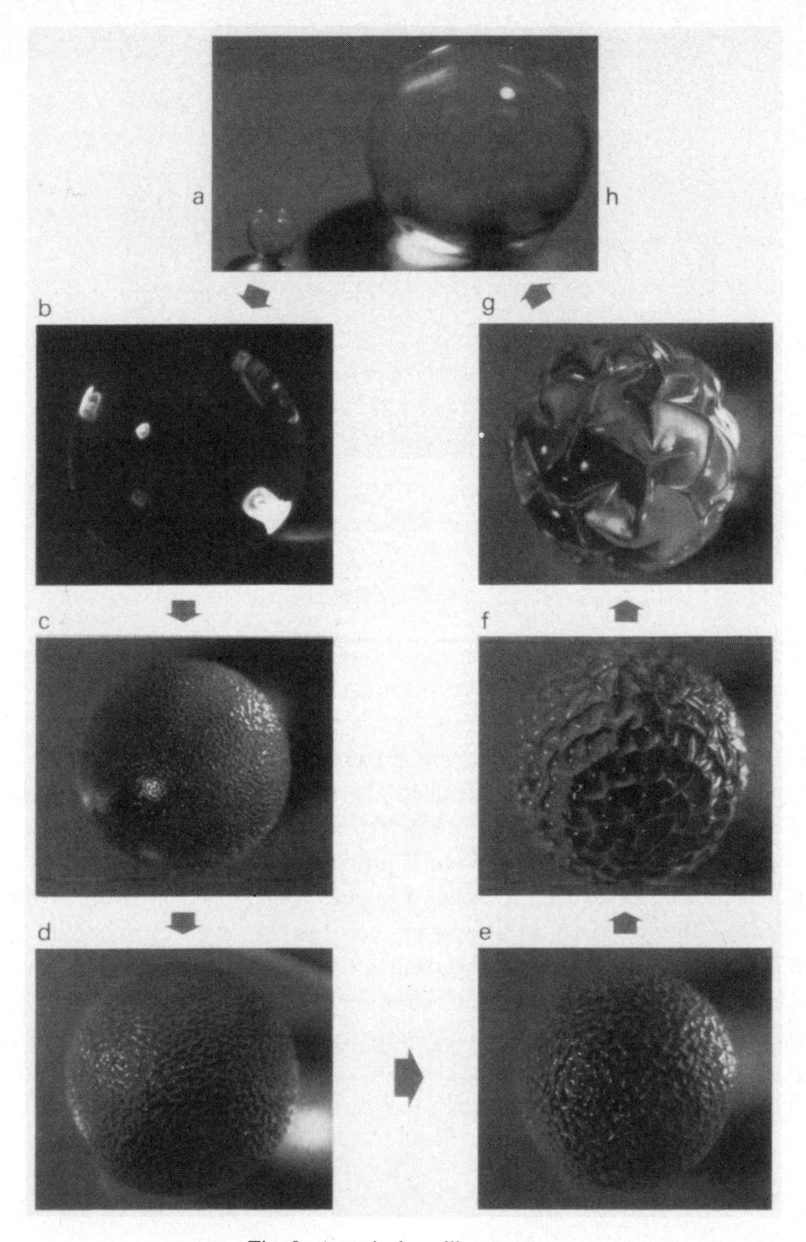

Fig. 5. A typical swelling process.

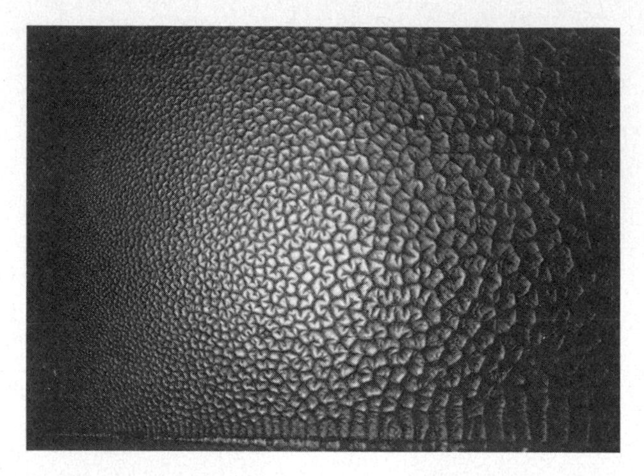

Fig. 6. A slab gels with thickness tapered from 0 to 1 mm with one surface mechanically fixed to a film is allowed to swell in water. The final equilibrium pattern shows gradual increase of unit size, which is proportional to the local thickness of the gel.

appear on the surface and within the gel, which depend on the gel structure, and initial and final conditions of the transition. A typical swelling process is shown in fig. 5. A spherical gel has a perfectly smooth surface before swelling. At the instant the gel is placed in water, an extremely fine pattern appears on the surface, which has the texture of a frosted glass[16]). As time goes on, the unit size of the pattern becomes larger due to coalescence of units into a larger unit, which is accomplished by conversion of side lines into thorns followed by disappearance of thorns. Eventually the size of the unit becomes comparable to that of the gel radius, then the pattern disappears completely. Even at this time, the gel volume is not yet fully swollen. It continues to swell to the final volume until it regains the spherical shape and totally homogeneous density.

Careful observation shows that the pattern consists of numerous line segments of cusps into the gel. At each stage of swelling, the pattern appears to be a hexagonal lattice, but it is quite irregular and contains triangles, squares and hexagonal cells as well as thorns.

6. Qualitative account for the pattern formation

The pattern formation may be qualitatively explained in the following way. The kinetic process of swelling and shrinking of a gel in response to an excess osmotic pressure is governed by the collective diffusion. Because of this, the gel starts to swell from the surface, and the swollen layer becomes thicker as time goes on. This "swollen layer" is under a mechanical constraint, namely, the outer surface of the layer is free to expand, whereas the internal surface of the layer is fixed to the core of the gel, which is not yet able to swell. Thus, the layer is under contradictory demands on the upper and lower surfaces, one to expand and the other to remain unswollen. When the osmotic pressure is small, this contradiction is resolved by stretching the gel unidirectionally perpendicular to the surface. When the osmotic pressure is large, the outer surface buckles itself to increase its surface area to resolve the contradiction. The buckling appears in the form of a semi-regular hexagonal pattern.

The characteristic wavelength of the pattern is proportional to the thickness of the swollen layer, as it should be, since that is the only relevant length scale appearing in the early stage of the phenomenon. As time goes on, the thickness of the swollen layer increases; so does the wavelength of the pattern. When the osmotic pressure front reaches the center of the gel, the mechanical constraint decreases and finally diminishes, at which time the pattern disappears.

7. Permanent patterns on mechanically constrained gel

The hypothesis presented above has been tested by several demonstrative experiments. A gel slab is prepared whose one surface is free, but the other surface is covalently, and thus permanently cross-linked to a film whose surface is specially treated to retain polymerization-active chemical groups. Such slab gels developed patterns, whose evolution eventually stopped and never disappeared. The final wavelength of the pattern was proportional to the thickness of the gel slab.

There is a critical osmotic pressure for pattern formation. Near this critical osmotic pressure the pattern appears to be quite irregular. The critical condition for pattern formation is independent of the gel thickness, but uniquely determined by the internal osmotic pressure.

8. Theoretical considerations

We present a theoretical explanation for the pattern formation using a similar logic first introduced by Euler to analyze the mechanical instability of a rod under a compressional force[17]). The theory predicts the critical osmotic pressure P_c and the critical wavelength λ_c: $P_c \approx E$ and $\lambda_c \approx h$. Here E is an averaged Young's modulus of the gel and h is the thickness of the gel. The theoretical predictions that the wavelength of the pattern is proportional to the thickness of the gel, and that the critical pressure is independent of the gel thickness agree with the experimental results.

The hexagonal pattern can be explained in a mathematical formulation similar to that of the Benard' cell pattern: once a mode becomes unstable and its amplitude increases, it is necessary to take into account the higher order terms. The third order terms play an important role in producing the hexagonal pattern.

9. Patterns on gel under shrinking transition

We also present studies on the kinetics of shrinking transition. It is interesting to observe that various kinds of patterns are formed in this reverse process, but the formation and evolution of patterns are totally different from that of swelling. In fact, the shrinking patterns have much more variety than that of swelling. The patterns can be explained in terms of combination of mechanical instability and phase transition under internal osmotic gradient.

The understanding of the kinetics of phase transition of gels may not only be important from purely the physics point of view, but also for biology, rich with a wide variety of polymorphism.

References

1) T. Tanaka, Phys. Rev. Lett. **40** (1978) 820–823.
2) T. Tanaka, D.J. Fillmore, S-T. Sun, I. Nishio, G. Swislow and A. Shah, Phys. Rev. Lett. **45** (1980) 1636–1639.
3) M. Ilavsky, Macromolecules **15** (1982) 782–788.
4) T. Tanaka, I. Nishio, S-T. Sun and S. Ueno-Nishio, Science **218** (1982) 467–469.
5) S. Katayama et al., Macromolecules **18** (1985) 2781–2782.
6) Y. Hirokawa, T. Tanaka and E. Sato, Macromolecules **18** (1985) 2782–2783.
7) T. Amiya and T. Tanaka, submitted for publication (1986).
8) T. Tanaka, S. Ishiwata and C. Ishimoto, Phys. Rev. Lett. **38** (1977) 771–774.
9) T. Tanaka, Phys. Rev. A **17** (1978) 763–767.
10) A. Hochberg, D. Nicoli and T. Tanaka, Phys. Rev. Lett. **43** (1979) 217–219.
11) J.C. Bacri and R. Rajaonarison, J. Phys. Lett. (Orsay, France) **40** (1979) L5–L8.
12) T. Tanaka and D.J. Fillmore, J. Chem. Phys. **70** (1979) 1214–1218.
13) T. Tanaka, L.O. Hocker and G.B. Benedek, J. Chem. Phys. **59** (1973) 5151–5159.
14) P-G. DeGennes, Scaling Concepts in Polymer Physics (Cornell Univ. Press, Ithaca, NY, 1979), pp. 214–218.
15) T. Tanaka et al., Phys. Rev. Lett. **55** (1985) 2455–2458.
16) Y. Hirokawa, J. Kucera, S-T. Sun and T. Tanaka, in Physical Optics of Dynamic Phenomena and Processes in Macromolecular Systems, B. Sedlacek, Ed. (de Gruyter, Berlin, 1985), pp. 197–204.
17) L.D. Landau and E.M. Lifschitz, Teoria Uprogosti (Nauka, Moscow, 1967).
18) H. Haken, Synergetics, An Introduction (Springer-Verlag, Berlin, 1983).

Physica **140A** (1986) 269–277
North-Holland, Amsterdam

EFFECTIVE RIGIDITY OF MEMBRANES

L. PELITI

Dipartimento di Fisica, Università "La Sapienza", Piazzale Aldo Moro 2, I-00185 Roma, Italy and GNSM-CNR, Unità di Roma*

The role of thermal fluctuations of shape (undulations) in reducing the effective rigidity of membranes is reviewed. The consequences of this effect on vesicle size distribution and on the structure of microemulsions, as well as on other physical phenomena, are sketched.

Physicists always have the habit of taking the simplest example of any phenomenon and calling it "physics", leaving the more complicated examples to become the concern of other fields.

R.P. FEYNMAN

1. Introduction

I hope that Israelachvili's contribution to this conference has convinced most of the audience that not all physicists shun the complications of actual systems in the way I do. For I am going to fall at once into the bad habit mentioned by Feynman and to forget most of what we know about actual membranes of biological interest, in order to entertain you with the deeds of an extremely purified model which has only a faint resemblance with them. I shall try to convince you at the end that this exercise is not unworthy of our efforts, for it may help us in the solution of a few interesting physical puzzles.

As a first idealization, we represent a membrane as consisting of two homogeneous layers of lipids oriented with their polar heads towards the exterior of the membrane. Such a lipid bilayer can be in a fluid state, which means that the molecules diffuse freely within the layers[1]). Real biological membranes possess in fact a quite complicated structure, including heterogeneities in their composition, small intercalated molecules such as cholesterol, and many intramembrane proteins.

*Present address: Dipartimento di Fisica, Università di Napoli, Mostra d'Oltremare, Pad. 19, I-80125 Napoli, Italy.

0378-4371/86/$03.50

As a second idealization, we shall consider a simple phenomenological model in which the membrane is treated as a continuous, fluid, bidimensional system whose free energy depends only on its shape[2-4]. In this context one speaks of *elastic curvature energy* of the system, which is a function of two phenomenological parameters: the *rigidity* κ and the *spontaneous curvature* H^s. What is peculiar of membranes with respect to other two-dimensional systems like fluid interfaces is that the surface tension r, if it does not vanish, is very small. This feature is shared by the water–oil interface in microemulsions[5]), because of the presence of a large concentration of surfactant (soap molecules). Since the stability of the system is only controlled by the elastic curvature energy, wild out-of-plane fluctuations take place. They are called *undulations* and are the first characters of our drama.

This model has been applied at the mean field level to reproduce the shapes of closed membranes, such as those of red blood cells. The success of this program is remarkable[6]). If one takes undulations into account, a few interesting phenomena arise. One witnesses e.g. the "flicker phenomenon", i.e. the enhanced light scattering of active red blood cells[4]). The crumpling of the membrane due to undulations makes its area appear smaller than its true value[7]). The reduction factor increases with increasing membrane size, since more and more modes enter the game.

Helfrich[8]) has recently suggested that undulations should also reduce the effective rigidity (measured at large scales) with respect to the bare rigidity (microscopically measured), again by a factor which increases with increasing membrane size. This effect has been confirmed by renormalization group calculations[9-11]). As a consequence, normals to the membrane have their directions correlated only up to a *persistence length* ξ [12]), which increases exponentially as $T \to 0$. Beyond this length scale, shape fluctuations become so large that the membrane tends to fill all space. The model we are describing, which does not take into account the effects of close encounters between different parts of the membrane, then loses its validity.

It is possible however to draw from it a few interesting consequences. On the one hand the exponential distribution of vesicle size which one would obtain at constant rigidity is modified by a power law[13]). On the other hand it is tempting to relate the crumpling of the membrane at length scales larger than ξ to the formation of microemulsions. A model of microemulsion incorporating the size dependence of κ has indeed been proposed[14]). Impurities coupling with the local curvature can also affect the rigidity via their fluctuations[15]). This may lead to a *curvature instability* and to the formation of spikes or of other modulations.

Because of undulations, nearby membranes frequently bump into each other. The corresponding excluded volume constraint produces a repulsive interaction which competes with the long attractive tail of Van der Waals interactions[16]).

One cannot consider this effect by simply adding one more potential to the interaction among membranes. This makes the problem of the stability of *multilamellar phases*, where several membranes are piled up, a delicate affair. One can apply to it a renormalization group treatment resembling that applied to the wetting transition[17]). It turns out that the membranes unbind via a continuous transition (although other investigators[18]) maintain the possibility of a first order phase transition, hence of a tricritical behavior).

I introduce and discuss in section 2 the model of the membrane and its renormalization treatment. Section 3 is dedicated to a brief discussion of vesicle size distribution and of curvature instability, and section 4 to an outline of some recent results on microemulsions and multilamellar phases. An outlook is contained in section 5.

2. Renormalization of membrane rigidity

We consider a two-dimensional membrane embedded in a three-dimensional space. The free energy $\mathscr{F}$ per unit area of a fluid membrane can be written[3]) as follows:

$$\mathscr{F} = r_0 + \tfrac{1}{2}\kappa_0 (H - H_0^s)^2 + \bar{\kappa}_0 K. \tag{1}$$

Here H is the *mean* curvature of the membrane, i.e. the sum of its inverse curvature radii, whereas K is the *Gaussian* curvature, i.e. the corresponding product. It is well known[3]) that it is not necessary to take the last term into account if one considers only fluctuations of the membrane shape which do not change its topology, since the integral of K over the whole area of the membrane is a topological invariant. The coefficient κ_0 is the *bare* rigidity, and H_0^s is the *bare* spontaneous curvature. The *bare* surface tension is r_0. We are considering systems whose *effective* (macroscopic) surface tension r is very small. If F is the total free energy and S is the membrane area, we have:

$$\mu = \left(\frac{\partial F}{\partial S}\right)_{\rm eq} \approx 0. \tag{2}$$

This does not imply $r_0 = 0$, since thermal fluctuations contribute to the effective surface tension r. We shall only consider symmetric membranes, and set $H_0^s = 0$. We have therefore, for any configuration of the membrane:

$$F = \int \mathrm{d}S \left(r_0 + \tfrac{1}{2}\kappa_0 H^2\right). \tag{3}$$

The probability of any configuration is proportional to the Boltzmann factor

$\exp(-F/T)$. The integral in eq. (3) runs over the whole area of the membrane and dS is the area element.

When we sum over fluctuations, we face the delicate problem of *counting* the membrane configurations, i.e. of assigning a functional measure to the space of all membrane shapes. Several proposals have been made[8–11,13,19–21]. In the first group[9–11,20] one uses the same functional measures which have been introduced in string theory or in the theory of fluctuating interfaces[22], although in different disguises. In the second group[8,13,19,21], which was represented single-handedly by Helfrich till very recently, one considers the peculiar effects of the constraint that the lipid film is of *constant density*. Therefore, whenever a fluctuation locally increases the membrane area, there should be a net inflow of lipids from the environment. I have long been an exponent of the first group, but I must confess that the arguments put forward by Helfrich[19] and Förster[21] are troubling. I hope that this Conference will give us the opportunity of settling the controversy.

Membrane shapes are identified by giving the position $\boldsymbol{r}$ of a membrane element (a three-dimensional vector) as a function of two parameters ζ^1, ζ^2 identifying the membrane element itself. If we neglect the Gaussian curvature term eq. (3) represents the only functional of the membrane shape which only depends on derivatives of $\boldsymbol{r}$ with respect to the ζ's up to second order and is invariant with respect to Euclidean and up-down symmetries. It only remains to constrain the parametrization $\boldsymbol{r}(\zeta^1, \zeta^2)$ in order not to count as different shapes different representations of the same shape, and most of all to define a functional measure in the space of shapes. The choice we made[9] was to use the Monge form for $\boldsymbol{r}(\zeta^1, \zeta^2)$, i.e.

$$r_x = \zeta^1, \quad r_y = \zeta^2, \quad r_z = u(\zeta^1, \zeta^2). \tag{4}$$

and to use as a functional measure simply $\mathrm{D}u$, since Wallace and Zia[22] have shown that it is invariant with respect to Euclidean transformations. On the other hand, if one considers the effect of the *longitudinal* motion of the molecules induced by the fixed density constraint, one finds with Helfrich[8,13,19] and Förster[21] a measure equivalent to $\mathrm{D}u\,(\det A)^{1/2}$, where A is a matrix related to the variation in membrane area due to the deformation. When $(\det A)^{1/2}$ is exponentiated, it induces a *long range* effective interaction which gives an extra contribution to the renormalization of κ and r. This appears to be the origin of the discrepancy among the results of the different groups. Yet I should like to be sure that the extra interaction does not wreck the renormalizability of the theory.

Whatever the measure, one imagines giving the membrane a certain average configuration at long wavelengths and computes its free energy by summing over short wavelength ripples. If the measure is Euclidean invariant the resulting functional will have the *same* dependence on the average configuration as F has on the actual configuration. We thus obtain an expression similar to eq. (3), but with different coefficients r and κ, which are identified with the effective surface tension and rigidity. The result of the calculation is most conveniently expressed by renormalization group equations, which represent the change in r and κ consequent to a change in a renormalization wavenumber μ at fixed bare parameters. For the rigidity this equation reads to lowest order:

$$\mu\left(\frac{\mathrm{d}x}{\mathrm{d}\mu}\right)_0 = \kappa\left(\frac{\alpha T}{4\pi\kappa}\right), \tag{5}$$

where $\alpha = 3$ for the first group of works[9–11,20]) and $\alpha = 1$ for the second group[8,13,19,21]). One can also consider the two-dimensional membrane embedded in a D-dimensional space[11,20]). By using a measure of the "first group" one obtains $\alpha = D$, which is confirmed by an exact calculation[20]) in the $D \to \infty$ limit. The $r = 0$ state appears[20]) as a kind of tricritical point with a relevant coupling (the surface tension) and a marginally relevant one (the inverse rigidity). There is a stable renormalization group trajectory leading from this point to a fixed point describing the transition studied by Wallace and Zia[22]). Since the inverse rigidity is marginally relevant, correlations of the normal to the membrane decay *exponentially* over a length scale ξ which is given by:

$$\xi \sim \exp\left(4\pi\kappa_0/(\alpha T)\right) \tag{6}$$

and may be identified with the persistence length of de Gennes and Taupin[12]). Membranes appear rigid at length scales shorter than ξ, and flexible and crumpled at length scales larger than ξ.

3. Vesicle size distribution and curvature instability

By integrating eq. (5) (identifying μ with the inverse linear size of the membrane) we obtain a logarithmic reduction of κ with increasing size L of the membrane:

$$\kappa = \kappa_0 + \frac{\alpha T}{4\pi}\ln\left(\frac{a}{L}\right), \tag{7}$$

where a is some microscopic length. This formula only holds to first order in the inverse rigidity. The effect is rather weak, but may be highlighted by looking at the size distribution of a fixed number of vesicles[13]). The bending energy of a sphere can be found by integrating eq. (1) and (for $H_0 = 0$) is given by:

$$E = 8\pi\kappa_0 + 4\pi\bar{\kappa}_0. \tag{8}$$

Labeling the vesicle size by the number N of lipid molecules, the probability of a given size N is given by $w(N) \propto \exp(-E/T)$. Taking eq. (7) with constant coefficients κ_0, $\bar{\kappa}_0$ we obtain:

$$\omega(N) = \frac{1}{\overline{N}} \exp\left(-\frac{N}{\overline{N}}\right), \tag{9}$$

where the exponential decrease is due to the presence of a chemical potential of the lipid molecules. If we had instead a size dependent rigidity according to eq. (7) we would have:

$$\omega(N) \propto N^{\alpha} \exp\left[-(1+\alpha)\frac{N}{\overline{N}}\right]. \tag{10}$$

This equation is valid if the Gaussian rigidity $\bar{\kappa}_0$ is not renormalized, what is also controversial. The form (10) should make it possible in principle to measure the coefficient α, but should fail at very small sizes because of the effects of higher curvature terms[23]) neglected in eq. (1).

Another interesting effect arises when one adds annealed impurities to the membrane which couple to the mean curvature H [15]). This produces in general a spontaneous curvature H_0^{s}. A less trivial effect is the reduction of the rigidity κ. We introduce an interaction term in eq. (1) of the form:

$$\mathscr{F}_{\mathrm{int}} = \lambda \phi H, \tag{11}$$

where $\phi = c - c_0$, c is the concentration of impurities and c_0 is its average value. Let us choose c_0 to cancel exactly the spontaneous curvature term in eq. (1). By integrating upon the (Gaussian) concentration free energy of the form of eq. (3), but with a renormalized rigidity given by:

$$\kappa = \kappa_0 - 3\gamma\lambda^2, \tag{12}$$

where $\gamma = \langle \phi^2 \rangle$. We see that if the concentration fluctuations are sufficiently large κ may even become *negative*. This is the curvature instability. The regions with higher concentrations of impurities will bend strongly. The final form reached by the membrane will depend on several factors which are not taken into account in this simple calculation, but one might expect the formation of spikes or of other modulated shapes. One can have a better understanding of the resulting shapes by reintroducing the surface tension[24]). Similar phenomena may arise in mixtures of lipids of different properties or in different fluidity states.

4. Microemulsions and multilamellar phases

I wish to draw your attention on the very interesting work of Safran et al.[14]), which shows how the concept of a size dependent effective rigidity helps in understanding the origin of middle phase microemulsions. The structure of microemulsions is discussed at length in Widom's contribution to this conference, so I shall be rather brief. Widom[5]) proposed some time ago a kind of mean field model of microemulsions based on concepts introduced by Talmon and Prager[25]) and de Gennes and Taupin[12]). The microemulsion is considered as an interwoven structure of cells of size l filled with either oil or water and separated by an interface filled with surfactant molecules. While the first investigators[25,12]) considered quite naturally l to be of the order of the persistence length ξ, Widom[5]) treats l as a variational parameter in a free energy which takes into account the entropy of mixing of the different phases and the curvature energy of the interface. He is able to find in this way rather sensible phase diagrams, with the drawback that the characteristic size l of the microemulsion becomes unrelated to the persistence length – and the sensitivity of the experimental phase diagram to the properties of the surfactant film remains quite unexplained. Safran et al.[14]) take into account the spontaneous curvature term in the bending energy and the size dependence of the rigidity (as given e.g. by eq. (7)). As a result they obtain middle phase microemulsions with structures of length l comparable with the persistence length ξ, in equilibrium with both dilute surfactant/oil and surfactant/water phases. It is the renormalization of the rigidity due to thermal undulations which is responsible for the stability of the middle phase. The volume fraction of surfactant in the middle phase is $\phi_s \sim \exp[-4\pi\kappa_0/(\alpha T)]$, which explains why microemulsions are so sensitive to changes in κ_0 induced e.g. by the presence of cosurfactants, i.e. of impurities acting according to the mechanism described in the previous section.

Flexible membranes often cohere to form multilayer lamellar phases. The cohesion necessary for the stability of these phases is provided by long-range Van der Waals forces. At short membrane separation these attractive interactions are balanced by strong repulsions of molecular or electrostatic origin. One might expect two membranes always to cohere on the basis of their Van der Waals attraction. However, both uncharged and charged bilayers have been found which seem to separate indefinitely far when allowed to equilibrate in excess water. These observations may be explained in terms of an effective steric interaction[16]) decaying like l^{-2}, where l is the membrane separation, resulting from thermally excited undulations. This repulsion would dominate at long distances, producing a first order transition. This effective interaction was derived under the assumption that the membranes do not interact through other forces. If the direct interaction mentioned above are taken into account one obtains a *continuous* phase transition[15]) from a state where the membranes are bound to one where they are completely separated. The properties of this transition cannot be obtained simply by superposing the steric repulsion with the direct interaction. One has to face instead the renormalization of a Hamiltonian of the type:

$$F = \int \mathrm{d}^{D-1}x \left\{ \tfrac{1}{2}\kappa(\nabla^2 l)^2 + V(l) \right\}, \tag{13}$$

where l is the membrane separation and $V(l)$ is the direct interaction potential (comprehensive of Van der Waals forces). This Hamiltonian is similar to that used to investigate the wetting transition[26]), except for the presence of a *curvature* elastic energy $\frac{1}{2}\kappa(\nabla^2 l)^2$ instead of a surface tension term $\frac{1}{2}r(\nabla l)^2$. One obtains as a result three regimes: for $D > d_1 = 5$, mean field theory is valid and the transition is first order. For $d_1 > D > d_2 = 11/3$, one is in a weak fluctuating regime and a continuous transition takes place when the amplitude of the Van der Waals forces W (Hamaker's constant) vanishes. For $d_2 > D$ (hence at $D = 3$), the transition is still continuous, but takes place with a nonzero value of W.

5. Outlook

Helfrich's observation[8]) on the reduction of membrane rigidity due to thermal undulations has been at the origin of an intense research activity. Subtle effects have been highlighted by research methods of increasing sophistication. It would be probably interesting in the future to look more closely inside the membrane to understand better the role of its internal degrees of freedom and to introduce methods by which the "bumping" of far parts of the membrane into one another and the effects of fluctuations which change its topology could be treated.

Acknowledgement

I am grateful to Dr. S. Leibler for having shared with me many of his insights on this fascinating field.

References

1) See e.g. E. Sackmann, in D. Chapmann, ed., Biomembranes (Academic Press, New York, 1985).
2) P.B. Canham, J. Theor. Biol. **26** (1970) 61.
3) W. Helfrich, Z. Naturforsch. **28c** (1973) 673.

4) F. Brochard and J.-F. Lennon, J. Physique **36** (1975) 1035.
5) B. Widom, J. Chem. Phys. **81** (1984) 1030.
6) H.J. Deuling and W. Helfrich, Biophys. J. **16** (1976) 861.
7) W. Helfrich and R.H. Servuss, Nuovo Cimento **2D** (1984) 137.
8) W. Helfrich, J. Physique **46** (1985) 1263.
9) L. Peliti and S. Leibler, Phys. Rev. Lett. **54** (1985) 1960.
10) D. Főrster, Phys. Lett. A **114** (1986) 115.
11) A.M. Polyakov, Nucl. Phys. **B268** (1986) 406.
12) P.G. de Gennes and C. Taupin, J. Phys. Chem. **86** (1982) 2294.
13) W. Helfrich, J. Physique **47** (1986) 321.
14) S.A. Safran, D. Roux, M.E. Cates and D. Andelman, Phys. Rev. Lett. **57** (1986) 491.
15) S. Leibler, J. Physique **47** (1986) 507.
16) W. Hefrich, Z. Naturforsch. **33a** (1978) 305.
17) R. Lipowsky and S. Leibler, Phys. Rev. Lett. **56** (1986) 2541; S. Leibler and R. Lipowsky, to be published.
18) D. Sornette, to be published.
19) W. Helfrich, to be published.
20) F. David, to be published.
21) D. Főrster, to be published.
22) D.J. Wallace and R.K.P. Zia, Phys. Rev. Lett. **43** (1979) 808.
23) M.D. Mitov, C.R. Acad. Bulgare Sci. **31** (1978) 513.
24) S. Leibler, private communication.
25) Y. Talmon and S. Prager, J. Chem. Phys. **69** (1978) 2984.
26) E. Brézin, B.I. Halperin and S. Leibler, Phys. Rev. Lett. **50** (1983) 1387.

Physica **140A** (1986) 278–284
North-Holland, Amsterdam

LIQUID STRUCTURE AND SHORT-RANGE FORCES BETWEEN SURFACES IN LIQUIDS

J.N. ISRAELACHVILI
Physical Chemistry I, Chemical Center, University of Lund, S-221 00 Lund, Sweden

and

H.K. CHRISTENSON
Department of Applied Mathematics, Res. School Physical Sciences, Australian National University, Canberra ACT 2601, Australia

A review is given of direct measurements of the forces between surfaces in liquids. The results provide a unique insight into liquid structure and its role in determining the fundamental interactions in liquids.

1. Introduction

There are four major physical forces that occur between surfaces and particles in liquids[1]). These are (i) attractive *Van der Waals* forces, (ii) repulsive electrostatic *double-layer* forces, which arise when surfaces carry a net charge, (iii) *solvation* forces, which arise from the structuring or ordering of the liquid (solvent) molecules at the surfaces (these can be repulsive or attractive), and (iv) repulsive *entropic* or *steric* forces, which arise from the interactions of protruding mobile surface groups such as polymers.

An understanding of these forces, especially their magnitude and range, is essential for understanding many complex but everyday phenomena at the molecular level. For example, the occurrence of landslides, biological organization and the interactions of biological cells, the properties of soaps, soils and paints, and many technological and industrial processes. Until recently only the attractive Van der Waals and repulsive double-layer forces were believed to be important. Thus, if the former dominated, two particles or surfaces would come together, while if the latter dominated they would be kept apart. And there were (apparently) rigorous theories to account for these forces[1]), which predicted (i) a monotonically attractive Van der Waals force that varied with some inverse power of the surface separation, and (ii) a monotonically repulsive double-layer force having an exponential distance dependence. However, with increasing theoretical attention and the advent of highly sensitive techniques to measure these forces it became clear that the situation is not quite so simple and that very subtle and unexpected interactions can occur in liquids, especially at small surface separations. Here we shall first describe one such technique for directly measuring the force laws (force vs. distance) between two surfaces, and then proceed to review some of the more interesting results.

2. The force measuring apparatus

Fig. 1 shows a force-measuring apparatus with which the force between two surfaces in liquid or vapour can be directly measured[2]). The distance resolution is about 0.1 nm and the force sensitivity is about 10^{-8} N (10^{-6} g). The apparatus contains two curved molecularly smooth surfaces of mica (of radius $R \sim 1$ cm) between which the interaction forces are measured. The mica surfaces can be bare, or they can have metal films, dielectric coatings, surfactants, polymers, or other macromolecules adsorbed (deposited) on them.

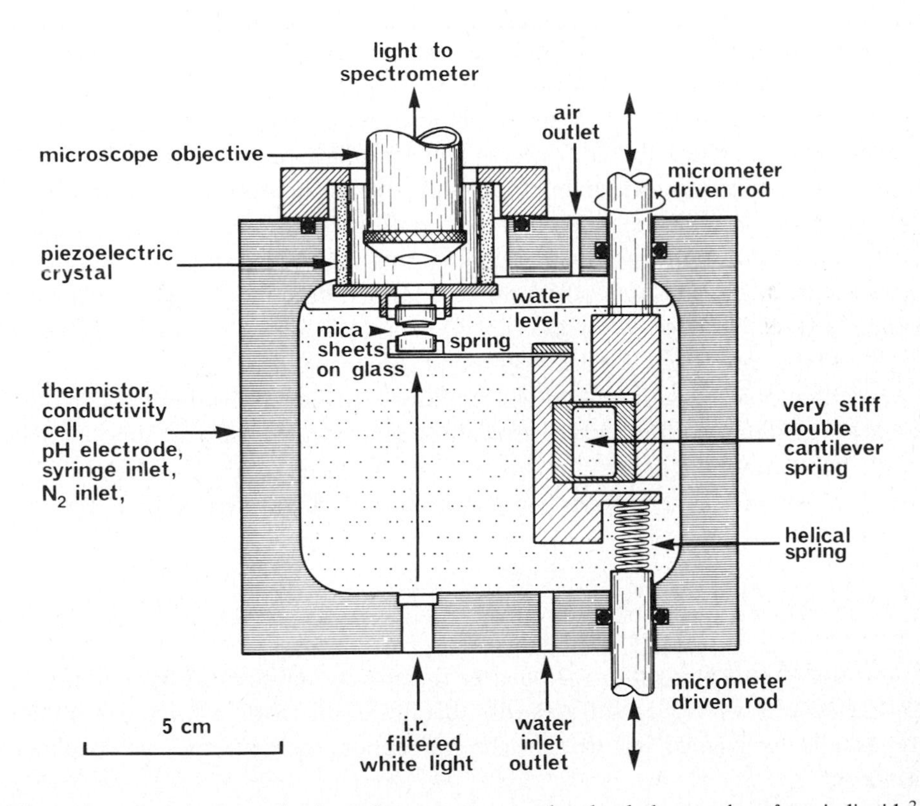

Fig. 1. Apparatus for measuring forces between two curved molecularly smooth surfaces in liquids[2]).

The separation between the two surfaces is measured by use of an optical technique using multiple beam interference fringes called 'fringes of equal chromatic order' or FECO. Here the two transparent mica sheets (each about 2 μm thick) are first coated with a semi-reflecting 50 nm layer of pure silver before they are glued onto the curved silica disks (silvered sides down). Once in position in the apparatus, as shown in fig. 1, white light is passed vertically up through the two surfaces and the emerging beam is then focussed onto the slit of a normal grating spectrometer. From the positions and shapes of the coloured FECO fringes seen in the spectrogram the distance between the two surfaces can be measured, usually to better than 0.1 nm (1 Å), as can the shapes of the two surfaces and the refractive index of the liquid (or material) between them.

The distance between the two surfaces is controlled by use of a three-stage mechanism of increasing sensitivity: the coarse control (upper rod) allows positioning to within ~ 1 μm, the medium control (lower rod, which depresses the helical spring and which in turn bends the much stiffer double cantilever spring by 1/1000 of this amount) allows positioning to about 1 nm, and the piezoelectric crystal tube – which expands or contracts vertically by ~ 0.5 nm per volt applied axially across its cylindrical wall – is used for the final positioning to ~ 0.1 nm. Given the facility for moving the surfaces towards or away from each other and, independently, of measuring their separation (each with a sensitivity or resolution of about 0.1 nm), the force measurements themselves now become straightforward. The force is measured by expanding or contracting the piezoelectric crystal by a known amount and then measuring optically how much the two surfaces have actually moved; any difference in the two values when multiplied by the stiffness of the force-measuring spring gives the force difference between the initial and final positions. In this way, both repulsive and attractive forces (especially adhesive forces) can be measured with a sensitivity of about 10^{-6} g and a force law obtained over any distance regime.

Once the force F as a function of separation distance D is known for the two surfaces (of radius R) the force between any other curved surfaces simply scales by R. Furthermore, it is easy to show[1]) that the corresponding interaction free energy E per unit area between two *flat* surfaces is simply related to F by

$$E(D) = F(D)/2\pi R. \tag{1}$$

It is therefore a straightforward matter to compare the results obtained for the forces between two curved surfaces with theories of intersurface forces since these are usually formulated in terms of the interaction energy between two planar surfaces.

3. Forces in simple liquids

Fig. 2 shows a typical short-range force-law measured between two molecularly smooth surfaces in any simple liquid. The decaying oscillations have a periodicity of the diameter of the liquid molecules and correspond to discrete (but diffuse) layers of molecules between the surfaces. The attractive minima occur at separations where the molecules can pack comfortably in layers while the repulsive (energetically unfavourable) walls is where each layer opposes being squeezed or pushed out (as must occur as the two surfaces approach each other). Such oscillatory force-laws are now well understood theoretically[1]); they are primarily due to the *geometry* of molecules, and they represent the most general type of solvation force between surfaces in liquids.

It is instructive to consider why the oscillatory force between two *flat* surfaces is not smeared out when measured between two *curved* surfaces. The reason becomes apparent when referring to eq. (1) where we see that the one is simply the derivative of the other, i.e. since the derivative of $\sin\theta$ is $\cos\theta$ we may conclude that the oscillations will persist between two surfaces with any curved geometry. However, this is not true for rough surfaces, where now a surface distribution of *random* hills and valleys will result in a smearing out of the oscillations (if these surface asperities are comparable to the molecular diameter σ) and indeed this has been found experimentally[6]).

In addition to the oscillatory solvation force component there can also be a purely monotonic contribution to the total solvation interaction. This monotonic

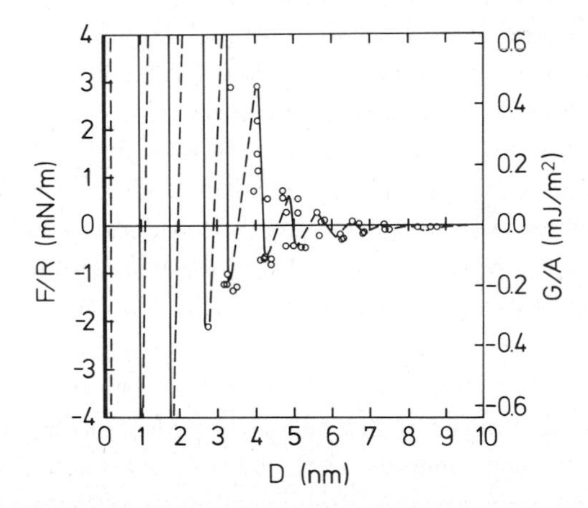

Fig. 2. Measured force between two mica surfaces in octamethyl-cyclo-tetrasiloxane[3]), a nonpolar liquid with nearly spherical molecules of diameter $\sigma \approx 0.85$ nm. Very similar force-laws are obtained in other liquids, including polar and hydrogen-bonding liquids[4,5]). Note that $F > 0$ corresponds to repulsion, while $F < 0$ to attraction.

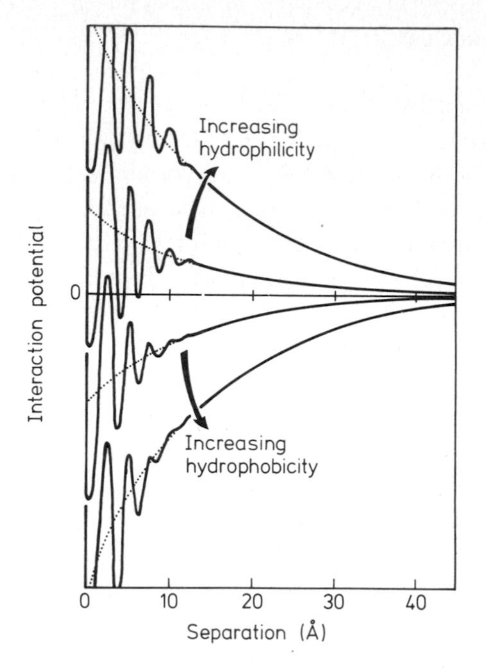

Fig. 3. Schematic short-range solvation (hydration) force-laws in water between surfaces of varying degrees of hydrophilicity[1]) and hydrophobicity[7,8]).

component appears to occur only in water (or in aqueous electrolyte solutions, where water is the solvent[1]) as illustrated in fig. 3. Repulsive solvation forces, or 'hydration' forces, occur between hydrophilic (water-loving) surfaces such as certain clay-mineral surfaces (leading to landslides), silica, and surfactant-water surfaces (stabilizing soap films). Attractive 'hydrophobic' forces in water occur between hydrophobic (water-fearing) compounds such as oils, hydrocarbon and other inert (non-polar) surfaces and groups. These latter forces are responsible for the self-assembly and organization of biological macromolecules into membranes and proteins.

Finally, we must not forget the long-range monotonically attractive Van der Waals force, and the repulsive double-layer force especially between charged surfaces in aqueous solutions[1]). When these are added to the net solvation force in water (fig. 3) we can imagine that the net interaction will be quite a complicated function of distance. A little reflection should make it clear how such interaction potentials can lead to metastable conditions, irreversible processes, time-dependent phenomena and hysteresis effects in many real systems of practical importance.

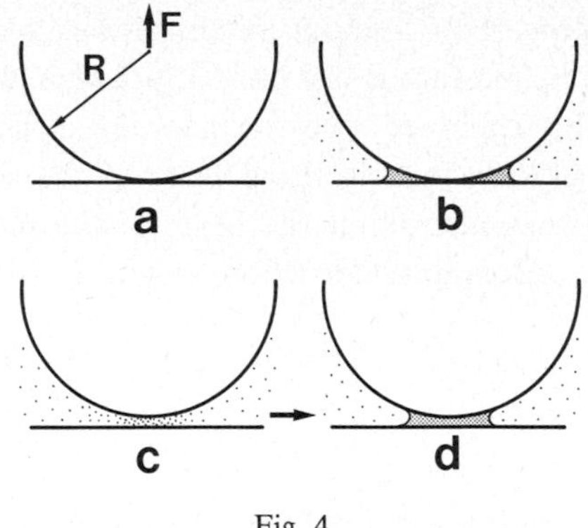

Fig. 4.

4. Adhesion forces

For two surfaces in contact (at $D = 0$) their surface energy (or surface tension) γ_S is defined by $\gamma_S = \frac{1}{2}E(0)$. Thus from eq. (1) the adhesion force (or pull-off force) of a sphere on a flat surface (fig. 4a) will be given by

$$F = 4\pi R\gamma_S \tag{2}$$

which has been verified by direct force measurements[1]). If the vapour consist of molecules which spread on (i.e. wet) the surfaces, then this vapour will 'capillary condense' into a bulk liquid annulus when the surfaces are in contact (fig. 4b), and the resulting adhesion force is now given by

$$F \simeq 4\pi R\gamma_L, \tag{3}$$

where γ_L is the surface tension of the liquid. The interested reader may easily derive this result by multiplying the Laplace pressure γ_L/r of the curved liquid meniscus (of radius r) by the area of the annulus in contact with the surfaces. Eq. (3) has been found to be valid for meniscus radii as small as 1 nm[1,9]).

5. Capillary condensation and surface-induced phase changes

More recently, Christenson and Israelachvili have observed the *spontaneous* (capillary) condensation of liquid bridges between two surfaces even before they come into contact (fig. 4c, d). This phenomenon appears to be quite general and was found to occur even in mixtures of immiscible liquids (e.g. water or methanol condensing out of hydrocarbon liquids). It suggests a type of surface-induced nucleation or phase transition that is not only of practical importance but also of theoretical interest. The effect of two surfaces in close proximity (but not touching) on increasing or decreasing the density of the solvent–solute mixture between them (fig. 4c) may also be at the heart of the monotonically repulsive and attractive solvation forces in water whose origin is still a mystery.

References

1) J.N. Israelachvili, Intermolecular and Surface Forces (Academic Press, London, 1985).
2) J.N. Israelachvili and G.E. Adams, J. Chem. Soc. Faraday Trans 1 **74** (1978) 975.
3) R.G. Horn and J.N. Israelachvili, J. Chem. Phys. **75** (1981) 1400; C.E. Blom and H.K. Christenson, submitted.
4) H.K. Christenson, J. Chem. Phys. **78** (1983) 6906.
5) H.K. Christenson and R.G. Horn, Chem. Phys. Lett. **98** (1983) 45; Chemica Scripta **25** (1985) 37.
6) H.K. Christenson, J. Phys. Chem. **90** (1986) 4.
7) J.N. Israelachvili and R.M. Pashley, Nature **300** (1982) 341.
8) R.M. Pashley, P.M. McGuiggan, B.W. Ninham and D.F. Evans, Science **229** (1985) 1088; P.M. Claesson, C.E. Blom, P.C. Herder and B.W. Ninham, J. Colloid Interface Sci., in press.
9) H.K. Christenson, J. Colloid Interface Sci. **104** (1985) 234.

Physica **140A** (1986) 285–290
North-Holland, Amsterdam

PHASE TRANSITIONS AND METASTABLE STATES IN THE PRESENCE OF RANDOM FIELDS

R.A. COWLEY
Department of Physics, University of Edinburgh, Mayfield Road, Edinburgh, UK

R.J. BIRGENEAU
Department of Physics, Massachusetts Institute of Technology, Cambridge, MA 02139, USA

and

G. SHIRANE
Brookhaven National Laboratory, Department of Physics, Upton, NY 11973, USA

The application of a uniform magnetic field to a diluted Ising antiferromagnet produces a random staggered field and so can be used to study the random field Ising model. In this paper we describe neutron scattering measurements on $Mn_xZn_{1-x}F_2$. They show that at low temperatures there are many metastable states which are all stable for experimentally accessible times. There is a sharp boundary in the H/T plane which separates metastability from ergodicity. The critical phenomena close to this metastability boundary is discussed.

The random field Ising model (RFIM) is one of the simplest problems with competing interactions. The exchange interactions act so as to produce long range order, but the magnetic fields, which are randomly directed at each site, act so as to destroy the order. Ever since the work of Imry and Ma[1]), it has been realized that random fields can produce dramatic effects on phase transitions and ordering. The RFIM has proved, however, to be a more complicated problem than was initially anticipated by either the theorists or the experimentalists, and despite a large body of work, it is still not fully understood.

A very convenient experimental realization of the random field Ising model is obtained by the application of a uniform field to a diluted antiferromagnet[2]); here the uniform field produces a random staggered field due to the randomness in the Zeeman energy. Over the past 6 years a large number of experiments employing a variety of experimental techniques have been performed using $d = 2$ and $d = 3$ antiferromagnets. The progress up until 1983 was reviewed at the preceding Statistical Physics Conference[3]). In this brief review we discuss the progress made since then with emphasis on results obtained using neutron

0378-4371/86/$03.50

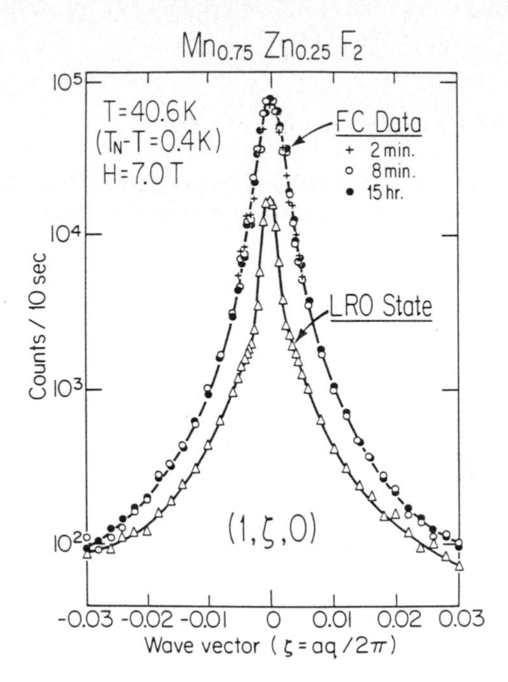

Fig. 1. The wavevector dependence[4]) of the neutron scattering from $Mn_{0.75}Zn_{0.25}F_2$. The fits are to a Lorentzian squared profile FC or to a Lorentzian squared and a Gaussian to describe the Bragg reflection, ZFC, or LRO state.

scattering techniques with the $d = 3$ system $Mn_xZn_{1-x}F_2$ [4]). Very similar results have also been observed with $Fe_xZn_{1-x}Fe_2$ [5]).

In all of the neutron scattering experiments on the diluted $d = 3$ systems, it is found that when the samples are cooled in a field, FC, long range order is not established at low temperature, but the scattering profiles have a characteristic Lorentzian squared form as a function of wavevector. In contrast, if the samples are cooled in the absence of the field, and the field applied, ZFC, long range order is retained until the sample is heated above a well defined temperature. Fig. 1 illustrates this difference at a temperature just 0.4 K below this metastability boundary, $T_M(H)$. Much of the recent work has been performed with a view to elucidating the behavior close to this metastability line. Differences in the FC and ZFC behavior are also observed in macroscopic measurements such as thermal expansion and magnetization[6]).

It is now generally believed that in equilibrium the lower critical dimension of the random field Ising model is 2 [7]). Consequently systems with $d = 3$ are predicted to have a long range ordered ground state, and there have been many theories[8]) of the critical phenomena describing the possibly continuous phase transition into this long range ordered phase at $T_N(H)$. Because none of these theories explain the observed metastability, models have also been developed[9]) to explain this metastability as arising from anomalously large barriers to the motion of domain walls between the antiferromagnetic domains. These theories predict that the size of the domains should increase with time as $\ln(t/\tau)$ where τ is a microscopic time. It is important to emphasize that $Mn_xZn_{1-x}F_2$ has only a very small spin wave energy gap: ~ 1 K at 7 T and 40.0 K. The metastability cannot, therefore, arise from a thermal freezing of the system as there are very many spin waves excited at the relevant temperatures. To probe the predicted time dependence of the domain size the scattering was measured as a function of time after cooling as rapidly as possible from above to below the metastability boundary. Typical results are shown in fig. 1. It was found that the size of the domains at 900 min was 1.01 ± 0.03 times that at 8 min, whereas with any reasonable microscopic time, τ, the logarithm would predict a minimum ratio of 1.14. Similar experiments have been performed by quenching the field as well as the temperature in $Mn_{0.5}Zn_{0.5}F_2$ with similar results[4]). We conclude that the FC state does not decay logarithmically with time.

In an attempt to understand the observed FC behavior in the critical region Villain[10]) and Fisher[10]) have suggested that the barrier heights for domain wall motion diverge just above $T_N(H)$. Further, the barrier height varies sufficiently rapidly with $T - T_N(H)$, in these theories, that one expects a quite well-defined metastability boundary. Close to this metastability boundary King et al.[11]), have indeed found that the ac susceptibility is frequency dependent at anomalously low frequencies. We conclude that anomalously slow dynamics can account for the metastability boundary.

Despite this result, the properties of the metastable phase are not yet fully understood. Fig. 2 shows that the FC inverse correlation length continues to decrease on cooling below the boundary; thus the system is not completely frozen. Secondly, if the field is reduced in the metastable region the system rapidly relaxes to the same state as that achieved by the FC procedure in the new field. In contrast, when the field is raised no contraction of the domains occurs. This asymmetry in the behaviour is possibly consistent with the theories, but it is surprising that the domain size is dependent on temperature, and that the same FC state is always reached from different more disordered configurations.

If the metastability transition at $T_M(H)$ occurs above a continuous phase transition at $T_N(H)$ to long range order, the behaviour of the scattering above the metastability may give information about the exponents to be expected at the continuous phase transition. Fig. 3 shows the inverse correlation length for various fields as a function of temperature. For $H = 0$ the correlation length exponent is $\nu = 0.71$ in agreement with theory for the random exchange Ising model. However fits to the data for higher fields give exponents between 1.0 and 1.9 depending on the field and range over which the data are fitted. The results

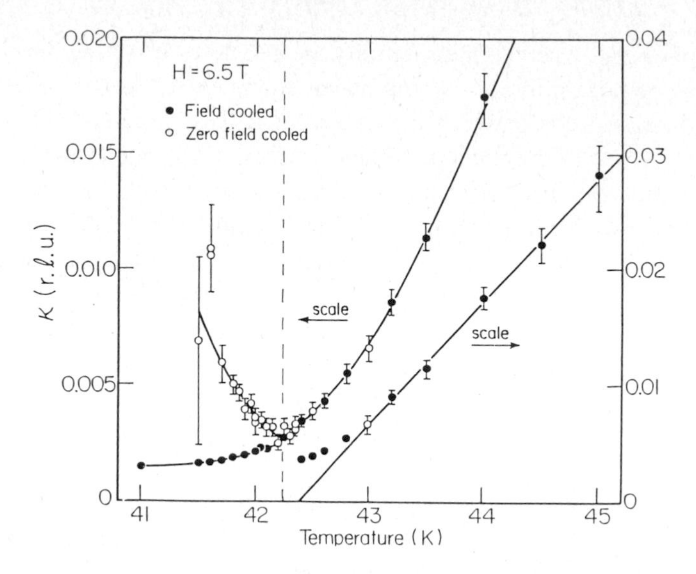

Fig. 2. The inverse correlation length obtained from neutron scattering measurements on $Mn_{0.75}Zn_{0.25}F_2$ in a field of 6.5 T. The dotted line is the metastability boundary.

obtained[5]) in $Fe_xZn_{1-x}F_2$ are very similar to those of $Mn_xZn_{1-x}F_2$. It is therefore very difficult if not impossible to extract these exponents reliably, without a theoretical prediction for the relevant crossover functions.

The amplitude of the Lorentzian squared, shown in fig. 4, increases steadily as the sample is cooled to the metastability temperature. This behavior also occurs at other fields. If there is a continuous phase transition, this Lorentzian squared

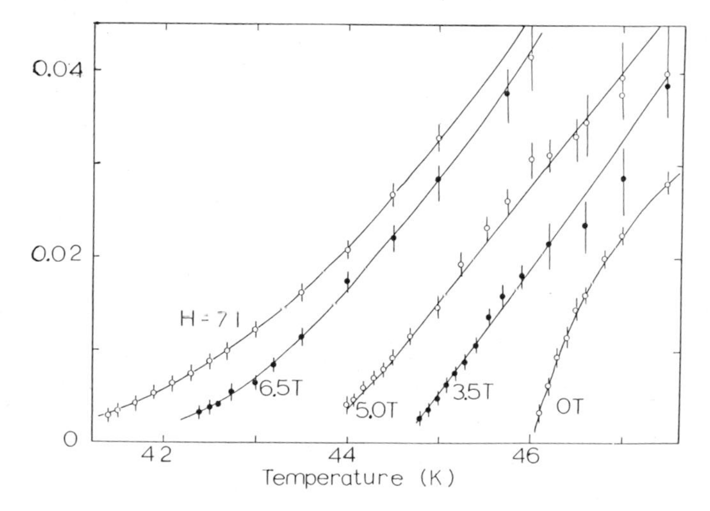

Fig. 3. The inverse correlation length above the metastability boundary for various fields in $Mn_{0.75}Zn_{0.25}F_2$.

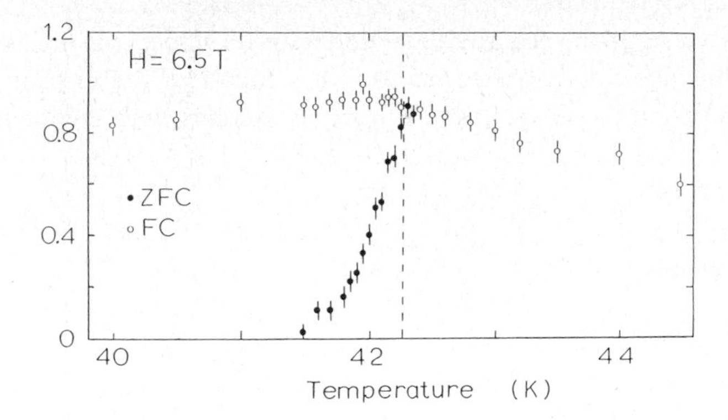

Fig. 4. The amplitude of the Lorentzian squared derived from neutron scattering measurements on $Mn_{0.75}Zn_{0.25}F_2$ in 6.5 T. The dotted line gives the metastability boundary.

term must evolve continuously into the Bragg reflection from the long range order. If $\beta > 0$, this requires that the amplitude vary continuously to zero at the phase transition temperature (that is $\bar{\eta} > 1$) in disagreement with the trend evident in fig. 4.

Alternatively the amplitude could tend to a constant at T_N in which case $\beta = 0$. On the other hand a conventional critical point analysis of the data above the metastability temperature suggests that the amplitude diverges at $T_N(H)$ ($\bar{\eta} < 1$), which is inconsistent if the transition is continuous. We conclude that our data suggest that if there is a transition to long range order it is probably of first order[12]), even if very weakly so.

We have studied the behavior of the ZFC state close to the metastability boundary. With increasing temperature just below the boundary there is a Lorentzian squared component to the scattering whose intensity increases, while the inverse correlation length decreases until the metastability boundary is reached. At that point they have the same values as obtained in the FC experiments. The long range order decreases to zero over a narrow, ~ 0.4 K, temperature interval at the metastability boundary. We conclude that ZFC state undergoes an abrupt, possibly first order transition at the metastability boundary to a state identical with that obtained by the FC procedure. It should be emphasized that in our $Mn_{0.75}Zn_{0.25}F_2$ sample the rounding of $T_N(H = 0)$ due to inhomogeneities is less than 0.025 K.

In conclusion, the random field Ising model has been found to have many unexpected features. At low temperatures there are many metastable states and the theory of these states is not yet completely convincing. It is predicted that if the system remained in equilibrium and the metastability did not occur, there would be a transition to a state with long range order. Our results suggest that this transition, if it occurred, would be of first order. In the future we can expect to see further improvements and more detailed experiments over many decades of frequency close to the metastability boundary. These experiments will hopefully help elucidate not only the metastability found in the random field Ising model, but also similar effects found in other systems with competing interactions.

Acknowledgements

R.A.C. is grateful for the hospitality of Brookhaven National Laboratory where much of this work was done. We are grateful to our collaborators, Prof. H. Yoshizawa and Dr. Y.J. Uemura.

Work at Brookhaven supported by the Division of Materials Sciences, U.S. Department of Energy under contract DE-AC02-76CH00016 and at MIT by the U.S. National Science Foundation – Low Temperature Physics program, contract No. DMR 85-01856 and at Edinburgh by the SERC.

References

1) Y. Imry and S.K. Ma, Phys. Rev. Lett. **35** (1975) 1399.
2) S. Fishman and A. Aharony, J. Phys. C **12** (1979) L729.
3) R.J. Birgeneau, R.A. Cowley, G. Shirane and H. Yoshizawa, J. Stat. Phys. **34** (1984) 817.
4) R.J. Birgeneau, R.A. Cowley, G. Shirane and H. Yoshizawa, Phys. Rev. Lett. **54** (1985) 2147. R.J. Birgeneau, Y. Shapira, G. Shirane, R.A. Cowley and H. Yoshizawa, Physica **137B** (1986) 83 and R.A. Cowley, G. Shirane, H. Yoshizawa, Y.J. Uemura and R.J. Birgeneau (to be published).
5) R.A. Cowley, H. Yoshizawa, G. Shirane and R.J. Birgeneau, Z. für Physik **58** (1984) 15. D.P. Belanger, A.R. King and V. Jaccarino, Phys. Rev. B **31** (1985) 4538. H. Yoshizawa, R.A. Cowley, G. Shirane and R.J. Birgeneau, Phys. Rev. B **31** (1985) 4548.
6) Y. Shapira, N.F. Oliveira and S. Foner, Phys. Rev. B **30** (1984) 6639.
7) J.Z. Imbrie, Phys. Rev. Lett. **53** (1984) 1747; Physica **140A** (1986) 291 (these proceedings).
8) For example: A. Aharony, Y. Imry and S.K. Ma, Phys. Rev. Lett. **37** (1976) 1364. G. Parisi and N. Sourlas, Phys. Rev. Lett. **43** (1978) 744. M. Schwartz, J. Phys. C **18** (1985) 135. Y. Shapir, Phys. Rev. Lett. **54** (1985) 154. A.T. Bray and M.A. Moore, J. Phys. C **18** (1985) L927. A.N. Berker and S.R. McKay, Phys. Rev. B **33** (1986) 4712.
9) J. Villain, Phys. Rev. Lett. **54** (1984) 1543. G. Grinstein and J. Fernandez, Phys. Rev. B **29** (1984) 6389.
10) J. Villain, J. de Physique **46** (1985) 1843. D.S. Fisher, Phys. Rev. Lett. **56** (1986) 416.
11) A.R. King, J.A. Mydosh and V. Jaccarino, Phys. Rev. Lett. **56** (1986) 2525.
12) A.P. Young and M. Nauenberg, Phys. Rev. Lett. **54** (1985) 429.

Physica 140A (1986) 291–297
North-Holland, Amsterdam

ON THE ISING MODEL IN A RANDOM MAGNETIC FIELD

John Z. IMBRIE*
Departments of Mathematics and Physics, Harvard University, Cambridge, MA 02138, USA

We discuss recent progress on the lower critical dimension question for the random-field Ising model, from early domain wall arguments and dimensional reduction to a mathematical proof that the ground state is ferromagnetically ordered in more than two dimensions. In settling this controversial point, a coarse-graining method is used to control the entropy of domain walls, and a "contour gas" representation for the ground state is used to control the loss of ordering at large length scales.

1. Introduction

Our understanding of how random magnetic fields affect Ising systems has evolved considerably over the last decade. Beginning with the argument of Imry and Ma[1]) for long-range order in dimensions $d > 2$, the central question has been to determine the borderline dimension for ferromagnetic order. The lower critical dimension, d_1, is the dimension above which long-range order is possible. Its value has at times been controversial[2]), with 2 and 3 being the leading candidates. Recently, we obtained a rigorous proof that long-range order occurs above two dimensions[3,4]). This result rules out $d_1 = 3$ and definitively settles the controversy. In this article we present some of the physical issues involved in the lower critical dimension question. We explain how the new ideas underlying the proof lead to a more complete understanding of the ordered states and why it exists.

2. The random-field Ising model

We begin with the usual Hamiltonian for spins $\sigma_i \neq 1$:

$$H = \tfrac{1}{2} \sum_{\langle i, j\rangle} (1 - \sigma_i\sigma_j) - \tfrac{1}{2}\sum_i h_i\sigma_i .$$

*Alfred P. Sloan Research Fellow. Supported in part by the National Science Foundation under grants PHY-85-13554 and PHY-84-13285.

0378-4371/86/\$03.50 © Elsevier Science Publishers B.V.
(North-Holland Physics Publishing Division)

Here, i, j are sites in $\mathbb{Z}^d$ and $\langle i, j\rangle$ denotes a nearest-neighbor pair. The random magnetic field h_i varies from site to site on the lattice, normally with independent fields at each site. Since we are interested in questions of symmetry breaking, it is important that the field have a symmetric distribution. For a Gaussian distribution, this means we assume each h_i has mean zero. This ensures that while the field locally favors plus spins or minus spins, there is no average tendency to favor either state. Symmetry breaking can only be induced by the boundary conditions (say $\sigma_i = +1$ for sites i outside Λ, a finite subset of $\mathbb{Z}^d$). If the spin at the origin tends to follow the boundary conditions even in the limit as Λ fills all of $\mathbb{Z}^d$, then we say there is long-range order (LRO). The width of the distribution for h_i is a measure of the strength of the disorder. We suppose that each h_i is Gaussian with variance $\varepsilon^2 \ll 1$. Strong disorder ($\varepsilon \gg 1$) destroys LRO whatever the dimension[5,6]).

The magnetization is computed in a fixed magnetic field configuration h:

$$\langle \sigma_0 \rangle(h) = \frac{1}{Z} \sum_{\{\sigma_i\}} \sigma_0 \, \mathrm{e}^{-\beta H}.$$

Second, this quantity is averaged over the random field h. This is denoted with a bar, so that the magnetization is $M = \overline{\langle \sigma_0 \rangle}$. This quenched prescription carries over to zero temperature ($\beta = \infty$) in which case the thermal average is replaced by a single configuration $\sigma^{\min}$, the one minimizing H. In this case $M = \overline{\sigma_0^{\min}}$. In either case, long-range order is present if $M > 0$ uniformly in Λ. By analogy with the ordinary (non-random) Ising model, LRO should occur only for low or zero temperature and for sufficiently many dimensions, i.e. for $d > d_1$.

Our rigorous result is that LRO occurs at zero temperature and weak disorder for any dimension $d > 2$ [3,4]). The result strongly suggests that the lower critical dimension is 2. Also for small, nonzero temperatures one should expect LRO for $d > 2$ – from the ordinary Ising model we know that the thermal fluctuations are not strong enough to break up LRO unless $d \leqslant 1$. While the behavior at nonzero temperature is not settled rigorously, we believe that the basic tools needed to treat $T > 0$ are already at hand. (Because of the problem of Griffiths singularities[7]), an improved perturbation expansion as in ref. 5 would be needed in addition to the methods of ref. 4.)

3. Random fields in nature

It would be difficult to arrange a random magnetic field in the laboratory. Instead, random fields are mimicked experimentally by placing a randomly diluted antiferromagnet in a uniform magnetic field. The approximate equivalence with the random-field Ising model was observed first by Fishman and

Aharony[8]). A simplified Hamiltonian for such a system is

$$H = \sum_{\langle i, j \rangle} -\tfrac{1}{2}(1 - \sigma_i \sigma_j) - h \sum_i \mu_i \sigma_i, \tag{1}$$

where h is a uniform field, and μ_i is a random magnetic moment. The moments might be either 0 or 1, depending on whether a magnetic or a non-magnetic atom is present at the site. The minus sign in the exchange term makes this system antiferromagnetic. (There should also be a random exchange coupling, but we suppress it for simplicity.) The idea is that the staggered nature of the ground states compensates for the lack of symmetry in the distribution of the μ_i's.

These systems have been studied extensively in the laboratory[9]). Initially the results led to conflicting interpretations on the question of LRO in $d = 3$. Subsequently it became clear that there was a failure to reach equilibrium, so the experiments did not settle the lower critical dimension controversy. Dynamic properties of these systems are still being actively investigated. (See Cowley's contribution to these proceedings for references and for a more complete discussion of these issues.)

It would be interesting to attempt to prove the existence of LRO in the diluted antiferromagnets directly. We would expect that the Hamiltonian (1) could be treated without too much additional work. Versions with random exchange interactions would require more thought, but should not be out of reach of present methods. Other systems that could presumably be treated are interfaces in dimensions larger than 3, either with random fields or with random exchange interactions.

4. Background

Imry and Ma[1]) first argued that $d_1 = 2$ as follows. They asked whether a predominantly plus state would be unstable due to formation of domains of minus spins. A domain of diameter L would cost an energy L^{d-1} from the antiparallel spins at its boundary. On the other hand, the typical fluctuation in magnetic energy inside such a domain is of the order of $L^{d/2}$. (The difference in magnetic energy is $\sum_{i \in \text{Domain}} h_i$, assuming the domain is all plus or all minus. If there are L^d spins in the domain, then by the law of large numbers, the fluctuations are typically $L^{d/2}$.) A comparison shows that if $d > 2$, the surface energy is more important, and ferromagnetic order is possible. In two or fewer dimensions, fluctuations will overcome the surface energy, destroying LRO.

More recently, a number of authors have noticed a correspondence between random field systems in d dimensions and pure systems in $d - 2$ dimensions[10]). This "dimensional reduction" was shown by Parisi and Sourlas[11]) to be a consequence of supersymmetry. A number of authors then questioned the $d_l = 2$ result of Imry and Ma. They argued that the lower critical dimension should be 3 because $3 - 2 = 1$ is the lower critical dimension for the pure (non-random) Ising model. The "failure" of the Imry–Ma argument was explained in terms of roughening of interfaces[12]) but this explanation was objected to by others who maintained that $d_l = 2$ [13]). The physics community was sufficiently divided that a vote was taken by M. Fisher at a conference in Geilo, Norway in 1983. The value $d_l = 3$ was strongly favored. At this point the mathematical physics community became involved, beginning with Fisher, Fröhlich, and Spencer[14]) and Chalker[15]). These authors showed how one of the main shortcomings of the Imry–Ma argument could be circumvented (the entropy problem – see below). Finally, a complete understanding of the ground state was obtained in our proof of LRO for $d > 2$. Mathematical physics deserves credit for a timely resolution of the lower critical dimension question.

5. The entropy problem

One serious problem with the Imry–Ma argument is the possibility that domain boundaries will tend to adjust so as to create domains with the lowest possible energy. Thus it is not sufficient to consider a "typical" domain and to show it is unlikely to be present in the ground state. Rather, we would show that it is unlikely for *any* domain to be present. It is easy to show that the number of domain boundaries (contours) of area A is $e^{\mathcal{O}(A)}$. A given domain with volume V can have its magnetic energy $\Sigma_{i \in V} h_i$ exceed A with probability $\exp(-A^2/\varepsilon^2 V)$. (Each h_i has variance ε^2, so the sum has variance $\varepsilon^2 V$.) Assuming that $A = L^{d-1}$, $V = L^d$, we find that the probability of a large field fluctuation is $\exp(-L^{d-2}/\varepsilon^2)$, which is small for $d > 2$ but not small enough to allow for all $\exp(L^{d-1})$ contours. The exponent L^{d-1} in the counting estimate should be thought of as the entropy for contours of area L^{d-1}.

In ref. 14, entropy is controlled by using coarse-grained contours. A given contour can be approximated on lattices with spacing 2^j, $j = 0, 1, 2, \ldots$. The balance between entropy and small probability (or "energy") is considered separately for each annular region separating successive coarse-grainings. These regions are bounded by contours on a lattice of spacing 2^j, so their entropy is reduced to $L^{d-1}2^{-2j}$. The volume is reduced to $L^{d-1}2^j$, which yields an improved estimate on probabilities. Specifically, the probability of a fluctuation in Σh_i of the order L^{d-1} is

$$\exp\left(- \frac{(L^{d-1})^2}{\varepsilon^2 L^{d-1} 2^j} \right) = e^{-\varepsilon L^{d-1} 2^{-j}} .$$

Evidently, this is sufficient to control $\exp(L^{d-1}2^{-2j})$ choices of regions. The reason for the improvement is the fact that all contours which look the same on the scale 2^j enclose essentially the same volume. We avoid imposing many conditions on h that are essentially the same, and instead look at "difference events" which are much less probable.

6. The contour gas

A second problem with the Imry–Ma argument is the implicit assumption that the configuration inside a domain is perfectly ordered. This was used to evaluate the magnetic energy of a domain as Σh_i. Actually, the magnetic energy is $\frac{1}{2}\Sigma h_i(\sigma_i^{(1)} - \sigma_i^{(2)})$, where $\sigma^{(1)}$ is the configuration without the domain, and $\sigma^{(2)}$ is the configuration with the domain. To the extent that these configurations are perfectly ordered ($+1$ and -1 respectively), the original approximation will be accurate. The problem is that we are interested in proving LRO, so what allows us to assume LRO in domains? For example, in two dimensions, there is presumably no LRO, and the magnetic energy is presumably not well approximated by Σh_i.

The short answer to this question is that one should prove LRO in domains of a certain size, and work inductively to larger and larger domains. At each step one should check the approximations needed for subsequent steps. This is the basic strategy of our proof.

A more complete answer is based on an understanding of the errors in the approximation and how they may grow with the size of the domain. To fully understand LRO in this model we are forced to consider the effect of domains within domains. To this end we obtain in ref. 4 an exact formula for the ground state energy which exhibits all the correction terms:

$$H(\sigma^{\min}) = -\tfrac{1}{2}\sum_i h_i + \sum_\gamma r_\gamma(h). \tag{2}$$

Here $\{r_\gamma\}$ is a collection of random variables (functions of h) labeled by contours γ. The sum runs over all connected contours (not just those in any particular configuration). These random variables are constructed in such a way that r_γ depends only on h in a neighborhood of γ (of diameter comparable to the diameter of γ). This property ensures that r_γ and $r_{\gamma'}$ are statistically independent, provided γ and γ' are separated by a distance comparable to their diameters. Independence is extremely useful when one tries to estimate the probability of a large fluctuation in domain energy.

While we cannot derive the formula here, it is possible to describe how it arises. We approximate the ground state first using only very small domains. A contour is included in the trial ground state if it is present in a local ground state in a neighborhood of γ. Next we get a better approximation using larger contours, possibly discarding contours which were included erroneously on the first pass. The incremental effect of including γ defines r_γ. Each r_γ depends only on h in a neighborhood of γ because the construction is in terms of local ground states in such neighborhoods. After all scales of contours have been tried, we arrive at the true ground state with an energy given by (2). The ground state has been represented as a gas of freely-roaming, almost independent contours.

Using (2) and related formulas for r_γ itself, it is possible to prove inductively estimates on the probability distribution of r_γ. (We express r_γ in terms of $r_{\gamma'}$ with γ' inside γ.) The induction maintains control over the tendency to become less ordered at larger scales. In this way we avoid the catastrophe of two dimensions, where at some large length scale, the contour gas ceases to be dilute, and LRO is lost.

Quantitatively, we can arrange contours so that after N induction steps, we are working on a length scale $\exp\left(\frac{3}{2}\right)^N$. This is because contours of scale l need not occur inside a contour of scale L if $l > L^{2/3}$. When they do occur, they are simply regarded as part of the larger contour. Using the fact that inner contours typically have an area $l^{d-1} \geqslant L^{2(d-1)/3}$, we estimate their positional entropy by

$$\left(\log L^d\right) \cdot \left(\text{number of small contours}\right) \leqslant \left(\log L^d\right) L^{d-2(d-1)/3} \leqslant \mathcal{O}\left(L^{d-1}\right),$$

which causes no difficulties. The probability that a contour of area A and volume V occurs in any local ground state is naively $\exp(-A^2/\varepsilon^2 V)$, but with each induction step we lose a factor of $(\log L)^{10}$ or so in the estimate from summations over scales $j = 1, \dots, \log L$). (L is the diameter of the contour.) After N steps we have lost

$$(\log L_1)^{10} \cdots (\log L_N)^{10} = \left(\frac{3}{2} \cdot \left(\frac{3}{2}\right)^2 \cdot \left(\frac{3}{2}\right)^3 \cdots \left(\frac{3}{2}\right)^N\right)^{10} \sim \mathrm{e}^{\mathcal{O}(N^2)}.$$

Thus the actual bound is

$$\operatorname{Prob}\left(r_\gamma \neq 0\right) \leqslant \exp\left(\frac{-A^2}{\varepsilon^2 V \cdot \exp \mathcal{O}(\log\log L)^2}\right),$$

and since $A^2/V \gtrsim L^{(d-2)/(d-1)}$ it improves as $L \to \infty$ despite the deterioration in $N \doteqdot \log\log L$.

We have tried to present a few of the physical ideas underlying the proof of LRO for $d > 2$. These ideas may be considered independently of the proof. The resulting semi-quantitative analysis treats the effects that were not considered in earlier work and should convince any reasonable physicist that $d_1 = 2$ in the random-field Ising model.

References

1) Y. Imry and S. Ma, Random-field instability of the ordered state of continuous symmetry, Phys. Rev. Lett. **35** (1975) 1399–1401.
2) T. Von Foerster, Random magnetic fields reduce critical dimensionality, Physics Today **36** (1983) 17–19.
3) J. Imbrie, Lower critical dimension of the random-field Ising model, Phys. Rev. Lett. **53** (1984) 1747–1750.
4) J. Imbrie, The ground state of the three-dimensional random-field Ising model, Commun. Math. Phys. **98** (1985) 145–176. The Ising model in a random field: Long range order in three dimensions, in: Critical Phenomena, Random Systems, Gauge Theories, Les Houches 1984. K. Osterwalder and R. Stora, eds. (North-Holland, Amsterdam, 1986).
5) J. Fröhlich and J. Imbrie, Improved perturbation expansion for disordered systems: Beating Griffiths singularities, Commun. Math. Phys. **96** (1984) 145–150.
6) A. Berretti, Some properties of random Ising models, J. Stat. Phys. **38** (1985) 483–496.
7) R.B. Griffiths, Nonanalytic behavior above the critical point in a random Ising ferromagnet, Phys. Rev. Lett. **23** (1969) 17.
8) S. Fishman and A. Aharony, Random field effects in disordered anisotropic antiferromagnets, J. Phys. **C12** (1979) L729–L733.
9) R. Birgeneau, R. Cowley, G. Shirane and H. Yoshizawa, Phase transitions in diluted magnets; Critical behavior, percolation, and random fields, J. Stat. Phys. **34** (1984) 817–847. R. Birgeneau, Y. Shapira, G. Shirane, R. Cowley and H. Yoshizawa, Random fields and phase transitions, in: Frontiers of Neutron Scattering, Physica **137B** (1986) 83–95.
10) A. Aharony, Y. Imry and S. Ma, Lowering of dimensionality in phase transitions with random fields, Phys. Rev. Lett. **37** (1976) 1346–1367. A. Young, J. Phys. **C10** (1977) L257–L262.
11) G. Parisi and N. Sourlas, Random magnetic fields, supersymmetry, and negative dimensions, Phys. Rev. Lett. **43** (1979) 744–745. Supersymmetric field theories and stochastic differential equations, Nucl. Phys. B **206** (1982) 321–332.
12) E. Pytte, Y. Imry and D. Mukamel, Lower critical dimension and the roughening transition of the random-field Ising model, Phys. Rev. Lett. **46** (1981) 1173–1177. D. Muhamel and E. Pytte, Interface fluctuations and the Ising model in a random field, Phys. Rev. B **25** (1982) 4779–4786.
13) G. Grinstein and S. Ma, Roughening and lower critical dimension in the random-field Ising model, Phys. Rev. Lett. **49** (1982) 685–688. Surface tension, roughening, and lower critical dimension in the random-field Ising model, Phys. Rev. B **28** (1983) 2588–2601. J. Villain, Commensurate–incommensurate transition with frozen impurities, J. Physique **43** (1982) L551–L558.
14) D. Fisher, J. Fröhlich and T. Spencer, The Ising model in a random magnetic field, J. Stat. Phys. **34** (1984) 863–870.
15) J. Chalker, On the lower critical dimensionality of the Ising model in a random field, J. Phys. **C16** (1983) 6615–6622.

Physica **140A** (1986) 298–305
North-Holland, Amsterdam

FORMATION OF QUASICRYSTALS

Leonid A. BENDERSKY* and Robert J. SCHAEFER
Metallurgy Division, National Bureau of Standards, Gaithersburg, MD 20899, USA

It has recently been discovered that "quasicrystalline" intermetallic phases can form which have point group symmetries not allowed in periodic crystals; icosahedral, with six five-fold axes, and decagonal, with one 10-fold axis. These phases have long-range orientational order, but they must have translational quasiperiodicity. They usually form at alloy compositions close to those of intermetallic compounds which have crystal structures containing icosahedral groups of atoms. The quasicrystal phases nucleate preferentially in supercooled melts, thus replacing the equilibrium phases.

1. Introduction

One can define two classes of solids according to their structure. One is glassy (amorphous) structures, where the distinct order of local atomic arrangement does not have a long-range correlation. The glassy structures are essentially frozen in undercooled liquids[1]). The other class is crystal structures where both short-range order and long-range translational periodicity are present[2]). The reciprocal space of a crystal is its reciprocal Bravais lattice, where nodes consist of Bragg reflection modulated according to the motif structure. Different deviations from ideal crystalline order (interfaces, chemical disorder, dislocations, etc.) appear as peak broadening and diffuse scattering.

The ordered structures which have Fourier transforms consisting of sharp peaks are not necessarily periodic; for example, incommensurate systems containing interacting subsystems with different lattice periods. These systems can be described as periodic only in higher than 3-dimensional space[3]) and usually have crystallographic point groups. Another class of structures with non-periodic Fourier transform is quasicrystals, in which incommensurability is a consequence of geometrical properties of their non-crystallographic 3-D point group[4]). Quasicrystals have an underlying quasiperiodic lattice which can be generated by cut and projection from higher-dimensional periodic lattice (see references in ref. 4). The existence of particular types of quasicrystal–icosahedral 3-D Penrose tiling

*Also with the Center for Materials Research, The Johns Hopkins University, Baltimore, MD 21218, USA.

0378-4371/86/$03.50

was first speculated by Mackay[5]) and later by Levine and Steinhardt[4]). The discovery of the Al–Mn icosahedral phase by Shechtman et al.[6]) seems to be the first experimental evidence that such quasicrystals exist. Additional support to the idea of quasicrystal came when another type of quasicrystal, decagonal phase, was found in Al–Mn and some other systems[7]).

2. Crystallography of icosahedral and decagonal phases

All selected area electron diffraction (SAD) patterns of the icosahedral phase[6]) are incommensurate with the golden mean being the parameter of incommensuration (fig. 1a–c). There is no 3-D periodic Bravais lattice, but the positions and intensities of the reflections correspond perfectly to the Fourier transform of the icosahedral quasilattice (e.g., a 3-D Penrose tiling) which is a special projection of a 6-D primitive cubic lattice (references in ref. 4). The SAD patterns of the decagonal phase (fig. 1d–f) have 1-D periodicity, except for one (fig. 1d) which has 10-fold rotation symmetry and can be described as the Fourier

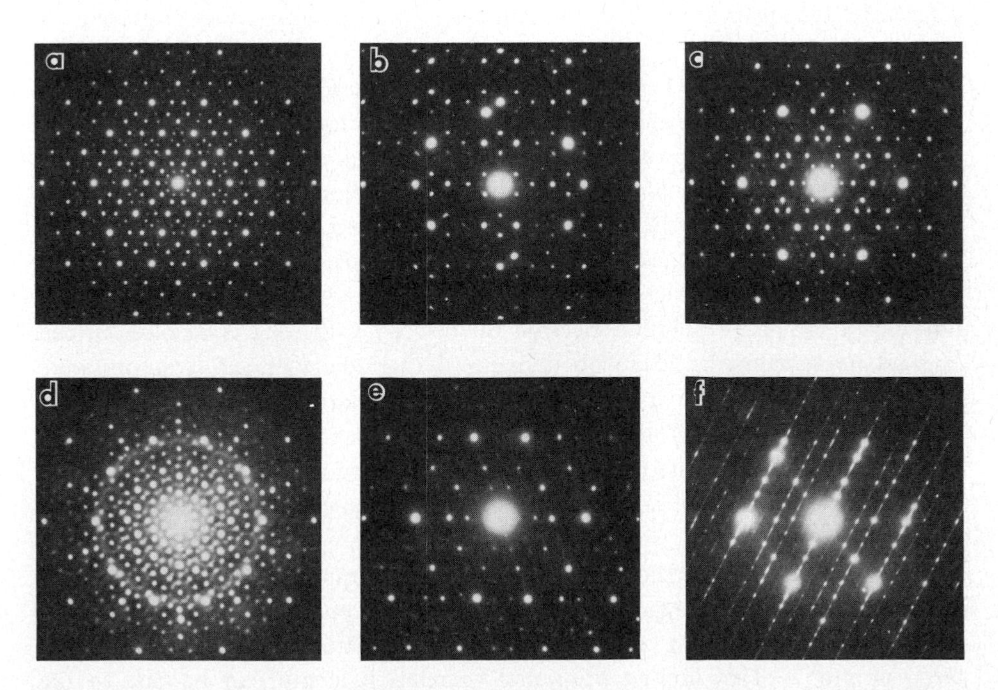

Fig. 1. Selected area electron diffraction patterns from icosahedral (a)–(c) and decagonal (d)–(f) phases. (a) Five-fold; (b) two-fold; (c) three-fold; (d) ten-fold; (e) pseudo-five-fold; (f) pseudo-three-fold.

transform of a 2-D Penrose tiling. The reciprocal lattice of the decagonal phase is thus a periodic stacking of reciprocal quasilattices of the 2-D Penrose tiling along the 10-fold axis. The results of powder X-ray and neutron diffraction are in perfect agreement with electron microscopy for the icosahedral phase[6,8–10]), whereas for the decagonal phase only the preliminary results are available[11]).

The point groups of both quasicrystals were determined by using convergent beam electron diffraction[7,12]): $m\overline{53}$ for the icosahedral phase and 10/m (or 10/mmm) for the decagonal phase. Neither of these two point groups is a subgroup of the other[13]); therefore a second order transformation (of ordering or displacements) between these two phases is not possible.

There is no space group for the icosahedral phase in 3-D, since a translational group does not exist. However, there are 6-D space groups of 5 possible cubic Bravais lattices, and therefore different projections of the 3-D icosahedral lattice. That of the icosahedral phase was shown to be the simple cubic[13]), according to the scaling properties of the diffraction pattern. Different decagonal phases can have different parameters of periodicity, e.g., ~ 1.24 nm for Al–Mn phase and ~ 1.65 nm for Al–Pd and Al–Fe phases (ref. 7 and its references). SAD patterns of both types of the decagonal phase show a systematic absence of odd reflections along the 10-fold c^* axis[7]) (also supported by kinematic X-ray diffraction[11]). Therefore, the space group $P10_5/m$ with a ten-fold screw axis was proposed for the decagonal phases[7]).

The described quasiperiodic features of the reciprocal space of both phases result from a truly quasiperiodic arrangement of atoms and not from either multiple twinning or approximate icosahedral symmetry of a crystal with a large unit cell. The conventional crystallographic explanations are unable to account for a large bulk of experimental evidence, as was demonstrated in ref. 14. The most direct evidence comes from field ion microscopy where definitely no twin boundary and no periodic structure are observed[15]). The same conclusions are obtained from high-resolution electron microscopy with ~ 2 Å resolution (ref. 16 and its references), where no domains of periodic twinned structure were observed. However, the images show a high regularity of its projected 3-D quasiperiodic structure. There is an excellent agreement between the experimental and computer-simulated structural images, where for the model we used the icosahedral quasiperiodic lattice with the nodes occupied by single and complex scatterers[16]).

If quasicrystals have ideal, continuously occupied quasilattices, the diffraction will have δ-function Bragg peaks. Real icosahedral phase materials show peak broadening corresponding to a mean positional correlation range on the order of ten's of nm[8,9]). This limited positional correlation length can be due to the presence of growth defects or to the intrinsic properties of the quasicrystals. A defect substructure with 40–100 nm domains was observed in icosahedral

Al–Mn–Si[9]). However, with this type of defect each diffracted peak should have the same width, which is not so[8]). An attempt has been made to correlate the width of peaks with $\boldsymbol{G}_{\perp}$ (the phason momentum, or the counterpart of the real scattering vector $\boldsymbol{G}$ of 6-D $\boldsymbol{G}$)[17]). The observed trend was found to be roughly consistent with the current theories either of phason strain (refs. 17, 20 and its references) or random icosahedral packing[18,19]). The anisotropic phason strain can also produce a shift of diffracted spots which was experimentally observed[20,21]).

3. Physical metallurgy of phase formation

3.1. *Relation between crystalline and quasicrystalline phases*

It has recently been reported[22]) that in the Al–Li–Cu system one of the equilibrium phases is icosahedral, and this phase can be formed by conventional slow solidification or solid state precipitation processes. In all other systems where icosahedral phases have been found, they have been metastable and were formed only by special processing methods, usually within particular groups of alloys thought to be particularly susceptible to their formation.

There are many intermetallic crystals in which atoms are arranged in local icosahedral groupings. These include, among others, the Frank–Kasper phases which contain combinations of polytetrahedral configurations of atoms with coordination numbers 12, 14, 15 and 16. Ramachandrarao and Sastry[23]) demonstrated that when one such alloy $[Mg_{32}(Al, Zn)_{49}]$ was rapidly solidified, a quasicrystal with long-range icosahedral symmetry was formed instead of the equilibrium phase, which is body-centered cubic with an extended icosahedral grouping of atoms at each lattice point. A similar approach has lead to the formation of icosahedral quasicrystals in Mg_4CuAl_6 [24]), Cd–Cu [25]), and $(Ti, V)_2Ni$ [26]). The other major group of alloys which has been formed into icosahedral and decagonal quasicrystals is the aluminum–transition metal series[6,20]).

In the binary aluminum–transition metal alloy systems, there are several cases where different icosahedral groups of atoms are present in equilibrium crystal structures. However, evidence from EXAFS[27]), Mossbauer spectroscopy[28]) and NMR[29]) all indicate that the icosahedral phase of Al–Mn is not formed from groups of atoms with a central Mn atom surrounded by a single icosahedral group of 12 aluminum atoms as found in the $Al_{12}Mn$ phase. Instead, such evidence points to larger icosahedral clusters of atoms such as are found in the ternary α phase of Al–Mn–Si.

The addition of Si to Al–Mn alloys is known to stabilize the icosahedral phase[30,31]), and possible similarities between the icosahedral phase and the α phase of Al–Mn–Si have been pointed out[32,33]). α(AlMnSi) phase has an almost body-centered cubic lattice with each lattice point being occupied by a large double-layered icosahedral cluster of atoms, usually described as a Mackay icosahedron. Thus formation of the icosahedral phase in Al–Mn–Si may be attributed to the same principle which was used to explain its formation in $Mg_{32}(Al,Zn)_{49}$.

There are also several Al–transition metal phases with crystal structures which are not known in detail, especially in the Al–Mn and Al–Cr systems. Recent diffraction evidence[34]) indicates that some of the Al–Mn phases of composition close to Al_4Mn contain icosahedral clusters. According to this work, it seems possible to describe the structure of quasicrystalline and crystalline Al–Mn phases roughly as an agglomeration of the same type of icosahedral clusters, joined in different ways such that either a periodic or a quasiperiodic lattice is formed. The cluster was shown to be oriented uniquely for the icosahedral phase, in two orientations for decagonal and in three for a hexagonal phase.

3.2. *Formation of quasicrystal phases*

In contrast to the formation of amorphous solids, the formation of quasicrystals is a first-order nucleation and growth process somewhat similar to that of conventional crystals. Thus the initial TEM images of icosahedral Al–Mn[7,19]) formed by rapid solidification of a molten alloy showed that the icosahedral phase had grown radially outward from a central point of origin with a branched or dendritic structure typical of diffusion-controlled crystal growth into an undercooled melt. Rapid solidification from the liquid state can be carried out by several different processes, but a common feature is that small volumes of liquid are used so that heat can be extracted rapidly and the probability of the presence of heterogeneous nucleants for the equilibrium phases is small. Under these conditions, the alloy melt can become highly supercooled and it is found that the nucleation rate of the icosahedral phase is large. Increasing cooling rates lead to greater rates of nucleation and correspondingly smaller grain sizes, until in extreme cases[35]) the size of the individual grains is so small that the structure is indistinguishable from glassy structures with certain local order. This material must be contrasted to true metallic glasses, however, in which an amorphous structure is formed by completely avoiding nucleation in the liquid; in the present case the structure is formed by an exceedingly high rate of nucleation. It has been speculated[35]) that preexisting icosahedral clusters in the undercooled melt can promote the observed homogeneous nucleation, being both centers for nuclei and reducing icosahedral phase/melt interface energy.

The growth of quasicrystals following their nucleation is in many ways similar to that of conventional crystals in a supercooled melt. In most cases the quasicrystals grow with a composition different from that of the melt, thus requiring solute diffusion to occur in the liquid. This is a slow process and leads to the formation of a dendritic (branched) structure, in which the dendrites grow preferentially in the directions of the three-fold symmetry axes[30]). The selection of the preferred dendrite growth direction must be attributed to the anisotropic characteristics of the solid–liquid interface, but there is no direct evidence to indicate whether this is an anisotropy of the surface energy or of the atomic attachment kinetics. Surface energy effects are expected to dominate with unfaceted solid–liquid interfaces while faceted interfaces indicate the importance of kinetic effects. Facets are only occasionally seen[9]) on the planes of five-fold symmetry. If these are the most densely packed planes of the solid–liquid interface, they would have the lowest surface energy and would also be the most difficult surfaces on which to add new layers of crystal. Either effect would result in preferred growth along the three-fold directions.

Quasicrystals have also been formed in the Al–Mn system by several processes other than rapid solidification of liquids, the success of which indicates that although the quasicrystal phases are not equilibrium phases of this system, they are not far from being stable and are frequently favored by kinetic factors. These processes include ion beam mixing[36]), ion irradiation[37]), ion implantation[38]), solid state interdiffusion[39,40]) and rapid pressurization[41]). There have been several cases in which icosahedral crystals are reported to form by devitrification of metallic glasses. This occurs within a very narrow composition range in the Pd–U–Si system[42]). It is also reported to occur in Al–Mn, using amorphous alloys formed by evaporation or ion beam mixing with cold substrates[36,37]). In such cases, extremely fine-grained structures are obtained, again indicating very high rates of nucleation.

4. Some conclusions (and more questions)

We presented experimental results on the structure and formation of quasicrystals, which lead to some conclusions but raise many questions.

1. The following empirical rules can be suggested for formation of quasicrystals:

(a) Suitable alloy systems contain intermetallic compounds where icosahedral clustering is extensively present.

(b) The alloy melt has to be undercooled to nucleate the usually metastable quasicrystals without forming more stable phases. This is most readily accomplished if the alloy composition is off stoichiometry of the intermetallic compounds.

(c) Phase diagrams with peritectic reactions are more favorable than eutectics, since condition (b) can be fulfilled without forming glass.

Questions. In the crystals of the Al–Mn system there are 3 types of icosahedral clusters; a shell of 12 aluminum with a central Mn in $Al_{12}Mn$ G phase, possible Mackay icosahedron in Al_4Mn and truncated icosahedra in Al_6Mn. Why is only one realized as quasicrystal? Do icosahedral phases of different metallurgical systems have the same local order?

2. The projection of a quasiperiodic lattice from higher dimension seems to be a well established description for the crystallography and structure of quasicrystals.

Questions. Should quasicrystals be considered as decorated tiles? Or does the quasilattice have to be rearranged (i.e., disordered) to allow symmetric grouping of atoms to dominate in a structure which is still quasiperiodic but not deterministic? What determines formation of quasicrystals: energetics of cluster formation and entropy of their packing, or long-range quasiperiodic interaction? Is the observed position disorder an intrinsic property of the structure, or is it due to metastable growth defects? The answers are expected from the presumably equilibrium Al–Li–Cu icosahedral phase.

3. Nucleation of the icosahedral phase is homogeneous and its subsequent growth is relatively slow and proceeds in the 3-fold directions.

Questions. Does the nucleation behavior of the icosahedral phase mean preexistence of the icosahedral clusters in liquid? How do atoms become attached to the growing quasicrystal surface? Is it an atom-by-atom transfer or cluster-by-cluster transfer? What is the critical nucleus of the icosahedral phase and how is it related to the cluster size?

References

1) S. Sachdev and D.R. Nelson, Phys. Rev. B **32** (1985) 1480 and references.
2) B.K. Vainstein, Modern Crystallography (Springer, Berlin, 1981), vol. 1.
3) A. Janner and T. Janssen, Acta Cryst. A (1980) 408.
4) D. Levine and P.J. Steinhardt, Phys. Rev. Lett. **53** (1984) 2477; Phys. Rev. B **34** (1986) 596.
5) A.L. Mackay, Acta Cryst. **15** (1962) 916; Physica **114A** (1982) 609.
6) D. Shechtman, I. Blech, D. Gratias and J.W. Cahn, Phys. Rev. Lett. **53** (1984) 1951.
7) L. Bendersky, Phys. Rev. Lett. **55** (1985) 1461; J. de Phys. Colloq. **C3** (1986) 457.
8) P.A. Bancel, P.A. Heiney, P.W. Stephens, A.I. Goldman and P.M. Horn, Phys. Rev. Lett. **54** (1985) 2422.

9) J.L. Robertson, M.E. Misenheimer, S.C. Moss and L. Bendersky, submitted to Acta Met.
10) B. Mozer, J.W. Cahn, D. Gratias and D. Shechtman, J. de Phys. Colloq. **C3** (1986) 351.
11) Chr. Janot, J. Dannetier and A. Pianelli, submitted to Phys. Lett. A.
12) L.A. Bendersky and M.J. Kaufman, Phil. Mag. B **53** (1986) L75.
13) J.W. Cahn, D. Shechtman and D. Gratias, J. Mater. Res. **1** (1986) 13.
14) J.W. Cahn, D. Gratias and D. Shechtman, Nature **319** (1986) 1478; P.A. Bancel, P.A. Heiney, P.W. Stephens and A.I. Goldman, ibid.
15) A.J. Melmed and R. Klein, Phys. Rev. Lett. **56** (1986) 1478.
16) M. Cornier, R. Portier and D. Gratias, Inst. Phys. Conf. Ser. **78** (1985), chap. 3, p. 91; M. Cornier, K. Zhang, R. Portier, and D. Gratias, J. de Phys. Colloq. **C3** (1986) 447.
17) W. Malzfeldt, P.M. Horn, D.P. Di Vincenzo, B. Stephenson, R. Gambino and S. Herd, J. de Phys. Colloq. **C3** (1986) 379.
18) P.W. Stephens and A.I. Goldman, Phys. Rev. Lett. **56** (1986) 1168.
19) D. Shechtman and I. Blech, Met. Trans. **16A** (1985) 1005.
20) P.A. Bancel and P.A. Heiney, J. de Phys. Colloq. **C3** (1986) 341; Phys. Rev. B **33** (1986) 7917.
21) M. Tanaka, M. Terauchi, K. Hiraga, and M. Hirabayashi, Ultramicroscopy **17** (1985) 127.
22) P. Sainfort and B. Dubost, J. de Phys. Colloq. **C3** (1986) 321.
23) P. Ramachandrarao and G.V.S. Sastry, Pramana **24** (1985) L225.
24) G.V.S. Sastry, P. Ramachandrarao and T.R. Anantharaman, Script Met. **20** (1986) 191.
25) L.A. Bendersky, unpublished work.
26) Z. Zhang, H.Q. Ye and K.H. Kuo, Phil. Mag. Lett. **52** (1985) L49.
27) E.A. Stern, Y.A. Ma and C.E. Bouldin, Phys. Rev. Lett. **55** (1985) 2172.
28) L.J. Swartzendruber, D. Shechtman, L. Bendersky and J.W. Cahn, Phys. Rev. B **32** (1985) 1383.
29) L.H. Bennett, J.W. Cahn, R.J. Schaefer, M. Rubinstein and G.H. Stauss, to be published.
30) R.J. Schaefer, L.A. Bendersky, D. Shechtman, W.J. Boettinger and F.S. Biancaniello, to appear in Met. Trans. A (1986).
31) C.H. Chen and H.S. Chen, Phys. Rev. B **33** (1986) 2814.
32) V. Elser and C.L. Henley, Phys. Rev. Lett. **55** (1985) 1883.
33) P. Guyot and A. Audier, Phil. Mag. **52B** (1985) L15.
34) L.A. Bendersky, submitted to J. Microscopy.
35) L.A. Bendersky and S.D. Ridder, J. Mater. Res. **1** (1986) 405.
36) D.M. Follstaedt and J.A. Knapp, J. Appl. Phys. **59** (1986) 1756.
37) D.A. Lilienfeld, M. Nastasi, H.H. Johnson, D.G. Ast and J.W. Mayer, Phys. Rev. Lett. **55** (1985) 1587.
38) J.D. Budai and M.J. Aziz, Phys. Rev. B **33** (1986) 2876.
39) D.M. Follstaedt and J.A. Knapp, Phys. Rev. Lett. **56** (1986) 1827.
40) K. Urban, J. Mayer, M. Rapp, M. Wilkens, A. Csanady and J. Fidler, J. Phys. Colloq. (1986).
41) J.A. Sekhar and T. Rajasekharan, Nature **320** (1986) 153.
42) S.J. Poon, A.J. Drehman and K.R. Lawless, Phys. Rev. Lett. **55** (1985) 2324.

Physica **140A** (1986) 306–311
North-Holland, Amsterdam

ORIGINS OF QUASICRYSTAL ORDERING

Christopher L. HENLEY*
Department of Physics, Cornell University, Ithaca, NY 14853, USA

I survey current models of the structure and formation of quasicrystals.

1. Introduction

Recently, metallic alloys were discovered exhibiting sharp diffraction patterns with icosahedral[1,2]) or decagonal symmetry groups[2]), which include fivefold symmetry axes. The sharpness of the diffraction peaks suggests long-range translational order, as in a periodic crystal, but fivefold axes are incompatible with periodicity[3]). This immediately poses two physical questions:

(1) How should we describe such a long-ranged order?

(2) By what mechanism or mechanisms is it engendered?

Considerable progress has been made on the first question, beginning with an independent theoretical work based on Penrose tilings[4]). The second question is still open. In this paper, I survey what is known (as of August 1986) about the possible origins of quasicrystal ordering. Space does not permit a real exposition of the various approaches which have been proposed, so I have tried to present a "road map" indicating their relationships for the statistical mechanician contemplating the rapidly expanding literature of quasicrystal theory. No comprehensive review has appeared; currently the best substitute is the Proceedings of the Les Houches Workshop on Aperiodic Crystals [(March, 1986), in J. Physique Colloq. **47** (1986) C3]. General or semi-popular articles may also be helpful[5,6]). The preceding paper[2]) discusses experiment – which is far ahead of theory at the moment! A more complete version of my survey is in preparation[7]).

Many intermetallic alloys with different types of local order[8]) form icosahedral structures. Quasicrystal ordering may have different origins in different systems. (There is no unique cause for cubic crystal ordering!) One scheme posits peculiar microscopic forces which have a quasicrystal as a ground (or at least metastable) state. A likelier scheme suggests that the quasicrystal ordering is due to entropy

*On leave from Department of Physics, Boston University, Boston, MA 02215, USA.

0378-4371/86/$03.50 © Elsevier Science Publishers B.V.
(North-Holland Physics Publishing Division)

rather than energy considerations and incorporates intrinsic disorder, somewhat like a glass[9-14]).

One should immediately distinguish between orientational order and translational order[5,6]). Orientational order means that there is some way we can construct a frame from the local arrangements of atoms and these frames are oriented equivalently in distant parts of the structure. Long-range translational order will be taken as equivalent to the existence of Bragg diffraction. Experimentally, orientational order is observed across grains which have diameters of ~ 10 μm (much larger for icosahedral AlCuLi), while X-ray diffraction indicates translational correlations to only 100 Å or so[2]). This is still large enough to suggest that a reasonable idealized model *might* develop true long-range translational order.

2. Abstract models: Penrose tilings and density waves

Fig. 1a illustrates a two-dimensional Penrose tiling[15,16]). It has perfect tenfold orientational order, by construction. Furthermore, it is "quasiperiodic": the Fourier transform is entirely composed of δ-functions. Thus, it has perfect long-range (yet *aperiodic*) translational order and can serve as a *geometrical* toy model of a quasicrystal[4,17]). Three-dimensional generalizations of the Penrose tiling exist[18,19]) with similar properties[10,19-21]).

A more trivial geometrical model is a sum of density waves[21-26]), e.g. (in $d = 2$),

$$\rho(\boldsymbol{x}) = \sum_{n=1}^{5} \rho_0 \, \mathrm{e}^{\mathrm{i} q_0 \hat{\boldsymbol{e}}_n \cdot \boldsymbol{x}} + (\text{higher harmonics}), \tag{1}$$

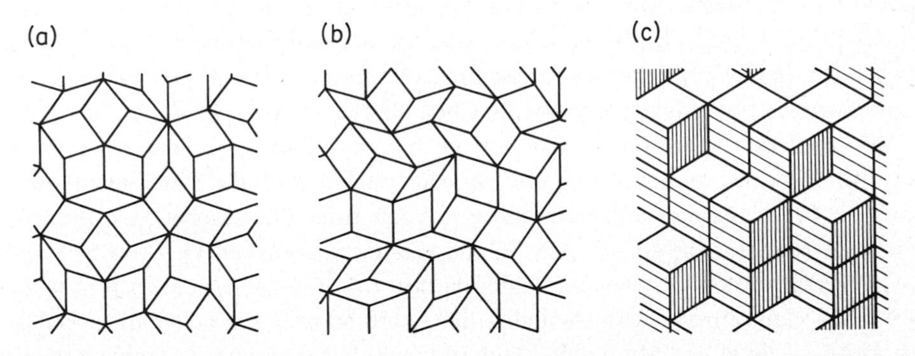

Fig. 1. (a) Perfect two-dimensional Penrose tiling by rhombi with 36° and 72° acute angles. (b) Random tiling by the same tiles. (c) Random tiling by rhombi with 60° acute angles: each orientation of tile is shaded differently.

where the basis vectors $\hat{e}_n$ point toward the vertices of a pentagon. The density $\rho(x)$ in (1) is manifestly quasiperiodic.

The first *physical* models to describe icosahedral ordering were based on Landau free-energy functionals of $\rho(x)$, which serve as the order parameters for the translational order[21-26]). Proper choices of functional had ground (or metastable) states of the form (1). These theories have been reviewed elsewhere[26]). Special note should be taken of Jarić's version[24]): his free energy had an additional *orientational* order parameter coupled to the density. Then an icosahedral orientational ordering could induce an icosahedral translational ordering – as in the microscopic models discussed below.

There is not yet any understanding of the connection between the density $\rho(x)$ and the microscopic structures. The density wave models are best viewed as nontrivial toy models which do order icosahedrally and conveniently describe properties (phonons, phasons, dislocations) which depend only on symmetry considerations[27]).

3. Atomic structure of quasicrystals

What are the actual atomic arrangements of atoms in the real icosahedral alloys, and how are these related to the diffraction pattern? The straightforward approach is to consider the density to be a cut through a truly periodic density in a higher (D) dimensional space[21,28,29]), as is valid for any incommensurate crystal. (Note how $\rho(x)$ in (1) is a periodic function in $\mathbb{R}^5$ of y with components $y_n \equiv \hat{e}_n \cdot x$.) Then the whole apparatus of crystallography (Bravais lattices, etc.) is applicable[29]). But this approach is unlikely to be useful for understanding the microscopic structure since quasicrystals surely have much disorder[30]).

The most successful approach is the "decoration" of tilings (or other rigid geometries) by atoms. This means that the same atoms are placed at equivalent places in each kind of tile, as in unit cells of a crystal structure. Generally, a decoration model contains two levels: first, the specification of a rigid geometry (this is trivial for ordinary crystals, but not here!); second, the decoration rule.

This approach is justified a posteriori in that crystalline structures exist which may be described as decorations of *periodic* packings of the same geometrical units[31]). From the crystalline structures we can infer the decoration rules, and guess the rules for the construction of the geometry. Icosahedral AlMn[31,32]) and AlCuLi[8]) correspond to two quite different decorations of the same geometry, the 3-dimensional Penrose tiling. Actually, the AlMn structure is better described as a packing of large $Al_{42}Mn_{12}$ units, each of which has icosahedral symmetry[31,32]). These icosahedral units are related by displacements along 3-fold symmetry axes in a staggered fashion, so that the neighboring units are oriented the same way.

Investigations using EXAFS[33]) and X-ray, neutron and electron diffraction[2]) are consistent with this model.

4. Quasicrystals: random geometries and growth

As a crystal grows, it maintains long-range order, not through long-range forces but merely because the ideal local structure propagates periodically. However, to produce a *perfect* quasiperiodic structure, the atoms must satisfy nonlocal constraints, which is unphysical. The resolution is to abandon perfect models for random models[9]).

Current structural models are all variations on the same theme: units of icosahedral symmetry which form "bonds" which ensure that all the units are oriented in the same way. The "bonds" are rigid in that they have well-defined lengths and well-defined orientations relative to the icosahedral units. Thus, we can draw a network, connecting neighboring units whenever they are related by a "bond" of the allowed type.

The models can be classified along a spectrum by how random and how well-connected this network is; this will determine the degree of translational order.

(1) Perfect quasicrystal: deterministic, quasiperiodic structure, such as the Penrose tiling.

(2) Random quasicrystal: random, rigid network which is well connected. That is, nearby units are connected by allowed bonds. An example is the *random* tiling[9,34]) of fig. 1(b). Such models are conjectured to have a nonzero configurational entropy, yet develop long-range quasiperiodic order in dimensions $d > 2$.

(3) "Pseudo-quasicrystal": broken network (not illustrated), which fills space, but contains long wall defects across which nearby units are *not* properly connected.

If we picture physical d-dimensional space as a (hyper)plane cutting through a D-dimensional cubic lattice $\mathbb{Z}^D$ at a particular incommensurate orientation, we find that each site occupied by a unit is the projection of a vertex of $\mathbb{Z}^D$; thus the bond network can be "lifted" to, and embedded in, $\mathbb{Z}^D$[10,14]). For the icosahedral case $d = 3$, $D = 6$. Just as fig. 1(c) is obviously the projection of (square) faces from $\mathbb{Z}^3$ (which form the surface of a cubic crystal), so fig. 1(b) is the projection of faces from $\mathbb{Z}^5$. It can be shown that the Fourier transform of the structure has Bragg spots (i.e., there is long-range order) if and only if the "height" $\boldsymbol{x}^\perp$ (displacement of network from the physical plane) has a finite r.m.s. fluctuation[10]), $\langle|\Delta\boldsymbol{x}^\perp|^2\rangle < \infty$. The spot positions $\boldsymbol{q}$ are projections of a D-dimensional reciprocal lattice; the intensities are largest for small values of the perpendicular component of wavevector, $\boldsymbol{q}^\perp$ (defined analogously to $\boldsymbol{x}^\perp$)[10,15,20,21]).

A perfect quasicrystal (class 1) corresponds to $|x^{\perp}| \simeq 0$, the "cut-and-project" method[10,20,21]) in which the embedded network approximates the physical plane as closely as possible. For class 2, by analogy to interface models (see fig. 1(c) again), we plausibly conjecture[14,34])

$$F \sim \int \mathrm{d}^d x \, |\nabla x^{\perp}(x)|^2,$$

which (like the harmonic crystal or XY ferromagnet in the spin–wave approximation) gives long-range order in $d > 2$. Besides the Bragg peaks, the diffraction pattern has a diffuse continuous background due to the randomness.

For the "pseudo-quasicrystal", class 3 (also known as the generalized Hendricks–Teller model[11]), the wall defects correspond to gaps ("tears") in the embedded network, across which $x^{\perp}$ changes discontinuously[14]). The diffraction peaks have nonzero, but surprisingly small widths[11,12]), which grow as a power of $|q^{\perp}|$.

A statistical–physics system of icosahedral units, interacting by potentials with angular dependences that favor (but do not constrain) the rigid "bonds", might exhibit a rich phase diagram: a high-temperature disordered phase, a "pseudo-quasicrystal" phase with only orientational[5]) icosahedral order, a quasicrystal phase (stabilized by its large entropy) and an ordinary crystal phase at low temperatures.

However, real quasicrystal formation may better be described by a nonequilibrium process of aggregation[9,11]). Growth instabilities would seem to make wall defects more likely, favoring the "pseudo-quasicrystal" case. Also, the growth direction breaks symmetry and forces a systematic linear "tilt" ($\Delta x^{\perp} \propto \Delta x$) of the network embedded in $\mathbb{Z}^6$. At least, that is an interpretation[12,14]) of recently observed distortions of the diffraction pattern from icosahedral symmetry[35]).

In conclusion, we have scarcely scratched the surface of the problem of the ordering (and its kinetics[2])). Large-scale simulations of the freezing are needed, as well as electronic structure calculations to explain the stability and "bonding" of the units. The ultimate importance of quasicrystals may be in the "spinoff" of methods applicable to many other complex structures.

Acknowledgements

This paper is suffused with concepts due to Veit Elser, who also gave detailed suggestions about a draft. I thank N.D. Mermin and S.M. Troian for comments and conversations. This work was supported by NSF grant DMR-8314625 and an IBM Postdoctoral Fellowship.

References

1) D. Shechtman, I. Blech, D. Gratias and J.W. Cahn, Phys. Rev. Lett. **53** (1984) 1951.
2) L.A. Bendersky and R.J. Schaefer, Physica **140A** (1986) 298 (preceding paper).
3) A material with 12-fold symmetry (also forbidden in crystals) may also exist; see Ishimasa et al., Phys. Rev. Lett. **55** (1985) 511.
4) D. Levine and P.J. Steinhardt, Phys. Rev. Lett. **53** (1984) 2477.
5) D.R. Nelson and B.I. Halperin, Science **229** (1985) 233.
6) D. Gratias, La Recherche **17** (1986) 788; D.R. Nelson, Scientific American **255** (Aug. 1986) 42.
7) C.L. Henley, Comments in Condensed Matter Physics, to be submitted to.
8) C.L. Henley and V. Elser, Phil. Mag. B **53** (1986) L59.
9) V. Elser, Phys. Rev. Lett. **54** (1985) 1730.
10) V. Elser, Phys. Rev. B **32** (1985) 4892.
11) P.W. Stephens and A.I. Goldman, Phys. Rev. Lett. **56** (1986) 1168.
12) A.I. Goldman and P.W. Stephens, unpublished.
13) P.M. Horn, W. Malzfeld, D.P. DiVincenzo, J. Toner and R. Gambino, preprint.
14) V. Elser, unpublished.
15) N.G. de Bruijn, Proc. Konink. Ned. Akad. Wetensch. **A84** (1981) 39, 53.
16) A.L. Mackay, Sov. Phys. Crystallogr. **26** (1981) 517 [Kristallografiya **26** (1981) 910]; Physica **114A** (1982) 609.
17) D. Levine and P.J. Steinhardt, Phys. Rev. B **34** (1986) 596.
18) P. Kramer and R. Neri, Acta Crystallogr. **A40** (1984) 580.
19) J.E.S. Socolar and P.J. Steinhardt, Phys. Rev. B **34** (1986) 617.
20) A. Katz and M. Duneau, J. Physique **47** (1986) 181.
21) P.A. Kalugin, A.Yu. Kitaev and L.C. Levitov, J. Physique Lett. **46** (1985) L601.
22) N.D. Mermin and S.M. Troian, Phys. Rev. Lett. **54** (1985) 1524.
23) P. Bak, Phys. Rev. Lett. **54** (1985) 1517.
24) M.V. Jarić, Phys. Rev. Lett. **55** (1985) 607.
25) S. Sachdev and D.R. Nelson, Phys. Rev. B **32** (1985) 4592.
26) S.M. Troian and N.D. Mermin, Ferroelectrics **66** (1986) 127.
27) T.C. Lubensky, S. Ramaswamy and J. Toner, Phys. Rev. B **32** (1985) 7444, **33** (1986) 7715.
28) P. Bak, Phys. Rev. Lett. **56** (1986) 861.
29) T. Janssen, Acta Cryst. **A42** (1986) 261.
30) D.M. Frenkel, C.L. Henley and E.D. Siggia, Phys. Rev. B, in press.
31) V. Elser and C.L. Henley, Phys. Rev. Lett. **55** (1985) 2883; C.L. Henley, J. Non-Cryst. Solids **75** (1985) 91.
32) P. Guyot and M. Audier, Phil. Mag. B **52** (1985) L15; M. Audier and P. Guyot, Phil. Mag. B **53** (1986) L43.
33) M.A. Marcus, H.S. Chen, G.P. Espinosa and C.L. Tsai, Solid State Communications **58** (1986) 227; E.A. Stern, Y. Ma, K. Bauer and C.E. Bouldin, J. Physique Colloq. **47** (1986) C3-371.
34) C. Henley, to be submitted to J. Phys. A.
35) M. Tanaka, M. Terauchi, K. Hiraga and M. Hirabayashi, Ultramicroscopy **17** (1985) 127; P.A. Bancel and P.A. Heiney, J. Physique Colloq. **47** (1986) C3-341.

Physica **140A** (1986) 312–318
North-Holland, Amsterdam

SPIN GLASSES THEORY

Giorgio PARISI
II Universitá di Roma "Tor Vergata", Dipartimento di Fisica and INFN, Sezione di Roma, Italy

We present the progress that has been made in the mean field theory of spin glasses and we explain the physical meaning of broken replica symmetry.

A spin glass is a disordered magnet: in a typical example 50% of the bonds (randomly chosen) between two spin are ferromagnetic while the others are antiferromagnetic. From the experimental point of view[1]) spin glasses are characterized by the presence of a low temperature phase (for any value of the magnetic field) in which the relaxation time toward equilibrium is very large (as it happens in real glasses): the magnetic susceptibility in the presence of a time dependent magnetic field also depends on the frequency in the region of a few hertz.

These properties may be connected to the practical impossibility of finding numerically the ground state of a spin glass (if the number of spins is not very small); indeed, according to the conventional wisdom, the number of operations, which any algorithm needs in order to find the ground state, increases exponentially with the volume (the problem is NP complete in more than 2 dimensions). There are an exponentially large number of spin configurations which are local ground states, in the sense that their energy increases if only one spin is flipped: it very easy to find efficient algorithms for finding local minima; on the contrary it is very difficult for the algorithm to find the global minimum (i.e., the true ground state).

This peculiar behaviour is a source of frustration: it is not possible that all pairs of spins which are connected by a ferromagnetic (antiferromagnetic) bond have the same (the opposite) sign. Different terms of the Hamiltonian push in different directions and the number of possible compromises is very high. For example in the Ising case the Hamiltonian is

$$H = -\tfrac{1}{2}\sum_{i,k} J_{i,k}\sigma_i\sigma_k - h\sum_i \sigma_i, \tag{1}$$

where h is the magnetic field and the J's are the coupling between the spins. It is clear that, if the product $J_{12}J_{23}J_{31}$ is negative, it is impossible to find a configuration of the σ's such that $\sigma_1 J_{12}\sigma_2$, $\sigma_2 J_{23}\sigma_3$ and $\sigma_3 J_{31}\sigma_1$ are all positive.

There are many systems in which different terms of the Hamiltonian (or different constraints) are in competition one with the others; this happens very frequently when the system is complex. Typical examples are real glasses, other NP complete problems like the traveling salesman or the matching problem, protein folding, biological organization, prebiotic evolution, neuron networks and so on. At the present moment many of the concepts that have been developed in the study of spin glasses, start to also be useful for these systems and it is quite possible that in the long run the applications of these ideas beyond solid state physics will be the most interesting ones.

Exhaustive numerical simulations on spin glasses of small size[2]) show that differences in the free energy (or energy at zero temperature) of the lowest local minima remain of $\mathcal{O}(1)$ also when the volume goes to infinity. This situation is rather unprecedented in statistical mechanics: different configurations with the same total free energy are normally present at the point where a first order transition happens (which is an isolated point); for spin glasses this phenomenon is present in a large range of temperatures and magnetic fields, where no first order phase transition is seen from the thermodynamical point of view. If a different phase of the system is associated to each of the local minima, we find that the Gibbs phase rule is violated. Moreover, in the usual system different phases can be easily distinguished macroscopically (e.g., water looks quite different from ice) or they are related one to the other by a symmetry operation (spin reversal in the Ising model at zero magnetic field).

In recent years much progress has been made and after much effort a self-consistent mean field approximation has been obtained[3]) (sometimes this construction goes under the name of broken replica theory); this theory should be exact for weak long range forces or when the dimensions of the space become very large (fluctuations are neglected). The ideas involved in this construction are rather different from those of the mean field approximation for other models and take care of the peculiar properties of spin glasses.

We start by analyzing the concept of pure equilibrium state. If we consider a simple three dimensional ferromagnetic Ising model (with zero magnetic field) at temperature greater than the critical one, there is only one pure equilibrium state and the spins are disordered. Below the critical temperature there are two translational invariant pure equilibrium states, one with positive magnetization, the other with negative magnetization.

At temperatures smaller than the roughening transition for the interface between regions of positive and negative magnetization, there are also pure equilibrium states, which are not translational invariant: half of the space has positive magnetization and half of the space has negative magnetization. These new pure equilibrium states have much greater total free energy than the translational invariant states, the difference in free energy being due to the presence of the interface; in the infinite volume limit the free energy density is the same, as it should be.

Generally speaking a pure equilibrium state is any state which is at local equilibrium and whose connected correlation functions go to zero at large distance; the expectation value in the state α will be denoted by $\langle\ \rangle_\alpha$. According to the conventional wisdom, a system, which is not at equilibrium at initial time, evolves toward a pure equilibrium state and the time needed to go from one equilibrium state to another equilibrium state is exponentially large.

From the technical point of view a pure equilibrium state is characterized by the clustering property: all the connected correlation functions go to zero at large distance; equivalently intensive quantities do not fluctuate. In a nutshell, a pure equilibrium state corresponds to our intuitive idea of a normal equilibrium state; sometimes a pure equilibrium state is called a phase of the system (e.g., liquid or solid phase). In order to avoid inconsistency with the thermodynamics, all equilibrium states must have the same free energy; states with higher free energy are metastable.

The previous considerations imply that an Ising spin glass has many more equilibrium states than a ferromagnetic Ising model[4]): the configurations which have lower free energy correspond to pure equilibrium states while those configurations which have an higher free energy density in the infinite volume limit correspond to metastable states.

We face the new problem of characterizing the ensemble of equilibrium states. We assign an a priori probability w_α for the state α to appear in the ensemble; the natural choice is:

$$w_\alpha \propto \exp(-\beta F_\alpha), \tag{2}$$

where F_α is the free energy of the state α.

In order to study how all these states differ one from the other, it is convenient to define a distance among them. A simple definition of the distance between the state α and the state β is the following:

$$d_{\alpha\beta}^2 = (1/N)\sum_{i=1}^{N}\left(m_i^\alpha - m_i^\beta\right)^2, \tag{3}$$

where N is the total number of spins, the sum over i is from 1 to N and m_i^α is the average magnetization of the spin i in the state α, i.e., $m_i^\sigma = \langle \alpha_i \rangle_\alpha$. In the same way we can define the overlap between two states α and β as

$$q_{\alpha\beta} = (1/N)\sum_{i=1}^{N} m_i^\alpha m_i^\beta. \tag{4}$$

If all the states have the same overlap with themselves (as should happen in spin glasses), i.e., $q_{\alpha\alpha} = q_{\beta\beta} = q_{\mathrm{EA}}{}^{5}$), the distance is very simply related to the overlap:

$$d^2_{\alpha,\beta} = 2(q_{\mathrm{EA}} - q_{\alpha\beta}). \tag{5}$$

It is possible to define the probability distribution of overlap $P(p)$:

$$P(q) = \overline{P_J(q)} = \overline{\sum_{\alpha\beta} w_\alpha w_\beta \delta(q_{\alpha\beta} - q)}, \tag{6}$$

where the bar denotes the average over different samples with different value of the couplings J.

In other words $P(q)$ is the average probability of finding two states with overlap q, weighting each state with its probability of appearing in the ensemble. Only states which have a nonzero probability ($w_\alpha \neq 0$) contribute to eq. (6).

In the ferromagnetic Ising model at nonzero magnetic field, $P(q)$ is a delta function, because there is only one equilibrium state; exactly at zero magnetic field, the function $P(q)$ contains two delta functions, one at $q = m^2$, the other at $q = -m^2$, where m is the spontaneous magnetization.

In spin glasses the function $P(q)$ at high temperatures has the same shape as in ferromagnetic materials; however, if many pure equilibrium states exist at low temperatures at nonzero magnetic field, $P(q)$ must have a more complex structure; in this case, where the function $P(q)$ is not a simple delta function, we say (for historical reasons) that the replica symmetry is broken.

When the replica symmetry is broken we are interested not only in computing the free energy, but in knowing which is the probability distribution $\mathscr{P}\{w, q\}$ of all the w's and q's. In other words we need to do a statistical mechanics of the phases of the system; this step is particularly difficult because we face a completely new problem that has never been studied before ($\mathscr{P}\{w, q\}$ is a function of an unbounded number of variables). However something of this kind must be done if we want to understand the behaviour of complex systems.

The problem of computing $\mathscr{P}\{w, q\}$ becomes simpler (or more intriguing) in the replica approach, where there is an one to one correspondence between the function $\mathscr{P}\{w, q\}$ and a matrix $Q_{a,b}$; Q is a zero by zero matrix and the group S_0 (the permutation group of zero elements) acts naturally on it: the form of the matrix Q (and consequently of the function $\mathscr{P}\{w, q\}$) may be selected by group theoretical arguments. Moreover in the infinite range model, the free energy can be exactly written as a functional of Q and uncorrect forms of Q can be disregarded using thermodynamical arguments. Up to this moment only one acceptable candidate has emerged for Q: the matrix Q has a hierarchical structure. In this hierarchical Ansatz, all possible probability distributions with more than one overlap can be computed from the function $P(q)$. The free energy can be computed as a functional of $P(q)$ and the function $P(q)$ can be found by looking for the maximum of free energy.

The first interesting result is that $P_J(q)$ fluctuates from sample to sample[6]:

$$\overline{P_J(q_1)P_J(q_2)} = \tfrac{1}{2}P(q_1)P(q_2) + \tfrac{1}{2}\delta(q_1 - q_2)P(q_1). \tag{7}$$

This result indicates that only few states give an important contribution to the function $P_J(q)$ and their relative weight and overlap fluctuate from sample to sample.

The most interesting and unexpected result concerns the probability distribution of the overlap of three or more states. One finds that the distance satisfies the ultrametricity inequality[6]:

$$d_{\alpha\beta} < \max\left(d_{\alpha\gamma}, d_{\alpha\gamma}\right) \quad \forall\gamma. \tag{8}$$

As is well known to mathematicians, this inequality implies that the space of states can be divided into clusters of states; each cluster may be further divided into subclusters and so on . . . ; a cluster can be characterized as the set of states whose distance from a given (but arbitrary) state of the cluster is smaller than a given distance.

This result is extremely important because it may be the basis for the existence of many different scales of relaxation times. Indeed many different times of relaxation are naturally present if the free energy landscape in configuration space contains many valleys which are separated by high mountains: the time going from one valley to another is the exponential of the height of the lower saddle among the two valleys. The division of configuration space in valleys is clearly ultrametric: if the distance between two configurations is defined as the height of the highest saddle one must cross doing the most convenient trip between the two configurations; the distance satisfies the ultrametric inequality[8]. The real surprise is that the space of the equilibrium state of the spin glass is ultrametric using the natural definition of the distance eq. (3).

Another very interesting result is that the specific form of the equilibrium states is very sensitive to external parameters: a small change of the external magnetic field, of order $1/N^{1/2}$, is sufficient to completely upset the microscopic details of the equilibrium states: the equilibrium states at two different magnetic fields have a very small overlap[4]. The magnetization as a function of the magnetic field has many small jumps, of the order of $1/N^{1/2}$; we could say pictorially that, when we change the magnetic field, the system undergoes a sequence of first order microtransitions.

All these results certainly hold when fluctuations are neglected; this approximation is good when the dimensions of the space are very large and the predictions of this approach compare well with the results of the numerical simulations in infinite dimensions, i.e., for the Sherrington and Kirkpatrick model[7]).

If we decrease the dimensions the situation is less clear: a likely scenario, although infinitely many different possibilities are open, is that the picture does not change for dimensions greater that D_c; at the lower critical dimension D_c, the critical temperature becomes zero and there is no transition to a spin glass phase. However, if the coherence length ξ is large and if the relaxation time is proportional to $\exp(\xi^{1/4})$, it is possible that the relaxation time becomes very large (seconds, days, years...) and for reasonably long observation times the system seems to be in the spin glass phase.

A different possibility would be that the ultrametricity property is correct only in infinite dimensions (numerical simulations seem to imply that it holds for the infinite range models) and violations of ultrametricity start to be present when we decrease the dimensions: more complex forms of the matrix Q would be needed in finite dimensions.

A very neat way of solving the problem consists in computing the corrections induced by fluctuations in the framework of the present theory and comparing them to the experimental and numerical data. Unfortunately such computation is very difficult and progress is very slow[8]). Moreover we should also investigate the possibility of violations of the ultrametricity. We should also remember that instantons have been completely neglected in the case of spin glass, while it is known that instantons are very important at low dimensions both for statics and dynamics: in all dimensions instantons are crucial for understanding Griffiths singularities in random systems[9]).

The task that remains before us, before we can obtain reliable predictions for the three-dimensional Ising spin glass, is gigantic: we must remember that sixty years separate the first mean field theory for ferromagnets from the computation of critical exponents and the problems we face are apparently much more complex.

References

1) For a nice short experimental review of spin glasses see J.A. Mydosh, in Disordered Systems and Localization, C. Castellani, C. Di Castro and L. Peliti, eds. (Springer-Verlag, New York, 1981). A theoretical review of spin glasses and related subjects can be found in M. Mezard, G. Parisi and M. Virasoro, Spin Glass Theory and Beyond (Word Scientific, Singapore, 1987).
2) A.P. Young, Phys. Rev. Lett. **43** (1983) 1206; N. Sourlas, LPTENS preprint 85/9 (1985).
3) G. Parisi, Phys. Rev. Lett. **43** (1979) 1754, J. Phys. A **13** (1980) 1101, 1887, L115.

4) G. Parisi, Phys. Rev. Lett. **50** (1983) 1946.
5) S.F. Edward and P.W. Anderson, J. Phys. **F5** (1975) 965; **F6** (1976) 1927.
6) M. Mezard, G. Parisi, N. Sourlas, G. Toulouse and M. Virasoro, Phys. Rev. Lett. **52** (1984) 1156; J. Physique **45** (1984) 843.
7) D. Sherringtong and S. Kirkpatrick, Phys. Rev. Lett. **35** (1975) 1792; Phys. Rev. B **17** (1978) 4384.
8) C. De Dominicis and I. Kondor, J. Physique Lett. **46** (1985) L1037 and references therein.
9) G. Parisi, in Field Theory and Statistical Mechanics, J.B. Zuber and R. Stora, eds. (North-Holland, Amsterdam, 1984), G. Parisi, in 60th birthday Festschrift for E.S. Fradkin (Adam Hilger, Bristol, 1986).

Physica **140A** (1986) 319–325
North-Holland, Amsterdam

HOW TO MEASURE SELF-GENERATED COMPLEXITY

Peter GRASSBERGER

Physics Department, University of Wuppertal, D-56 Wuppertal 1, Gauss-Strasse 20, Fed. Rep. Germany

In an increasing number of simple dynamical systems, patterns arise which are judged as "complex" in some naive sense. In this talk, quantities are discussed which can serve as measures of this complexity. They are measure-theoretic constructs. In contrast to the Kolmogorov complexity, they are small both for completely ordered and for completely random patterns. Some of the most interesting patterns have indeed zero randomness but infinite complexity in the present sense.

1. Introduction

The main aim of physics is (and always has been) to explain the complexity of nature surrounding us by a few, hopefully simple, laws. Thus, one might believe that nobody should be surprised to encounter simple models which lead to very complex behaviour. But for some reason, this is not true. One of the most fascinating developments of the last two decades is the enormously rich structure which is observed in very simple dynamical systems.

The simplest and most famous example is maybe the logistic map[1])

$$x_{n+1} = 1 - ax_n^2, \qquad x_n \in [-1,1]. \tag{1.1}$$

Here, we can digitize the itinerary by defining binary variables s_n with $s_n = L$ for $x_n < 0$ and $s_n = R$ for $x_n > 0$. The complexity of the behaviour of the system is then seen in the complexity of the string $\ldots s_n s_{n+1} s_{n+2} \ldots$. Similar strings encoding the history can be formed for other dynamical systems such as Lorenz's equations, Bernard cells or nonlinear driven oscillators.

Another simple class of systems are the simplest cellular automata (CA) in one dimension[2]). Here, we have an infinite array of "spins" $\{s_i\}$ with $s_i = 0$ or 1. Starting, e.g., with a random starting configuration, time evolution is defined by some deterministic rule $s_i' = f(s_{i-1}, s_i, s_{i+1})$. Notice that here time is discrete, and the updating is done simultaneously. The nontriviality of some of these models is seen in complex space–time patterns (see fig. 1b), or in strings of spins at fixed time or at fixed space. These CA are crude models of spatially distributed

0378-4371/86/$03.50 © Elsevier Science Publishers B.V.
(North-Holland Physics Publishing Division)

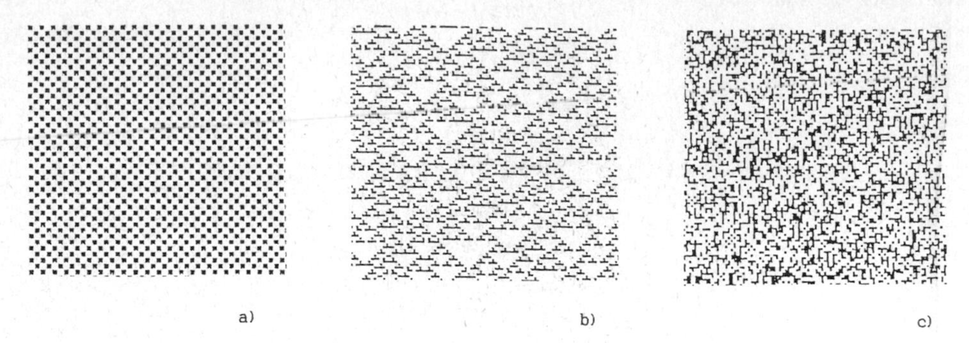

Fig. 1. Three patterns used to demonstrate that the pattern which one intuitively would call the most complex (panel b) is neither the one with lowest entropy nor the one with highest. Panel b is indeed obtained by cellular automaton No. 22[2]) from a random start, with time running from top to bottom.

systems like reaction-diffusion systems or turbulent flows, which often show similarly complex patterns.

Finally, as a last example, we might mention quasicrystals[3]) and other quasi-periodic structures[4]).

One characteristic common to all these examples is that the complexity is "self-generated" in the sense that the formulation of the problem is translationally invariant, and the observed structures arise from spontaneous breakdown of translational symmetry. In the following, we shall restrict ourselves to 1-dimensional such patterns. Higher dimensions are shortly discussed in ref. 5.

Also common to all these examples are order–disorder transitions. Indeed, the concept of complexity is closely related to that of randomness. One purpose of the present talk is to clarify that relationship. More generally, we shall discuss ways of defining an observable which could measure complexity in this context, just as entropy measures randomness.

Of course, this is not the first attempt to quantify complexity. Apart from the "computational complexity"[6]) which is not applicable in our case because it applies to functions (or algorithms computing these functions) and not to patterns, the best known concept is that of Kolmogorov (or "algorithmic") complexity[7]). A sequence of N symbols is said to be algorithmically complex if the length of its shortest complete description increases $\propto N$, for $N \to \infty$. Indeed, if we are interested in attributing a measure of complexity to any definite sequence, this seems to be the only way possible: a sequence is more complex the more information is needed to specify it (see note added in proof).

Clearly, this resembles closely Shannon's definition of information. A difference is seen when considering the string of digits of $\pi = 3.141592\ldots$. Its Kolmogorov complexity is zero since there exist efficient algorithms for computing π, but any statistical analysis would yield a nonvanishing Shannon information per digit[8]). But we claim that this example is not typical for the situations we are interested in. In the latter, Kolmogorov complexity and Shannon information are indeed equivalent. The reason is our assumption of translational symmetry. If we would ask for the shortest program to compute any randomly chosen sequence of N digits of (and not just the first N ones), it would in general consist of a listing of these digits, and would thus have length $\sim N$.

Thus we are faced with the puzzle[9]) that no accepted measure of complexity could, e.g., corroborate that music written by Bach is more complex than the random music written by a monkey. Another example is shown in the 3 panels of fig. 1. Panel a is clearly the least random and at the same time the least complex. But at least when told that the third panel is produced by a random number generator, most people would not call it complex, although it has the highest randomness and Kolmogorov complexity.

The way out of this conundrum relates to our well-known ability to distinguish intuitively between "important" and "irrelevant" features. Even if one seems to speak about an individual pattern, one has in mind an ensemble of patterns all having the same "important" structures. A particular piece of music produced by a monkey is considered as not complex since we recognize no such structure. We do see some structure in fig. 1b. The complexity we seem to attribute to this pattern is indeed the difficulty to describe the whole ensemble characterized by this structure. While we might first try to see structure also in panel c, we give up as soon as we are told that this picture is random, i.e., is in the trivial class of all possible pictures.

The most important problem would be to understand how our brain makes this classification. I shall leave this to the philosophers, and just keep the following lesson: complexity in the naive sense measures the information content of ensembles and not of individual patterns, contrary to what one might have believed. The ensembles in the following will be the sets of patterns produced, e.g., by the dynamical systems mentioned above.

Notice that I used the word "ensemble" and not "set", implying that the sets of patterns are endowed with some probability measure. This is an important difference to most previous discussions of complexity[6,7]).

One might question this distinction between individual patterns and ensembles in the case of ergodic systems like the logistic map. There, any infinite itinerary contains complete information about the set of all itineraries. Yet there is an important difference. Assume we are given part of an itinerary $\ldots s_{i-2}s_{i-1}s_i$, and we are asked how to continue it. If we are interested in the actual itinerary, the relevant question is "what is s_{i+1}?". The information missing in the past itinerary for answering that question is the Shannon information (which in this case is also the Kolmogorov–Sinai entropy[1])). But if we are only interested in the ensemble, the relevant question is "given the past history, what can s_{i+1} be, and with what probability?". The information about the past needed for that is called the true measure complexity (TMC) in the following.

Another example: we don't call Latin more complex than, e.g., Esperanto since it carries more information per letter, but since its grammar – needed to verify the correctness of a string of letters – is bigger. As a final example, take weather forecasting. One problem there (usually considered the most important) is how big is the uncertainty still remaining after an optimal prediction. For a practitioner, a more important problem might be "how much effort is needed (how much data have to be collected, stored and processed) to make this optimal prediction?". Obviously, this is closely related to the TMC.

Come back to the logistic map. When the constant a in eq. (1.1) is smaller than $\frac{3}{2}$, the x's converge to a fixed point and the itinerary becomes $\ldots RRR \ldots$. Thus both the metric entropy and the TMC are zero. When $a = 2$, all 2^N itineraries of length N occur with the same probability[1]). Thus, the metric entropy there is maximal (1 bit/iteration), but the TMC is again zero: no information about the past is needed for an optimal prediction since no prediction can be better than a simple guess. Finally, at the Feigenbaum point $a = 1.401155\ldots$ any itinerary is non-random[1]) since the Shannon information stored in a string of length N increases only as $\log N$[5]). But this increase implies (see section 2) that the TMC is infinite.

This last example of a pattern with zero randomness but infinite measure complexity is presumably the most interesting. In section 3 we shall give other similar examples. Before that, we have to clarify some technical points. This will be done in section 2.

2. Technicalities and variations of the theme

We have seen that naive complexity is some "information" needed to verify the "correctness" of a string during a scan. For the TMC, "information" meant Shannon entropy, and "correct" implied right statistical properties. Alternatives to both will be discussed now.

If one is only interested in the "grammatical" correctness, the needed Shannon information is called the "set complexity" SC. It is obviously a lower bound to the TMC. For its technical definition, let us first consider regular languages[6]). These are sets of strings whose possible continuations depend on a finite number N of states of a scanner. Otherwise said, there is a state transition graph with N nodes, and with directed links labelled by letters of the alphabet (by values of the spins). After the scanner has arrived at node i, the string is allowed to continue with letter (spin) s only if there is a suitable link which can be followed. To each ensemble of such strings corresponds a set of probabilities p_i to be at node i. The

SC is then defined as

$$\mathrm{SC} = \min_{\{\text{graphs}\}} - \sum_{i=1}^{N} p_i \log p_i . \tag{2.1}$$

If these probabilities are not given, one has to replace the Shannon information in eq. (2.1) by log N, and arrives at the "algorithmic complexity"[2])

$$\mathrm{AC} = \min_{\{\text{graphs}\}} \log N . \tag{2.2}$$

Non-regular languages can in general be represented by infinite transition graphs, so eq. (2.1) can still make sense even if eq. (2.2) is meaningless[5]).

At least when the connectivity is low, the number of nodes in a graph is roughly proportional (when N becomes large) to the number of bits needed to describe the entire graph. Thus, the AC is roughly equal to the Kolmogorov information of the grammar.

A drawback of all 3 complexities defined so far is that they are not observables. Any observation can only give an approximate probability distribution. The grammar could always be very different from the one seemingly observed, due to rules which are applied sufficiently rarely. This means that from any observation one cannot even estimate bounds for the SC and AC. Since the SC is a lower bound for the TMC, the latter cannot be estimated from above either. Also, we haven't yet given an operational definition for it.

In the remainder of this section, we shall thus discuss still another quantity, the "effective measure complexity" EMC. It is closely related to the TMC, to which it is a lower bound. Its main virtue is that it is operationally defined and can be estimated from observed data. We consider it as the most interesting measure of complexity within the present context.

Let us call $p_N(s_1 \cdots s_N)$ the probability of finding the string $s_1 \cdots s_N$ of length N in our ensemble. The information in such a string is

$$H_N = - \sum_{\{s_i\}} p_N(s_1 \cdots s_N) \log p_N(s_1 \cdots s_N) . \tag{2.3}$$

Given s_1 to s_N, the information needed to give s_{N+1} is then $h_N = H_{N+1} - H_N$. The Shannon entropy per letter (or spin) is $h = \lim_{N \to \infty} h_N$. Finally, the decrement $h_{N-1} - h_N$ is the amount by which the uncertainty of s_{N+1} decreases when learning about s_1, all spins between being already known. Since at least this information has to be stored during N steps for an optimal prediction, the TMC cannot be smaller than

$$\mathrm{EMC} = \sum_{N=0}^{\infty} N(h_{N-1} - h_N) = \sum_{N=0}^{\infty} (h_N - h) . \tag{2.4}$$

One might believe that the TMC is indeed equal to that, but counter examples show that this need not be so[5]).

Given all block entropies h_n with $n \leqslant N$ and nothing else, the minimal EMC is obtained when $h_n = \text{const} = h$ for all $n \geqslant N$. This would also correspond to maximal entropy h, and to a Markov ansatz of order N. Thus, a Markov ansatz corresponds to the hypothesis of maximal randomness but minimal complexity.

Using mutual entropies as measures of complexity or "structure" had been proposed previously in refs. 10 and 11.

3. Applications

(a) We have already discussed the logistic map. We might add that for typical chaotic parameter values the AC and SC should be infinite, though the EMC seems to be finite. The convergence of the block entropies has been studied for this case in ref. 12.

(b) *Quasiperiodicity*. The remarks about the logistic map at the Feigenbaum point apply mutatis mutandis also to the critical circle map at the golden winding number[4])

$$x_{n+1} = f(x_n) = x_n - \sin x_n + 3.81176\ldots \quad [\mathrm{mod}\, 2\pi]. \tag{3.1}$$

Defining a discrete itinerary by $s_n = R$ if $x_n \in [0, f(0)]$ and $s_n = L$ otherwise, the information in a string of length N again increases as $\log N$. Thus these itineraries are again not random but metrically complex.

A quasiperiodic sequence like this itinerary is the 1-dimensional analog of a quasicrystal[3]). The latter are known to have zero entropy (at zero temperature). It seems very likely that they also have infinite metric complexity.

(c) *One-dimensional cellular automata* (CA). Algorithmic complexities of spatial sequences obtained by iterating a CA t times were studied in ref. 2. For all automata judged naively as "complex", they increase rapidly with t. But they increase also for some automata which naively would be called simple. Taking probabilities into account by going from AC to SC, the results agree better with naive expectations[5]).

The limit ensembles of spatial patterns after infinitely many iterations seem to be simple in most cases. One notable exception is rule 22 (in Wolfram's notation[2])). A typical pattern is shown in fig. 1b. In spite of the seeming randomness of the pattern, entropies H_N both of spatial and of temporal strings are not proportional to the length N of the string, but are found numerically to increase like powers of N, with exponents ~ 0.94 (resp. ~ 0.82)[5,13]). Thus again these are non-random patterns with infinitely large complexity. For some other

CA, one finds temporal sequences with infinite EMC but with non-zero randomness[5]).

(d) *Spin glasses*. It is conjectured [A. Zippelius, private communication] that in some spin glasses at zero temperature the entropy is not an extensive quantity but increases like some power of the volume. This would make them very similar to CA 22. It agrees with the intuitive feeling that spin glasses are very complex.

Further examples of patterns with large complexity might be natural languages, hydrodynamic turbulence (including the weather) and DNA. In all these examples, it should be very useful to have objective measures of complexity. But before our concepts can be applied to these cases, one has to devise more efficient algorithms (straightforward computation of h_N is very tedious), in particular for higher-dimensional patterns.

Note added in proof

While writing this, I was unaware of C.H. Bennett's notion of "logical depth" (in Emerging Syntheses in Science, D. Pines, ed., pp. 297–313, 1985). This measures essentially the time required to run the shortest program describing completely the sequence.

References

1) P. Collet and J.-P. Eckmann, Iterated Maps on the Interval as Dynamical Systems (Birkhäuser, Boston, 1980).
2) S. Wolfram, Rev. Mod. Phys. **55** (1983) 601; Commun. Math. Phys. **96** (1984) 15.
3) D. Levine and P.J. Steinhard, Phys. Rev. Lett. **53** (1984) 2477.
4) M.J. Feigenbaum, L.P. Kadanoff and S.J. Shenker, Physica **5**D (1982) 370.
5) P. Grassberger, preprint WU B 86-9 (1986), to be published in Int. J. Theor. Phys.
6) J.E. Hopcroft and J.D. Ullman, Introduction to Automata Theory, Languages, and Computation. (Addison-Wesley, Reading, MA, 1979).
7) V.M. Alekseev and M.V. Yakobson, Phys. Rept. **75** (1981) 287.
8) S. Wagoner, Is pi normal?, Math. Intelligencer **7** (1985) 65.
9) T. Hogg and B.A. Huberman, Order, complexity, and disorder, Xerox Palo Alto preprint (1985).
10) M.H. van Emden, An Analysis of Complexity (Mathematical Centre Tracts, Amsterdam, 1975).
11) G.J. Chaitin, Towards a mathematical theory of 'life', in The Maximum Entropy Principle, R.D. Levine and M. Tribus, Eds. (MIT Press, Cambridge, 1979).
12) G. Györgyi and P. Szepfalusy, Phys. Rev. A **31** (1985) 3477; and to appear.
13) P. Grassberger, preprint WU B 86-10 (1986), to be published in J. Stat. Phys.

Physica **140A** (1986) 326–335
North-Holland, Amsterdam

LATTICE GAS AUTOMATA FOR FLUID MECHANICS

D. D'HUMIÈRES and P. LALLEMAND

CNRS, Laboratoire de Physique de l'Ecole Normale Supérieure, 24 rue Lhomond, 75231 Paris Cedex 05, France

A lattice gas is the representation of a gas by its restriction on the nodes of a regular lattice for discrete time steps. It was recently shown by Frisch, Hasslacher and Pomeau that such very simple models lead to the incompressible Navier–Stokes equation provided the lattice has enough symmetry and the local rules for collisions between particles obey the usual conservation laws of classical mechanics. We present here recent results of numerical simulations to illustrate the power of this new approach to fluid mechanics which may give new tools for numerical studies and build a bridge between cellular automata theory and complex physical problems.

1. Introduction

Fluid mechanics is one of the most important fields for practical applications: airplanes, weather prediction, etc., and one of the most challenging for theoretical studies. Despite a lot of theoretical work, a large part of the present knowledge comes from experiments. Fluid dynamics can be studied from two points of view. The first one consists of describing the fluid from a macroscopic point of view. In this case the equations of motion are derived as the most general equations for mass, momentum and energy conservation which satisfy the usual space invariance (translation, rotation and Galilean invariances). These equations have been known for more than one hundred years as the continuity equation,

$$\partial_t \rho + \operatorname{div}(\rho \boldsymbol{u}) = 0, \tag{1a}$$

the Navier–Stokes equation,

$$\rho \partial_t \boldsymbol{u} + \rho(\boldsymbol{u} \cdot \operatorname{grad})\boldsymbol{u} = -\operatorname{grad} P + \eta \Delta \boldsymbol{u} + \xi \operatorname{grad}(\operatorname{div} \boldsymbol{u}), \tag{1b}$$

and the heat equation; where ρ, $\boldsymbol{u}$ and P are the local fluid density, velocity and pressure and η and ξ are the shear and bulk viscosities[1]). Despite their apparent simplicity, the non-linear terms make these equations very difficult to use: the analytical solution is known only in a few simple cases. Generally, eqs. (1) are used under their dimensionless form in order to compare the results obtained for different fluids or geometries. Then, the isothermal flows depend on two im-

0378-4371/86/$03.50

portant dimensionless parameters: the Reynolds number: $\mathrm{Re} = \rho u L/\eta$ (where u and L are the typical velocity and length scales of the problem), which gives the influence of the non-linear terms in eqs. (1), non-stationary and turbulent flows occurring for high Reynolds numbers, and the Mach number: $M = u/c$ (where c is the sound velocity), which gives the influence of compressible effects, low Mach numbers implying incompressible flows.

The second approach, the molecular dynamics, consists of a statistical description of the fluid from its microscopic properties. The transport coefficients (sound velocity, viscosities,...) are derived from the molecular properties such as the interparticle potential. However, the full description of the microscopic world allows only to simulating the behavior of a small number of particles. A first attempt to get correct microscopic properties from a simplified description of the microscopic world was done by Broadwell and Gatignol[2]) when they studied a gas in which the possible molecular velocities are restricted to a finite set. Then, Hardy, de Pazzis and Pomeau (HPP)[3]) tried to use the simplest possible model: a gas on a square lattice with very simple collision and propagation rules. However, this model was too simple to give realistic macroscopic behavior.

More recently, the fast development of integrated circuits and computers has triggered a new interest in cellular automata, which are regular networks of simple finite state machines with local connections between cells. The most complete study of classes of behavior in cellular automata has used analysis of machine simulations of such systems in one spatial dimension[4]) and it was shown that they are capable of extremely complex behavior. Conjectures have been made that there exist cellular automata capable of simulating the solution space of partial differential equations of physical interest[5]). No non-trivial example of this conjecture was known until Frisch, Hasslacher and Pomeau (FHP) recently found a way to cure most of the pathology of the HPP model and were able to prove that this new lattice gas evolves according the Navier–Stokes equation in the limit of large lattice size and low velocity[6]).

We present here recent results of computer simulations of the FHP models and some of its variants. First, we give a brief description of the model. Next we present the general procedure for the computer simulations; then we describe the linear hydrodynamic regime, measure transport coefficients and simulate two dimensional flows at moderate Reynolds number: a flow at the inlet of a duct and a von Karman street behind a flat plate.

2. The models

We consider particles moving on a triangular lattice with unit velocity $\boldsymbol{c}_i$ in direction i between a node and one of its six neighbors ($i = 1,\ldots,6$). At each time step each particle stays on a node and interacts on it with the other incoming particles according to collision laws assumed to conserve the number of particles and the total momentum on the node. Additionally, there is an exclusion principle such that no two particles with the same velocity may occupy the same node at the same time. Then the particles propagate according their new velocity. We have also used a variant of this model with "rest particles" which allows an additional particle with zero velocity at each site ($i = 0$ for notational purposes); these particles may be considered to have an internal energy to satisfy energy conservation.

These lattice gas models can be implemented on a two dimensional cellular automata with a triangular lattice topology, where each cell has six neighbors and a seven bit state. The collision laws of the lattice gas give the transition rules for the internal states of each cell and the propagation rules of the particles give the correspondence between the internal states and the cell inputs and outputs.

The original FHP model uses only six particles and the collision rules given in fig. 1a (five possible collisions, model I), which is the minimal set of rules to prevent spurious mass or momentum conservation. In fact, most of the computer simulations were done with the model with resting particles and the additional rules shown in fig. 1b and those of fig. 1a with rest particles (twenty-two possible collisions, model II). More recently we used a model with rest particles and all the possible collisions conserving mass and momentum (seventy-six possible collisions, model III).

The macroscopic quantities, density ρ and momentum $\rho \boldsymbol{u}$, are related to the macroscopic local average populations N_i of the Boolean states by[7])

$$\rho = \sum_i N_i, \qquad \rho \boldsymbol{u} = \sum_i N_i \boldsymbol{c}_i. \tag{2}$$

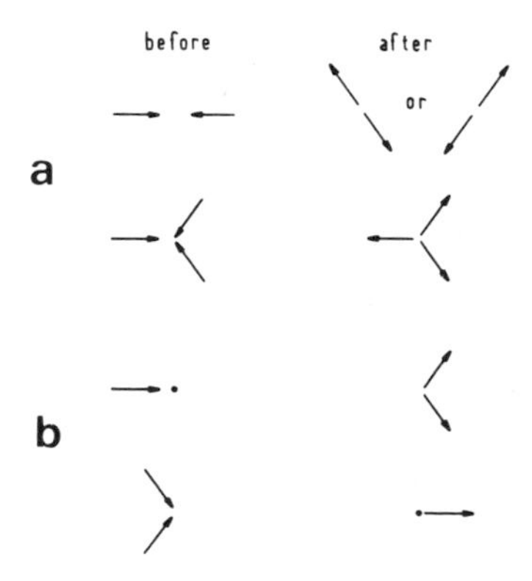

Fig. 1. Schematic representation of some collision rules; (a) no rest particles; (b) additional rules for rest particles.

As the particles satisfy an exclusion principle, at thermodynamic equilibrium the N_i are given by Fermi–Dirac distributions and can be expanded around small u by[6,8,9])

$$N_i = d\{1 + (\rho/3d)\boldsymbol{c}_i \cdot \boldsymbol{u}\}, \tag{3}$$

where the reduced density d is the average density per direction (without rest particles $\rho = 6d$, with rest particles $\rho = 7d$). Using a Chapman–Enskog expansion up to second order terms in velocity and gradients one gets

$$\partial_t(\rho u_\alpha) + \sum_\beta \partial_\beta[g(\rho)\rho u_\alpha u_\beta] = -\partial_\alpha P + \eta\Delta u_\alpha + \xi\partial_\alpha(\mathrm{div}\,\boldsymbol{u}), \tag{4}$$

where the Greek subscripts refer to the x and y components and $g(\rho)$ and P are given by

$$g(\rho) = (\rho - 3)/(\rho - 6), \qquad P = \rho/2, \tag{5a}$$

for model I without resting particles and by

$$g(\rho) = \tfrac{7}{12}(2\rho - 7)/(\rho - 7), \qquad P = 3\rho/7, \tag{5b}$$

for the models II and III with rest particles. The term $g(\rho)$ comes from the Fermi–Dirac distribution and restricts the use of lattice gas models to incompressible flows. In this case, $g(\rho)$ is a constant which may be absorbed in a rescaled velocity, the resulting Reynolds number being $\mathrm{Re} = \rho g(\rho)uL/\eta$.

The values of the viscosities can be computed in the Boltzmann approximation from the collision operator[10]). As noticed by Hénon[11]) an additional negative term comes from the discrete nature of the lattice gas and decreases the effective viscosities. Since the definition of the density on a triangular lattice is slightly different from the usual one and since the reduced density is the natural parameter for the computer simulations, in what follows we will use reduced viscosities $\eta_r = d\eta/\rho$ and $\xi_r = d\xi/\rho$. A straightforward extension of the results of ref. 10 gives

$$\eta_r = \frac{1}{12(1-d)^3} - \frac{d}{8}, \qquad \xi_r = 0, \tag{6a}$$

for model I,

$$\eta_r = \frac{1}{28(1-d)^3(1-4d/7)} - \frac{d}{8}, \qquad \xi_r = \frac{1}{98(1-d)^4} - \frac{d}{28}, \tag{6b}$$

for the model II and

$$\eta_r = \frac{1}{28(1-d)(1-8d(1-d)/7)} - \frac{d}{8},$$
$$\xi_r = \frac{1}{98(1-d)(1-2d(1-d))} - \frac{d}{28}, \tag{6c}$$

for the model III (units of time, length and mass are the time step, the link length and the particle mass, respectively).

3. Computer simulations

The present work was done by simulation of the lattice gas models on an FPS-164 using lattices of order 10^6 nodes with a typical speed of 10^6 updates per second[12-15]). The evolution of the lattice gas is computed according to a parallel iteration in two steps. During the first one, the post-collision state is computed using either a look-up table (in this case the states are coded with eight bits, seven for the particles and one to choose between the different possible head-on collisions, and eight nodes are packed in each of the 64 bits words of the FPS) or the combination of Boolean operators giving the collision rules (in this case 64 nodes are packed in a word and seven words are used to code the different particles, the choice between the head-on collisions is made randomly at each time step). During the second step the bits coding the different particles are moved from one node to the next one.

The lattice boundary conditions can be either periodic in both directions or of the "wind tunnel" type: particles are generated and absorbed at the lattice edges up and downstream according to eq. (3), $\boldsymbol{u}$ being the velocity of the wind tunnel. The boundary conditions on obstacles are "bounce-back" reflections: the particle velocity after the collision is the opposite of the incoming velocity. Initial flows were generated by a Monte Carlo procedure with average population N_i related to the local velocity by eq. (3). Macroscopic quantities are obtained by averaging the N_i according to eqs. (2) over rectangular regions with shapes and sizes adapted to the studied flow.

3.1. *Linear hydrodynamics*

The velocity of sound c, the shear viscosity η and the bulk viscosity ξ of the lattice gas models described above have been measured, using the relaxation of an initial periodic perturbation $(\boldsymbol{u}_\parallel + \boldsymbol{u}_\perp)\cos(\boldsymbol{k}\cdot\boldsymbol{r})$ of the velocity field[16]). $\boldsymbol{k}$ is

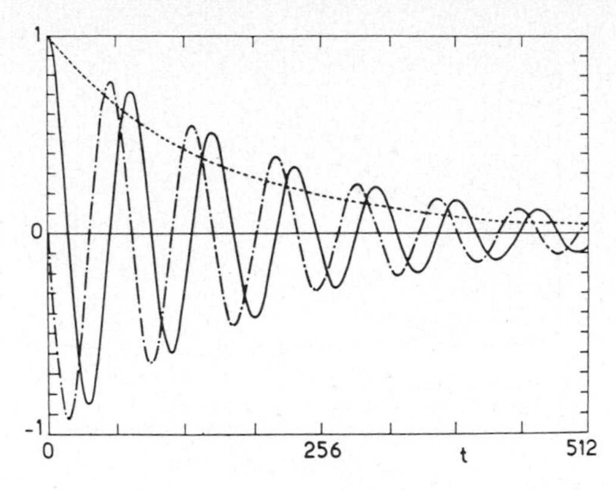

Fig. 2. Time evolution of $u_\perp$ (dashed line), $u_\parallel$ (solid line) and ρ (dot-dash line) normalized by the initial values of the velocity components.

the wave vector of the perturbation, and $\boldsymbol{u}_\parallel$ and $\boldsymbol{u}_\perp$ are the velocity components parallel and perpendicular to the wave vector.

The relaxation in time of the velocity $\boldsymbol{u}(\boldsymbol{r}, t)$ and of the density perturbation $\delta\rho(\boldsymbol{r}, t)$ are given by

$$\boldsymbol{u}(\boldsymbol{r}, t) = \big(\boldsymbol{u}_\parallel \cos(\omega t + \varphi) \exp(-k^2 t(\eta + \xi)/2\rho)$$

$$+ \boldsymbol{u}_\perp \exp(-k^2 t\eta/\rho)\big) \cos(\boldsymbol{k} \cdot \boldsymbol{r}), \tag{7a}$$

$$\delta\rho(\boldsymbol{r}, t) = (\rho u_\parallel / c) \sin(\omega t) \exp(-k^2 t(\eta + \xi)/2\rho) \sin(\boldsymbol{k} \cdot \boldsymbol{r}), \tag{7b}$$

with $\omega = ck$, $\tan\varphi = k(\eta + \xi)/2\rho c$.

Starting from the initial conditions, at each time step the momentum and the density are averaged along lines perpendicular to the wave vector. The result is Fourier transformed to get the component of the momentum and density corresponding to $\boldsymbol{k}$. Typical relaxation curves are given in fig. 2. From these curves, c, η and ξ are measured by least squares fits of eqs. (7) to the time evolution of $u_\perp(k)$, $u_\parallel(k)$ and $\rho(k)$.

The measured sound velocities are isotropic and agree with theoretical values $1/\sqrt{2}$ for model I and $\sqrt{3/7}$ for models II and III. The measured values of the viscosities are summarized in fig. 3, along with the theoretical curves computed from eqs. (6). These measurements were obtained on 256^2 lattices with periodic boundary conditions and for wavelength between 16 and 32 nodes; the size of the symbols corresponds roughly to the error bars. The measured viscosities agree with theoretical prediction for models II and III and confirm the presence of the negative lattice viscosities. Without rest particles the experimental values of η are

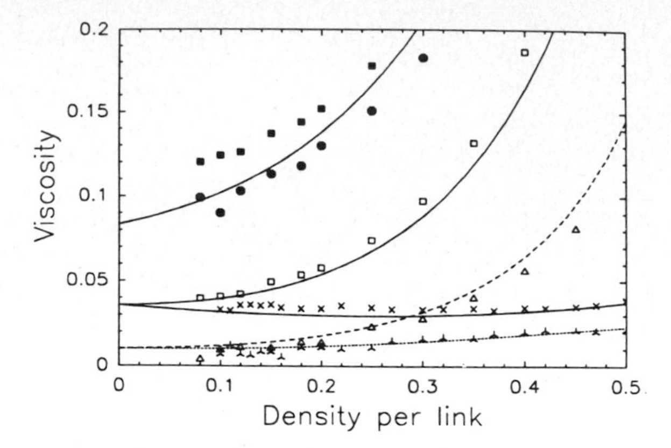

Fig. 3. Theoretical shear (solid lines) and bulk (dashed lines) reduced viscosities as a function of the reduced density, compared with the results of numerical simulations for different lattice gas models: original FHP model (solid squares and circles) model with rest particles and limited collision rules (open squares and triangles) and model with rest particles and all the possible collisions (oblique and triangular crosses).

too high and ξ is found negative. Thus, the presence of rest particles apparently improves the qualitative behavior of the lattice gas while decreasing the viscosity significantly. Fig. 4 shows the achievable Reynolds number for the different models for $uL = 1$; the maximum value is obtained with model III and is six times larger than the maximum value for the original FHP model. These results show that Reynolds numbers depend not only on the number of nodes as noted in refs. 6 and 17 but also on the lattice gas models. Further improvements on

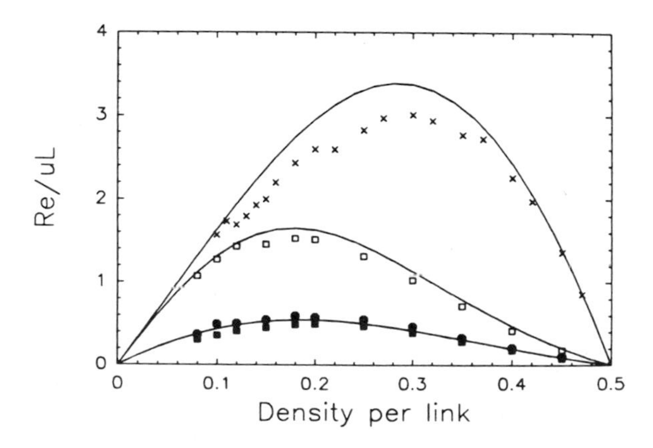

Fig. 4. Available Reynolds number as a function of the reduced density (solid lines: theory; symbols: measured values) for the different models with the same symbols as in fig. 3.

them may increase the maximum achievable Reynolds numbers which is presently of order of 10^3.

3.2. *Non-linear flow simulations*

The experimental points of fig. 4 are computed from the measured viscosity values and the theoretical values for $g(\rho)$. In order to test the non-linear terms in the lattice gas dynamics, we studied the flow at the inlet of a 2-D duct using a 512×3072 lattice[15]). The average density and velocity of the gas are 1.54 ($d = 0.22$) and 0.30 and the lattice gas model is model II. In fig. 5, we have compared the momentum obtained for the lattice gas flow to the values computed according to Slichting[18]) without any adjustable parameters, using the viscosity value from the relaxation measurements and the theoretical value for $g(\rho)$. The agreement between the two methods shows that the lattice gas dynamics is correctly described by eq. (4).

In fig. 6 is shown a von Karman street behind a "bounce-back" flat plate placed normal to a mean flow of velocity 0.51 in a 1024×512 wind-tunnel[14]). The lattice gas model is model II, the length L of the plate is 120 and the density $\rho = 1.4$ ($d = 0.2$). The corresponding Reynolds number is 90. The flow is

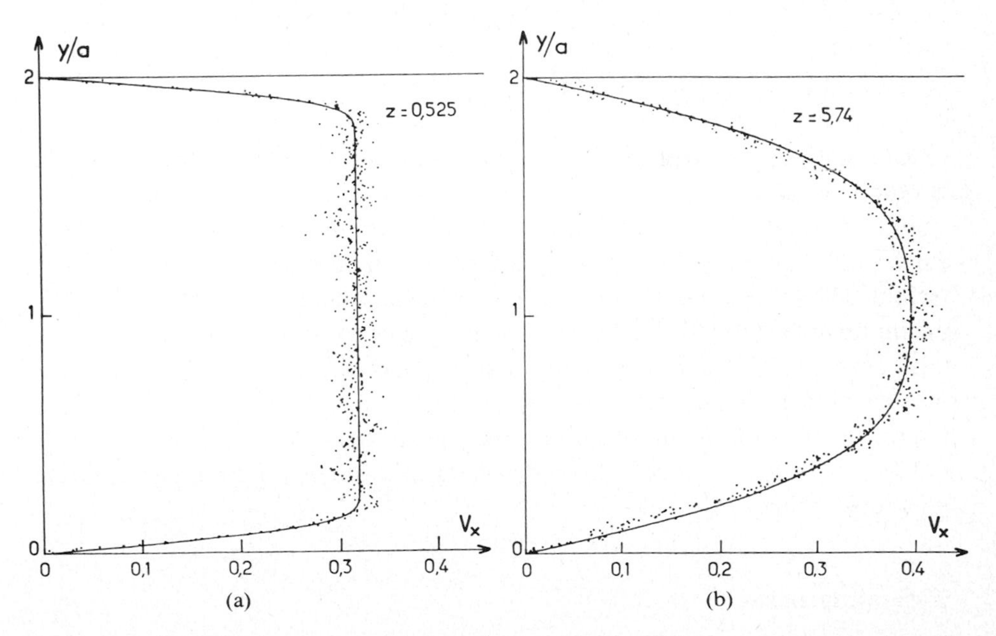

Fig. 5. Velocity component across a channel, relative distance from the inlet (a) 0.525, (b) 5.74. The dots are obtained by the lattice gas simulations, the solid lines are calculated using the Slichting method.

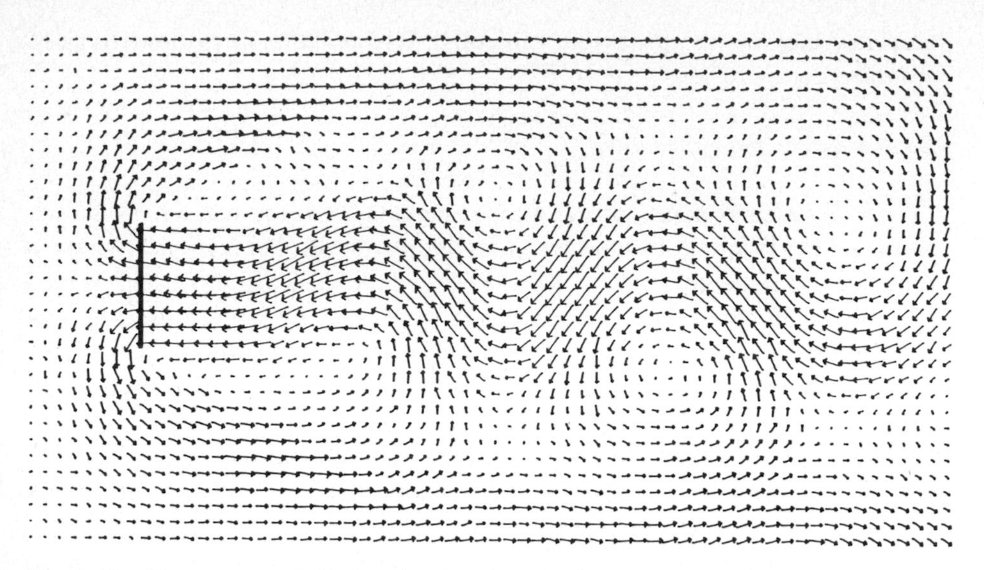

Fig. 6. Von Karman street behind a flat plate at 5000 time steps (Re ~ 90) with "wind-tunnel" boundary conditions; the mean flow has been subtracted.

time-dependent with a period $T = 1200$ time steps and a Strouhal number L/uT of approximately 0.2 close to the experimental value 0.16.

4. Conclusion

Quantitative agreement between theory and simulation has been demonstrated in both the linear and non-linear regimes, for moderate Reynolds numbers. It should be noted that the introduction of obstacles into the flow is particularly simple and represents, for the cases studied, a computational overhead of a few percent. The present models are limited to Reynolds numbers of order of 10^3 and incompressible flows. Moreover, since local equilibrium is a function of ρ and $\rho \boldsymbol{u}$ only, such models cannot simulate thermal phenomena. However, more complicated models derived from the FHP model [8,9]) may overcome most of these limitations in the near future. In this case, lattice gas simulations will be a new tool for experimental work in hydrodynamics, with the main advantage of being inherently stable.

Acknowledgements

We thank U. Frisch, B. Hasslacher, M. Hénon, Y. Pomeau and J.P. Rivet for helpful discussions and GRECO 70 "Expérimentation numérique" for support.

References

1) L.D. Landau and F.M. Lifshitz, Fluid Mechanics (Pergamon, London, 1959); the Navier-Stokes eq. (1b) is given for two dimensional flows, for D dimensions ξ must be replaced by $\xi + (D - 2)\eta/D$.
2) I.E. Broadwell, Phys. of Fluids **7** (1964) 1243; R. Gatignol. Théorie cinétique des gaz à répartition discrète de vitesse, Lecture Notes in Physics (Springer, Berlin, 1975).
3) J. Hardy, and Y. Pomeau, J. Math. Phys. **13** (1972) 1042; J. Hardy, O. de Pazzis and Y. Pomeau, J. Math. Phys. **14** (1973) 1746; Phys. Rev. A **13** (1976) 1949.
4) S. Wolfram, Rev. Mod. Phys. **55** (1984) 96.
5) Y. Pomeau, J. Phys. A **17** (1984) 415; G. Vichniac, Physica **10D** (1984) 96 and refs. therein; N.H. Packard and S. Wolfram, Two dimensional cellular automata, I.A.S. preprint, (Sept. 1984).
6) U. Frisch, B. Hasslacher and Y. Pomeau, Phys. Rev. Lett. **56** (1986) 1505.
7) Using these formulae, the density is defined as the number of particles per node; this is $\sqrt{3}/2$ of the density which is usually defined as the number of particles per square unit.
8) D. d'Humières, P. Lallemand and U. Frisch, Europhys. Lett. **2** (1986) 291.
9) S. Wolfram, Cellular automaton fluids 1: basic theory, Institute for Advanced Studies preprint (1986).
10) J.P. Rivet and U. Frisch, C.R. Acad. Sci. Paris II **302** (1986) 267.
11) M. Hénon, Calcul de la viscosité dans le réseau triangulaire and Viscosité d'un réseau, preprints Observatoire de Nice, B.P. 139, 06003 Nice Cedex, France; U. Frisch and J.P. Rivet, Lattice gas hydrodynamics: Green–Kubo formula, submitted to C.R. Acad. Sci. Paris II (1986).
12) D. d'Humières, Y. Pomeau and P. Lallemand, C.R. Acad. Sci. Paris II **301** (1985) 1391.
13) D. d'Humières, P. Lallemand and T. Shimomura, Computer simulations of lattice gas hydrodynamics, submitted to Phys. Rev. Lett. (1985).
14) D. d'Humières, Y. Pomeau and P. Lallemand, Two dimensional hydrodynamics calculations with a lattice gas, Innovative Numerical Methods in Engineering **241** (A Computational Mechanics Publication, Springer-Verlag, Berlin, 1986).
15) D. d'Humières and P. Lallemand, C.R. Acad. Sci. Paris II **302** (1985) 983.
16) R.D. Mountain, Rev. Mod. Phys. **38** (1966) 205.
17) S.A. Orszag and V. Yakhot, Phys. Rev. Lett. **56** (1986) 1691.
18) H. Slichting, Z.A.M.M. **14** (1934) 368; Boundary Layer Theory (Pergamon, London, 1955).

Physica **140A** (1986) 336–348
North-Holland, Amsterdam

A CRITICAL PHENOMENON APPROACH OF BIOGENESIS

Constantino TSALLIS
Centro Brasileiro de Pesquisas Fisicas / CNPq, Rua Xavier Sigaud 150,*
22290 Rio de Janeiro, RJ, Brazil
and Service de Physique Théorique-CEN-Saclay, 91190 Gif-sur-Yvette cedex, France

and

Ricardo FERREIRA
Departamento de Quimica Fundamental, Universidade Federal de Pernambuco,
50000 Recife, PE, Brazil

A fundamental prebiotic stage of the origin of life is the formation, from a random assembly of oligomers, of information-containing self-replicating polymers. By adopting a growth mechanism (already used by Anderson and others) essentially based in the complementarity of the Crick and Watson base-pairs, we show that this stage may have occurred as a critical phenomenon. Within a simple real-space renormalization group framework we calculate the relevant critical lines, which are different for different sequences of base-pairs. The picture incorporates in a natural way Darwinian-like evolution, is consistent with reasonable nucleotide ratios (A + T)/(C + G) and suggests that polymers like ADN double-chains are more primitive than proteins.

1. Introduction

Life seems to have started on Earth about 3.8×10^9 years ago. The Earth itself was formed about 5×10^9 years ago. Consequently, if we assume that life on our planet started spontaneously, there was a period of approximately 1.2×10^9 years during which a complex set of steps was overcome in order to arrive at the primitive biological systems. If life did not start spontaneously on Earth, but rather it came from elsewhere, then in that other place (or those other places) of the universe a similar set of steps has presumably occurred. Among the various prebiotic stages which probably occurred during the transition from inanimate to living matter, onc of great importance no doubt is the growth of codified self-replicating polymers starting from a random assembly of oligomers (dimers, trimers, etc). This is an important link of the chain which joins organic chemistry to biology, and its study constitutes the central scope of the present work. This

*Permanent address.

step remains, in spite of its obvious importance, insufficiently understood. This is due in part to the fact that the amount of related experimental work has not yet achieved the point where an enlightening and comprehensive view would be possible. This situation is, in some sense, in contrast with the present knowledge[1,2]) of a more primitive step, namely the formation of nucleotides, aminoacids, etc., starting from H_2O, methane, etc. (the transition from inorganic to organic chemistry, generally speaking); indeed, this step is now considered to be based on scientifically reliable grounds through the picture of violent non-equilibrium phenomena (electrical discharges, light and heat flashes, etc.) occurring in a relatively simple atmosphere.

The growth of codified self-replicating macromolecules (DNA- or RNA-like) has recently attracted quite intensive theoretical attention[3-10]) within thermodynamic and/or statistical mechanics frameworks. The basic growth mechanism is assumed to be autocatalysis relying on a Crick and Watson-like complementarity[11]), A and T as well as C and G thus constituting base-pairs. This is a convenient place for making clear that, although we shall use throughout the present work the notation A–T and C–G, *we do not necessarily refer to the well known nucleotides* (adenine, thymine, cytosine and guanine); the notation might as well refer to their *precursors*; even more, if the basic macromolecule is to be RNA-like[2,12-16]), rather than DNA-like, the notation A–T would then refer to the A–U pair.

Other growth mechanisms, using clay[17]), protein[18]) and aminoacid pairing[19]) as basic ingredients, have been proposed. However, autocatalysis based on A–T, C–G pairings is very appealing[2,6,8-10,12-16]), and this is the viewpoint we adopt in the present work.

We have developed since 1983 a thermal equilibrium critical phenomenon picture for understanding the growth of codified self-replicating polymers. Our first approach[20]) assumed a *single* base-pair; it provided polymeric growth consistent with *diversity*, but with *no selection* in it. Our second approach[21]) generalized the first one in the sense that the fugacity K_{AT} of the A–T hydrogen bridge might be *different* from the fugacity K_{CG} of the C–G hydrogen bridge; it showed that if *two* different base-pairs are assumed in the autocatalysis, *both diversity and selection* become possible, thus satisfying in a natural way the basic requirements for Darwinian evolution. The importance of having *four*, and *not two*, different monomers (capable consequently of forming *two*, and *not one*, complementary pairs) has also been emphasized by Anderson[10]).

Our critical phenomenon theoretical picture is developed within the renormalization group (RG) framework[22,23]), more specifically within a real space version of it, similar to those available in the literature[24-26]) for polymer problems. A comprehensive and pedagogical review of our approach[20,21]) of the above discussed prebiotic stage constitutes the central purpose of the present talk.

2. Model, formalism and results

The monomers A, T, C and G can form double strings through *intrachain* covalent bonds (noted – in the illustration which follows) and *interchain* (hydrogen-like) bridges (noted $\cdots$) as illustrated below:

$$\begin{array}{cccccc} A- & G- & T- & A- & C- & G \\ \vdots & \vdots & \vdots & \vdots & \vdots & \vdots \\ T- & C- & A- & T- & G- & C \end{array} \tag{1}$$

Let K_{AT} and K_{CG} be the *fugacities* (or bonding constants) respectively associated with the *interchain* A–T and C–G bridges (K_{AT}, $K_{CG} > 0$). K_{AT} and K_{CG} depend, in a complex unknown manner, on all the thermal equilibrium (or almost equilibrium) external parameters (temperature, pressure, humidity, various salts concentrations, etc.) which characterize the *primordial soup*, assumed to contain arbitrary amounts of randomly codified oligomers (dimers, trimers, etc.) like that of scheme (1). We further assume that oligomers can grow through the autocatalytic process illustrated below:

$$\text{(A–G–T–A)} + \text{(C–G)} + \text{(A–T–G–C)} \rightleftarrows$$

$$\begin{array}{l} \text{(A–G– T– A) (C– G)} \\ \quad\vdots \quad \vdots \quad \vdots \quad \vdots \\ \quad\text{(A– T – G– C)} \end{array} \rightleftarrows \begin{array}{l} \text{(A–G– T– A– C– G)} \\ \qquad\quad \vdots \quad \vdots \quad \vdots \quad \vdots \\ \qquad\quad\text{(A– T– G– C)} \end{array} \rightleftarrows$$

$$\text{(A–G–T–A–C–G)} + \text{(A–T–G–C)}. \tag{2}$$

Notice that we have obtained, as a final product, the *hexamer* of scheme (1), whereas at the initial stage, we had nothing longer than *tetramers*. In the present illustration, (A–G–T–A) and (C–G) play the role of *growing fragments*, and (A–T–G–C) plays the role of *catalysing fragment*. We are assuming that the intrachain condensation (characterized in our illustration by the fugacity J_{AC} of the A–C covalent bond) between the two growing fragments is greatly favoured ($J_{AC} \gg 1$) in the presence of the catalysing fragment bonded, to *both* growing fragments, through the interchain bridges.

In order to better understand the RG framework within which we shall perform calculations, let us first discuss the *single* base-pair particular case ($K_{AT} = K_{CG} \equiv K$; both A and C denoted by A; both T and G denoted by B). We perform the configurational analysis associated with the growth of a small oligomer (e.g., a *dimer* in fig. 1a, where procedure I has been illustrated), according to the following rules: (i) we consider all the growth-active configura-

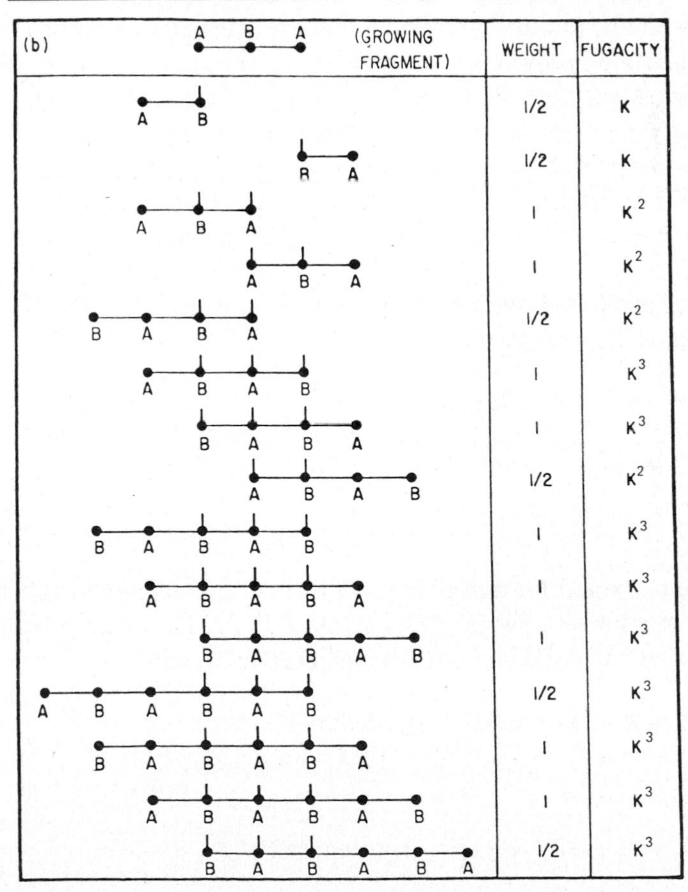

Fig. 1. Configurational analysis, within procedure I, of the catalysing fragments corresponding to the growth of a *dimer* (a) and of a *trimer* (b).

tions of all the catalysing fragments whose size is not longer than twice the growing fragment under consideration (we want to retain only the most probable mechanisms, and the probability of occurrence of catalysing fragments *much longer* than the growing fragment is rather poor); (ii) the "weight" equals 1 when the catalysing fragment is unambiguously associated with the growing fragment under consideration, equals $\frac{1}{2}$ when it can equally well be associated with the other growing fragment and equals 0 (and is therefore absent from the figure) when it is unambiguously associated with the other fragment (to be more precise, when the number of non-connected residues at any given end of the catalysing fragment exceeds the number of its residues actually connected to the growing fragment under consideration); (iii) the number of growth-active ends (1 or 2) of the catalysing fragment can be disregarded (procedure I) or taken into account (procedure II) by introducing a "growth efficiency" which equals the number of growth-active ends; (iv) the interchain bonds are assumed independent (hence the effective fugacity of a given set of simultaneous bonds is just the *product* of the corresponding fugacities); (v) multiple catalysing processes (involving more than one catalysing fragment) or similar complex processes are neglected because of a presumably low probability of occurrence. This set of rules obviously involves a certain degree of arbitrariness; however, we believe that any other "reasonable" set of rules would lead to results not essentially different from those we shall present.

Fig. 1a yields, through the sum of (weight) × (growing efficiency) × (fugacity), the following effective fugacity:

$$R_2^{\mathrm{I}}(K) = K + 4K^2 \quad \text{(procedure I)}. \tag{3}$$

Analogously we obtain, within procedure II,

$$R_2^{\mathrm{II}}(K) = K + 5K^2 \quad \text{(procedure II)}. \tag{3$'$}$$

The subscripts 2 stand for dimer. We now repeat the configurational analysis for the growth of a longer oligomer (e.g., a *trimer* in fig. 1b, where procedure I has been illustrated). We obtain the following effective fugacity:

$$R_3^{\mathrm{I}}(K) = K + 3K^2 + 8K^3 \quad \text{(procedure I)} \tag{4}$$

and also

$$R_3^{\mathrm{II}}(K) = K + 3K^2 + 11K^3 \quad \text{(procedure II)}. \tag{4$'$}$$

We can now write down the RG recursive equation, namely

$$R_2^{\alpha}(K') = R_3^{\alpha}(K) \qquad (\alpha = \mathrm{I} \text{ or } \mathrm{II}). \tag{5}$$

Both recurrences admit the trivial (stable) fixed points $K = 0$ (corresponding to lack of infinite growth, and characterizing the *finite growth* (FG) *phase*) and $K = \infty$ (characterizing the *infinite growth* (IG) *phase*). They also admit a critical (unstable) fixed point, namely $K^* = \frac{1}{8} = 0.125$ for procedure I and $K^* = \frac{2}{11} \simeq 0.18$ for procedure II. The present calculation provides further information: while approaching the critical value K^*, the *mean length* ξ of the growing fragment diverges as $\xi \propto (K^* - K)^{-\nu}$, where the critical exponent ν is given (within the present RG approximation) by

$$\nu = \frac{\ln(b/b')}{\ln(\mathrm{d}K'/\mathrm{d}K)_{K^*}} = \frac{\ln(b/b')}{\ln\left[(\mathrm{d}R_b(K)/\mathrm{d}K)/(\mathrm{d}R_{b'}(K)/\mathrm{d}K)\right]_{K^*}}, \tag{6}$$

where b (b') is the size of the original (renormalized) oligomer under analysis (in our present example, $b = 3$ for the trimer and $b' = 2$ for the dimer) and $R_b(K)$ ($R_{b'}(K')$) the corresponding effective fugacity. We obtain $\nu \simeq 7.0$ for procedure I and $\nu \simeq 3.3$ for procedure II. The smaller and more satisfactory (because of better consistency with related calculations in polymer physics) value of ν obtained through procedure II is to be attributed to the higher realism introduced by the growth efficiency. Anyhow, it is completely out of the scope of the present very crude approximations to obtain reliable numbers for K^* or ν: our arguments concern only the qualitative facts of the picture.

Summarizing, we have seen that the autocatalytic mechanism might lead, when approaching a critical value for the interchain fugacity, to the growth of codified self-replicating polymers. This already seems to us a very suggestive conclusion. However, if a *single* base-pair is assumed, all codes grow, and all do so at the same value of K: this is fine regarding diversity, *but*, from the biological standpoint, completely unsatisfactory in what concerns selection! We shall next see that the (realistic) assumptions of *two* (or more) different base-pairs will lead to a remarkable improvement.

The parameter space of our problem will now be a two-dimensional one, namely determined by K_{AT} and K_{CG} (all intrachain fugacities are assumed infinite at this level of approximation). The RG flow will now be determined by (explicit or implicit) recursive relations of the following type:

$$K'_{\mathrm{AT}} = f_{bb',\sigma}(K_{\mathrm{AT}}, K_{\mathrm{CG}}), \tag{7a}$$

$$K'_{\mathrm{CG}} = g_{bb',\sigma}(K_{\mathrm{CG}}, K_{\mathrm{AT}}), \tag{7b}$$

where $f_{bb',\sigma}$ and $g_{bb',\sigma}$ are functions which will in general depend on the respective sizes b and b' of the original and renormalized oligomers we have chosen to work with, *as well as on the particular code which is growing* (and which is denoted by the index σ). Examples of such codes are the following:

$\cdots K_{AT}K_{AT}K_{AT}K_{AT} \cdots$ ($\sigma = 1$), $\cdots K_{CG}K_{CG}K_{CG}K_{CG} \cdots$ ($\sigma = 2$), $\cdots K_{AT}K_{CG}K_{AT}K_{CG} \cdots$ ($\sigma = 3$), $\cdots K_{AT}K_{AT}K_{CG}K_{CG}K_{AT}K_{AT}K_{CG}K_{CG} \cdots$ ($\sigma = 4$) and $\cdots K_{AT}K_{AT}K_{CG}K_{AT}K_{AT}K_{CG} \cdots$ ($\sigma = 5$). The single base-pair particular case can be obtained through three different limits namely: (i) $K_{AT} = K_{CG} \equiv K$ and arbitrary σ, therefore $f_{bb',\sigma}(K, K) = g_{bb',\sigma}(K, K) \equiv F_{bb'}(K)$; (ii) $K_{AT} \equiv K$, arbitrary K_{CG}, and $\sigma = 1$, therefore $f_{bb',1}(K, K_{CG}) \equiv F_{bb'}(K)$; (iii) $K_{CG} \equiv K$, arbitrary K_{AT}, and $\sigma = 2$, therefore $g_{bb',2}(K, K_{AT}) \equiv F_{bb'}(K)$. Furthermore, for codes which are invariant through $K_{AT} \rightleftarrows K_{CG}$ permutation (e.g., $\sigma = 3, 4$, but not $\sigma = 5$), the following property must be satisfied: $f_{bb',\sigma}(X, Y) = g_{bb',\sigma}(X, Y)$ for arbitrary (X, Y). Several of the above properties can be verified on the following example concerning the growth of the $\cdots K_{AT}K_{CG}K_{AT}K_{CG} \cdots$ sequence ($\sigma = 3$). The RG equations are given by

$$R_5^{\text{II}}(K'_{AT}, K'_{CG}, K'_{AT}, K'_{CG}) = R_9^{\text{II}}(K_{AT}, K_{CG}, K_{AT}, K_{CG}) \tag{8}$$

and

$$R_5^{\text{II}}(K'_{CG}, K'_{AT}, K'_{CG}, K'_{AT}) = R_9^{\text{II}}(K_{CC}, K_{AT}, K_{CG}, K_{AT}), \tag{9}$$

where

$$R_5^{\text{II}}(K_{AT}, K_{CG}, K_{AT}, K_{CG}) = K_{AT} + 3K_{AT}K_{CG} + 5K_{AT}^2K_{CG} + 7K_{AT}^2K_{CG}^2 + 29K_{AT}^3K_{CG}^2 \tag{10}$$

and

$$R_9^{\text{II}}(K_{AT}, K_{CG}, K_{AT}, K_{CG}) = K_{AT} + 3K_{AT}K_{CG} + 5K_{AT}^2K_{CG} + 7K_{AT}^2K_{CG}^2 + 9K_{AT}^3K_{CG}^2 + 11K_{AT}^3K_{CG}^3 + 13K_{AT}^4K_{CG}^3 + 15K_{AT}^4K_{CG}^4 + 89K_{AT}^5K_{CG}^4. \tag{11}$$

See in fig. 2 the associated RG flow, which determines the corresponding critical line, and also exhibits that the two base-pairs case belongs to the same university class as the single base-pair case.

Eq. (10) has been established by making the configurational analysis associated with the sequence $\cdots K_{AT}K_{XX'}K_{YY'}K_{CG}K_{AT}K_{XX'}K_{YY'}K_{CG} \cdots$ (see fig. 3), calculating the effective fugacity $R_5^{\text{II}}(K_{AT}, K_{XX'}, K_{YY'}, K_{CG})$, and then taking $XX' = \text{CG}$ and $YY' = \text{AT}$. We have proceeded analogously to obtain eq. (11).

We have indicated in fig. 4 the critical lines corresponding to various typical sequences. We notice an important improvement with respect to the one base-pair

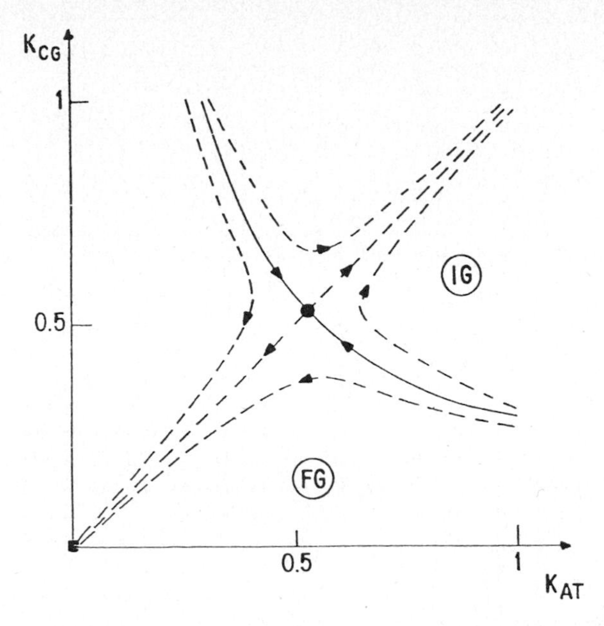

Fig. 2. Critical line (full line) in the (K_{AT}, K_{CG}) fugacity space, separating the finite growth (FG) phase from the infinite growth (IG) one of the sequence $\cdots K_{AT}K_{CG}K_{AT}K_{CG} \cdots$. Arrows and dashed lines indicate RG flow; the central dot indicates the single base-pair critical fixed point (responsible for the universality class of the whole critical line).

model: the picture presents now *both* diversity and selection! In other words, a microscopic basis for Darwinian evolution is now achieved. However, and in spite of this interesting achievement, the model is not yet free from two important limitations: (i) if we assume a reasonable time evolution of K_{AT} and K_{CG} (see fig. 4), the most privileged codes are those presenting either very low or very high (A + T)/(C + G) ratios, a fact which is not easily consistent with the values ($\frac{1}{2} \lesssim$ (A + T)/(C + G) $\lesssim 2$; see ref. 11) associated with modern living systems (at least in the biosphere); (ii) the critical line is one and the same for all sequences of nucleotides which correspond to a single sequence of bonds (e.g., $\cdots$ ACACAC $\cdots$, $\cdots$ AGAGAG $\cdots$, $\cdots$ ACTGTCTG $\cdots$, etc., correspond to the sequence $\cdots K_{AT}K_{CG}K_{AT}K_{CG} \cdots$), a fact which has no biochemical support. Both limitations disappear by considering the different intrachain fugacities (they are, within a nearest-neighbor picture, 10 in number, and will be denoted by J_{AA}, J_{AT}, J_{AC}, J_{AG}, J_{TT}, J_{TC}, J_{TG}, J_{CC}, J_{CG} and J_{GG}). In fact, our approach thus far corresponds to assigning to these 10 constants the value infinity. It is intuitive that finite values for these fugacities will make it *more difficult* to attain the point of infinite polymeric growth. We have indicated in fig. 5 the expected critical line assuming say that all the J's are equal among them

Growing fragment: A X Y C A — Catalysing fragments	Weight	Growth efficiency I	Growth efficiency II	Fugacity
G T	$\frac{1}{2}$	1	1	K_{AT}
T X'	$\frac{1}{2}$	1	1	K_{AT}
G T X'	1	1	1	$K_{AT} K_{XX'}$
G T X'	1	1	1	$K_{AT} K_{CG}$
Y' G T X'	$\frac{1}{2}$	1	1	$K_{AT} K_{XX'}$
G T X' Y'	1	1	1	$K_{AT} K_{XX'} K_{YY'}$
Y' G T X'	1	1	1	$K_{AT} K_{YY'} K_{CG}$
G T X' Y'	$\frac{1}{2}$	1	1	$K_{AT} K_{CG}$
Y' G T X' Y'	1	1	1	$K_{AT} K_{XX'} K_{YY'}$
G T X' Y' G	1	1	1	$K_{AT} K_{XX'} K_{YY'} K_{CG}$
X' Y' G T X'	1	1	1	$K_{AT} K_{XX'} K_{YY'} K_{CG}$
Y' G T X' Y'	1	1	1	$K_{AT} K_{YY'} K_{CG}$
X' Y' G T X' Y'	$\frac{1}{2}$	1	1	$K_{AT} K_{XX'} K_{YY'}$
Y' G T X' Y' G	1	1	1	$K_{AT}^2 K_{XX'} K_{YY'} K_{CG}$
G T X' Y' G T	1	1	1	$K_{AT}^2 K_{XX'} K_{YY'} K_{CG}$
T X' Y' G T X'	1	1	1	$K_{AT}^2 K_{XX'} K_{YY'} K_{CG}$
X' Y' G T X' Y'	1	1	1	$K_{AT} K_{XX'} K_{YY'} K_{CG}$
Y' G T X' Y' G	$\frac{1}{2}$	1	1	$K_{AT} K_{YY'} K_{CG}$
X' Y' G T X' Y' G	1	1	1	$K_{AT} K_{XX'} K_{YY'} K_{CG}$
Y' G T X' Y' G T	1	1	1	$K_{AT} K_{XX'} K_{YY'} K_{CG}$

Growing fragment: A X Y C A — Catalysing fragments	Weight	Growth efficiency I	Growth efficiency II	Fugacity
G T X' Y' G T X'	1	1	2	$K_{AT}^2 K_{XX'} K_{YY'} K_{CG}$
T X' Y' G T X' Y'	1	1	1	$K_{AT}^2 K_{XX'} K_{YY'} K_{CG}$
X' Y' G T X' Y' G	1	1	1	$K_{AT} K_{XX'} K_{YY'} K_{CG}$
T X' Y' G T X' Y' G	$\frac{1}{2}$	1	1	$K_{AT} K_{XX'} K_{YY'} K_{CG}$
X' Y' G T X' Y' G T	1	1	1	$K_{AT}^2 K_{XX'} K_{YY'} K_{CG}$
Y' G T X' Y' G T X'	1	1	2	$K_{AT}^2 K_{XX'} K_{YY'} K_{CG}$
G T X' Y' G T X' Y'	1	1	2	$K_{AT}^2 K_{XX'} K_{YY'} K_{CG}$
T X' Y' G T X' Y' G	1	1	1	$K_{AT}^2 K_{XX'} K_{YY'} K_{CG}$
X' Y' G T X' Y' G T	$\frac{1}{2}$	1	1	$K_{AT} K_{XX'} K_{YY'} K_{CG}$
T X' Y' G T X' Y' G T	1	1	1	$K_{AT}^2 K_{XX'} K_{YY'} K_{CG}$
X' Y' G T X' Y' G T X'	1	1	2	$K_{AT}^2 K_{XX'} K_{YY'} K_{CG}$
Y' G T X' Y' G T X' Y'	1	1	2	$K_{AT}^2 K_{XX'} K_{YY'} K_{CG}$
G T X' Y' G T X' Y' G	1	1	2	$K_{AT}^2 K_{XX'} K_{YY'} K_{CG}$
T X' Y' G T X' Y' G T	1	1	1	$K_{AT}^2 K_{XX'} K_{YY'} K_{CG}$
G T X' Y' G T X' Y' G T	$\frac{1}{2}$	1	1	$K_{AT}^2 K_{XX'} K_{YY'} K_{CG}$
T X' Y' G T X' Y' G T X'	1	1	2	$K_{AT}^2 K_{XX'} K_{YY'} K_{CG}$
X' Y' G T X' Y' G T X' Y'	1	1	2	$K_{AT}^2 K_{XX'} K_{YY'} K_{CG}$
Y' G T X' Y' G T X' Y' G	1	1	2	$K_{AT}^2 K_{XX'} K_{YY'} K_{CG}$
G T X' Y' G T X' Y' G T	1	1	2	$K_{AT}^2 K_{XX'} K_{YY'} K_{CG}$
T X' Y' G T X' Y' G T X'	$\frac{1}{2}$	1	1	$K_{AT}^2 K_{XX'} K_{YY'} K_{CG}$

Fig. 3. Configurational analysis, within both procedures I and II, of the catalysing fragments corresponding to the growth of a *pentamer* (sequence $\cdots K_{AT} K_{XX'} K_{YY'} K_{CG} K_{AT} K_{XX'} K_{YY'} K_{CG} \cdots$).

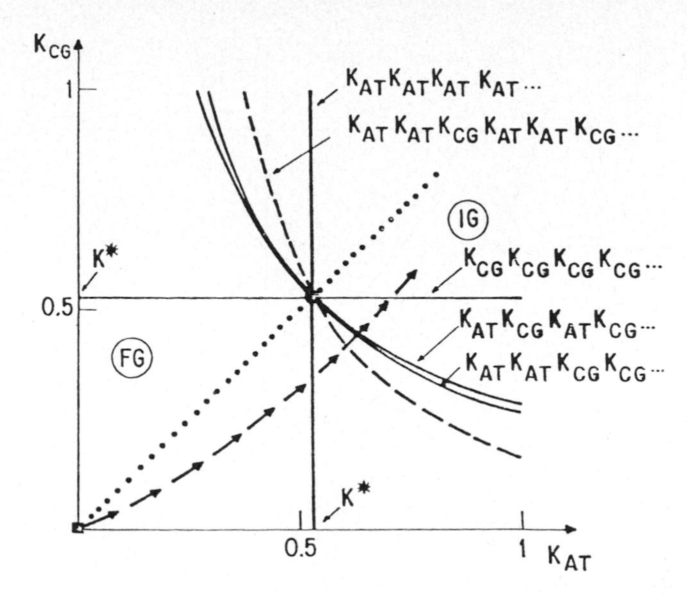

Fig. 4. Critical lines (in the (K_{AT}, K_{CG}) fugacity space) corresponding to the growth of selected sequences (the dashed line is indicative); FG (IG) denotes the finite (infinite) growth phase. The point at $K_{AT} = K_{CG} = K^*$ reproduces the fixed point of fig. 2; the dotted line is a symmetry axis of some of the sequences (e.g., $\cdots K_{AT}K_{CG}K_{AT}K_{CG}\cdots$ and $\cdots K_{AT}K_{AT}K_{CG}K_{CG}K_{AT}K_{AT}K_{CG}K_{CG}\cdots$). The arrows indicate a plausible (slow) time evolution of K_{AT} and K_{CG}.

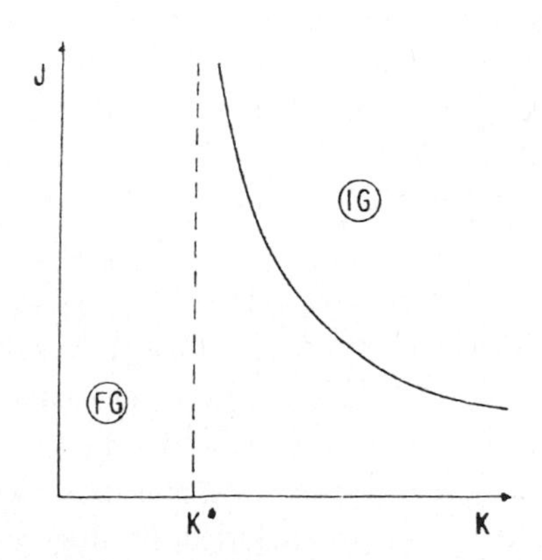

Fig. 5. Indicative FG–IG critical line (one and the same for all sequence types) corresponding to $K_{AT} = K_{CG} \equiv K$ and all J's equal among them (and equal to J). K^* refers to the single base-pair critical point of fig. 2.

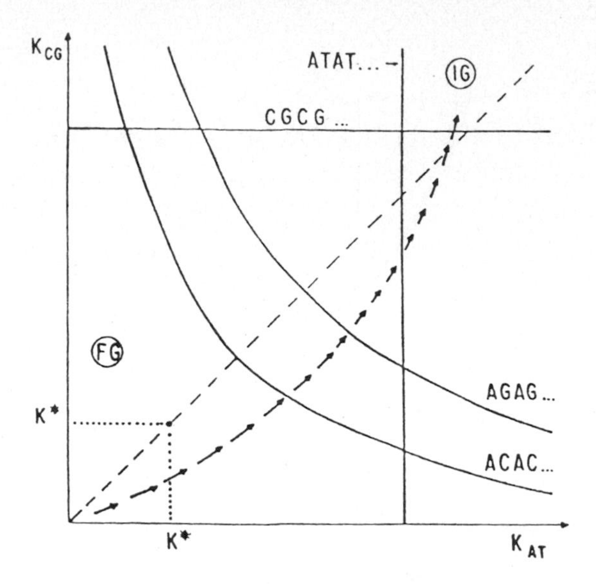

Fig. 6. Indicative FG–IG critical lines corresponding to the growth of different *nucleotide sequences* (not only different sequences of interchain links). The dashed line is a symmetry axis of some sequences (e.g., $\cdots$ACAC$\cdots$, $\cdots$AGAG$\cdots$); the dotted lines indicate the value K^* of all previous figures. The arrows indicate a plausible (slow) time evolution of K_{AT} and K_{CG}.

(and equal to J), and that $K_{AT} = K_{CG} \equiv K$; note that K approaches K^* when J diverges. The fact that the actual J's are *finite and different from one another*, will make all the critical lines (of fig. 4) shift toward *higher* values of K_{AT} and K_{CG}. This shift is in general *different for differing sequences of nucleotides*, *even if they preserve the same sequence of* K_{AT}'s and K_{CG}'s. The result is indicated in fig. 6, by arbitrarily choosing $J_{CC} \simeq J_{CG} \simeq J_{GG} < J_{AA} \simeq J_{AT} \simeq J_{TT} < J_{AG} \simeq J_{TC} < J_{AC} \simeq J_{TG} < \infty$. We have not carried actual RG calculations corresponding to finite J's. They are in principle tractable, though burdensome because of the large number of RG parameters. It is clear, in any case, that this is a realistic path for overcoming the two limitations mentioned previously.

Let us add that if we were to pursue calculations along this line, a further realistic element would have to be included in the model, namely a certain amount of cross-links which do exist between different points of the (folded) double-chain. The effect of these cross-links would presumably be (as observed, for instance, in hemoproteins) to make the (intrinsic) fractal dimensionality of the macromolecule higher than one. This would be necessary for having phase transitions at finite (not vanishing) temperatures in the presence of finite (not infinite) J's.

3. Conclusions

Within a critical phenomena approach we have studied the prebiotic stage concerning the growth of codified self-replicating polymers (DNA- or RNA-like) starting from small oligomers (dimers, trimers, etc.). The growth mechanism that has been adopted essentially is autocatalysis through Crick and Watson-like pair complementarity. The picture which emerges can be synthesized as follows:

(i) A *single* base-pair ($K_{AT} = K_{CG}$) yields critical growth consistent with *diversity* but not with *selection*.

(ii) Two (or more) base-pairs ($K_{AT} \neq K_{CG}$) yield polymeric growth consistent with *both diversity and selection*; these fundamental ingredients naturally come into the theory, thus providing a microscopic basis for Darwinian evolution; life would have then appeared from a certain amount of self-replicating codes, and not from a single one (those different codes would have grown at different, though close, moments of the Earth's evolution).

(iii) The role played by *finite* values for the interchain covalent fugacities (J_{AA}, J_{AC}, etc.) is to make possible realistic values for the nucleotide ratio $(A + T)/(C + G)$ (roughly between $\frac{1}{2}$ and 2).

(iv) In the old querelle "which came first: nucleic acids or proteins?", our picture suggests a more primitive role for the nucleic acids.

(v) The polymeric growth we are concerned with *can* be thought as essentially being a *thermal equilibrium* stage (no assumptions of processes far from equilibrium are needed).

As a final remark we might add that, if analogies with spin $\frac{1}{2}$ magnetic systems are to be done for the prebiotic stage under study, the roughest "reasonable" model seems to be that in which *the binary code* ($S = \pm \frac{1}{2}$) *refers to the base-pairs* (A–T or C–G) *and not to the nucleotides* (purine or pirimidine).

Acknowledgement

One of us (C.T.) acknowledges hospitality received at the Service de Physique Théorique, CEN-Saclay, where the present review was written.

References

1) A.E. Oparin, The Chemical Origin of Life (Thomas, Springfield, 1964).
2) S. Miller and L.E. Orgel, Origin of Life on Earth (Prentice Hall, Englewood Cliffs, NJ, 1974); J. Ninio, Molecular Approaches to Evolution (Princeton Univ. Press, Princeton, 1983).

3) M. Eigen and P. Schuster, Naturwiss. **64** (1977) 541.
4) M. Eigen and P. Schuster, Naturwiss. **65** (1978) 341.
5) F. Dyson, J. Mol. Evol. **18** (1982) 344.
6) P.W. Anderson, Proc. Nat. Acad. Sci. USA **80** (1983) 3386.
7) L. Demetrius, Proc. Nat. Acad. Sci. USA **81** (1984) 6068.
8) P.W. Anderson and D. Stein, Proc. Nat. Acad. Sci. USA **81** (1984) 1751.
9) P.W. Anderson and D. Stein, Self-organizing Systems: The Emergence of Order, F.E. Yates, ed. (Plenum, New York, 1985).
10) P.W. Anderson, Computer modeling of prebiotic evolution: general theoretical ideas on the origin of biological information, preprint (1985), to appear in Comments on Molecular and Cellular Biophysics (Gordon and Breach, New York).
11) J.D. Watson, Molecular Biology of the Gene, 2nd. ed. (Benjamin, New York, 1970).
12) M. Eigen, Naturwiss. **58** (1971) 465.
13) H. Kuhn, Angew. Chem. **84** (1972) 838.
14) R. Lohrmann and L.E. Orgel, J. Mol. Biol. **12** (1979) 237.
15) H. Kuhn and J. Waser, Angew. Chem. **93** (1981) 495.
16) H. Kuhn and J. Waser, in Biophysics, Hoppe, Lohrmann, Markel and Ziegler, eds. (Springer, Berlin, 1983), pp. 830–847.
17) A.G. Cairns-Smith, J. Theor. Biol. **10** (1966) 53.
18) S.W. Fox and K. Dose, Molecular Evolution and the Origin of Life (Dekker, New York, 1977).
19) R.S. Root-Bernstein, J. Theor. Biol. **94** (1982) 895.
20) C. Tsallis and R. Ferreira, Phys. Lett. **99A** (1983) 461.
21) R. Ferreira and C. Tsallis, J. Theor. Biol. **117** (1985) 303.
22) K.G. Wilson and J. Kogut, Phys. Rep. **12C** (1974) 75.
23) G. Toulouse and P. Pfeuty, Introduction au Groupe de Renormalisation et à ses Applications (Presses Univ. Grenoble, Grenoble, 1975).
24) S.L.A. de Queiroz and C.M. Chaves, Z. Phys. B **40** (1980) 99.
25) H. Gould, F. Family and H.E. Stanley, Phys. Rev. Lett. **50** (1982) 686.
26) H.E. Stanley, F. Family and H. Gould, J. Poly. Sci.: Polymer Symposium **73** (1985).
27) G. Nicolis and I. Prigogine, Self-organization in Non-equilibrium Systems (Wiley, New York, 1977).

Physica **140A** (1986) 349–358
North-Holland, Amsterdam

SPIN SLIPS AND LATTICE MODULATIONS IN HOLMIUM: A MAGNETIC X-RAY SCATTERING STUDY

Jakob BOHR and Doon GIBBS

Brookhaven National Laboratory, Upton, NY 11973, USA

and

D.E. MONCTON and K.L. D'AMICO

Exxon Research and Engineering Co., Annandale, NJ 08801, USA

The analysis of data obtained in magnetic X-ray scattering experiments on rare earth metals, particularly holmium, has led to a phenomenological model for one-dimensional spatially propagating magnetic structures. Based on the concept of spin slips or discommensurations, this model explains the observed lock-in transitions in the magnetic spirals of the rare earths in terms of simple commensurate structures. Further, the anomalous intensities of previously observed higher harmonic magnetic satellites, as well as the qualitative behavior of the magnetic wave vector in the presence of a magnetic field, are understood directly within the spin slip description. We discuss how the presence of spin slips in the magnetic structure can lead to modulations of the crystal lattice as demonstrated by a recent magnetic X-ray study of holmium.

1. Introduction

Magnetic X-ray scattering with synchrotron radiation has recently been demonstrated to be a fruitful technique for magnetic structure and phase transition studies in a set of experiments beginning with the rare earth metal holmium[1–2]). These experiments led to the discovery of localized spin defects in holmium, which were previously undetected. We believe these defects, which we call spin slips, occur generally in the magnetic structure of rare-earths and perhaps in other magnets as well. In this paper we will review our understanding of the magnetic structure of holmium, including previously unpublished data and analysis, with emphasis on the concept of spin slips. A description of our magnetic X-ray scattering methods has been published elsewhere[2]).

The magnetic structure of holmium has been extensively studied by neutron diffraction[3–6]) and the basic structure is well established. Below the Néel temperature (132 K) the structure is an incommensurate spiral with the propagation wave vector in the direction perpendicular to the basal planes of the hcp crystal structure. The magnetic moments of the atoms within the basal planes are all

0378-4371/86/$03.50 © Elsevier Science Publishers B.V.
(North-Holland Physics Publishing Division)

parallel and confined to the plane. At low temperature (20 K), the structure becomes conical with a small component of the moments along the direction of the propagation of the spiral (cone angle 10°). In the early and comprehensive studies by Koehler et al.[3,4]), it was found that for temperatures below 19 K the conical spiral is commensurate with a wave vector of 0.167 = 1/6 (units of c^*). In this phase the magnetic spiral repeats itself every six crystallographic unit cells (or 12 atomic layers). For increasing temperatures the wave vector apparently increased continuously. Hysteresis was present near the lock-in transition. Felcher et al.[5]), in a later neutron diffraction study of Ho, found that at low temperatures their sample locked to 0.185 (or 5/27) rather than 1/6. In a sample studied more recently by Pechan and Stassis[6]), the low temperature period was again observed to be 1/6. The presence of higher harmonics of the magnetic satellites in the neutron diffraction patterns shows that at low temperature the spiral deviates from uniform spatial propagation. Koehler et al.[3]) pointed out that 5th and 7th harmonic magnetic satellites of equal intensity, which they observed in the low temperature 1/6 structure, implied that the magnetic moments in this 12 atomic layer unit cell are distorted toward the six easy directions of the basal plane crystal field.

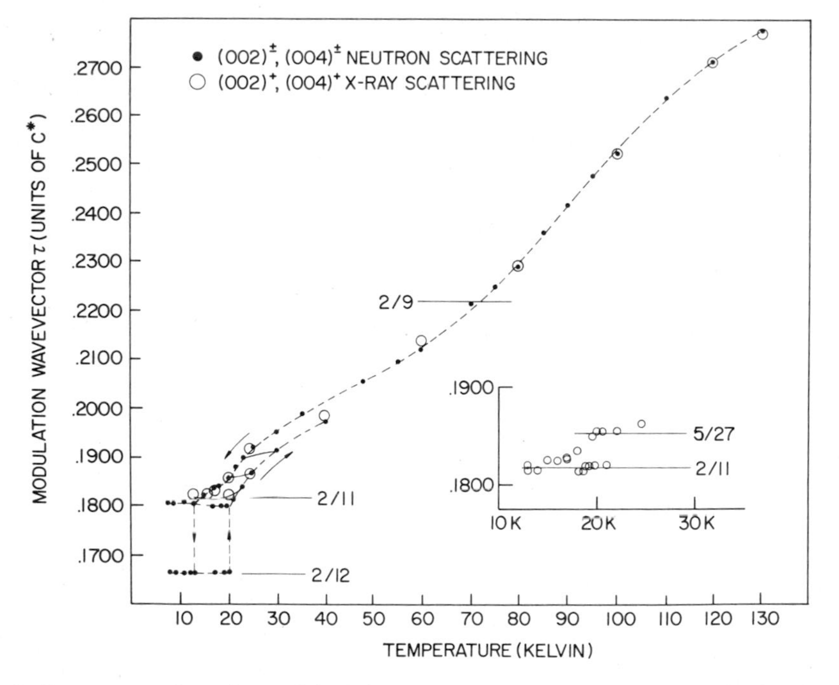

Fig. 1. Temperature dependence of the holmium magnetic wave vector τ as obtained with synchrotron X-ray diffraction (open circles) and with neutron diffraction (dots).

Recently, we reported an X-ray study of the magnetic spiral of holmium[1]). Fig. 1 shows the magnetic wave vector as a function of temperature. By analyzing the polarization[2]) of the scattered X-rays, we demonstrated that the scattering is magnetic in origin. The main features of the temperature dependence of the magnetic wave vector are: an inflection point around 75 K or $\tau_m = 2/9$, hysteresis below 50 K, and coexistence of phases with different wave vectors. At the lowest temperatures, there is a first order transition between two commensurate periods, namely, $0.182 = 2/11$ and $0.167 = 1/6$, and there is an indication of a lock-in at $0.185 = 5/27$. As has been mentioned, the 1/6 and 5/27 phase have been observed previously in other samples.

2. Spin-slip model: application to holmium

An understanding of these rational fractions arises naturally from the concept of discommensurations[7]). The magnetic spiral of holmium is an example of a structure where the local configuration can be described by a phase parameter $\phi(i)$ which depends on the index i of the atomic layer. Let the phase space be f-fold degenerate as defined by the easy directions of the crystal field ($f = 6$ in the basal plane of holmium). The crystal field therefore provides a periodic potential in the phase space with a period $2\pi/f$. Hence, for a simple structure where $\phi(i+1) = \phi(i) + C$ (C is an arbitrary constant), there exists an integer k such that the number of sequential atomic-layers for which $\phi(i)$ falls within the same period of the crystal field potential is always either k or $k-1$. Therefore, k is the smallest integer which satisfies the inequality $fk\tau_m \geqslant Q$, where τ_m is the magnetic wave vector and Q the wave vector of the interatomic layer distance. The sequence of blocks of k and $k-1$ atomic layers is a unique property of the wave vector τ_m and defines a translational symmetry (commensurate structures) or shows a lack of translational symmetry (incommensurate structures). We will introduce a shorthand notation for this sequence using integers and dots, for example ($\cdot p$). The integer p gives the number of blocks of length k which come in sequence and a dot represents a block of length $k-1$ in the sequence. Whenever a block of length $k-1$ appears in the sequence we will say that there is a spin slip in the structure. The phase parameter advances additionally by $2\pi/fk$, as if there were an extra layer of atoms in the structure. The above described sequence will be referred to as the spin slip sequence. The average number of atomic layers x between spin slips can be found from the equation $(x+1)2\pi/fk = x2\pi\tau_m/Q$, i.e.,

$$x = 1/(fk\tau_m/Q - 1). \tag{1}$$

For holmium $\tau_m \geqslant 1/6$ (see fig. 1), $Q = 2$, and $f = 6$. Consequently, $k = 2$ and

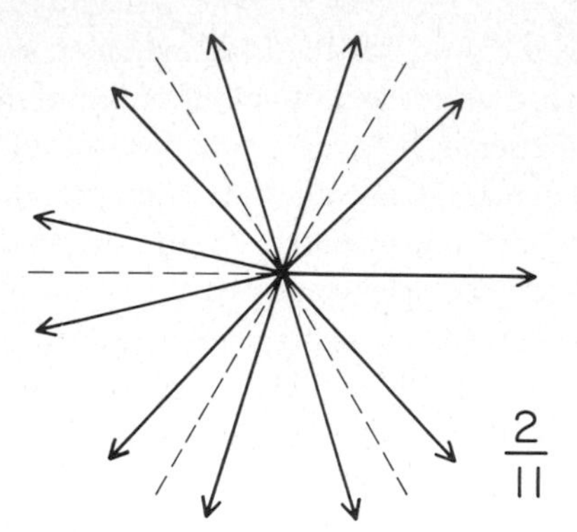

Fig. 2. A schematic and simplified drawing of the directions of the atomic moments in the 11 atomic layer commensurate structure •5. The dotted lines indicate the 6 easy directions in the basal plane of the hcp crystal structure.

we say that the spiral of holmium consists of doublets and singlets. The low temperature phase with $\tau_m = 1/6$ has a commensurate structure with 6 doublets in the unit cell, as was recognized by Koehler et al.[3]) The first order transition in fig. 1 takes place between $\tau_m = 1/6$ and $\tau_m = 0.182$ or 2/11. The 2/11 structure results from removing one spin from the 1/6 structure per magnetic unit cell (see fig. 2). The spin slip notation •5 describes the 2/11 structure. The 5/27 structure ($\tau_m = 0.185$) has an equally simple spin slip configuration, namely a spin slip every 9 atomic layers rather than one for every 11 layers. The spin slip sequence is •4 • 4 • 4 • 4 • 4 • 4, or 6 spin slips distributed over 5 magnetic periods[8]).

2.1. *Lattice modulations*

If the spin slips are uniformly distributed in the magnetic spiral, or if the spin slip pattern is periodic, then interplanar lattice modulations with a period of the spin slip pattern may arise from magneto-elastic interactions. For the $\tau_m = 2/11$ structure shown in fig. 2 the wave vector of the interplanar lattice modulation is also 2/11. For the $\tau_m = 5/27$ structure (•4 • 4 • 4 • 4 • 4 • 4) there is one spin slip for every 9 atomic layers and, consequently, the corresponding interplanar lattice modulations have a wave vector $\tau_s = 2/9$. Fig. 3 shows the X-ray satellites above the (004) Bragg point for temperatures decreasing from $T = 25$ K to $T = 17$ K. Clearly, for $T \geqslant 23$ K additional, and initially broad, scattering is observed above the magnetic satellites. As the temperature is lowered and the magnetic wave vector approaches 5/27, then the additional scattering from the interplanar lattice modulations sharpens up and approaches 2/9. A polarization analysis of the diffraction pattern at $T = 17$ K showed that this additional peak

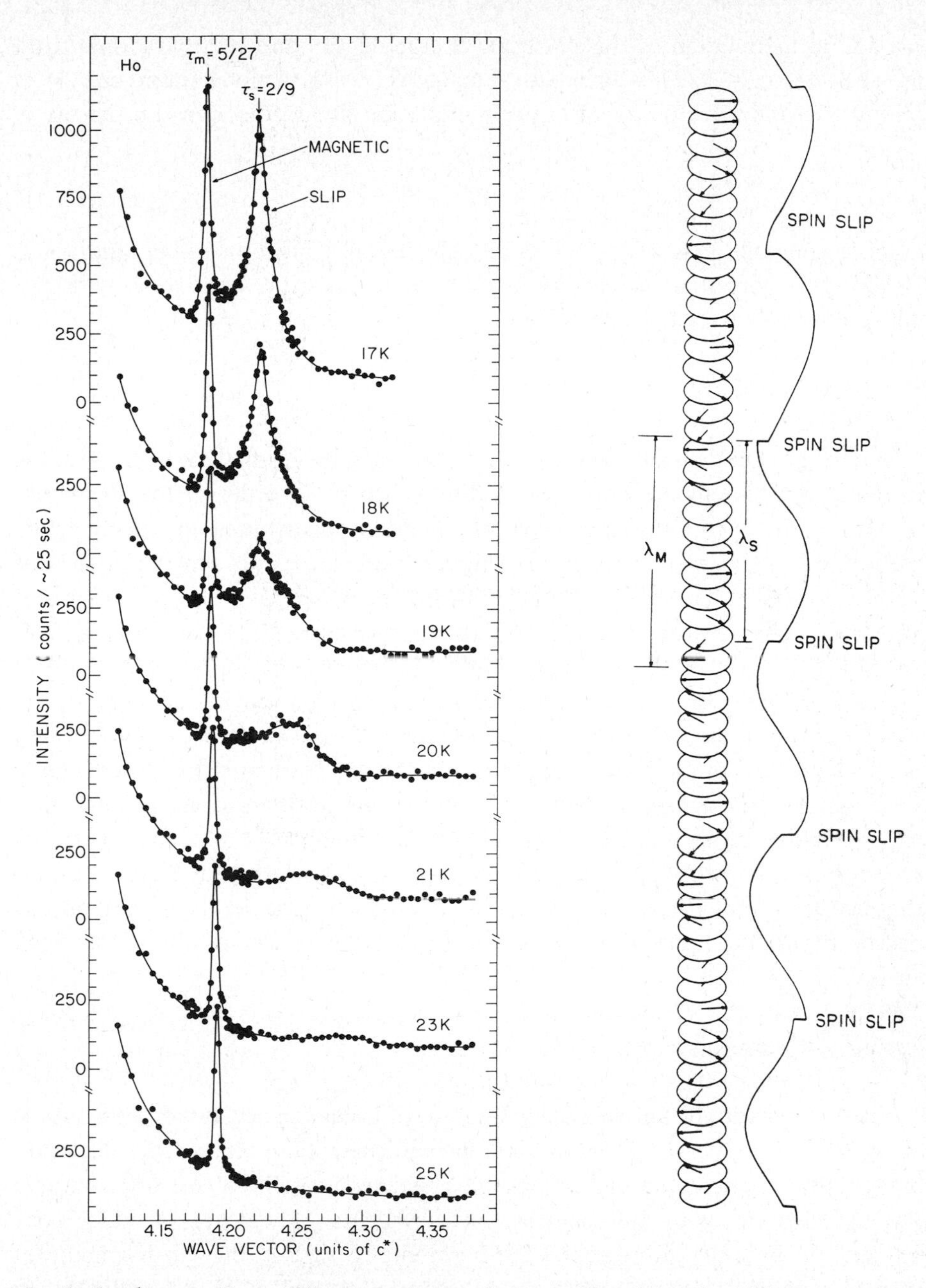

Fig. 3. Synchrotron X-ray diffraction patterns of the satellite above the (004) Bragg point of holmium, studied for decreasing temperature form $T = 25$ K to 17 K. Note that in addition to the sharper magnetic satellite, a second broad satellite appears at the slip position $\tau_s = 12\tau_m - 2$. When the temperature is lowered and τ_m approaches 5/27 then the inter planar lattices modulation sharpens up and approaches $\tau_s = 2/9$.

was due to modulation of the electronic charge density and clearly distinct from the peak at $\tau_m = 5/27$ which had a magnetic cross section. From eq. (1) it follows that the wave vector of lattice modulation due to the spin slip density is

$$\tau_s = Q/x = 12\tau_m - 2, \tag{2}$$

or more generally $\tau_s = kf\tau_m - Q$. Although the peak position of the additional scattering is somewhat uncertain due to its width, it is consistent with the magnetic wave vector through eq. (2).

2.2. *Disorder effects*

Due to the fact that the phase advance of a spin slip is small, $\phi_0 = 2\pi/12$, the magnetic correlations are only weakly diminished by disorder in the slip structure. In particular, if the spin slips are randomly distributed in the magnetic spiral, then the magnetic correlation length is still substantial, on the order[1]) of $2\pi/\phi_0^2\tau_s = 100$ Å. The width of the additional peak at τ_s is a direct measure of the correlations between spin slips. We therefore see (fig. 3) that as the temperature is lowered and as the commensurate 5/27 structure is approached, the spin slip correlations continuously increase. In fact, such behavior is qualitatively consistent with a model in which a slightly incommensurate structure is obtained by occasionally replacing the •4 sequences in the 5/27 structure by •3 sequences. If we assume no correlations between the •3 sequences then we have found that the slip correlations increase continuously to infinity as the commensurate structure is approached. As can be seen from fig. 3, however, the peak at 2/9 is not instrumentally narrow implying that additional disorder is needed in the model to produce quantitative agreement with the observed diffraction peak widths.

2.3. *Higher harmonics*

At low temperature the magnetic structure of holmium deviates from a simple spiral. The spin slip description gives specific predictions about this deviation and, therefore, about the higher harmonic magnetic satellites. We will compare these predictions with the neutron measurements of Felcher et al.[5]) They reported an intensity ratio of the 5th harmonic to that of the 7th harmonic of about 2. The wave vector underwent a discontinuity at $T = 25$ K and locked to $\tau_m = 0.185$. The authors state the fraction 5/27 for the possible commensurability. This corresponds to the •4 • 4 • 4 • 4 • 4 • 4 spin slip structure. At $T = 6$ K they measured the intensity of the 5th and 7th harmonics to be $0.65 \pm 0.02\mu_B^2$ and $0.38 \pm 0.02\mu_B^2$, respectively, where μ_B is one Bohr magneton. We simplify

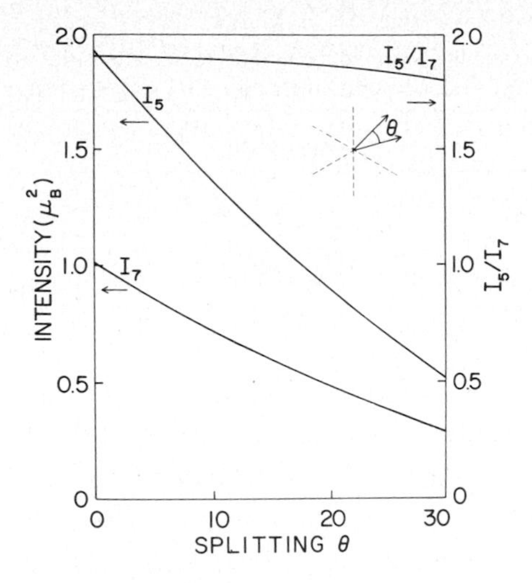

Fig. 4. The two curves I_5 and I_7 give the calculated neutron intensities of the 5th and 7th harmonic satellites of the •4 • 4 • 4 • 4 • 4 • 4 structure as a function of splitting angle θ (see inset upper right). The I_5/I_7 curve gives the intensity ratio of I_5 to I_7 as function of θ.

the calculation of structure factors by assuming a one-parameter model. In order to emphasize not only the spin slips, but also the general influence of the crystal field on the detailed reconstruction of the spiral, we choose a structure where the singlets are taken to be precisely aligned with the easy directions. For the doublets the sum of the two spins is taken to be aligned with the easy directions, but a splitting angle θ between the spins is allowed. Fig. 4 shows the calculated intensities of the 5th and 7th harmonics satellites of the •4 • 4 • 4 • 4 • 4 • 4 structure as a function of the splitting angle θ. The intensities were calculated as:

$$I(\boldsymbol{\tau}) = \left| \frac{M}{m} \sum \frac{1}{|\boldsymbol{\tau}|} \boldsymbol{\tau} \times \boldsymbol{S}_j \exp\{i\boldsymbol{\tau} \cdot \boldsymbol{r}_j\} \right|^2, \tag{3}$$

where m is the number of atomic layers in the unit cell (i.e. 54), $\boldsymbol{\tau}$ is the momentum transfer, and $\boldsymbol{r}_j$ is the position of the jth spins $\boldsymbol{S}_j$ in the unit cell. M is the atomic moment and is taken to be $10.3\mu_B$. At $T = 6$ K the sample is in the conical phase so for the calculation we have assumed that all spins have the same out-of-plane angle, namely, $\chi = 10°$. Effectively this enters with the small correction term $[\sin 80°]^2 = 0.97$. Fig. 4 shows the calculated ratio I_5/I_7 as a function of the splitting angle θ. We conclude that, independent of the splitting angle, the spin–slip model accounts for the observed ratio of I_5 to I_7. Furthermore, for a splitting $\theta = 26°$, we calculate $I_5 = 0.66\mu_B^2$ and $I_7 = 0.38\mu_B^2$ which is in excellent agreement with the experimental values (see table I). We further note

TABLE I
Compares the experimentally observed higher harmonics intensities as reported by Felcher et al.[3]) with our calculation assuming a •4 • 4 • 4 • 4 • 4 • 4 spin slip structure

	Experiment	$\lvert F_{5/27} \rvert^2$
I_2, I_3, I_4, I_6	0	0
I_5/I_7	1.7 ± 0.1	$1.7–1.9(\theta)$
I_5 (μ_B^2)	0.65 ± 0.02	0.66 $\theta = 26°$
I_7 (μ_B^2)	0.38 ± 0.02	0.36 $\theta = 26°$

that for the (•4 • 4 • 4 • 4 • 4 • 4) structure the 3rd (as well as the 2nd, 4th, 6th, etc.) harmonic is forbidden. This agrees with the fact that no 3rd harmonic intensity was observed along the (001) direction[5]).

3. Ferrimagnetic spin slip structures

The deviation of the magnetic structure from a simple spiral will in some commensurate spin slip structures lead to spontaneous in-plane ferrimagnetism with its accompanying macroscopic magnetic moment. The •5 structure of fig. 2 is a specific example. In contrast, rigorously incommensurate structures cannot be ferrimagnetic. Therefore, spontaneous ferrimagnetism indicates a commensurate magnetic periodicity. The c-axis modulated magnetic structure of thulium becomes ferrimagnetic at about $T = 32$ K and a neutron diffraction study[9]) has shown that at this temperature the magnetic wave vector locks to $\tau_m = 2/7$ (i.e., 3 spins up followed by 4 spins down). In the spin slip notation this is a •1 structure with $k = 4$. Magnetic susceptibility measurements have revealed three intermediate temperatures where erbium becomes ferrimagnetic[10]). We have recently completed an X-ray study where the magnetic structure was found to undergo first order lock-in transitions to simple commensurate ferrimagnetic spin slip structures at the same temperatures[11]). Åström and Benediktsson have recently confirmed the existence of at least two of these transitions by observing a latent heat[12]).

We now consider a sample in an external magnetic field. In general, the moments rotate away from their zero-field positions. As a result, ferrimagnetism may be induced even in incommensurate structures. More dramatically, an external magnetic field may provide an additional lock-in energy and enhanced wave vector stability for certain commensurate structures. Such behavior has been observed in a neutron scattering study of holmium in a magnetic field by Koehler et al.[4]) In a high magnetic field (oriented parallel to a hard direction in the basal plane) the wave vector $\tau_m = 0.182$ or $2/11$ was found to be stable over a wide temperature range from 75 K to 40 K. A more comprehensive analysis of the wealth of available data for holmium in a magnetic field is a challenging problem for future study.

4. Conclusions

The main discussion of this paper has been concerned with the magnetic structure of holmium. The accompanying lattice modulations are a reflection of the spin slip density (blocks of $k - 1$ spins) and provide a means, using X-ray scattering, to study the spin slip correlation directly. We note, that for a general magnetic structure the blocks of k spins may be equally important in generating interplanar lattice modulations. The wave vector for this modulation is $\tau_{s'} = Q - (n - 1)f\tau_m$. Moreover, for a strong crystal field, interplanar lattice modulations also appear at $\tau_f = f\tau_m$. These three wave vectors are related by: $\tau_s = \tau_f - \tau_s$. In our recent study of erbium additional scattering was present at all of these wave vectors[11]). Finally, we point out that lattice modulations can appear with a longer period if the spin slip pattern has a more complicated period. This is the case for the 5/21 basal plane phase of erbium where there are three slips every seven atomic layers, but the spin slip pattern repeats every seven layers,

$$(\cdot\cdot 1 \cdot 1 \cdot\cdot 1 \cdot 1 \cdot\cdot 1 \cdot 1 \cdot\cdot 1 \cdot 1 \cdot\cdot 1 \cdot 1 \cdot\cdot 1 \cdot 1).$$

In summary, we have shown how the concept of spin slips as periodic perturbations of the simple "bunched" spiral naturally explains the observed commensurate periods in holmium. It also accounts for the observed ratios of the harmonic intensities and qualitatively explains the behavior of holmium in a magnetic field. From an experimental viewpoint it is the existence of the interplanar lattice modulations induced by the spin slips which has allowed their detailed study by X-ray diffraction. Presumably, further study will lead to a more detailed understanding of magnetostriction and related phenomena. We believe that these concepts apply, not only to holmium, but to rare earth metals in general. Finally, we note a simple but remarkable feature of magnetic structure factors: unless the magnetic structure is ferrimagnetic the magnetic scattering at τ_s vanishes. This accounts for the fact that the spin slip has remained uncovered for so many years.

Acknowledgements

We would like to thank J.D. Axe for inspiring discussions as well as for his careful reading of this manuscript. Stanford Synchrotron Radiation Laboratory is supported by the U.S. Department of Energy Office of Basic Energy Sciences and the National Institute of Health Biotechnology Resource Program. Work performed in the Brookhaven Physics Department was supported by the Division of Materials Research, U.S. Department of Energy under contract DE-AC02-76CH00016.

References

1) D. Gibbs, D.E. Moncton, K.L. D'Amico, J. Bohr and B.H. Grier, Phys. Rev. Lett. **55** (1985) 234.
2) D.E. Moncton, D. Gibbs and J. Bohr, Nucl. Inst. and Meth. A **246** (1986) 839.
3) W.C. Koehler, J.W. Cable, H.R. Child, M.K. Wilkinson and E.O. Wollan, Phys. Rev. **151** (1966) 414.
4) W.C. Koehler, J.W. Cable, H.R. Child, M.K. Wilkinson and E.O. Wollan, Phys. Rev. **158** (1967) 450.
5) G.P. Felcher, G.H. Lander, T. Arai, S.K. Sinha and F.H. Spedding, Phys. Rev. B **13** (1976) 3034.
6) M.J. Pechan and C. Stassis, J. Appl. Phys. **55** (1984) 1900.
7) For a review article on commensurate phases, see for example, P. Bak, Rep. Prog. Phys. **45** (1982) 587.
8) Alternatively, for commensurate structures we may express the magnetic wave vector τ_m as the ratio of the number n of 2π rotations of the moments in the unit cell to the number of layers: $\tau_m = nQ/(fkn - s)$ where s is the number of the spin slips in the unit cell. Then the wave vector for the average spin slip density is $\tau_s = sQ/(fkn - s)$. For the $\tau_m = 5/27$ structure $n = 5$ and $s = 6$.
9) T.O. Brun, S.K. Sinha, N. Wakabayashi, G.H. Lander, L.R. Edwards and F.H. Spedding, Phys. Rev. B **1** (1970) 1251.
10) See data of W.A. Taylor, B.C. Gerstein and F.H. Spedding, in article by K.A. McEwen on p. 428, Handbook of the Physics and Chemistry of Rare Earths, Vol. 1, K.A. Gschneidner, Jr and L.R. Eyring, eds. (North-Holland, Amsterdam, 1978).
11) D. Gibbs, J. Bohr, J.D. Axe, D.E. Moncton and K.L. D'Amico, to be published.
12) H.U. Åström and G. Benediktsson, Abstract presented at the 6th EPS Conf. on Cond. Matter, Stockholm (1986).

Physica **140A** (1986) 359–367
North-Holland, Amsterdam

EXOTISM AROUND THE SMECTIC A STATE IN ASSOCIATED LIQUID CRYSTALS

F. HARDOUIN*

Centre de Recherche Paul Pascal, Université de Bordeaux I, Domaine Universitaire, 33405 Talence Cedex, France

Of greatest current interest is a class of thermotropic mesomorphic compounds containing a long aromatic core and a strongly polar end group (CN or NO_2). A considerable variety of phase diagrams have been experimentally and theoretically described by the Bordeaux group[1–3]). The first conclusion was to challenge the uniqueness of the smectic A state and consequently to divide the smectic A phases into different basic groups depending on the layering arrangement. Another point of interest of these polymorphisms was that multi-reentrant nematics or intermediate 2D states often separate two smectic A phases. Although these 2D mesophases possess a long range modulation inside the layers, they remain smectics with liquid-like ordered layers at short range. Finally, such anomalies of periodicity have been recently observed in other LC classes: 1) peculiar electron donor–acceptor systems with a polymorphism of smectic A phases at low temperature; 2) nonpolar systems with liquid crystalline side-chain polymers.

1. Introduction

The nematic phase (N) is characterized by long range orientational order in the sense that the molecules have their axis, on average, parallel to a preferred direction (with unit vector, $\boldsymbol{n}$, called the director). In the smectic A phase (S_A) the translational symmetry is restricted in the direction of the director where a mass density modulation takes place. From X-ray analysis one can measure the underlying wave vector $\boldsymbol{q} = 2\pi/d \cdot \boldsymbol{n}$ of this density wave. Thus smectic A liquid crystals are layered phases (with a layer thickness d) in which there is no order within each layer and in which the director is perpendicular to the plane of the layers.

Prior to 1978, the distance d between layers corresponded either to the molecular length for terminal-nonpolar compounds or to 1.4 times the molecular length for associated systems consisting of diaromatic compounds which possess

*Present address: Laboratoire Léon Brillouin – CEN Saclay; 91191 Gif-sur-Yvette Cedex, France.

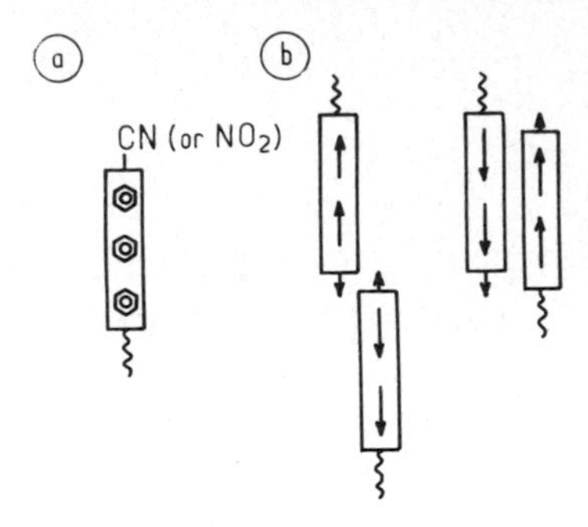

Fig. 1.

a strongly polar end group. But in both cases the uniqueness of the smectic A state is not questioned.

It has been shown since 1978 that the thickness of the layers could actually range from one to two molecular lengths in associated thermotropic systems including triaromatic polar compounds (fig. 1a). A competition between two possible layering periods hence takes place for the formation of the smectic A phase, namely, one connected to the molecular length and the other connected to the length of antiparallel "pairs". The latter is very much dependent on both temperature and dipole delocalization on the molecule (fig. 1b). The main consequences concerning these systems (pure compounds or binary mixtures) are: (i) the evidence for direct S_A–S_A transitions indicating that they are indeed distinct phases; (ii) the existence of multi-reentrant N–S_A behavior at atmospheric pressure (more symmetric nematic phase reoccurring once or twice at lower temperature than the less symmetric SA phases).

So far, the physico-chemical works of characterization[1]) joined to an important effort in organic synthesis[2]) have led to classification of six phases (directly derived from S_A phases) resulting from these anomalies of periodicity.

Simultaneously a phenomenological model[3]) which describes most of these experimental results has been developed, based on a general Landau–Ginzburg free energy expression, with two coupled smectic order parameters (generally characterized by incommensurate wavelengths), this Prost model introduces the idea of "frustrated" smectics with competing length scales. In an unrelated molecular theory[4]) which reproduces the multi-reentrant sequences the frustration is inherent in a triangular close packing of the dipolar molecules in the layers.

In the following we present an experimental overview of the variety of the structural changes proceeding from stacking of asymmetric molecules in smectic A layers.

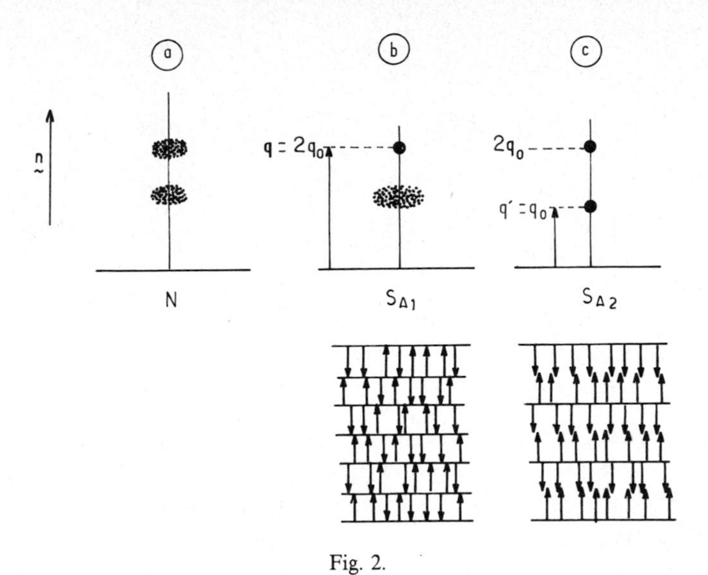

Fig. 2.

2. S_{A1}–S_{A2} transitions

The first evidence for a line of transitions separating two smectic A regions was obtained from DSC measurements in a binary diagram of a triaromatic polar compound with a nonpolar one[5]). In the HT S_{A1} phase there is a monolayer quasi-Bragg spot at wave vector $q = 2q_0 \simeq 2\pi/l$. In the LT S_{A2} an additional bilayer quasi-Bragg spot wave vector appears at $q' = q_0 \simeq 2\pi/2l$. This doubling of the periodicity[6]) is described as a long range antiferroelectric array of the dipolar heads (fig. 2). Locally, bilayers with antiferroelectric order are still present in the S_{A1} phase and even in the nematic phase (see diffuse scattering spots in fig. 2a).

Consistent high-resolution X-Ray scattering[7]) and calorimetric studies[8,9]) indicate that the S_{A1}–S_{A2} phase boundary can be second order beyond a tricritical point. Although the corresponding observed critical exponents disagree with predicted Ising model[10,11]), they are in rather good agreement with the so-called "Fisher renormalized" values of the simple Ising values.

3. S_{Ad}–S_{A2} changes

With increasing aliphatic tail in the triaromatic polar compound another smectic A phase is observed between the nematic and the bilayer S_{A2} phase. This intermediate S_A phase with $l < d < 2l$ is called partial bilayer S_A phase (S_{Ad}).

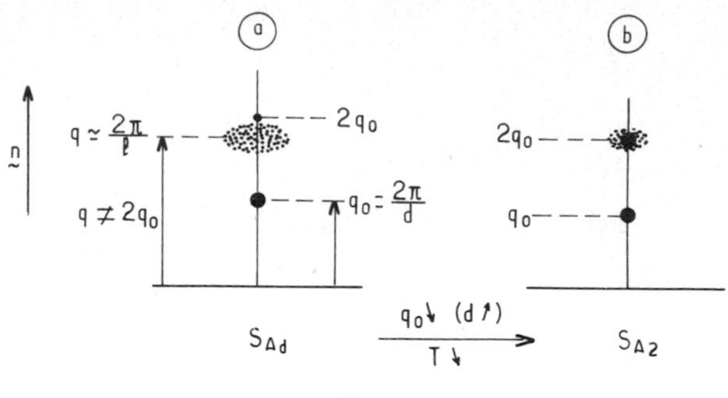

Fig. 3.

In the S_{Ad} phase two incommensurate wave vectors can coexist[12]), namely a quasi-Bragg spot at $q_0 = 2\pi/d$ and a diffuse scattering centered at $q < 2q_0$, the periodicity of this second wave vector corresponding to the molecular length $q \simeq 2\pi/l$ (fig. 3a). Generally the intensity of the 2nd order quasi-Bragg spot at $2q_0$ is weak and it undergoes a dramatic increase at the S_{Ad}–S_{A2} transition[13]). This is the signature of the commensurate lockin $2q_0 = q$ obtained with decreasing temperature (fig. 3b). Indeed, as the wave vector of the long range modulation q_0 decreases with decreasing temperature (i.e., the layer spacing d increases), the incommensuration in the system decreases because the wave vector q of the diffuse scattering is temperature independent.

Of particular interest is the thermodynamic evolution of the character of the S_{Ad}–S_{A2} transition in a (T, molecular length) phase diagram. Especially the DSC recordings clearly indicate the lack of latent heat contribution for the higher homologues[14]). High resolution experiments are now running to resolve the questions of the localization and the nature of a critical S_{Ad}–S_{A2} point predicted as similar to the liquid–gas critical point[15]).

4. S_{Ad}–S_{A1} changes

This evolution can be discontinuous through a first order S_{Ad}–S_{A1} transition. In such a case the layer spacing increases in the S_{Ad} phase as the temperature is decreased and at the $S_{Ad} - S_{A1}$ transition temperature the long range periodicity suddenly exhibits a dramatic change. As a result the low temperature smectic A phase has a quasi-Bragg spot corresponding to the monolayer ordering and an incommensurate diffuse scattering corresponding to the partial bilayer periodicity (fig. 4)[16]).

Apart from a first order coexistence line the S_{Ad} to S_{A1} evolution is observed either continuous without any transition or with an intermediate reentrant nematic phase. A remarkable illustration of this latter situation is the very exotic

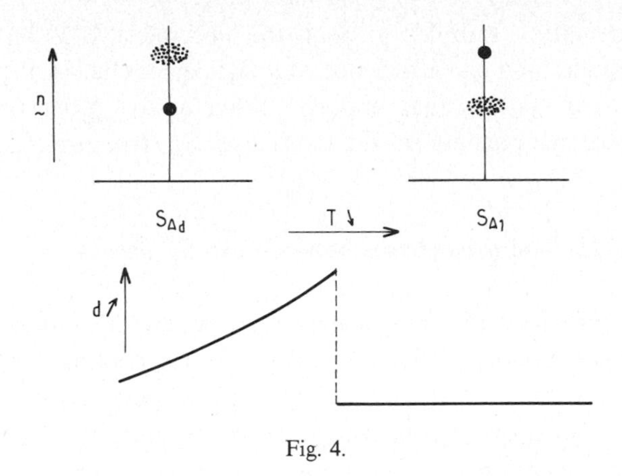

Fig. 4.

triple reentrant behavior[18,1]) found in a pure nitro derivative: $N-S_{Ad}-N_{re}-S_{Ad}-N_{re}-S_{A1}$.

A more general behavior is the shortness of the first order $S_{Ad}-S_{A1}$ transition line due to the vicinity of a reentrant nematic phase and to the occurrence of a two-dimensional $S_{\tilde{C}}$ mesophase (fig. 5). In the phenomenological model for frustrated smectics[3]) the $N_{re}-S_{Ad}-S_{A1}$ intersection (fig. 5a) is predicted as a peculiar bicritical point. On the experimental side, we show that such a triple

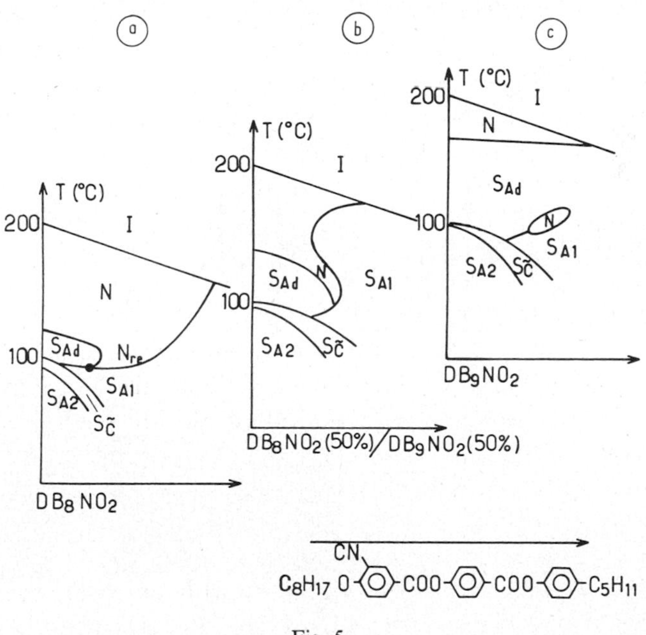

Fig. 5.

point and a nematic "bubble"[17,19]) at the end of a S_{Ad}–S_{A1} line are two connected topologies in a given thermodynamical space (fig. 5). Fig. 5c is actually direct evidence for systems that undergo either a first order transition or no transition between the partial and the monolayer S_A structures.

5. Intermediate 2D fluid mesophases between two S_A phases

Most polar systems with two coexisting wavelengths induce non-collinear modulations in the monolayer S_{A1} phase (fig. 6b). On cooling, a phase transition can occur such that the off-longitudinal axis diffuse spots get condensed (fig. 6c).

As a result a superlattice long range order expands in a lateral direction. Locally the two-dimensional (2D) arrangement remains smectic with liquid-like disordered layers (i.e., fluid mesophases).

Different 2D mesophases were discovered[20-22]). They are close to the bilayer smectic A2 but the $S_{\tilde{A}}$ (fluid antiphase) and the $S_{\tilde{A}cre}$ (crenelated phase) are periodically broken by normal defect walls (fig. 6c, d) whereas the $S_{\tilde{C}}$ structure (fig. 7) is periodically broken by oblique defect walls involving tilted molecules with respect to the normal to the layers.

Thus these mesophases possess a long range in-plane polarization modulation superimposed on the bilayer smectic structure. On the theoretical side[3,23]), the 2D biaxial fluid smectics are described as responses, in two dimensions, to the frustration due to the incommensurate wavelengths observed at higher temperature.

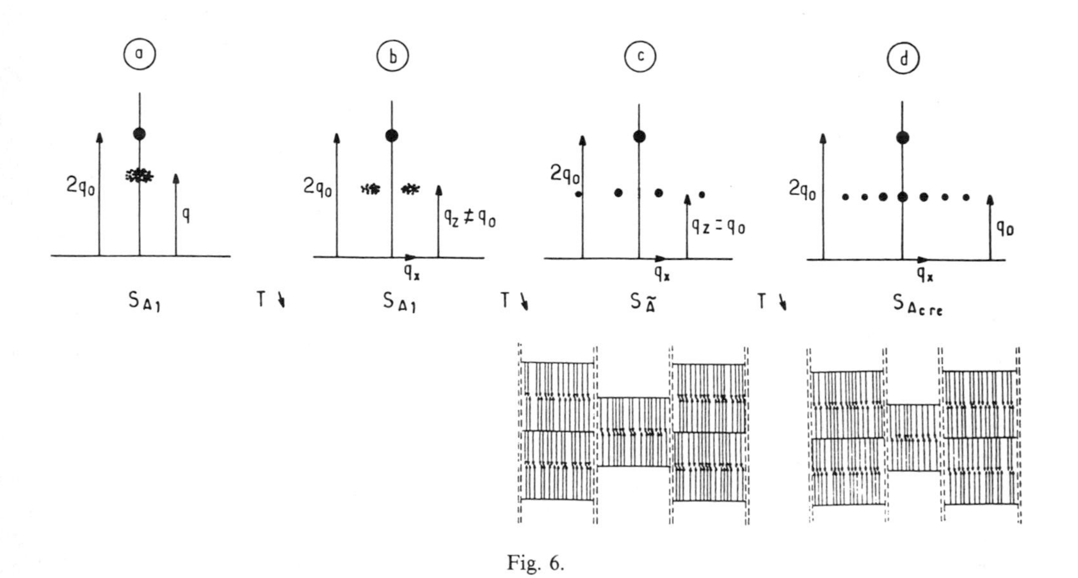

Fig. 6.

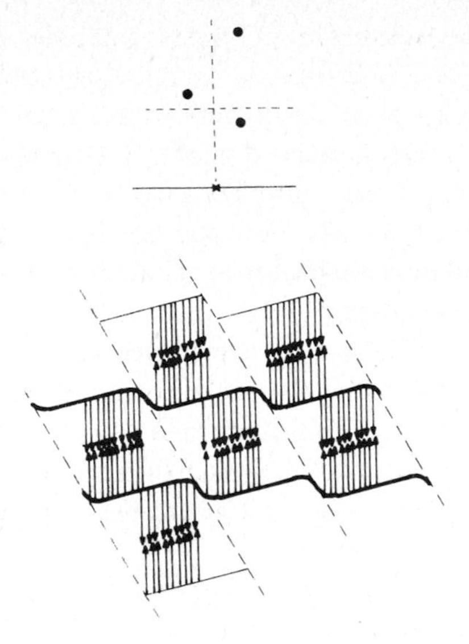

Fig. 7.

Contrary to the $S_{\tilde{A}}$ phase, which is a centered rectangular lattice (fig. 6c), the $S_{\tilde{A}cre}$ phase is a simple rectangular lattice (fig. 6d). Indeed, the in-plane polarization is symmetrically modulated in the $S_{\tilde{A}}$ phase, but asymmetrically modulated in the $S_{\tilde{A}cre}$ [22,24]). Thus two adjacent domains of the antiphase structure are of unequal size in the $S_{\tilde{A}cre}$ phase signaling an approach to the uniform polarization of the S_{A2} phase (fig. 2c). At any rate, this bilayer S_A phase is the final step for uniaxial disordered smectics at low temperature.

Finally the evolutions from $S_{\tilde{C}}$ to $S_{\tilde{A}}$ have faint thermodynamic and structural changes. Up to now a textural criterion only allowed us to identify the $S_{\tilde{C}}$–$S_{\tilde{A}}$ phase transition which lead to the following generalized sequence S_{A1}–$S_{\tilde{C}}$–$S_{\tilde{A}}$–$S_{\tilde{A}cre}$–S_{A2} decreasing temperature[25]).

6. Conclusion

Should this remarkable richness of polymorphism be restricted to a rare class of long polar rods? In fact the phenomenological descriptions[3]) still apply whatever the molecular causes for the anomalies of periodicity. Moreover, some microscopic models incorporate stabilizing interactions of various kinds[26,27]).

We propose the outline of two experimental advances which extend this polymorphism to systems in which the interactions are modified.

In the first, the strengthening of the attractive interactions by some associations among cyano-amino compounds leads to induced S_{A1}, $S_{\tilde{A}}$ and S_{A2} phases[28]).

In the second, some side-chain polymers combine the opposite tendency (reinforcement of repulsive interactions, decrease of dipole–dipole interactions) with a strong asymmetry of the mesogenic moiety inherent to the chemical structure itself since one end is attached to the polymer backbone. Free of highly polar parts, such systems are apt to provide smectic structures in close connection with our previous descriptions. We note the observations of partial bilayer S_{Ad} phases[29,30]), of reentrant nematic phases in mixtures of monomer and polymer[31]) and of antiphase-like fluctuations[32,33]).

At last, we have eluded here – the occurrence of a true incommensurate S_A phase (double collinear quasi-Bragg spots coexisting at wave vectors with incommensurate ratio)[34,35]) not yet confirmed in a pure compound – a part of the polymorphism in associated LC systems which is related to biaxial and/or ordered smectics, connected to the same anomalies of periodicity.

References

1) F. Hardouin, A.M. Levelut, M.F. Achard and G. Sigaud, J. de Chimie Phys. **80** (1983) 53.
2) Nguyen Huu Tinh, J. de Chimie Phys. **80** (1983) 83 and the references therein.
3) J. Prost and P. Barois, J. de Chimie Phys. **80** (1983) 65 and the references therein.
4) J.O. Indekeu and A.N. Berker, Phys. Rev. A **33** (1986) 1158.
5) G. Sigaud, F. Hardouin, M.F. Achard and H. Gasparoux, Int. Liq. Cryst. Conf. Bordeaux (1978); J. Phys. **40** (1979) C3–356.
6) F. Hardouin, A.M. Levelut, J.J. Benattar and G. Sigaud, Solid State Commun. **33** (1980) 337.
7) K.K. Chan, P.S. Pershan, L.B. Sorensen and F. Hardouin, Phys. Rev. Lett. **54** (1985) 1694.
8) C. Chiang and C.W. Garland, Mol. Cryst. Liq. Cryst. **122** (1985) 25.
9) C.W. Garland, C. Chiang and F. Hardouin, Liq. Cryst. **1** (1986) 81.
10) J. Prost, J. Phys. **40** (1979) 581.
11) J. Wang and T.C. Lubensky, Phys. Rev. A **29** (1984) 2210.
12) F. Hardouin, A.M. Levelut and G. Sigaud, J. de Phys. **42** (1981) 71.
13) K.K. Chan, P.S. Pershan, L.B. Sorensen and F. Hardouin, Phys. Rev. A **34** (1986) 1420.
14) F. Hardouin, M.F. Achard, Nguyen Huu Tinh and G. Sigaud, J. Phys. Lett. **46** (1985) L-123.
15) P. Barois, J. Prost and T.C. Lubensky, J. Phys. **46** (1985) 391.
16) A.M. Levelut, R.J. Tarento, F. Hardouin, M.F. Achard and G. Sigaud, Phys. Rev. A **24** (1981) 2180.
17) F. Hardouin, M.F. Achard, Nguyen Huu Tinh and G. Sigaud, Mol. Cryst. Liq. Cryst. Lett. **3** (1986) 7.
18) Nguyen Huu Tinh, F. Hardouin and C. Destrade, J. Phys. **43** (1982) 1127.
19) P.E. Cladis and H.R. Brand, Phys. Rev. Lett. **52** (1984) 2261.
20) G. Sigaud, F. Hardouin, M.F. Achard and A.M. Levelut, J. Phys. **42** (1981) 107.
21) F. Hardouin, Nguyen Huu Tinh, M.F. Achard and A.M. Levelut, J. Phys. Lett. **43** (1982) L-32.
22) A.M. Levelut, J. Phys. Lett. **45** (1984) L-603.
23) J. Wang and T.C. Lubensky, J. Phys. **45** (1984) 1653.
24) L.G. Benguigui, Phys. Rev. A **33** (1986) 1429.
25) G. Sigaud, M.F. Achard and F. Hardouin, J. Phys. Lett. **46** (1985) L-825.
26) L. Longa and W.H. de Jeu, Phys. Rev. A **28** (1983) 2380.
27) F. Dowell, Phys. Rev. A **31** (1985) 3214.

28) G. Sigaud, M.F. Achard, F. Hardouin and H. Gasparoux, J. Phys. Lett. **46** (1985) L-321.
29) R.M. Richardson and N.J. Herring, Mol. Cryst. Liq. Cryst. **123** (1985) 143.
30) M. Mauzac, F. Hardouin, H. Richard, M.F. Achard, G. Sigaud and H. Gasparoux, Eur. Polym. J. **22** (1986) 137.
31) G. Sigaud, F. Hardouin, M. Mauzac and Nugyen Huu Tinh, Phys. Rev. A **33** (1986) 789.
32) P. Davidson, P. Keller and A.M. Levelut, J. Phys. **46** (1985) 939.
33) P. Keller, B. Carvalho, J.P. Cotton, M. Lambert, F. Moussa and G. Pepy, J. Phys. Lett. **46** (1985) L-1065.
34) B.R. Ratna, R. Shashidar and V.N. Raja, Phys. Rev. Lett. **55** (1985) 1476.
35) P. Barois, Phys. Rev. A **33** (1986) 3632.

Physica **140A** (1986) 368–375
North-Holland, Amsterdam

MOLECULAR TAIL LENGTHS, DIPOLE PAIRINGS, AND MULTIPLE REENTRANCE MECHANISMS OF LIQUID CRYSTALS

J.O. INDEKEU*

Laboratorium voor Technische Natuurkunde, Technische Universiteit Delft, 2600 GA Delft, The Netherlands

and

A. Nihat BERKER

Department of Physics, Massachusetts Institute of Technology, Cambridge, MA 02139, USA

Employing the "spin–gas" model which incorporates the inherent microscopic frustration of dipoles, observed sequences of reentrant nematic and smectic phases are reproduced. In particular, the quadruple reentrance (nematic ↔ smecticA_d ↔ nematic ↔ smecticA_d ↔ nematic ↔ smecticA_1) is explained. The molecular tail length theoretically required for this phenomenon agrees with experiment. In the theory, associations of correlated triplets of molecules propagate smectic order. By contrast, dipole pairings giving rise to dimers of molecules frustrate smectic order and favor the nematic phase. Calculated dimer concentrations are insensitive to $N \leftrightarrow A_d$ transitions, consistently with dielectric measurements, whereas dimers break up at the $N \leftrightarrow A_1$ transition, causing the large transition enthalpy.

1. Introduction

Current studies of thermotropic liquid crystals reveal a remarkable sensitivity of thermodynamic phases and phase transitions to details of molecular structure[1–3]). In particular, less ordered phases can sometimes "reenter" as temperature is lowered from more ordered phases. Thus, in some liquid crystals of molecules with strongly dipolar heads (–CN or –NO_2), the nematic phase (orientational order) and the smectic phases (orientational and partial positional order) reenter[4–9]). This unexpected behavior is nevertheless consistent with thermodynamics[10]). Phenomenological Landau–Ginzburg theories[11–14]) discuss reentrance in terms of two competing lengths: the molecular length l and a

*Permanent address: Department of Physics, Katholieke Universiteit Leuven, Celestijnenlaan 200 D, B-3030 Leuven, Belgium.

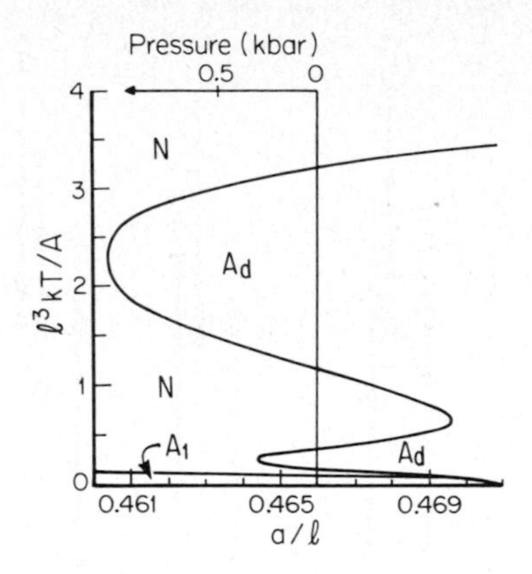

Fig. 1. Calculated quadruply reentrant phase diagram. The molecular parameters are $B/A = 1.455$, $n = 5$, $m = 3$, $\delta = 0.015\ l/n$. For comparison with experiment (fig. 3 in ref. 9), a tentative pressure scale is added.

dipole-pair length d, wavelengths, respectively, of the density and polarization modulations associated with smectic fluctuations. Our current paper presents a possible microscopic explanation of the delicate stability and reentrances of nematic and smectic phases in polar liquid crystals[15–19]).

Several reentrance sequences have been encountered as temperature is lowered, involving nematic (N), interdigitated partial-bilayer smectic (A_d), and monolayer smectic (A_1) phases. These sequences are "simple reentrance"[4,5,15]) ($N \to A_d \to N$), "double reentrance"[6,15]) ($N \to A_d \to N \to A_1$), and, most intriguingly, "quadruple reentrance"[7,8,9,16]) ($N \to A_d \to N \to A_d \to N \to A_1$), which is shown in fig. 1. In our microscopic theory, these sequences are reproduced and explained, while monomers (free molecules), dimers (free pairs of molecules with antiparallel dipoles), and triplets of molecules propagating smectic order are identified[15,16,19]).

The molecules of concern have the shape of rods approximately 30 Å long and 5 Å thick. They have a longitudinal electric dipolar head, a rigid aromatic core, and a semi-flexible aliphatic tail. Representations are shown in fig. 2. A micro-

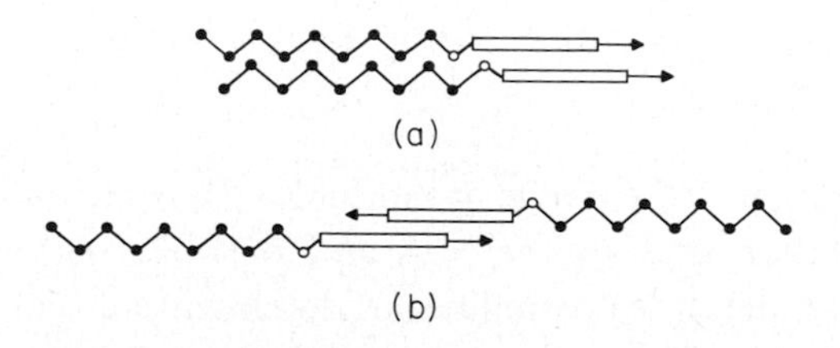

Fig. 2. Schematic representation of molecules of the compound DB_9ONO_2. The molecular pairs have dominant tail–tail attraction (a) or steric hindrance (b).

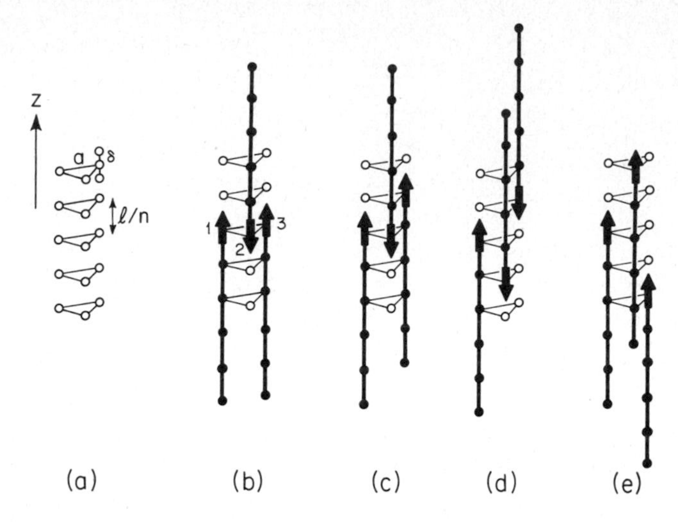

Fig. 3. Examples of configurations of a triplet of molecules. (a) The atomic permeation positions of the dipole heads. Librational permeation positions are illustrated only in the upper right-hand corner. (b) A frustrated configuration: A zero net force is felt by either of dipoles 1 and 3. (c) Another frustrated configuration: A zero net force is felt by dipole 3. Configurations (b) and (c) are thus not conducive to layering. On the other hand, (d) and (e) are configurations in which frustration is relieved by permeation, respectively conducive to interdigitated partial-bilayer (A_d) and monolayer (A_1) smectic layering.

scopic theory should embody intermolecular steric hindrance, Van der Waals attraction, and dipole–dipole forces. The latter are important since liquid crystal reentrance is almost exclusively seen in dipolar systems. In these systems, dipolar frustration[20]) plays a key role (figs. 3b, c).

2. The frustrated spin–gas model

Neighboring molecules are taken to interact via the pair potential[15,16])

$$V(\boldsymbol{r}_1, \hat{\boldsymbol{s}}_1, \boldsymbol{r}_2, \hat{\boldsymbol{s}}_2) = \left[A\hat{\boldsymbol{s}}_1 \cdot \hat{\boldsymbol{s}}_2 - 3B(\hat{\boldsymbol{s}}_1 \cdot \hat{\boldsymbol{r}}_{12})(\hat{\boldsymbol{s}}_2 \cdot \hat{\boldsymbol{r}}_{12})\right] / |\boldsymbol{r}_{12}|^3, \tag{1}$$

where $\boldsymbol{r}_i$ is the position of the dipole of molecule i, $\hat{\boldsymbol{s}}_i$ is the unit vector describing the dipolar orientation, and $\boldsymbol{r}_{12} = \boldsymbol{r}_1 - \boldsymbol{r}_2$ and $\hat{\boldsymbol{r}}_{12} = \boldsymbol{r}_{12}/|\boldsymbol{r}_{12}|$. For purely dipolar forces, $A = B$. The alternate possibilities of dominant tail–tail attraction (fig. 2a) or steric hindrance (fig. 2b) are incorporated[15,16]) in the ratio B/A: $B < A$ for net hindrance (favoring the antiferroelectric term) and $B > A$ for net entanglement (disfavoring the antiferroelectric term). Fluctuations toward the isotropic phase are ignored, namely the molecules are taken aligned along the z direction: $\hat{s}_i = \pm\hat{z}$, or simply $s_i = \pm 1$. Screening suppresses interactions between further neighbors. In the close-packing of a liquid, the potential in eq. (1) inherently causes frustration due to substantial cancellations of forces between a molecule and its nearest neighbors (figs. 3b, c).

Molecular tails play a crucial role. Apart from the free energy of their entanglement embodied in B/A, their lengthwise corrugation is essential, creating energetically preferred positions ("notches") of mutual permeation, i.e. positional fluctuation along the z direction, for a nearest-neighbor pair (cf. fig. 3a). Additionally to these discrete "atomic" permeations on a length scale l/n (a few Å), where $4 \lesssim n \lesssim 8$, small oscillations are of course allowed on a subscale $m\delta \ll l/n$. These "librational" permeations can be calculationally approximated to occur in m discrete subnotches within each notch. (Quantitative calculational stability under the continuum limit $m \to \infty$ with $m\delta \to$ constant has been verified[19]).

Within a three-dimensional system, a reference layer of molecules can be identified by the coordinates z_i being within the interval $z_0 \pm l/2$. The smallest unit of possible frustration is a triplet of molecules in the reference layer. First, the strengths are obtained for effective orientational couplings between the spins s_i by thermally averaging over the positional fluctuations $\boldsymbol{r}_i$. We found that the positional fluctuations that underly reentrance are permeations, the molecular motions in the z direction. Possible permeations are shown in fig. 3a. Displacements normal to z do not change the results qualitatively, and have generally been left out for calculational ease. [Thus, the degree of in-plane (xy) order is not directly related to smectic order along z. This prediction[15]) of our theory has been confirmed by subsequent X-ray scattering experiments[21]).] Accordingly, our calculations proceed by considering three nearest-neighbor molecules on a triangular prism of side a (fig. 3). After one molecule is fixed, e.g., in notch 3 for $n = 5$, there are n^2m^3 positional configurations per orientational configuration $\{s_i\}$. With a prefacing transformation, the average-strongest (K_S), average-intermediate (K_I), and average-weakest (K_W) antiferroelectric couplings are extracted:

$$\exp\left(K_S s_1 s_2 + K_I s_2 s_3 + K_W s_3 s_1 + G\right) = \sum_{\boldsymbol{r}_{2,3}}^{n^2m^3} \exp\left[-\beta V(\boldsymbol{r}_1, \hat{s}_1, \boldsymbol{r}_2, \hat{s}_2) - \beta V(\boldsymbol{r}_2, \hat{s}_2, \boldsymbol{r}_3, \hat{s}_3) - \beta V(\boldsymbol{r}_3, \hat{s}_3, \boldsymbol{r}_1, \hat{s}_1)\right], \tag{2}$$

where the molecule labels (12), (23), and (31) respectively span the strongest, intermediate, and weakest antiferroelectric couplings specific to each positional configuration. The resulting couplings $\{K_S, K_I, K_W\}$ depend sensitively on temperature, which is given as $l^3 k_B T/A$ in dimensionless units, on pressure and concentration in a mixture, which affect the ratio l/a of effective molecular length to average lateral separation, and on molecular properties, which are reflected in B/A, n, and $m\delta$. The next step in the treatment is to gauge, approximately, whether smectic order is supported, by referring to Houtappel's ordering condition[22]) of a uniformly distorted Ising model with the couplings $\{K_S, K_I, K_W\}$:

$$\sinh(2\bar{K}_S)\sinh(2\bar{K}_I) + \sinh(2\bar{K}_I)\sinh(2\bar{K}_W) + \sinh(2\bar{K}_W)\sinh(2\bar{K}_S) > 1, \tag{3}$$

where $\bar{K}_\alpha$ are K_α with or without any pairwise change of signs. Fig. 1 shows a phase diagram calculated as described.

3. Quadruple reentrance and molecular tail length

In order to expose the details of the reentrance mechanisms, a labeling scheme is introduced for the microscopic configurations. Consider $n = 5$, i.e. with notches 1, 2, 3, 4, 5, and $m = 3$, i.e. with subnotches a, b, c at each notch. The configuration in fig. 3b is labeled $(3b, \bar{3}b, 3b)$ if all dipoles are in subnotches b. The overbar signifies $s_i = -1$. Hereafter, only relevant subnotch labels will be given. The configurations in figs. 3c, d, e are $(3b, \bar{3}b, 4)$, $(3, \bar{2}, \bar{4})$, $(3, 5, 1)$, respectively.

Consider the system corresponding to fig. 1. At high temperature, the nematic phase occurs through ordinary thermal disorder, namely through the equal importance of all configurations.

Lowering temperature at an appropriate fixed l/a, two segments of smectic A_d phase are encountered, separated by a segment of (reentrant) nematic phase. On the one hand, if the local-antiferroelectric configurations that relieve frustration through atomic permeation, such as $(3, \bar{2}, \bar{4})$, are eliminated from the prefacing transformation of eq. (2), the upper A_d segment disappears[19]). On the other hand, if the local-antiferroelectric configurations that relieve frustration through librational permeation, such as $(3a, \bar{3}b, 3c)$, are eliminated from the prefacing transformation, the lower A_d segment disappears[19]). Evidently the upper and lower smectic A_d phases are due to local-antiferroelectric configurations that relieve frustration by, respectively, atomic and librational permeations, whereas in the intervening nematic phase, the frustrated configurations such as $(3b, \bar{3}b, \ldots)$ prevail. Thus, these smectic phases are characterized by antiparallel molecular orientation and are therefore composed of interdigitated partial bilayers. This is also reflected in the quantitative calculations of layer thickness[16]), in good agreement with experiment.

Upon lowering temperature below the two A_d segments, a second reentrant nematic segment is encountered. However, directly monitoring[19]) the importance of all types of configurations via their Boltzmann weights and multiplicities we see that the frustrated configurations have declined in importance in this region of the phase diagram. This nematic reentrance occurs through the simultaneous importance, competition, and mutual cancellation of local-antiferroelectric and local-ferroelectric, such as $(3, 5, 1)$, configurations. The latter configurations dominate in importance upon further lowering temperature, causing the smectic A_1 phase. This phase is characterized by parallel molecular orientation.

Fig. 1 compares satisfactorily with the experimental[9]) phase diagram of DB_9ONO_2. Interestingly, quadruple reentrance is hard to find both experimentally and in our model where it occurs for the limited cases and narrow intervals of $n = 4$ and $1.856 < B/A < 1.896$, $n = 5$ and $1.445 < B/A < 1.466$, or $n = 6$ and $1.468 < B/A < 1.479$. We have not found quadruple reentrance for $n = 3, 7, 8, 9$. This is equivalent to the experimental finding[8]) that the phenomenon depends crucially on the tail length of the molecule. Moreover, $n = 4$ or 5 are indeed the most reasonable notch numbers that can be deduced from considering DB_9ONO_2 with its tail of nine carbon atoms (fig. 2).

4. Molecular dimers and the molecular polymer

It has become fashionable to discuss reentrance in polar liquid crystals in terms of a competition between molecular monomers and molecular dimers[23]). We believe this is inappropriate. Molecular dimers are identified in our theory as the correlated antiparallel pair of molecules in a frustrated triplet such as $(3b, \bar{3}b, \ldots)$. Our model and calculations suggest that smectic order is propagated by correlated triplets [such as $(3, \bar{2}, \bar{4})$ or $(3, 5, 1)$] which form a molecular polymer that extends into the xy plane. This is an annealed polymer, in the sense that molecular units abandon it and join it continuously under thermal fluctuations. Conversely, molecular dimers, unlike the polymer, can pack uniformly along the z direction. Therefore, when they preponderate and the polymer cannot form, the nematic phase occurs. In the smectic phases, the dimers coexist with the polymer and form the nematic background to the smectic modulation underpinned by the polymer.

We have calculated the concentration of molecular dimers as a function of temperature[19]). This concentration is insensitive to $N \leftrightarrow A_d$ phase transitions, which is compatible with the temperature variation of the measured dielectric anisotropy[3]). By contrast, the dimers break up at the onset of the smectic A_1 phase, causing the large transition enthalpy as calculated in our model and observed in experiment[17]).

Acknowledgements

We are indebted, along with the statistical physics community, to Professor H.E. Stanley for his extraordinary ability and efforts in bringing about the Statphys-16 Conference, where this paper was presented.

We thank Professor C.W. Garland for many useful discussions and an enjoyable collaboration (ref. 17). One of us (J.O.I.) was a Postdoctoral Research Fellow of the Dutch "Stichting F.O.M.". Research at M.I.T. was supported by the National Science Foundation under Grant No. DMR84-18718.

References

1) G.W. Gray, Phil. Trans. R. Soc. Lond. A **309** (1983) 77.
2) J. Prost, Adv. Phys. **33** (1984) 1.
3) S. Chandrasekhar, Mol. Cryst. Liq. Cryst. **124** (1985) 1.
4) P.E. Cladis, Phys. Rev. Lett. **35** (1975) 48.
5) P.E. Cladis, R.K. Bogardus and D. Aadsen, Phys. Rev. A **18** (1978) 2292.
6) F. Hardouin and A.M. Levelut, J. Phys. (Paris) **41** (1980) 41.
7) N.H. Tinh, F. Hardouin and C. Destrade, J. Phys. (Paris) **43** (1982) 1127.
8) F. Hardouin, A.M. Levelut, M.F. Achard and A. Sigaud, J. Chim. Phys. **80** (1983) 53.
9) R. Shashidhar, B.R. Ratna, V. Surendranath, V.J. Raja, S. Krishna Prasad and C. Nagabhushan, J. Phys. (Paris) **46** (1985) L445.
10) N.A. Clark, J. Phys. (Paris) **40** (1979) C3.
11) P.S. Pershan and J. Prost, J. Phys. (Paris) **40** (1979) L27.
12) J. Prost and P. Barois, J. Chim. Phys. **80** (1983) 65.
13) P. Barois, J. Prost and T.C. Lubensky, J. Phys. (Paris) **46** (1985) 391.
14) P. Barois, Phys. Rev. A **33** (1986) 3632.
15) A.N. Berker and J.S. Walker, Phys. Rev. Lett. **47** (1981) 1469.
16) J.O. Indekeu and A.N. Berker, Phys. Rev. A **33** (1986) 1158.
17) J.O. Indekeu, A.N. Berker, C. Chiang and C.W. Garland, M.I.T. preprint (1986).

18) A.N. Berker and J.O. Indekeu, in Structural Incommensurability in Crystals, Liquid Crystals, and Quasi-Crystals, J.F. Scott, ed. (Plenum, New York, 1987).
19) J.O. Indekeu and A.N. Berker, to be published.
20) For an alternate approach and mechanism, see F. Dowell, Phys. Rev. A **28** (1983) 3526; **31** (1985) 2464; 3214.
21) A.R. Kortan, H. von Känel, R.J. Birgeneau and J.D. Litster, J. Phys. (Paris) **45** (1984) 529.
22) R.F.M. Houtappel, Physica **16** (1950) 425.
23) See, for example, L. Longa and W.H. de Jeu, Phys. Rev. A **26** (1982) 1632.

Physica 140A (1986) 376–389
North-Holland, Amsterdam

SYNCHROTRON X-RAY STUDIES OF LIQUID–VAPOR INTERFACES

J. ALS-NIELSEN
Risø National Laboratory, DK-4000 Roskilde, Denmark

The density profile $\rho(z)$ across a liquid–vapor interface may be determined by the reflectivity $R(\theta)$ of X-rays at grazing angle incidence θ. The relation between $R(\theta)$ and $\rho(z)$ is discussed, and experimental examples illustrating thermal roughness of simple liquids and smectic layering of liquid crystals are presented.

1. Introduction

The variation of density across the liquid–vapor interface from essentially zero density far out in the vapor phase to a homogeneous density deep in the liquid phase can be determined by X-ray reflectivity measurements[1]) at grazing angles θ well beyond the critical angle for total reflection, θ_c. We shall first derive the relation between the density variation, $\rho(z)$, and the reflectivity $R(\theta)$. Next we consider as an experimental example the roughness caused by thermal excitations of capillary waves on simple liquids like water and carbon-tetrachloride[2]). Then we consider the smectic layering of liquid crystal molecules of the free surface, first in a system with a spontaneous first order phase transition from the isotropic phase to a smectic A phase[3]), and afterwards in a system with a second order phase transition from the nematic phase to the smectic A phase[4–6]).

2. Reflectivity and density profile

We consider in fig. 1 a monochromatic X-ray beam (wavelength λ, wavevector $\boldsymbol{k}_{in}$) incident on the liquid surface at grazing angle θ and specular reflected to wavevector $\boldsymbol{k}_{out}$. Specular reflection means that the wavevector transfer

$$\boldsymbol{Q} \equiv \boldsymbol{k}_{out} - \boldsymbol{k}_{in}, \quad |\boldsymbol{Q}| = (4\pi/\lambda)\sin\theta \qquad (1)$$

is along the surface normal.

0378-4371/86/$03.50 © Elsevier Science Publishers B.V.
(North-Holland Physics Publishing Division)

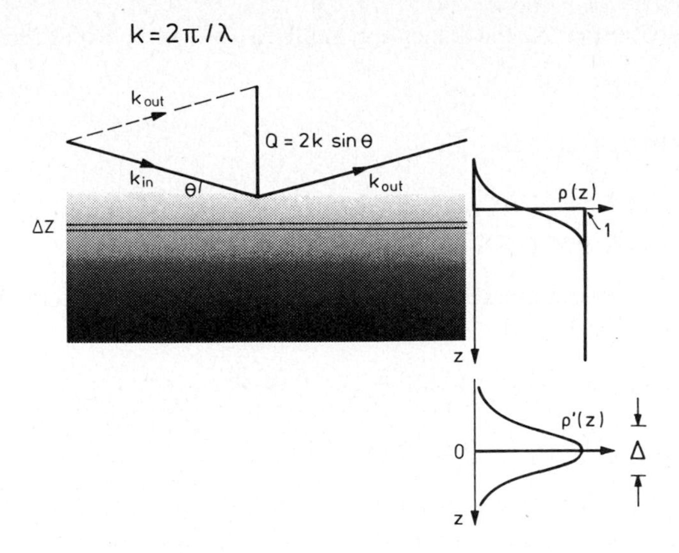

Fig. 1. The density variation across a surface is indicated by shading in the left part and more quantitatively by the function $\rho(z)$ in the right part. The reflectivity versus wavevector-transfer Q is related to the Fourier transform of the gradient of the density, $\rho'(z)$.

The dimensionless density profile, $\rho(z)$, is normalized to unity for large z where the electron density is ρ_{el}. The gradient of $\rho(z)$ is denoted $\rho'(z)$. In the ideal case of $\rho(z)$ being a step function at $z = 0$, standard derivation from optics obtains the Fresnel reflectivity $R_F(Q)$ as briefly outlined below. In the real case where $\rho(z)$ is a continuous function we shall see that the reflectivity $R(Q)$ is related to $\rho(z)$ by

$$\frac{R(Q)}{R_F(Q)} = \left| \int \rho'(z) \exp[iQz] \, dz \right|^2 . \tag{2}$$

We observe that eq. (2) has the correct limiting behaviour when $\rho(z)$ approaches a step function since $\rho'(z)$ then approaches a delta-function and the Fourier-transform therefore unity.

As far as the Fresnel reflectivity is concerned we recall Snell's law relating the grazing angles θ and θ' for $\theta > \theta_c$, the critical angle of total reflection:

$$\cos\theta / \cos\theta' = n = 1 - \lambda^2 \rho_{el} r_0 / 2\pi . \tag{3}$$

Here n is the index of refraction which for X-rays is given by the second equation in terms of λ, ρ_{el} and r_0, the electron radius or the scattering length of

Thompson scattering. As the angles are small eq. (3) can be expanded to

$$\theta^2 = \theta'^2 + \theta_c^2, \quad \theta_c^2 = \lambda^2 \rho_{el} r_0/\pi. \tag{4}$$

The reflectivity $R_F(\theta)$ is

$$R_F(\theta) = \left\{\frac{\theta - \theta'}{\theta + \theta'}\right\}^2. \tag{5}$$

For $\theta \gg \theta_c$, $R_F(\theta)$ approaches $(\theta_c/2\theta)^4$ or in terms of wavevector transfer Q relative to Q_c

$$Q_c = (4\pi/\lambda)\theta_c = 4\pi[\rho_{el} r_0/\pi]^{1/2}, \tag{6}$$

$$R_F(Q) \simeq (Q_c/2Q)^4, \quad Q \gg Q_c. \tag{7}$$

Anticipating eq. (2) we can estimate that Q must be of the order of the inverse width Δ of $\rho'(z)$ (or the period of $\rho'(z)$ if its oscillating) before an appreciable deviation of $R(Q)$ from $R_F(Q)$ is obtained. The corresponding value of Q/Q_c is

$$Q/Q_c \sim [\Delta \cdot Q_c]^{-1} \simeq \left[\mathscr{R} 4\sqrt{\pi}\,(\rho_{el} r_0)^{1/2}\right]^{-1}, \tag{8}$$

where we have assumed Δ to be of order the molecular radius $\mathscr{R}$. In a molecule with Z electrons, $\rho_{el} \sim Z/(4/3\pi\mathscr{R}^3) \sim Z/(\pi\mathscr{R}^3)$, so $\Delta \cdot Q_c \simeq 4(Z\mathscr{R}r_0)^{1/2}$ which numerically is typically around 10. The approximate simple expression for $R_F(Q)$ in eq. (7) is therefore useful, and refraction effects can usually be neglected. Absorption, which we have neglected for simplicity in eq. (5), modifies the reflectivity around the critical angle as it rounds the kink at $\theta = \theta_c$, but for $Q \gg Q_c$ the effects are negligible.

The X-ray reflectivity which contains the information about $\rho(z)$ is thus quite low since for $Q \sim 10Q_c$, $R_F(Q) \sim 10^{-5}$ and furthermore $R(Q) < R_F(Q)$. That is one reason for using the intense X-ray beams obtainable from synchrotron radiation in this kind of experiments, in particular because the beam height h must be small (0.1 mm) as the "footprint" on the sample, h/θ, must be limited to say 20–30 mm. Another reason is the narrow $\boldsymbol{Q}$-resolution which is necessary to distinguish clearly between intensity from specular reflection and intensity due to scattering from the illuminated bulk, which typically has a depth of several microns even for grazing angles of the order of milli-radians.

In order to derive eq. (2) we consider the reflected wave as a superposition of waves reflected from infinitesimal planes at varying depth z implying the phase factor $\exp[iQz]$. In accordance with the order-of-magnitude discussion given above we shall neglect refraction and absorption effects. We only need to know the reflectivity of a thin plate with thickness Δz, cf. fig. 1. The reflected wave is the result of Thompson scattering of the incident photon wave by the individual electrons. The reflected amplitude ΔA_r must be proportional to the incident amplitude A_i, to the scattering length r_0 of a single electron and to the number of electrons per unit area perpendicular to the incident beam, $\rho(z)\cdot(\rho_{el}\,\Delta z/\sin\theta)$. Since $\Delta A_r/A_i$ is dimensionless and the dependence on quantities with length dimensions such as r_0, ρ_{el} and Δz is exhausted by their product the only additional length in the problem, λ, must enter linearly, i.e.

$$\frac{\Delta A_r}{A_i} = c\rho(z)[\lambda r_0 \rho_{el}/\sin\theta]\,\Delta z, \tag{9}$$

where c is a dimensionless, complex number. The quantity in the square-bracket is by eq. (6) the same as $Q_c^2/(4Q)$. By superposition of all infinitesimal layers and squaring the amplitude ratio to get the reflectivity we find

$$R(Q) = |c|^2[Q_c^2/(4Q)]^2\left|\int \rho(z)\exp[iQz]\,dz\right|^2 \tag{10a}$$

$$= |c|^2[Q_c/(2Q)]^4\left|\int \rho'(z)\exp[iQz]\,dz\right|^2. \tag{10b}$$

Here eq. (10b) follows from eq. (10a) by partial integration. We note that the squared bracket by comparison to eq. (7) is the Fresnel reflectivity $R_F(Q)$ in the present approximation of neglecting refraction effects, i.e. $Q \gg Q_c$. By noting that $R(Q)$ must equal $R_F(Q)$ for a step function $\rho(z)$, or delta function $\rho'(z)$, it follows that $|c|^2 = 1$, and the master formula eq. (2) is obtained within the ab initio assumptions.

3. Experiments

The general experimental set-up is shown in fig. 2. A suitable aperture S_1 is placed in the horizontal synchrotron radiation beam incident from the left in fig. 2. A monochromatic beam is extracted from the "white" spectrum by Bragg reflection from a monochromator crystal, typically perfect Ge in the (1, 1, 1) orientation. By tilting the normal to the reflecting planes out of the horizontal plane, the monochromatic beam can be bent down to any glancing angle with the horizontal liquid surface. The sample is mounted on an elevator so that the liquid surface always intersects the monochromatic beam. A narrow slit S_2 in front of the sample defines a suitable footprint of the beam on the liquid surface, and a beam monitor after S_2 ensures that intensities measured at different slit heights and at different currents in the storage ring are all normalized to the same number of incident photons. A slit S_3 defines the angular resolution of the reflected beam. This slit together with the detector, a NaJ scintillation counter, is

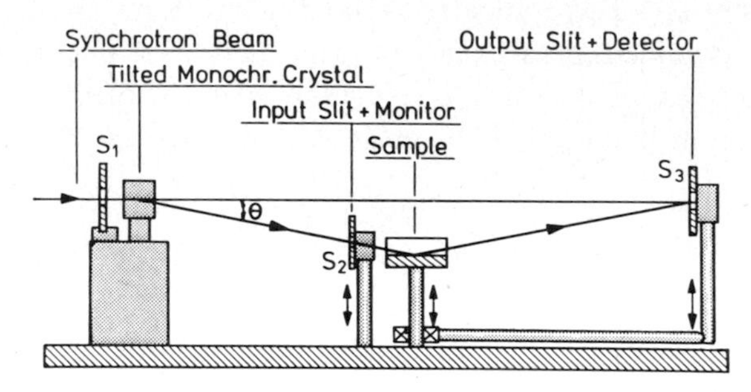

Fig. 2. Side view of the vertical, scattering plane. Beam directions are defined by slits. The monochromatic beam is bent down towards the sample by tilting the monochromator crystal. The incident beam intensity is monitored after slit S_2.

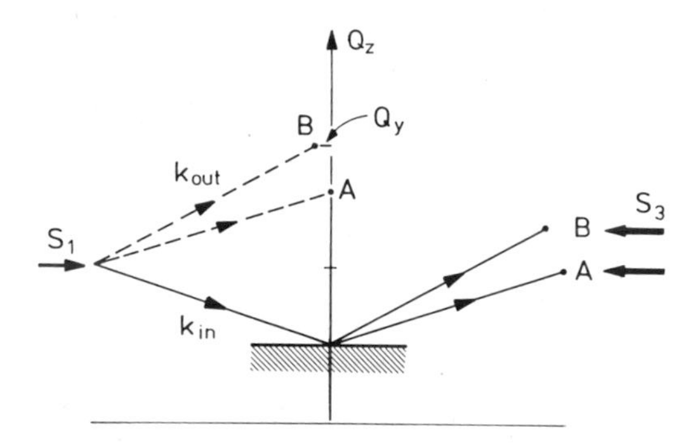

Fig. 3. A transverse wavevector component in the vertical plane may be obtained without rotating the sample by having the incident glancing angle different from the reflected angle.

mounted on an elevator situated on a spectrometer arm which is pivoted around a vertical axis through the sample center. In this way one has enough freedom to obtain an arbitrary kinematically allowed wavevector transfer $\boldsymbol{Q}$ with its three components: Q_z along the surface normal, Q_y transverse to Q_z in the vertical plane and Q_x transverse to Q_z in the horizontal plane, cf. fig. 3.

3.1. *Thermal roughness of simple liquids* [2])

A "quiet" surface of H_2O or CCl_4 was obtained by wetting an area of approximately 60 mm diameter of a clean glass flat. The corresponding liquid thickness is a few tenth of a millimeter. Reflectivities of more than 96% was obtained for $\theta < \theta_c$, and the reflected line shape was essentially identical to the incident beam line shape. The reflectivity, relative to the Fresnel reflectivity,

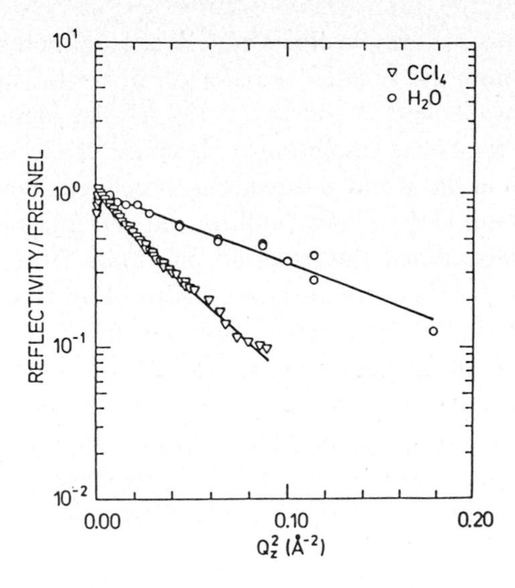

Fig. 4. Reflectivity data versus the squared wavevector transfer for water (open circles) and carbon tetrachloride (triangles). The slope expresses the thermal mean squared displacement of the atoms from the ideal flat surface.

turned out to be a Gaussian function of Q_z as shown in fig. 4, where the logarithm of $R(Q)/R_F(Q)$ is plotted versus Q_z^2. The slope of the lines for both H_2O and CCl_4 reflects essentially the thermal roughness which is three times bigger for CCl_4 than for H_2O because the surface tension γ for CCl_4 is only a third of that of H_2O. Thermal roughness means that the density gradient $\rho'(z)$ is a Gaussian, $\exp[-z^2/2\langle u^2\rangle]$; the Fourier transform is then also Gaussian, $\exp[-Q^2\langle u^2\rangle/2]$, and $R(Q)/R_F(Q)$ by eq. (2) equals therefore $\exp[-Q^2\langle u^2\rangle]$. We shall now express $\langle u^2\rangle$ in terms of the surface tension γ. To this end consider a sinusoidal perturbation of a flat square of side length L. The displacement at position $\boldsymbol{r}$ is $u_q \sin(\boldsymbol{q}\cdot\boldsymbol{r})$. The excess area over the flat surface is $\frac{1}{4}u_q^2q^2L^2$ and the associated energy of this mode therefore $\gamma\cdot\frac{1}{4}u_q^2q^2L^2$. By equipartition follows that the thermal average value of u_q^2, which is denoted $\langle u_q^2\rangle$, must be $\frac{1}{2}kT/[\frac{1}{4}\gamma q^2L^2]$. Finally, $\langle u^2\rangle$ is obtained by summing over all possible $\boldsymbol{q}$ values ranging from $q_{\min}$ to $q_{\max}$:

$$\langle u^2\rangle = \frac{1}{2}\sum_q \langle u_q^2\rangle = \frac{1}{2}(L/2\pi)^2\int_{q_{\min}}^{q_{\max}}\langle u_q^2\rangle\,\mathrm{d}^2q \tag{11a}$$

$$= (2\pi)^{-1}(kT/\gamma)\ln[q_{\max}/q_{\min}]. \tag{11b}$$

The last equality presumes that the volume element d^2q can be written as $2\pi q\,\mathrm{d}q$.

We shall now discuss the wavevector limits. Since the derivation is within the continuum assumption, the smallest wavelength of disturbance must be of the order of a molecular diameter d and $q_{max} \approx 2\pi/d$. The minimum value, q_{min}, is determined by the transverse resolution in $\boldsymbol{Q}$ space of the instrumental set up. Since the resolution in the x and y-directions as defined above are different, the assumption behind eq. (11b) is not fulfilled and the integral in (11a) must be evaluated numerically. When this is done one finds $\langle u^2 \rangle_{H_2O}^{1/2} = 2.8$ Å which added in quadrature to a root-mean-square radius of an H_2O molecule of 1.9 Å gives an effective $\langle u^2 \rangle_{eff}^{1/2}$ of about 3.4 Å. The slope for H_2O in fig. 3 corresponds to $\langle u^2 \rangle_{eff}^{1/2} = 3.2$ Å. Similar good agreement with no adjustable parameters is found for CCl_4 which has a significantly larger thermal mean squared displacement due to its lower surface tension. We conclude that the thickness of interface of simple liquids far from the critical point is simply determined by the combined effects of molecular size and thermal roughness due to excitation of capillary waves.

3.2. *Free surface of liquid crystals*

In the following two sections we shall review recent experimental studies of the density profile across the surface of liquid crystal materials in different phases. This section provides the minimum background knowledge on liquid crystals. For a general reference, the reader is referred to "The Physics of Liquid Crystals" by de Gennes[7]).

Liquid crystals consist of long molecules with a typical length-to-diameter ratio of 5 to 1. In describing the structures, the molecules are considered as rigid rods. The variety of structures or phases is due to the combination of order/disorder between the *position* of molecules and their *orientation*. For the present purpose it suffices to recall three liquid phases: The *isotropic* phase where both position and orientation are disordered as in an ordinary simple liquid, the *nematic* phase where the position is disordered but all molecules have the same spontaneous average direction, and the *smectic* A phase where the common orientation is maintained, but in addition the molecules are positioned in layers perpendicular to their long axis with a well-defined repetition distance between layers, but with positional disorder of molecules within the same layer.

Different sequences of transitions between these phases may occur. The high temperature isotropic phase may be followed directly by the smectic A phase, or a nematic phase at intermediate temperatures may intervene between the isotropic and smectic A phases. The transition from the isotropic phase is always discontinuous of first order, whereas the nematic to smectic A phase may be first order or continuous (second order). In the latter case critical fluctuations of short range order smectic A regions in the nematic matrix become more and more pronounced as the nematic to smectic A phase transition temperature is approached.

Examples of the molecular structure are the so-called nCB molecules

$$C_nH_{2n+1}—\langle\bigcirc\rangle—\langle\bigcirc\rangle—C—N$$

$$C—N—\langle\bigcirc\rangle—\langle\bigcirc\rangle—C_nH_{2n+1}$$

The upper molecule is shown with its aliphatic tail to the left and the polar cyano head to the right. The intramolecular interactions are so strong that this molecule pairs with one of opposite orientation as shown, and the unit rod should be considered as this pair of molecules.

The sequence of phases versus temperatures depends on n, the aliphatic tail length. The phase diagram of nCB is shown in fig. 5. In the following we shall describe the smectic layering at the free surface in two distinctly different cases: (i) in the isotropic phase near a strong first order phase transition to the smectic A phase ($n = 12$), (ii) in the nematic phase near a second order phase transition to the smectic A phase ($n = 8$).

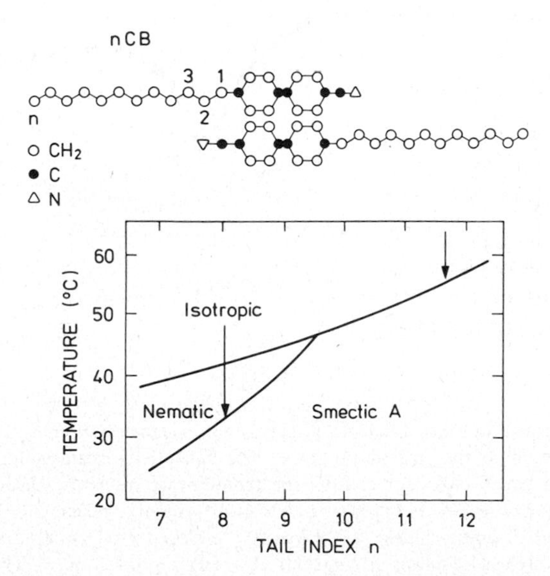

Fig. 5. Phases of nCB at various tail length n and temperature T. Experiments for $n = 12$ and $n = 8$ are described in the following.

3.3. *Quantized layer growth at free surface of liquid crystal in isotropic phase*

Reflectivity curves at different temperatures for 12 CB are given in fig. 6. Note the logarithmic scale of reflectivity and the wavevector scale relative to Q_0, the reciprocal lattice vector in the smectic A phase. The high temperature data, curve f, are quite similar to the data for H_2O or CCl_4. The root-mean-square displacement, which we prefer here to denote σ_s, is 5.5 Å. As the temperature is lowered toward T_{IA}, structure develops in the spectra. The full lines represent a simple model with a set of temperature independent parameters and only one parameter,

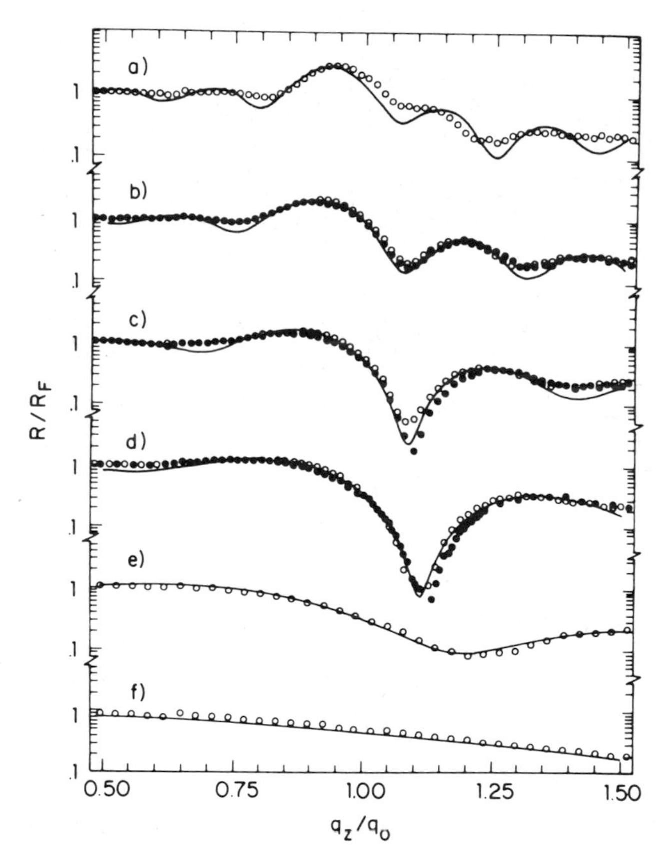

Fig. 6. Reflectivity relative to Fresnel reflectivity versus wavevector at different temperature intervals ΔT above the transition to the smectic A phase. The reflectivity scale is logarithmic and each spectrum is displaced two decades for clarity. The temperature intervals relative to the transition temperature T_{AI} are a) $t = 3 \times 10^{-5}$; b) $t = 8 \times 10^{-5}$ (open circles), $t = 1.4 \times 10^{-4}$ (closed circles); c) $t = 3 \times 10^{-4}$ (open circles), $t = 8.3 \times 10^{-4}$ (closed circles); d) $t = 1.1 \times 10^{-3}$ (open circles), $t = 3 \times 10^{-3}$ (closed circles); e) $t = 1.9 \times 10^{-2}$; f) $t = 6.1 \times 10^{-2}$. The solid line is for a density model with a sinusoidal modulation terminated after an integral number (a) 5, (b) 4, (c) 3, (d) 2, (e) 1 and (f) 0 of periods.

the number of smectic layers, varying with temperature. Specifically the model is as follows: The density is decomposed into two parts. The first is a step function to liquid density smeared by a Gaussian accounting for thermal roughness etc. The corresponding Fourier transform of gradient density is

$$\Phi_1(Q) = \exp\left[-Q^2\sigma_s^2/2\right].$$

This part alone accounts for the curve labelled f in fig. 6. The second part is a sinusoidal density wave of amplitude A_0 between $z = 0$ and $z = NL$, where L is the layer spacing in the smectic phase and N an integer number of layers. Also this part is smeared by a Gaussian and the corresponding Fourier transform of gradient density is

$$\Phi_2(Q) = A(x)\exp\left[-Q^2\sigma_m^2/2\right]\sin[\pi N]\exp[i\pi N]$$

with $A(x) = 2A_0 x/[(x+1)(x-1)]$ and $x \equiv Q/(2\pi/L) = Q/Q_0$.

In the model is allowed for a phase factor $\exp[iQz_0]$ between the two contributions so the model reflectivity is

$$\frac{R(Q)}{R_F(Q)} = |\Phi_1(Q)\exp[iQz_0] + \Phi_2(Q)|^2.$$

The full lines in fig. 6 all have the same values of $\sigma_s = 5.5$ Å, $\sigma_m = 4.5$ Å, $z_0/L = -0.35$, $A_0 = 0.12$ whereas N varies from 0 (curve f) to 5 (curve a). The discreteness of layer by layer growth as the temperature approaches T_{IA} is strikingly apparent in fig. 7 showing the intensity variation with reduced temperature at $Q/Q_0 = 0.93$.

3.4. *Layer growth in nematic phase*

All it takes in nCB molecules to obtain a nematic phase between the isotropic and smectic A phases is a shortening of the aliphatic tail from $n = 12$ to say $n = 8$. The surface layering is now quite different and so is the reflectivity data. In the isotropic phase a few smectic layers are formed on the surface, but now with a decreasing amplitude, as shown in the top of fig. 8, in contrast to 12CB where the model calculation assumed a sinusoidal density variation of constant amplitude up to N periods[8]). The most remarkable difference is however in the nematic phase. The reflected intensity versus Q_z/Q_0 is shown in the bottom left part of fig. 8 and the corresponding model density in the right bottom part of fig. 8.

The total picture of smectic layering in the nematic phase is given in fig. 9. The surface layers are of infinite lateral extent and essentially perfect, with an

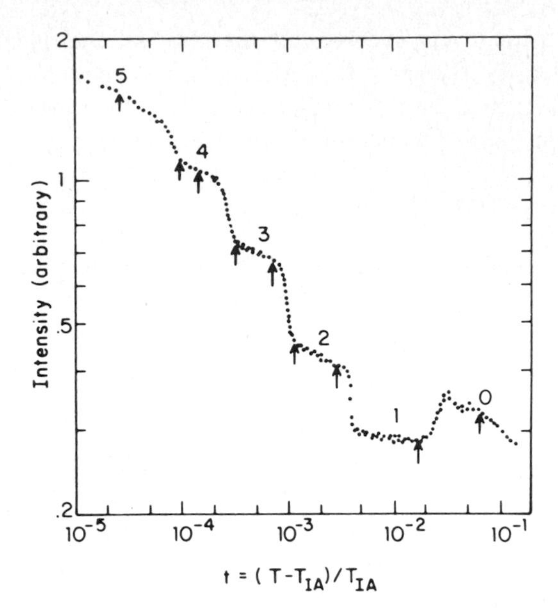

Fig. 7. The reflected intensity from 12CB in the isotropic phase changes in discrete steps as the temperature is lowered towards the transition temperature to the smectic A-phase. The numbers indicate the number of smectic A layers at the surface.

exponentially decaying amplitude into the bulk of penetration depth ξ_s. Deep in the bulk smectic fluctuations in the nematic matrix have one correlation range $\xi_{\|}$ along the molecular axis and another, finite, correlation range $\xi_{\perp}$ in the lateral direction. The top part of fig. 9 recapitulates the scattering diagram for separating bulk and surface scattering. The reflected intensity in fig. 8 displays a typical interference lineshape with constructive interference between the ordinary Fresnel wave and the wave scattered from the surface layers for $Q < Q_0$, and destructive interference for $Q > Q_0$. As the temperature approaches the transition temperature T_{NA} to the smectic A phase the correlation ranges $\xi_{\|}$ and $\xi_{\perp}$ diverge and also the penetration depth ξ_s increases, because the nematic phase becomes more and more susceptible to the layering imposed by the surface field. Most remarkably, it was found that ξ_s is identical to $\xi_{\|}$ not only in its temperature dependence but also in its numerical value[4]). The reason may be understood from the Landau theory of phase transitions and, since the argument is quite general, we shall outline it here. The basic quantity in the Landau theory is the order parameter ψ, which in the present case is the smectic density wave, but for a ferromagnet for instance would be the local magnetization. The order parameter varies in space and we shall here only be concerned with the variation along the z-axis. The average value is denoted $\langle\psi\rangle$ and, for a second order phase

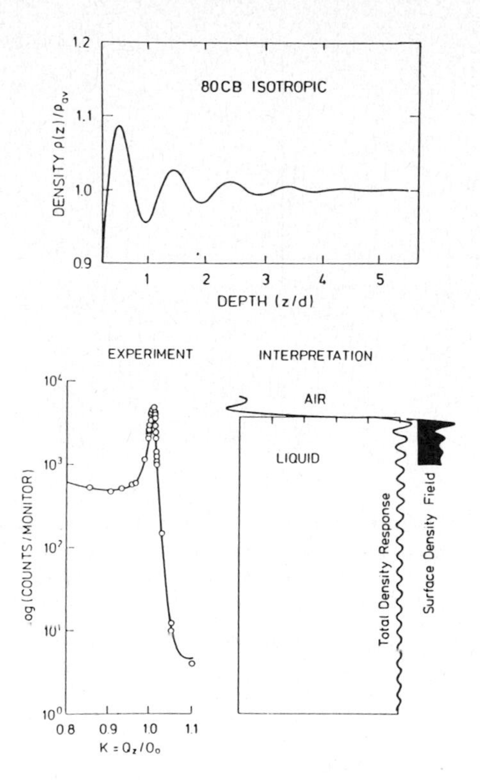

Fig. 8. Top: Density profile in the isotropic phase slightly above the transition temperature to the nematic phase of the liquid crystal 80CB. Bottom: Reflected intensity versus Q_z/Q_0 in the nematic phase (left) and the corresponding density (right). This can be considered as the response in the nematic phase to smectic layering imposed by the first few top layers.

transition, $\langle\psi\rangle = 0$ for $T \geqslant T_c$ but is finite for $T < T_c$. The phenomenological Landau free energy density $f = a\langle\psi\rangle^2 + b\langle\psi\rangle^4$ will lead to such a phase transition if one assumes $a = a_0(T - T_c)$ and $b > 0$. However, spatial fluctuations will cost energy and we therefore add a term of the form $c(\partial\psi/\partial z)^2$ to the free energy density. With this expression for f one readily finds critical fluctuations for $T > T_c$ with a correlation range $\xi_{\parallel} = (c/a)^{1/2}$, assuming an infinite system. Now we consider the penetration from the surface, requiring a finite value ψ_0 of the order parameter at $z = 0$. The order parameter must decay as z increases to reach its bulk average value of zero. On the other hand any spatial change costs energy, cf. the term $c(\partial\psi/\partial z)^2$, so there must be an optimal way for ψ to decay in the sense that the total energy obtained by integrating the energy density becomes minimal, i.e. a typical problem of variational calculus. Explicitly one finds that the optimal decay is exponential with a decay length $\xi_s = (c/a)^{1/2}$, i.e. $\xi_s = \xi_{\parallel}$.

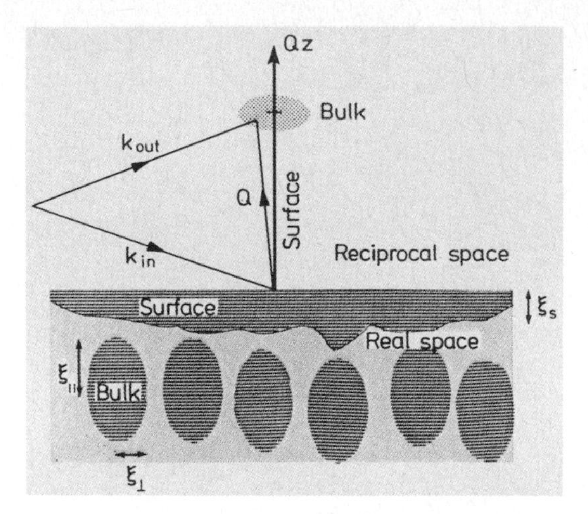

Fig. 9. Smectic layering in the nematic phase. The boundary conditions at the surface impose smectic layering of infinite lateral extent at the surface, decaying exponentially in going into the bulk. Here spontaneous critical fluctuations have correlation ranges $\xi_{\parallel}$ and $\xi_{\perp}$ along and perpendicular to the molecular axis, respectively. The top part shows scattering geometries to separate bulk and surface signals.

4. Conclusions

The three examples of X-ray reflectivity data from liquid surfaces illustrate the degree of detail that the density profile across the surface can be determined. A recent example on antiferroelectric surface layering on a liquid crystal material[9]) further emphasizes this point. The liquid crystal materials have a rich variety of surface layering which is interesting in itself but in the present context can also be viewed as the testing grounds for development of a new methodology for exploring the liquid–vapor interface in general, and we believe that this method will be widely used in the future.

Acknowledgements

The work described in this review has been carried out during the past 5 years in close cooperation with Professor P.S. Pershan and his group from Harvard University both as concerns methodology, interpretation and choice of problems. The experiments were carried out at HASYLAB, DESY in Hamburg.

References

1) For a review see J. Als-Nielsen in Handbook of Synchrotron Radiation, Vol. 3 (North-Holland, Amsterdam), to be published.
2) A. Braslau, M. Deutsch, P.S. Pershan, A.H. Weiss, J. Als-Nielsen and J. Bohr, Phys. Rev. Lett. **54** (1985) 114.
3) B. Ocko, A. Braslau, P.S. Pershan, J. Als-Nielsen and M. Deutsch, Phys. Rev. Lett. **57** (1986) 94.
4) J. Als-Nielsen, F. Christensen and P.S. Pershan, Phys. Rev. Lett. **48** (1982) 1107.
5) P.S. Pershan and J. Als-Nielsen, Phys. Rev. Lett. **52** (1984) 759.
6) P.S. Pershan, A. Braslau, A.H. Weiss and J. Als-Nielsen, Phys. Rev. (1986), to be published.
7) P.G. de Gennes, The Physics of Liquid Crystals (Clarendon, Oxford, 1974).
8) The data actually derive from the material 8OCB, not 8CB. The extra oxygen atom is placed between the aliphatic tail and the first benzene ring of the polar head.
9) E.F. Gramsbergen, W.H. de Jeu and J. Als-Nielsen, J. de Physique **47** (1986) 711.

Physica **140A** (1986) 390–396
North-Holland, Amsterdam

SYSTEMATIC DENSITY EXPANSION OF TRANSPORT PROPERTIES OF RANDOM LATTICES

Matthieu H. ERNST*

Institute for Physical Science and Technology, University of Maryland, College Park, MD 20742, USA

Using kinetic theory methods we develop systematic expansions in powers of the concentration of impurities (scatterers, super/non-conducting bonds/sites) and apply them to calculate explicitly to linear and quadratic order in the concentration, moments of displacement, velocity autocorrelation functions, etc., of random walks on square and simple cubic lattices with site or bond disorder (lattice Lorentz gas, ant and termite models, site and bond percolation).

Special attention is given to the velocity autocorrelation function (cage effect, long time tail) and its comparison with results of computer simulations.

The simplest model for diffusion of particles or conduction of charge carriers is perhaps the model of a random walker (RW) on a regular lattice (with N sites), taking a step to a nearest neighbor site, chosen at random, at every tick of a clock. Everything is known about it: the mean square displacement of the RW is $\langle(\Delta x)^2\rangle = 2D_0 t$ with a diffusion coefficient $D_0 = a^2/(2d\tau_0)$ for a simple cubic lattice in d dimensions with lattice distance a and τ_0 the period of the clock; the velocity autocorrelation function (VACF), $\langle v_x(0)v_x(t)\rangle = D_0\delta_+(t)$, shows no memory ($\delta_+(t)$ is a Dirac delta function normalized as $\int_0^\infty \mathrm{d}t\, \delta_+(t) = 1$); the probability $p_n(t)$ for a displacement $n = \{n_x, n_y, \ldots, n_d\}$ approaches a Gaussian. The evolution equation, governing $p_n(t)$, may also be interpreted as Kirchhoff's laws for an electric network, where the bonds correspond to resistors. There exists also a close connection with random elastic networks.

The above problems are mostly part of standard physics courses. However, what happens to the conductivity as a function of c, if a random fraction c of bonds or sites in the uniform lattice are removed, or replaced by superconductors (percolation phenomena)? What happens to the macroscopic sound velocity if a random fraction of springs (masses) is replaced by sloppier (heavier) or stiffer (lighter) ones? If the model is viewed as a Lorentz gas, what happens to the long time behavior of the mean square displacement or the VACF of a RW on the lattice with a fraction c of bonds or sites blocked by hard impurities? Due to

*Permanent address: Inst. for Theor. Physics, RUU, Utrecht, The Netherlands.

their presence the scattering of the RW fluctuates from site to site, which is the mechanism responsible for long time tails.

The purpose of this paper is to present a systematic method for calculating the static and frequency dependent macroscopic properties of these systems through a systematic expansion in powers of the impurity concentration c where I use many results from refs. 1–3. The results obtained here are exact to $\mathcal{O}(c^2)$ at low concentrations; the standard effective medium theories yields only results, exact to $\mathcal{O}(c)$ [4]).

In the *models* to be considered I mainly use the language of RW-models, consider only d-dimensional simple cubic lattices with unit lattice distance $a = 1$, use the so-called continuous-time-RW-models with exponential waiting time distributions, and refer to refs. 4 and 5 for the connection between discrete and continuous time RW's on disordered lattices. Here the distribution of displacements $p_n(t)$ of the RW is described by the master equation

$$\dot{p}_n = \sum_\rho [w_{n,n+\rho} p_{n+\rho} - w_{n+\rho,n} p_n], \tag{1}$$

where $w_{n'n}$ is the transition probability per unit time to jump from site n to its n.n.-site $n' = n + \rho$. In a uniform lattice $w_{n+\rho,n} = (2d\tau_0)^{-1}$, in a disordered lattice a random variable is assigned to the transition rate $w_{n'n}$. How that is done depends on the model considered. Here I restrict myself to the most simple cases, where a random variable assumes only two different values $\{\sigma_0, \sigma\}$ with $\sigma_0 = 1$ for host lattice sites or bonds, and $\sigma = 1 - b$ $(b \leqslant 1)$ for impurity sites or bonds, and I consider the following models:

The *random barrier model* (RBM) is an example of bond-disorder, where the random variable $c_{n,n+\rho} = 1$ is assigned to the bond $(n, n + \rho)$. It has the values $c_{n,n+\rho} = 1$ with probability c (impurity bond) and $c_{n,n+\rho} = 0$ with probability $1 - c$ (host lattice bond). The RBM is defined through ref. 4:

$$w_{n+\rho,n} = w_{n,n+\rho} = (1/2d)\psi_{n,n+\rho} = (1/2d)\{1 - bc_{n,n+\rho}\}. \tag{2}$$

As $b \uparrow 1$ (ant-in-a-labyrinth problem) the model shows a bond percolation transition, where the diffusion coefficient vanishes for c larger than a threshold value c_p. Large negative b-values (superconducting impurities, termite problem) describe a percolation transition from a normal to a superconducting state at the same threshold value c_p as above.

In *models with site disorder* the random variable c_n is assigned to site n, and has the value $c_n = 1$ with probability c (impurity site) and $c_n = 0$ with probability $1 - c$ (host lattice site). Here we consider models where the jump rate $w_{n'n}$ depends only on the site of arrival n', viz.

$$w_{n,n+\rho} = (1/2d)\psi_n = (1/2d)\{1 - bc_n\}. \tag{3}$$

This model with $b = 1$ is the usual site-percolation model. In the random jump rate model $w_{n'n} = (2d)^{-1}\psi_n$ depends only on the site of departure.

The *kinetic theory* approach[1,2]) is only illustrated for the RBM, where the master equation (1) can formally be written as $\dot{p}_n = -(\hat{L}p)_n$. The quantity of main interest is the average probability for a displacement from site m to site n, viz. $P_{n-m}(t) = N^{-1}\langle p(nt|m0)\rangle$, averaged over the disorder. Here $p(nt|m0) = (\exp - t\hat{L})_{nm}$ is the solution of the master equation with initial condition $p(n0|m0) = \delta_{nm}$.

A resolvent formalism is introduced for the response function $F(q, z)$, which is the Fourier–Laplace transform of $P_n(t)$, viz.

$$F(q, z) = N^{-1} \sum_{n,m} e^{iq(n-m)}\langle (z + \hat{L})^{-1}\rangle_{nm} \equiv \langle (z + L)^{-1}\rangle_{qq}, \tag{4}$$

where $A_{qq'}$ denotes the Fourier-representation of a matrix $\hat{A}_{nm}$. Once $F(q, z)$ is calculated (the Laplace transform of) the moments of displacements, $\langle n_x^l\rangle(z)$, can be calculated by taking q_x-derivatives and setting $q = 0$ afterwards.

To make eq. (4) suitable for kinetic theory analysis, I split L into $L = L^0 - \delta L$, where L^0 describes the uniform lattice and δL is linear in the impurity variables $\{c_{n,n+\rho}\}$. Then I make a formal perturbation expansion in powers of δL, yielding $F(q, z) = g(q) + g^2(q)M(q, z)$. Here $g(q) = (z + \omega(q))^{-1}$ is the response function of the uniform lattice with $\omega(q) = D_0\Sigma_\alpha(1 - \cos q_\alpha)$, where $\alpha = x, y, \ldots, d$, and $D_0 = (2d)^{-1}$ is its diffusion coefficient. Furthermore $M(q, z) = \Sigma_l\langle(\delta L g)^l\delta L\rangle_{qq}$ and $g_{qq'} = g(q)\delta_{qq'}$. Next, I sum the repeated visits ("repeated ring collisions") of the RW to a single impurity by introducing the T-operator,

$$T(z) = (b/2d)(1 - bJ(z))^{-1},$$
$$J(z) = d^{-1}\int_q \omega(q)/[q + \omega(q)]. \tag{5}$$

Here $J(z)$ is the ring-collision integral, well-known in the kinetic theory of the Lorentz gas, and $\int_q$ denotes an average over the first Brillouin zone of the simple cubic lattice with weight $(2\pi)^{-d}$.

To obtain a *density expansion* of $M(q, z)$ we observe that $M(q, z)$ contains repeated sums over different impurities. The terms in this sum involving the same impurity variable $c_{n,n+\rho}$ account for all possible visits of the blind ant to the same impurity, and give the exact $\mathcal{O}(c)$-contribution. The $\mathcal{O}(c^2)$-contributions are obtained by summing all possible walks between impurity pairs, symbolically denoted by [12], [121], [1212], [12121], ...

From $M(q, z)$ and the response function one can extract all desired information. The most important one is the lattice analog of the VACF $\phi(t)$, defined as the second derivative of the mean square displacement. Its Laplace transform, $\Phi(z) = \frac{1}{2}z^2\langle n_x^2\rangle(z)$, is proportional to the frequency dependent conductivity $\sigma(\omega) \sim \mathrm{Re}\,\Phi(z = \mathrm{i}\omega)$, and reads

$$\Phi(z) = D_0\Big[1 - 2dcT(z)\Big\{1 + c\sum_{n\neq 0} R_{xx}(n, z)/[1 - R_{xx}^2(n, z)] + c{\sum_{n\beta}}' R_{x\beta}^2(n, z)/[1 - R_{x\beta}^2(n, z)]\Big\} + \mathcal{O}(c^3)\Big], \tag{6}$$

where the prime on the sum indicates that the pair label $(n, \beta) = (0, x)$ is excluded. The integral $R_{\alpha\beta}(n, z) = G_{\alpha\beta}(n, z)T(z)$ contains a single crossing $(G_{\alpha\beta}(n, z))$ from impurity 1 to impurity 2 (where n is their relative position), followed by all possible returns $(T(z))$ to impurity 2. Here $G_{\alpha\beta}(n, z)$ is essentially the probability for a displacement n on a uniform lattice, and is defined as

$$G_{\alpha\beta}(n, z) = \int_q \mathrm{e}^{-\mathrm{i}qn} g(q)(1 - \mathrm{e}^{-\mathrm{i}q_\alpha})(1 - \mathrm{e}^{\mathrm{i}q_\beta}). \tag{7}$$

We further note that the dependence on the parameter $\sigma = 1 - b$ enters only through the single impurity T-operator (5), which is quite different in the "pure ant" limit ($b = 1$) and in the "pure termite" limit ($b \to -\infty$). In the remaining part of this paper some typical results will be discussed.

The *diffusion coefficient* for the RW-models is found by taking the static limit ($z \to 0$) in (6). It has the form of an expansion in powers of the impurity concentration, viz. $D(c, \sigma) = D_0\{1 + \alpha_1 c + \alpha_2 c^2 + \mathcal{O}(c^3)\}$. For the RMB, $\alpha_1(\sigma) = -db/(d - b)$ with $\sigma = 1 - b$, and $\alpha_2(\sigma)$ is an expression in terms of lattice sums, explicitly given in ref. 2. In the one-dimensional case one verifies that $R_{xx}(n, 0) \sim \delta_{n0}$ and the density expansion is in agreement with the exact result $1/D(c, \sigma) = \langle 1/\psi\rangle = 1 + cb/(1 - b)$ for $b < 1$. For the 2-D case the expansion coefficients are seen to satisfy the relations $\alpha_1(1/\sigma) = -\alpha_1(\sigma)$ and $\alpha_2(1/\sigma) = \alpha_1^2(\sigma) - \alpha_2(\sigma)$. These relations are equivalent to Straley's[6]) relation $D(c, \sigma)D(c, 1/\sigma) = D_0^2$ derived from self-duality of the square lattice.

The lattice sums for $\alpha_2(\sigma)$ (which converge rapidly in the RW-models with bond-disorder) have been evaluated numerically for the square lattice[2]) and for the simple cubic lattice[3]) at different σ-values. For instance, in the case of bond-percolation ($b = 1$) one finds $\alpha_2(\sigma = 0) = -0.21075$ ($d = 2$) and $\alpha_2(\sigma = 0) = -0.0162$ ($d = 3$). Hence, the approximate expression for $D(c, \sigma = 0)$, quadratic in c, vanishes at $c \simeq 0.48$ ($d = 2$) and $c \simeq 0.66$ ($d = 3$), where the exact percolation thresholds for the bond problem are $c_\mathrm{p} = \frac{1}{2}$ ($d = 2$) and $c_\mathrm{p} = 0.75$ ($d = 3$). The $\mathcal{O}(c^2)$-expression for the 3-D diffusion coefficient at $\sigma = 0$ is not noticeably different from the effective medium theory[4]).

From the symmetry relation $D(c, \sigma) = \sigma D(1 - c, 1/\sigma)$, based on homogeneity and interchange of the two types of bonds, the low-density expansion can be transformed into an expansion in powers of $(1 - c)$.

For *site-percolation* on a square lattice (where $b = 1$) the diffusion coefficient was found[1]) as $D(c, \sigma = 0) = \{1 - c(\pi - 1) + \alpha_2 c^2\}/4$ with $\alpha_2 \simeq -0.856$. This result, exact to $\mathcal{O}(c_2)$, vanishes at $c \simeq 0.40$, close to the threshold $c_p \simeq 0.41$ for site percolation on a square lattice. The result is also reasonably close[1]) to the intuitive result of Watson and Leath[7]), $D(c, \sigma = 0) = \{1 + (1 - \pi)c + (1 - \pi/2)c^2\}/4$. The latter result is in excellent agreement with existing computer simulations[8]).

The VACF $\phi(t)$ for the RBM is given by the inverse Laplace transform of (6). Since $\Phi(\infty) \equiv D_E$ is non-vanishing, the VACF has the general form $\phi(t) = D_E \delta_+(t) + \phi_+(t)$, where $\phi_+(t)$ is a *negative* function for all $t > 0$ (so called "cage effect") and is of $\mathcal{O}(c)$ for small densities. $D_E = D_0 \langle \psi \rangle$ is the short time (or Enskog) diffusion coefficient, and $\langle \psi \rangle = 1 - bc$ is the effective free volume fraction or porosity of the disordered lattice. The *short time behavior* of $\phi_+(t)$ can be calculated exactly to all orders in the concentration[2]). The *long time behavior* $\phi(t) \simeq \beta^{(d)}(c, \sigma) t^{-1-d/2}$ is typical for Lorentz gases and diffusive systems with static disorder. For later reference I note that $\beta^{(d)}(c, \sigma)$ is proportional to the local fluctuations in the diffusion coefficient. This result is based on mode-coupling theory[9]), exactly solved 1-D models[10]) and kinetic theory calculations[11]).

For small c-value the long time tail of the VACF is determined by the dominant small-z-singularity in the ring-collision integral $J(z)$ in (5) and given by $\beta^{(d)}(c, \sigma) = -(\frac{1}{2})[c\pi b^2/(d - b)^2](4\pi D_0)^{-1-d/2}$. To $\mathcal{O}(c^2)$ two lattice sums have to be evaluated numerically. This has only been carried out for the RBM on a square lattice[2]).

The coefficient of the long time tail on the square lattice exhibits the symmetry $\beta^{(2)}(c, \sigma) = \beta^{(2)}(c, 1/\sigma)$, which seems to be a consequence of self-duality. The reason for this symmetry in the density expansions is that $\beta^{(d)}(c, \sigma)$ depends only on σ through $|T(0)|$, defined in (5). Only for $d = 2$ is this quantity invariant under the substitution $\sigma \to \sigma' = 1/\sigma$. The same mathematical property guarantees the self-duality relation for the diffusion coefficient in the bond problem on the square lattice, mentioned above. Further support for the validity of $\beta^{(2)}(c, \sigma) = \beta^{(2)}(c, 1/\sigma)$ at general densities derives from the work of Wright et al.[12]). These authors use the self-duality of the square lattice to show that the local fluctuations in the resistivity, i.e., in the diffusion coefficient, satisfies the same symmetry relation.

Of course, there is also the general symmetry $\beta^{(d)}(c, \sigma) = \beta^{(d)}(1 - c, 1/\sigma)$ based on homogeneity and interchange of the two types of bond variables. Combination of the above symmetries yields for the square lattice, $\beta^{(2)}(c, \sigma) = \beta^{(2)}(1 - c, \sigma)$.

A *long standing problem* is the lack of quantitative agreement in Lorentz gases between computer simulations[13]) and the theoretical predictions[9]). In dense fluids, however, the agreement is satisfactory. Recent computer simulations by Fenkel[14]) of the site percolation model (with $b = 1$) on the square lattice together with detailed calculations[1]) of the VACF indicate the solution to this problem: in *the time interval below about 60 collision times* (which covers the time intervals of existing computer simulations) *one cannot see the pure t^{-2}-tail,* because the VACF has not reached its pure asymptotic behavior. However, inclusion of the first subleading long time corrections $(\ln t)/t^3$ in the $\mathcal{O}(c)$-terms leads to quantitative agreement.

These subleading long time corrections have been calculated analytically for the bond-and-site disordered RW-models on a square and a simple cubic lattice[1-3]), but not for a deterministic Lorentz gas.

At *finite times the VACF for termite models* has an interesting structure already to lowest order in c. This is shown in fig. 1 where the VACF (solid curve, labeled a) is plotted for the RBM on a square lattice for $\sigma = 1 - b = 11$. The result was obtained from (6) and (5) by numerically inverting the Laplace transform of

$$\Phi_+(z) = \Phi(z) - \Phi(\infty) \simeq -\tfrac{1}{4}cb^2J(z)/[1 + |b|J(z)]. \tag{8}$$

This formula represents the repeated ring collisions of the termite with a "superconducting" bond. On the short time scale of $1/\sigma$ sec the termite hops

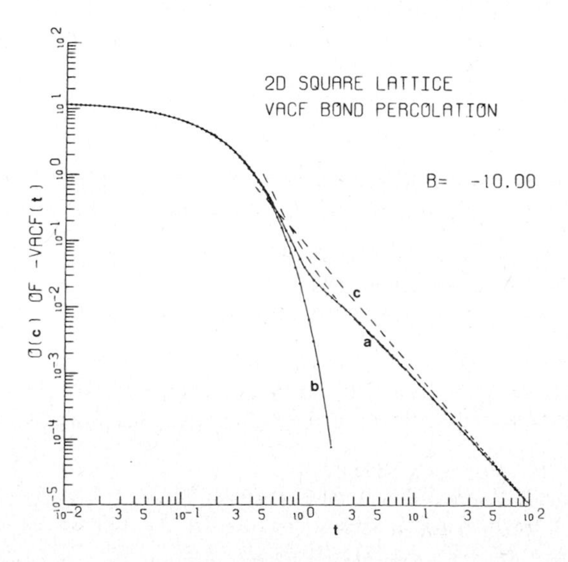

Fig. 1. VACF versus time for a termite model with a small concentration c of "superconducting" bonds with $\sigma = 1 - B = 11$. The labels (a), (b) and (c) are explained in the text.

essentially back and forth across the superconducting impurity bond with only a small probability $\sim 1/\sigma$ per hop of wandering off into the normal conducting region. This behavior is described by a pole in (8), leading to exponential behavior (solid curve, labeled b) and completely determines the VACF for $t < 1$. For larger times the contributions come from a branch cut in (8). The dashed curve, labeled c, represents the t^{-2}-tail. The unmarked dashed line, virtually coinciding with the VACF for $t > 2$, is the long time tail including subleading corrections $(\ln t)/t^3$.

In *summary*, the kinetic theory methods enable one to calculate d.c. and a.c. transport properties of disordered lattices, exactly to $\mathcal{O}(c^2)$ and $\mathcal{O}((1-c)^2)$ in the impurity concentrations c, and lead in many cases to significant corrections on the effective medium theory. The results may also be useful for constructing effective medium theories[4]) or renormalization group calculations[15]), exact to the same order in c.

Acknowledgement

It is a pleasure to thank Th. Nieuwenhuizen, P. van Velthoven and B. van Velzen, with whom I have collaborated on different subjects covered in this review. It is a pleasure to thank B. Bagchi, J. Brey, T. Kirkpatrick and R. Zwanzig for interesting discussions and IPST of the University of Maryland for its hospitality during the summer and fall of 1986. This work was supported in part by National Science Foundation Grant No. DMR-83-09449.

References

1) Th.M. Nieuwenhuizen, P.F.J. van Velthoven and M.H. Ernst, Phys. Rev. Lett., submitted.
2) M.H. Ernst, P.F.J. van Velthoven and Th.M. Nieuwenhuizen, J. Phys. A: Math. Gen. (1986), to appear.
3) B. van Velzen and M.H. Ernst, to be published.
4) J.W. Haus and K.W. Kehr, Phys. Rep., to appear.
5) D. Bedeaux, K. Lakatos-Lindenberg and K.E. Shuler, J. Math. Phys. **12** (1971) 2116.
6) J.P. Straley, Phys. Rev. B **15** (1977) 5733.
7) B.P. Watson and P.J. Leath, Phys. Rev. B **9** (1974) 4893.
8) A.B. Harris and S. Kirkpatrick, Phys. Rev. B **16** (1977) 542.
9) M.H. Ernst, J. Machta, J.R. Dorfman and H. van Beijeren, J. Stat. Phys. **34** (1984) 477, **35** (1984) 413.
10) H. van Beijeren, Rev. Mod. Phys. **54** (1982) 195.
11) A.K. Harrison and R. Zwanzig, J. Stat. Phys. **42** (1986) 935.
12) D.C. Wright, D.J. Bergman and K. Kantor, Phys. Rev. B **33** (1986) 396.
13) B.J. Alder and W.E. Alley, Physica **121A** (1983) 523.
14) D. Frenkel, Phys. Rev. Lett., submitted.
15) U.M.S. Costa, C. Tsallis and G. Schwachheim, Phys. Rev. B **33** (1986) 510.

Physica 140A (1986) 397–404
North-Holland, Amsterdam

FRACTONS

S. ALEXANDER

The Racah Institute of Physics, The Hebrew University, Jerusalem, Israel and Department of Physics, University of California, Los Angeles, CA 90024, USA

The fracton description of vibrational excitations in the strong scattering regime, above the Ioffe–Regel limit is reviewed. The reasons why this is a plausible model for vibrations in amorphous materials are explained and some predictions discussed. It is shown how such a model arises in the context of scattering theory when the Ioffe–Regel length is large and a new self-consistency argument suggesting that the relevant fracton dimension for such vibrations has a universal value of 4/3 is described.

1. The fracton model

The concept of fractons was introduced in the context of the study of the dynamics of self-similar fractal objects[1]). When one tries to map a Laplacian or a second-order difference operator on a fractal geometry one finds anomalous behavior[2]). For the low frequency vibrations one finds[1]) a density of states

$$N(\omega) \propto \omega^{\bar{\bar{d}}-1} \tag{1}$$

and a dispersion relation

$$\lambda \approx \omega^{-D/\bar{\bar{d}}}, \tag{2}$$

where D is the fractal (Haussdorf) dimension and $\bar{\bar{d}}$ is a new intrinsic fracton (or spectral) dimension. The scaling form of these expressions follows from the self-similarity (dilation symmetry) of the fractal. The fact that $\bar{\bar{d}}$ is, in general, different from D results from the fact that quantities like the diffusion constant[3]) or the elastic rigidity are not related to the scaling of the mass in the same way they would be in a Cartesian geometry.

For many purposes the relevant dimension is this intrinsic fracton dimension. Thus, e.g., the generalization of standard arguments shows that $\bar{\bar{d}}$ is the relevant dimension for Anderson localization[4]) so that, in the presence of disorder, all states are localized when $\bar{\bar{d}} \leqslant 2$ independent of the value of the fractal dimension (D). Moreover, most connected fractals of interest are weakly connected and

0378-4371/86/$03.50

have a fracton dimension between one and two. In particular, $\bar{\bar{d}} \equiv 1$ for all configurations of linear polymers[1,5]). More interesting is the fact that $\bar{\bar{d}} \simeq 4/3$ for percolation clusters generated in spaces of dimension $d \geqslant 2$, within experimental accuracy. Mostly one is of course interested in disordered fractals so that all eigenstates must be localized[4]).

An interesting question is the meaning of the scaling length (λ) in (2). The scaling is of course that one usually finds for wavelengths and that is indeed a correct interpretation for a folded linear chain (say a SAW) with no internal disorder ($\bar{\bar{d}} = 1$)[1]). In the other cases of interest the disorder is an intrinsic property of the *geometry* (e.g., for percolation or DLA clusters). We have conjectured[6,7]) that on such fractals λ is a *unique* length describing *both* the wavelength and the localization length. One notes that the localization edge is at $\omega = 0$ on the fractal, so that the localization length also scales with a power of ω. One must therefore have

$$\lambda_l \propto \omega^{-x}. \tag{3}$$

Comparing this with (2) one notes that $x > D/\bar{\bar{d}}$ is obviously inconsistent because it implies $\lambda_l < \lambda$ at small ω. $x < D/\bar{\bar{d}}$ only contradicts one's intuition because it implies that $\lambda_l(\omega)/\lambda(\omega)$ *increases* with a power of frequency. I will show below that a more careful argument[8]) shows that $x = D/\bar{\bar{d}}$ implies $\bar{\bar{d}} = 4/3$.

If one accepts this conjecture ($\lambda_l \approx \lambda$ for all ω) one has very peculiar eigenstates described by a *single* length scale ($\approx \omega^{-D/\bar{\bar{d}}}$). This has many implications. It should be contrasted with the situation near the Anderson localization edge where one has three length scales – the wavelength (λ), the scattering length (λ_s) and the localization length (λ_l), which are all different. Because of this special property we[1,6,7]) gave these excitations a name and called them fractons. In a different terminology one can say that fractons are always at the Ioffe–Regel limit. They remain there at all frequencies.

It is fairly straightforward to apply these ideas to a percolation model of quenched disorder[6,7]). The main new feature is that one has a crossover to phonons at the correlation length (ξ) and that one has to take into account a Stauffer[9]) distribution of finite clusters to make sure that the overall density is uniform. The model one obtains has a lot of predictive power. One can for example calculate a density of states and the temperature dependence of the specific heat[7,10]), the high-temperature thermal conductivity[11]), scattering form factors ($S(q, \omega)$), relaxation rate distributions[12]) and inelastic scattering off fractons[13]). One is also forced[6]) to introduce two-level systems[14]).

It is of course tempting to try to apply these results to real amorphous materials without worrying too much how one would map realistic microscopic models on this type of scaling model. One notes that percolation is, by far, the best understood model for quenched disorder and that the physical properties we are talking about seem, experimentally, to be surprisingly universal. One also notes that the only generally accepted universal feature of amorphous materials, the two-level systems[14]), has still, after many years, not been connected to microscopic models in any satisfactory way.

The results are surprisingly, sometimes embarrassingly, encouraging. This is of course particularly true when they are not very sensitive to the detailed scaling indices which are not really accessible. Some of these results are far from obvious. Thus one predicts a crossover in the density of states to a convex frequency dependence with an excess density of state in the crossover region[10]). The prediction of a plateau in the thermal conductivity when the vibrations become strongly localized is sort of built into the model[7]). The prediction of a linear temperature dependence of this quantity[11]) at high temperatures is much more striking and we had considerable difficulty in getting any sort of experimental results (new or old but all unpublished) on this regime.

None of this can of course be regarded as conclusive proof of the validity of the scaling, fracton description of the high frequency vibrations of amorphous materials. Published results also usually have other explanations related to the specific properties of the materials involved. Thus the most common explanation of the universal plateau in the thermal conductivity in (transparent "window") glass is not localization but scattering from the torsional modes of SiO_4 tetrahedra. This may certainly be the correct explanation combined with a series of material-specific explanations for all other materials for which such a plateau is observed.

It is nevertheless surprising to read a recent didactic Phys. Rev. Letter[15]) so enthusiastically acclaimed (apparently) by all referees that it was published within four weeks of submission, whose only point is that a convex density of states can arise in many ways and does not imply the validity of the fracton model (which is of course correct). The conclusion of this paper is (in a free translation from the German) that the only merit of the fracton model is that it is a "beautifully complicated" explanation of phenomena which have been well understood for a long time. The shopping list of options offered in ref. 15 is rather long, but as far as I have checked they are mostly valid mechanisms for changing a density of states – when applicable. The editor of Nature found this letter sufficiently important to merit an editorial[16]) calling it "a gentle warning on fractal fashions". I may be forgiven for refraining from commenting in detail on this criticism in spite of the implied compliment that we have indeed managed to convince the scientific community that amorphous materials should be considered "fractal at short distances"[16]). We have certainly tried, but I must confess so far with very limited success.

2. Scattering and strong scattering

I would like to devote the rest of this paper to a discussion of the reasons why a fractal (fracton) description of this (relatively) high frequency vibrational mode in amorphous solids is reasonable, what it means and why it implies a fracton dimension $\bar{\bar{d}} = 4/3$. The last part of this discussion is very recent work[8]). It also contains the most important new results in this paper. Thus A. Aharony O. Entin-Wohlman, and R. Orbach should certainly be considered authors for the parts of this paper which are really new. I have not included their names above solely because I am not sure they would agree to accept responsibility for the preceding section of this paper.

I want to consider the vibrational spectrum of amorphous materials. Macroscopically such materials behave as isotropic solids. The low frequency modes are therefore phonons with a dispersion

$$\lambda(\omega) = C/\omega, \tag{4}$$

where C is the velocity of sound and a density of states

$$N(\omega) = \frac{\omega^{d-1}}{C^d}. \tag{5}$$

At low frequencies the scattering is weak and can be described by a Rayleigh scattering[17]) time

$$\frac{1}{\tau_s} = \frac{\omega^{d+1}}{\omega_{IR}^d}, \tag{6}$$

where we have introduced the Ioffe–Regel[18,19]) frequency (ω_{IR}) for which

$$\frac{1}{\tau_s} = \omega. \tag{7}$$

The scattering time (τ_s) in (5) introduces a second length scale

$$\lambda_s(\omega) = C\tau_s(\omega) = \lambda(\omega)\left(\frac{\omega_{IR}}{\omega}\right)^d. \tag{8}$$

Below ω_{IR} one is in the weak (elastic) scattering regime. The (Born) scattering matrix elements are small. One result is that the density of states is not modified by the scattering and is still given by its free form [eq. (5)]. Somewhere in this regime Anderson localization sets in so that the phonons become localized. This has been discussed in great detail in a number of recent papers, particularly those by John, Sompolinsky and Stephen[20]), by Akkerman and Maynard[21]) and by Anderson[22]).

It is useful to look at the structure of the eigenfunctions or the Green's function [$G(0, r)$ *before* averaging]. There are three length scales involved with an inequality

$$\lambda \leqslant \lambda_s \leqslant \lambda_l, \tag{9}$$

where λ_l is the length scale associated with multiple scattering Anderson localization. One has free particle ballistic motion for $r < \lambda_s$ and anomalous diffusion in the range $\lambda_s < r < \lambda_l$. For scales large compared to λ_l one has normal diffusive behavior below the localization edge ($\omega < \omega_A$) with a renormalized diffusion constant. Above ω_A the diffusion constant ($D(\omega)$) vanishes in this limit.

Now all these three length scales decrease above ω_A. λ_s decreases faster than λ [eq. (8)] and they become equal at the Ioffe–Regel frequency (ω_{IR}). In considering the behavior of the localization length one has to remember that the scattering length (λ_s), or equivalently the scattering coupling parameter, is an ingredient in the multiple scattering theory. Thus on frequency scales for which λ_s changes the critical behavior of λ_l basically multiplies the noncritical frequency dependence of λ_s [eq. (8)]. Thus $\lambda_l(\omega)$ must approach both $\lambda_s(\omega)$ and $\lambda(\omega)$. We shall claim that the three scales all coincide for $\omega \approx \omega_{IR}$.

A heuristic argument is to consider the Thouless localization length[23]) (λ_T) which neglects the anomalous character of the diffusion between λ_s and λ_l. This should not matter when $\lambda_s \approx \lambda_l$.

One has

$$\frac{\lambda_T^2(\omega)}{D(\omega)} = N(\omega)\lambda_T^d, \tag{10}$$

and, using $D(\omega) = C\lambda_s(\omega)$ and (5) and (8), one finds

$$\lambda_T = \lambda(\omega_{IR})\left(\frac{\omega}{\omega_{IR}}\right)^{2/(d-2)}, \tag{11}$$

so that the three lengths become comparable at ω_{IR}. The eigenfunctions at ω_{IR} thus have a single length scale, $\lambda \approx \lambda_s \approx \lambda_l$, as we assumed for fractons. They are of course localized. One expects this single length scale feature to be a characteristic of the eigenfunctions at higher frequencies. It seems implausible that even stronger scattering can be renormalized so that λ and λ_s regain their separate identity.

This assumes that the basic length scale for the disorder is a ($\ll \lambda_{IR}$) – so that such a strong scattering regime has physical significance. The latter is not a trivial assumption. We note that the treatment of the disorder in refs. 20 and 21 implicitly assumes $\lambda_{IR} \approx a$. Fractons live in this strong scattering range. Recent experimental evidence[24]) also shows that $\lambda_{IR} \gg a$. We therefore consider the meaning of having a physical model for which $\lambda_{IR} \gg a$ in detail.

The basic quantity one wants is clearly some local index of refraction or velocity of sound. One is interested in the averages relevant to an excitation of characteristic length l. A simple procedure would be to divide space into cells of size l. For each cell we determine the lowest frequency (with some reasonable boundary conditions). We then define a local (cell) velocity of sound

$$C_l^2 = \Omega_l^2 \cdot l^2. \tag{12}$$

In a disordered system one will find a distribution with an average

$$\langle C_l^2 \rangle = \langle C^2 \rangle_l \tag{13}$$

and a variance

$$\frac{\langle C_l^4 \rangle}{\langle C^2 \rangle_l^2} - 1 = \gamma_l^2, \tag{14}$$

where $\gamma_l(\omega)$ is the scattering parameter governing the effect of the disorder for these waves. In particular the Ioffe–Regel condition [eq. (4)] is equivalent to some critical value

$$\gamma_l = \gamma^* \approx 1. \tag{15}$$

Consider now the statistical distribution. For a normal distribution one would have

$$\langle C^2 \rangle_l = C^2, \tag{16}$$

independent of l and

$$\gamma_l^2 \approx \frac{\gamma_1^2}{l^d}, \tag{17}$$

where γ_1^2 is the variance at the atomic level ($l = 1$). This is only consistent with a large Ioffe–Regel length

$$\lambda_{IR} \approx l(\gamma^*) \gg 1, \tag{18}$$

if $\gamma_1^2 \gg 1$. While one cannot exclude this option it is rather difficult to think of (mechanically stable) microscopic models for which it would make sense. An obvious alternative is suggested by percolation,

$$\langle C^2 \rangle_l \propto l^{-x}, \tag{19}$$

$$\gamma_l^2 \equiv \gamma_*^2. \tag{20}$$

The meaning of eq. (19) is that the coupling of cells of size l into larger cells (of size bl) involves not only their internal structure (expressed by the C_l^2) but also the coupling between the cells. Eqs. (19) and (20) are a scaling description of the result if these couplings are sufficiently random. For a percolation model, for which only a fraction of the potentially possible bonds exists, this means that space is uniformly covered by clusters, of any size l, but only a fraction ($b^{D-d} = b^{-\beta/\nu}$) of these clusters is coupled into larger clusters (of size bl). Variations on this scenario are of course possible[25]).

Finally I want to show that the consistency of this argument leads in the strong scattering Ioffe–Regel limit to a fracton spectral dimension of 4/3. I follow ref. 8.

Consider vibrations *on* a fractal. We want to compute the scattering width in the Born approximation analogous to Rayleigh scattering [eq. (6)], assuming weak disorder at some small length scale. This requires the definition of an effective strain on the scale of the "wave length" ($\lambda_{\mathrm{fr}}(\omega)$)

$$\frac{1}{\tau} \approx V^2 N(\omega) \approx \left(\frac{\phi}{R(\lambda)}\right)^4 N(\omega), \tag{21}$$

where ϕ is the vibrational amplitude and $R(\lambda)$ an effective distance on the fractal corresponding to a distance λ in cartesian space. It can be seen[26]) that $R(\lambda)$ is the distance along the bonds, and therefore *proportional* to the resistance between two points on the fractal at a distance λ in space when there is a unique connecting path. More generally $R(\lambda)$ scales like the point to point resistance (or force constant) one would need if one were to coarse grain the fractal to scale λ. In both situations one finds[26])

$$R \propto \lambda^{\bar{\zeta}}, \qquad \bar{\zeta} = (2 - \bar{\bar{d}})(D/\bar{\bar{d}}), \tag{22}$$

and using $\phi/R \sim [\omega^{1/2}R]^{-1}$ and eq. (2) one finds $V^2 \sim \omega^{6-4\bar{\bar{d}}}$ and finally, using (1) in (21)

$$1/\tau \propto \omega^{5-3\bar{\bar{d}}}. \tag{23}$$

This forces an effective "quantum" fractal dimension of 4/3 if one wants the scattering to remain marginally strong (or weak) and obey the Ioffe–Regel condition [eq. (4)] at all frequencies. The philosophy of this calculation is of course that one tries to take account of the strong disorder and multiple scattering by "channeling" the vibrations on a fractal on which the scattering is (relatively) weak.

I have tried to show what a scaling model for the strong scattering regime in amorphous materials means, why it seems reasonable, and to discuss some implications.

Acknowledgement

I would like to thank H.E. Stanley for insisting on my writing this manuscript for these Proceedings.

References

1) S. Alexander and R. Orbach, J. de Physique Lett. **43** (1982) L625.
2) S. Alexander, Phys. Rev. B **27** (1983) 1541.
3) Y. Gefen, A. Aharony and S. Alexander, Phys. Rev. Lett. **50** (1983) 77.
4) R. Rammal and G. Toulouse, J. de Physique Lett. **44** (1983) L13.
5) This well-known result was recently rediscovered in ref. 15. It was certainly not new in ref. 1, where it is used as a didactic illustration.
6) S. Alexander, Ann. Isr. Phys. Soc. **5** (1983) 144.
7) S. Alexander, C. Laermans, R. Orbach and H.M. Rosenberg, Phys. Rev. B **28** (1983) 4615.
8) A. Aharony, O. Entin-Wohlman, S. Alexander and R. Orbach, Phys. Rev. Lett., submitted.
9) D. Stauffer, Phys. Rept. **54** (1979) 3.
10) A. Aharony, S. Alexander, O. Entin-Wohlman and R. Orbach, Phys. Rev. B **31** (1985) 2565.
11) S. Alexander, O. Entin-Wohlman and R. Orbach, Phys. Rev. Lett., submitted; Phys. Rev. B **34** (1986) 2726.
12) S. Alexander, O. Entin-Wohlman and R. Orbach, J. de Phys. Lett. **46** (1985) L549, L555; Phys. Rev. B **32** (1985) 6447; **33** (1986) 3935.
13) O. Entin-Wohlman, S. Alexander and R. Orbach, Phys. Rev. B **32** (1985) 8007.
14) P.W. Anderson, B.I. Halperin, and C.M. Varma, Phil. Mag. **25** (1972) 1; W.A. Phillips, J. Low Temp. Phys. **7** (1972) 351.
15) J.A. Krumhausl, Phys. Rev. Lett. **56** (1986) 2696.
16) J. Maddox, Nature **322** (1986) 303.
17) Lord Rayleigh, The Theory of Sound (Macmillan, London, 1986), vol. II.
18) A.F. Ioffe and A.R.-Regel, Prog. in Semiconductors **4** (1960) 237.
19) N.F. Mott, Phil. Mag. **13** (1974) 93.
20) S. John, H. Sompolinsky and M.J. Stephen, Phys. Rev. B **27** (1983) 5592.
21) E. Akkermans and R. Maynard, Phys. Rev. B **32** (1985) 7850.
22) P.W. Anderson, in T. Holstein Memorial Volume, R. Orbach, Ed. (Los Angeles, 1986).
23) D. Thouless, Phys. Repts. **13** (1974) 93.
24) J.E. Graebner, B. Golding and L.C. Allen, preprint.
25) In particular viscoelastic effects can become important. Limitations of time and space prevent me from exploring this here.
26) A. Aharony, Y. Gefen and Y. Kantor, J. Stat. Phys. **36** (1984) 795; Ann. Isr. Phys. Soc. **5** (1983) 301.

Physica **140A** (1986) 405–409
North-Holland, Amsterdam

WHAT STATISTICAL MECHANICS HAS TO SAY TO COMPUTER SCIENTISTS

P.W. ANDERSON

Department of Physics, Joseph Henry Laboratories, Princeton University, Jadwin Hall, P.O.B. 708, Princeton, NJ 08544, USA

We discuss the relationship between the large N limit of statistical mechanics and the large N limit of computer complexity theory. The "cost function" of many computer optimization problems may be treated as a Hamiltonian whose ground state is the solution of the problem, and statistical mechanics gives us many insights into the structure of possible algorithms or solutions.

Many problems of computer science involve of order N bits of data where N is large: for instance $N = n^2$ intercity distances in the famous traveling salesman problem for n cities, or N propositions to be simultaneously satisfied in a typical satisfiability problem, or $N = n^2$ matrix elements in a diagonalization. The question is then, roughly, the order of magnitude of time necessary to solve the problem, specifically whether it is polynomial in N or exponential. If the latter, one normally assumes that the problem is impossible by computers for any reasonable large N and must be attacked by "heuristic algorithms".

A statistical mechanic immediately recognizes this as like a "thermodynamic limit" $N \to \infty$, and wonders if there are any relationships between the two. Some 10 years ago, such relationships began to be known, and by now there has grown up rather a large interface between the two fields.

In computer science there is a large class of so-called "NP complete" problems with roughly the following properties:

(i) A proposed answer can be evaluated in $\sim N^P$ steps, and simple enumeration of all possible answers will take $\sim e^N$ steps, but can be carried out.

(ii) No polynomial algorithm for solution has been found for any of ~ 300 different problems, each of which can be transformed into the other in $\sim N^P$ steps. This class of problems became of interest to statistical mechanics when it was realized that the ground state of the Sherrington–Kirkpatrick spin glass is an NP complete problem.

Ever since the late 70's Palmer, Kirkpatrick and others, had known that the spin glass ground state problem was hard, and they had been finding ground states of the SK model by an algorithm called "simulated annealing": using

0378-4371/86/$03.50

Mitropolis–Teller or other molecular dynamics technique, start above T_c and gradually change the effective temperature to 0. At $T = 0$ you hope to find the ground state, and indeed often find a very low energy state.

This suggested to Gelatt and Kirkpatrick trying to solve other NP complete optimization problems like TSP by SA, and indeed this algorithm has proved very useful as a heuristic for some types of problems. The scheme is to treat the optimization function as a Hamiltonian by appropriately scaling its variables to make its value $\mathcal{O}(N)$. Thence the space is explored by a molecular dynamics algorithm using a slowly decreasing effective temperature. We will speculate later on why this works.

It was the idea of Stein, Fu and myself, and independently of the Toulouse–Mézard group in France, to further exploit those parallels. Our idea was to approach such problems as though they were statistical mechanics problems of random, quenched systems such as the spin glass. In contrast to the computer scientists, we learn only *generic*, *highly probable* results, where the problem parameters are chosen randomly from some given probability distribution, not the exact solution in any one case; but from that we gain the fact that we are not tyrannized by the worst case if that is a highly improbable one. We expect to gain

(a) a general understanding of the nature of the problem;

(b) values of the optimization function itself valid to order N for most cases.

(a) With regard to general understanding: Stein and I recognized that the mere existence of a transition temperature in the spin glass, graph partition (according to Fu and Anderson), and in the TSP (according to Kirkpatrick and the French), tells us a great deal about the structure of the Hamiltonian = cost function. A transition normally signals the phenomenon of "broken ergodicity" as Palmer describes it. Using the thermodynamic identity $U = F + TS$ and $S = \partial F/\partial T$, the energy U is a unique function of T, $U(T) = F(T) + TS(T)$. Thus a transition temperature T_c corresponds to an energy (="cost") $E_c(T_c)$ at which ergodicity breaks down: the states of the system at energies below E_c are no longer available to each other via conventional thermal fluctuation, generalized isoenergetic trajectories, or what have you.

In the specific case of the SK spin glass, the work of Parisi, Toulouse and collaborators has given us a clear picture of the multifurcation of states below T_c and the geometry of the energy surface in configuration space. They show that the states form an *ultrametric* manifold, like an evolutionary tree, with no closer relationship between any states at any given energy E than their possession or not of a common "ancestor" state at a higher energy $E + \Delta E$ (in particular, there are no "clusters" or "domains" in this case, which appears to be quite different from finite-dimensional spin glasses, most of which are also NP complete). These authors show that there are a large – but probably only polynomial (and even linear) – number of different "ground states" within $< N$ in energy of each other, a fact which had been known only empirically to the computer simulators. What our analysis of the energy surface tells us is that some of these states are separated by energy barriers of $\mathcal{O}(N)$; hence an algorithm starting at a false minimum has to reverse $\mathcal{O}(N)$ spins to correct itself, which for thermal-like algorithms implies an exponential time scale.

This is the reason for a conjecture due to Stein: that problems with finite T_c are exponentially hard in the general case, which presumably implies NP completeness. "Short cuts" from the region of some false minimum to another are denied by the facts that (a) the algorithms can really not know anything about a state but its value of $\mathscr{H}$; (b) in ultrametric systems, there are no local correlations, in fact no locality properties at all. No relationships exist except ancestry.

The only possible weakness of this conjecture lies in the success of simulated annealing; can we anneal our way to a ground state in less than exponential time? This success is caused, as Virasoro has suggested, by another thermodynamic identity:

$$\frac{\mathrm{d}S}{\mathrm{d}E} = \frac{1}{T}.$$

Integrating this up along our particular ultrametric branch, one finds

$$S = \int_{E_{\min}}^{E_c} \frac{\mathrm{d}E}{T(E)}$$

which implies that the entropy at T_c in a deep basin with $E_{\min}$ very negative is bigger than that in a shallow one, hence one is more likely to find oneself in a deep basin: deep basins are also broad ones. This is probably specious as far as NP completeness is concerned since the minima differ only by $\mathcal{O}(< N)$ and thermodynamics is only correct to $\mathcal{O}(N)$, but it does give a clean reason for the success of SA. This amounts to a strongly-supported conjecture (the "moral equivalent of a proof", I have called it) that the SK spin glass in the general case is exponentially hard; if that were so it would be a profound theorem of computer complexity theory, since it would settle the fate of the class NP.

In any case, from this example we obtain analytic demonstrations of two very important facts which were only empirically understood by the computer scientists: (1) the existence of an E_c and the splitting up of configuration space for energies below it; (2) the existence of many solutions at nearly the minimum E or cost. On the other hand, there are some difficulties if one takes too naive a view. The famous case is the two-dimensional short range spin glass. Apparently $\geqslant 3d$ is NP complete, as expected from the existence of T_c, and $2d$ pure is not. But a

2-layer $2d$ spin glass is NP complete. I believe this to be an example of the tyranny of the special case, since special periodic versions of this case do have complicated ground states.

Baum has shown also that at least one NP complete problem transforms into a spin glass with a "golf course" potential, with only one hole and exponentially many identical states. This case has a T_c, all right, where $\Delta E = T_c S$, but no broken ergodicity.

(b) A number of NP complete problems have been "solved" as instances of spin-glass like statistical mechanics problems. The Parisi solution of the spin glass is itself one of these.

A very simple and well-studied example is partition of a random graph of N vertices: N sites i having ij bonds with probability p. The average number of bonds to a given site is then $\alpha = Np$. The problem is to find the cost in bonds of dividing this graph into 2 equal sets.

Let us write

$$H = -\sum_{i>j} J_{ij} S_i S_j,$$

$$S_i = \pm 1, \quad J_{ij} = J \text{ if there is a bond}$$

$$= 0 \text{ if not.}$$

Then if $\Sigma_i S_i = 0$,

$$C = \frac{\langle H \rangle}{2J} - \frac{N(N-1)p}{4}.$$

If desired, we may enforce $\Sigma S_i = 0$ by adding $J'(\Sigma_i S_i)^2$ which gives a conventional spin glass. There are two cases:

(1) if $p \sim \mathcal{O}(1), \alpha \sim N, J \sim N^{-1/2}$, we get a system equivalent to SK which can be solved by Parisi's methods, giving

$$C = \frac{N^2}{4} p - 0.38 N^{3/2} [p(1-p)]^{1/2}.$$

This agrees with all known limits, and improves on them; it also agrees well with simulations.

(2) if $\alpha \sim \mathcal{O}(1)$, $p \sim N^{-1}$, $J \sim 1$, the above expression is incorrect, as can be seen from the fact that $C = 0$ for $\alpha < \alpha_c = 2 \ln 2 = 1.38$, by a calculation of Erdös. This case is more like the finite-dimensional spin glass, where conventional mean field theory fails. By including more order parameters in replica space Liao has improved somewhat on the above answer but has not found a

good solution. Similar difficulties have been encountered by the French with the TSP.

In conclusion, we have made definite inroads into the field of complexity theory and, more important, into the understanding of these complex optimization problems involving quenched, random, frustrated functions of many variables. Since such optimization problems are ubiquitous in the real world, from economics to genetics to brain function to protein folding, this is a field of the utmost importance.

Physica **140A** (1986) 410–415
North-Holland, Amsterdam

HIERARCHICAL MODEL OF MEMORY

Viktor S. DOTSENKO

Landau Institute for Theoretical Physics, Academy of Sciences of the USSR, Moscow, USSR

A layered hierarchical memory model is constructed. The model can store families of correlated images ordered in the form of a hierarchical tree. The structure of the model is similar to that of the visual cortex of the brain.

Memory models or statistical (Ising) spin systems which can store information and function as content-adressable memory systems are an interesting development in the theory of spin-glasses, and are considered as a certain analogue of neural networks[1]). Only recently born, the problem now is an extensive field of activity comprising statistical properties of memory models storing uncorrelated patterns[2–5]), dynamical properties[6,7]), effects of non-symmetric couplings[8]), and finally the problem of storing correlated patterns[9–11]). Apart from the discussion of the reliability of memory models as models for neural networks, I'll concentrate in this paper on purely statistical–mechanical aspects of the problem concerning the storing of correlated patterns.

The idea of storing memories by spin models could be briefly described as follows. Consider a lattice system of N Ising spins $\{\sigma_i\}$ ($\sigma = \pm 1$) described by the Hamiltonian:

$$H = -\sum_{ij} \mathcal{J}_{ij} \sigma_i \sigma_j . \tag{1}$$

The given p uncorrelated spin configurations (images) $\{\sigma_i^{(s)}\}$, $s = 1, 2, \ldots, p$ will be memorized if the (infinite-ranged) interactions $\mathcal{J}_{ij}$ are determined as[1])

$$\mathcal{J}_{ij} = \frac{1}{N} \mathcal{J}_0 \sum_{s=1}^{p} \sigma_i^{(s)} \sigma_j^{(s)} . \tag{2}$$

In a certain range of temperatures and for $\alpha = p/N$ not very large ($\lesssim 0.14$) the model (1) described by ordinary relaxation dynamics serves as a system of content-adressable memory[2–4]). Namely, depending on which "image" out of $\{\sigma_i^{(s)}\}$ $(s = 1, 2, \ldots, p)$ the initial spin configuration resembles more, the relaxation process will recover one of the stored "images".

With an increasing number of stored memories their effective interference results in an increasing number of local spurious minima of the free energy, which does not correspond to any of memorized configurations $\{\sigma_i^{(s)}\}$. Since these spurious memories could also be evoked by the relaxation process, the effective size of attraction space of the given true memorized configuration becomes smaller and the "quality of functioning" of the content-adressable system worsens. For a sufficiently large number of memories, the system breaks down into chaotic spin-glass states where no memory is possible. There is a critical curve $\alpha(T)$ on the plane (T, α) separating the region of "good memory" from the disordered spin-glass region[2,3]). Naively speaking, for a large number of memories the spin–spin interactions (2) are effectively random and the model (1) appears to be the Sherrington–Kirkpatrick spin-glass model[12]).

In some sense the simplest situation is when all memorized spin configurations are completely uncorrelated, since in this case the corresponding free energy minima are independent (this is possible, e.g., for finite p and $N \to \infty$). For finite α the memories are unavoidably correlated. In this case one can try to classify a large number of correlated configurations into the hierarchical tree in terms of spin-glass hierarchical classification[13,14]) and then try to recover what interactions $\mathcal{J}_{ij}$ are needed to get this tree of states. This is hardly possible in a general case but it proved to be possible when the choice of memorized configurations was to a certain extent limited to give an ordered tree of states.

Such a hierarchical tree with an exponential number of correlated spin states can be memorized by a special learning algorithm. The idea is to divide the system into a hierarchy of spin clusters so that at each hierarchical level the clusters should act as independent Hopfield models memorizing certain parts of spin configurations of the respective hierarchical level[11]).

Here we will consider the hierarchical memory model constructed by means of a system of interacting spin layers. The layered structure permits us in a direct and transparent way to "see" the spin configurations of each hierarchical level and simplifies the learning algorithm also. Among other interesting features, this model has a striking similarity to the layered hierarchical structure of the cortex of the brain.

Consider a system of M layers, each having N_m Ising spins $(m = 1, 2, \ldots, M)$. Divide each layer into spin clusters $\{\Omega_{i_m}\}$ so that each cluster of the mth layer has Ω_m spins. Interactions of the layers are arranged in such a way that all spins of one cluster of the $(m - 1)$th layer are interacting with one and only one spin of the next mth layer (fig. 1). For this reason the number of spins in each subsequent layer is reducing: $N_m/N_{m+1} = \Omega_m$. The interaction of the spin of the

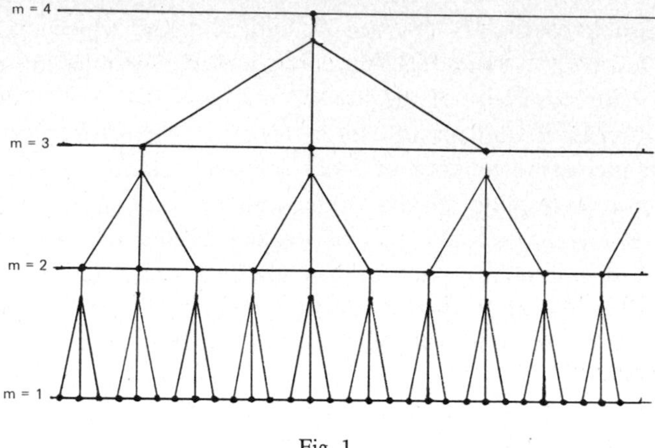

Fig. 1.

mth layer with total magnetization of the spin cluster of the $(m-1)$th layer is ferromagnetic and fixes their signs to be equal:

$$\mathrm{Sign}\left(\sum_{i\in\Omega_{m-1}} \sigma_{i_{m-1}}\right) = \sigma_{i_m}. \tag{3}$$

The spin configurations to be memorized should be classified in the form of the hierarchical tree (fig. 2)[11]). A given family of spin configurations of the mth layer will be assumed to belong to one common ancestor state in the next hierarchical level if the signs of magnetization of the spin clusters $\{\Omega_{i_m}\}$ in the mth layer are the same for all these configurations. Therefore the ancestor state of this family is given according to (3) by one spin configuration in the $(m+1)$th layer.

This way climbing from layer to layer we can construct a whole hierarchical tree of states (fig. 2). The number of levels of the tree is equal to the number of

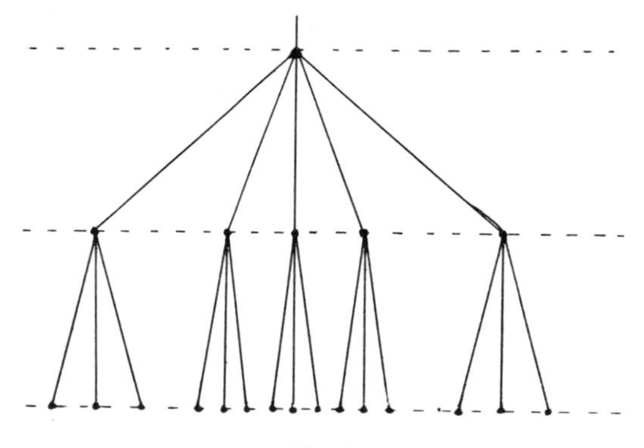

Fig. 2.

layers of the spin system. The number of the trees is equal to the number of the spin configurations in the last Mth layer. The learning algorithm of all these spin states is the following. Spin interactions in the layers are introduced according to the Hopfield model inside clusters only:

$$\mathscr{J}_{ij}^{(m)} = \frac{1}{\Omega_m} \mathscr{J}_0 \sum_{s=1}^{p_m} \sigma_{i_m}^{(s)} \sigma_{j_m}^{(s)}. \tag{4}$$

Here $\{\sigma_{i_m}^{(s)}\}$ $(s = 1, 2, \ldots, p_m)$ are cluster spin configurations which compose the global spin configurations of the layer. Note that all spin configurations memorized this way are "constructed" by some limited number of "blocks" – fixed spin states of the clusters. These "blocks" could of course be different in each cluster and in each layer.

The number of the "blocks" p_m in each cluster should be limited: $p_m/\Omega_m < \alpha(T)$ where $\alpha(T)$ is the abovementioned critical curve restricting the region of "good memory"[2,3]).

The total number of spin configurations in the family belonging to one ancestor state of the $(m + 1)$th hierarchical level is obviously

$$(p_m)^{N_m/\Omega_m}.$$

The total number of states of the whole tree is

$$\prod_{m=1}^{M} (p_m)^{N_m/\Omega_m} \sim \exp\left\{ \frac{N_1}{\Omega} \ln p \right\},$$

where p and Ω are characteristic values of p_m and Ω_m.

Spin configurations of the upper layer where there are no clusters are memorized in the same way as in the Hopfield model (2). Here the restriction should be imposed on the number of memorized trees: $p_M/\Omega_M < \alpha(T)$.

The evoking process in the present memory model proceeds as follows. After the initial spin configuration is fixed in the first layer, it automatically fixes all spins in all other layers. Evoking begins in the upper layer where the most general information (the most coarse image) is stored. Here the correct tree is chosen. Then in each subsequent layer the corresponding branch of the tree is chosen. Finally in the last layer the exact configuration is evoked.

Note that in an analogous way the hypothesized process of "learning" by this memory model could be imagined. After a new image to be learned is exposed to the first layer of the model which already has a lot of "knowledge" in its memory, it will eventually enter into some family of coarse images at some hierarchical level. Then the spin interactions should be fixed only below this level and this will create a new branch of a tree.

The model described above is interesting first of all in the respect that it can be useful for recognizing and memorizing images as part of the hierarchical family of similar images.

Another important feature of the model is its scaling property. If there are two images of the same object but of different size described by $\{\sigma_i^{(1)}\}$ and $\{\sigma_i^{(2)}\}$, these two spin configurations are not close in a usual (spin-glass) sense (since the distance $q^{12} \equiv (1/N)\Sigma_i \sigma_i^{(1)}\sigma_i^{(2)}$ could be close to zero). They are unavoidably different for the Hopfield model. The model presented here (with some modifications) can provide some sort of scale invariance to recognize the images as equivalent although on different hierarchical levels.

It is noteworthy that similar hierarchical organization is a characteristic property of the cortex of the brain or at least of the visual cortex[15]). The visual cortex has six layers ($M = 6$). Neurons of each subsequent layer receive and sum up the signals from a group of neurons of the previous layer (like in fig. 1). In each subsequent layer, however, the image becomes not just more coarse, but also there exists some sort of decomposition into components. For example a particular neuron of the second layer receives information from such a "cluster" of neurons of the first layer that it can be excited only by the image of a line with definite size and definite direction. In the next layer neurons are excited by the same kind of line but without strictly determined position[15]). The purpose of such an arrangement is not quite clear yet (it is also the way to catch some general features of the image) but it could also be made by the layered spin model described in this paper.

In conclusion, I want to say that I was not trying to give any kind of a review of the problem so I'm not discussing the attempts by Parga and Virasoro[9]), Toulouse, Dehaene and Changeaux[10]) or Ioffe and Feigelman[8]). But I would like to stress that, although the present hierarchical memory models have obvious drawbacks, any *practical* memory model (even without pretending to model neural networks) should possess in some form a hierarchical structure of its memory.

References

1) J.J. Hopfield, Proc. Natl. Acad. Sci. USA **79** (1982) 2554.
2) D.J. Amit, H. Gutfreund and H. Sompolinsky, Phys. Rev. Lett. **55** (1985) 1530.
3) L.B. Ioffe and M.V. Feigelman, Europhys. Lett. **1** (1986).
4) D.J. Amit, Proc. Physics in Environmental and Biomedical Research, Rome (1985).
5) W.Z. Kinzel, Phys. B **60** (1985) 205.
6) J.J. Hopfield, D.I. Feinstein and R.G. Palmer, Nature **304** (1983) 158.
7) S. Shinomoto, preprint RIFP-645 (1986).
8) L.B. Ioffe and M.V. Feigelman, to be published.

9) N. Parga and M.A. Virasoro, Triest preprint, IC/85, 186.
10) G. Toulouse, S. Dehaene and J-P. Changeux, Paster Institute, Paris, preprint.
11) V.S. Dotsenko, J. Phys. C **18** (1985) L1017.
12) D. Sherrington and S. Kirkpatrick, Phys. Rev. Lett. **32** (1975) 1792.
13) M. Mezard, G. Parisi, N. Sourlas, G. Toulouse and M. Virasoro, Phys. Rev. Lett. **52** (1984) 1156.
14) M. Virasoro and M. Mezard, preprint (1985).
15) D.H. Hubel and T.N. Wiesel, Proc. Roy. Soc. London Ser. B **198** (1977) 1.

Physica **140A** (1986) 416–420
North-Holland, Amsterdam

AGGREGATION PHENOMENA AND FRACTAL STRUCTURES

M. KOLB

Laboratoire de Physique des Solides, Bat. 510, Université de Paris-Sud, 91405 Orsay, France

General interest paper

In disordered aggregation processes fractal structures appear in a natural way. Many of their properties can be characterised in terms of scaling and numerous models have been introduced to describe different physical situations. "Clustering of Clusters" is a model applicable to several growth processes in colloid science and elsewhere. The basic results of a scaling analysis for this model is reviewed.

Aggregation processes play an important role in many areas of science and technology, and in particular in the field of aerosols and colloids[1]). Several years ago, it was proposed[2]) that the geometrical structures that result from such processes can be described in terms of fractals and scaling[3]). Since, there has been an enormous activity[4]) in a) interpreting experimental results in terms of fractals and b) building models that characterise specific types of aggregation mechanisms.

'Clustering of Clusters' (ClCl) is one of the basic aggregation models[5]). For simplicity, let us define it on a square lattice: Initially one deposits N_0 particles randomly on the sites of a periodic square lattice. No two particles can occupy the same site. Any two particles sitting on nearest neighbour sites belong to the same cluster. The clusters now start to move by hopping randomly and independently one lattice spacing at a time. Whenever a particle of one cluster is nearest neighbour to a particle of another cluster, the two clusters merge to form a single larger cluster. This process goes on until one large cluster contains all N_0 particles.

There are a number of parameters in this model, which can be adjusted to describe different physical situations and which lead to different scaling properties. First, we define a parameter which controls the diffusivity or mobility of the clusters relative to their size. Each cluster of size or mass m is moved with a probability proportional to m^{α} to one of its neighbouring sites per unit time. For $\alpha < 0$ the small clusters diffuse faster than the large ones; this is the physically relevant case. A second parameter is the sticking or bonding probability p_b.

0378-4371/86/$03.50

Whenever two particles belonging to different clusters sit on neighbouring sites, a bond forms between them with a probability equal to p_b; this models the fact that besides the attractive Van der Waals interaction, which is responsible for the strong bonding, an ionic repulsion of variable strength is present which prevents the clusters from getting close to each other. $p_b = 1$ corresponds to aggregating at first contact. Still another parameter is the fragmentation probability p_f which controls the stability of the bonds: each bond breaks with a probability p_f per unit time. Hence the lifetime of each bond is $\tau = 1/p_f$. For $p_f = 0$ the process is strictly irreversible.

The properties that are used to characterise this growth process are the radius of the clusters $R(m)$ or $R(t)$ as a function of mass m resp. the time t, the density correlation function within a cluster $C(r)$, and the cluster size distribution function $N(m, t)$, i.e. the number of clusters of a given mass m at time t. Scaling postulates that

$$
\begin{aligned}
&R \propto m^{1/D}, \\
&C(r) \propto r^{D-d}, \qquad (1)\\
&N(m,t) = \overline{m}^{-2} p(m/\overline{m}),
\end{aligned}
$$

where $\overline{m} = N_0/N$ and $N = N(t)$ is the total number of particles at a given time, $N(t) = \Sigma_m N(m, t)$. $p(x)$ is the scaling function. These relations are valid in the scaling regime, i.e. at large times and for large cluster sizes.

The fractal dimension is first and foremost determined by the spatial dimension d. For example, for the irreversible diffusive aggregation process where large clusters move more slowly than small ones $D = 1.43, 1.75, 2.02, 2.30, 2.60$ for dimension $d = 2, 3, 4, 5, 6$ as determined from simulations at low concentration. Varying the parameters permits to cross over into other growth regimes.

The mobility exponent separates a 'flocculation like' from a 'gelation like' regime. When $\alpha > 0$ the large clusters dominate the growth and they grow by absorbing small particles. This is in contrast to flocculation where two aggregating particles have about the same size. Fig. 1 shows the scaling function $p(x)$ for flocculation[6]). The corresponding curve for gelation ($\alpha > 0$) is monotonically decreasing instead of being bell-shaped. The process for $\alpha > 0$ at low concentration corresponds asymptotically to DLA[7]) (diffusion limited particle aggregation), which is used to model dielectric breakdown, fluid–fluid displacement and electrodeposition[4]). The fractal dimension changes from $D = 1.75$ for $\alpha < 0$ to $D = 2.50$ for $\alpha > 0$ in $d = 3$ dimensions.

The bonding probability permits the model to cross over from chemical or reactive ClCl to diffusive ClCl. When $p_b \to 0$ the clusters collide many times before sticking to one another, which reduces the effect of the diffusive motion.

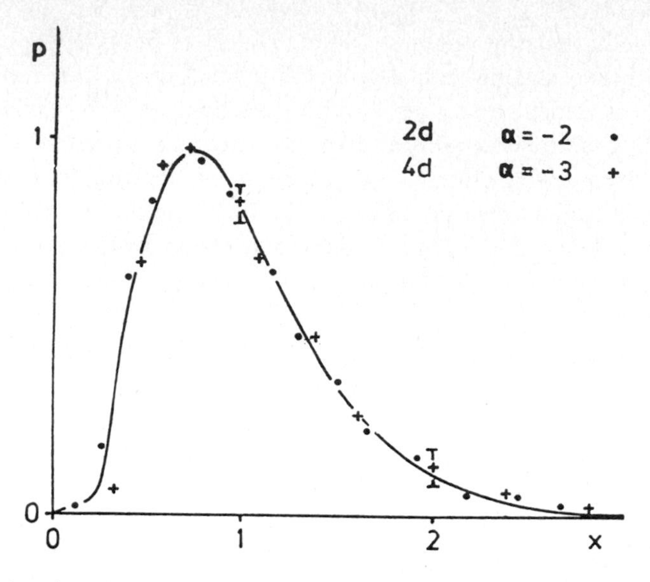

Fig. 1. Scaling function p for irreversible diffusion limited ClCl at low concentration as a function of the scaling variable $x = m/\overline{m}$. The data points are from simulations in $d = 2$ and $d = 4$ dimensions with the values of α indicated and the curve represents the solution of the Smoluchowski equation[6]) with a kernel satisfying eq. (2).

The fractal dimension increases, as the clusters penetrate more deeply into each other: $D = 2.02$ compared to $D = 1.75$ in $d = 3$. This prediction has been observed in experiments of gold[8]) and silica[9]) colloids.

The fragmentation probability p_f reduces the irreversibility of the aggregation process. For small, but nonzero values of p_f the growth process initially has the characteristics of irreversible aggregation and then crosses over to a state of dynamic equilibrium. This has been predicted theoretically from simulations[10]). There are aggregation experiments[11,12]) where reversibility plays a role and which show a change in the fractal dimension with time. The reversible clustering model has $D = 2.0$ in $d = 3$ which makes it impossible to distinguish it from chemical clustering solely on the basis of the fractal dimension.

The last parameter that we shall consider here is the cluster density. It describes the crossover from the flocculation regime at low concentration to the gelation regime at high concentration. At high concentration the clusters are so close together that again the effect of the diffusive motion is diminished. The fractal character of the clusters (at low concentration) invariably causes a crossover into the gelation regime, no matter how low the initial particle concentration as the volume occupied by fractal clusters grows with time like $NR^d \propto \overline{m}^{(d-D)/D}$. As a consequence the structure compactifies and its effective fractal dimension gradually changes from D to d. The gel time for completely

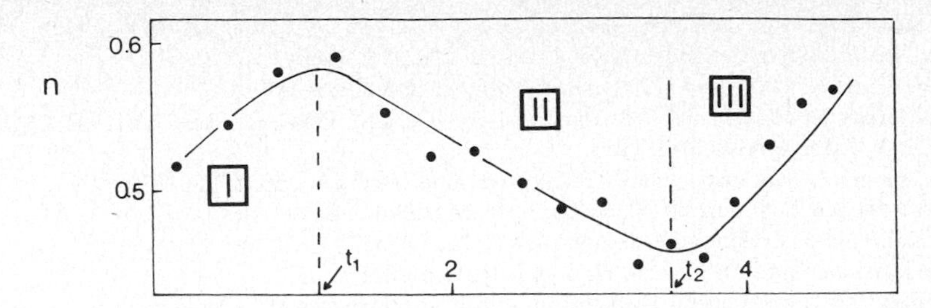

Fig. 2. The average number of moves n before a cluster sticks to another one (which is directly related to the surface-to-surface distance between neighbouring clusters) for irreversible ClCl as a function of time. Regime I corresponds to flocculation, regime II to gelation and regime III to the post gelation regime for a finite system[14]). At time t_2 the largest cluster spans the system.

irreversible gelation ($\alpha < 0$) is estimated to be infinite[13]). Three different regimes can be identified very clearly when plotting the surface-to-surface distance between neighbouring clusters: Initially the distance increases in the flocculation regime, then it decreases in the gelation regime and eventually one enters the finite size gelation regime where it grows again. This is shown in fig. 2.

The scaling hypothesis allows one to determine the cluster size distribution and the dynamic scaling exponent z in $R \propto t^{1/z}$ as a function of D and α [7]):

$$z = D(1 - \alpha) - (d - 2), \qquad \lambda = \alpha + (d - 2)/D. \tag{2}$$

The second equation relates ClCl to the kinetic approach by means of the Smoluchowski equation[15]). λ is the degree of homogeneity of the kernel, which determines the shape of $p(x)$. In fig. 1 the results of the Smoluchowski approach are compared with the simulations. The numerical results indicate that the two approaches agree.

The interrelation of the models presented above and the fractal structures that characterise each mechanism are illustrated in a film based on simulations of the growth processes[16]).

References

1) S.K. Friedlander, Smoke, Dust and Haze: The Fundamentals of Aerosol Behaviour (Wiley, New York 1977).
2) S.R. Forrest and T.A. Witten, J. Phys. **A12** (1979) L109.
3) B.B. Mandelbrot, The Fractal Geometry of Nature (Freeman, San Francisco, 1982).
4) On Growth and Form, H.E. Stanley and N. Ostrowsky, eds. (Martinus Nijhoff, Groningen, 1986). Fractals in Physics, L. Pietronero and E. Tosatti, eds. (North-Holland, Amsterdam, 1986).
5) P. Meakin, Phys. Rev. Lett. **51** (1983) 1119; M. Kolb, R. Botet and R. Jullien, Phys. Rev. Lett. **51** (1983) 1123.

6) M. Kolb, Phys. Rev. Lett. **53** (1984) 1653.
7) T.A. Witten and L.M. Sander, Phys. Rev. Lett. **47** (1981) 1400.
8) D.A. Weitz, N.Y. Lin and C.J. Sandroff, Surf. Sci. **158** (1985) 147.
9) D.W. Schaefer, J.E. Martin, P. Wiltzius and D.S. Cannell, Phys. Rev. Lett. **52** (1984) 2371.
10) M. Kolb, J. Phys. **A19** (1986) L263.
11) Ph. Mangin, B. Rodmacq and A. Chamberod, Phys. Rev. Lett. **55** (1985) 2899.
12) C. Aubert and D.S. Cannell, Phys. Rev. Lett. **56** (1986) 738.
13) M. Kolb and H.J. Herrmann, J. Phys. **A18** (1985) L435.
14) H.J. Herrmann and M. Kolb, J. Phys. A Lett., to appear.
15) M. Ernst, in: Fundamental Problems in Statistical Mechanics IV, E.G.D. Cohen, ed. (Elsevier, Amsterdam, 1985).
16) Aggregation, film by M. Kolb on growth processes and fractal structures. For information contact the author.

Physica **140A** (1986) 421–427
North-Holland, Amsterdam

SPECIAL PURPOSE COMPUTERS IN STATISTICAL PHYSICS

Hans J. HERRMANN

SPhT, CEN Saclay, 91191 Gif-sur-Yvette Cedex, France

General interest paper

A new trend in physics, namely the building of special purpose computers (SPC), is reviewed. Special emphasis is given to the following questions: Why does one build SPC's? When is it worthwhile to build an SPC? How does one proceed if one wants to build an SPC? Finally the most important results that have been obtained for statistical physics through SPC's up to now will be sketched.

Imagine a theoretical physicist choosing a transceiver for an address bus. His Gradshteyn Ryzhik is deeply buried under electronic design sheets and data books from Texas Instruments. The phone rings and a salesman from Analog Devices gives him the price of a new sequencer.

This scenario is becoming a reality for the increasing number of physicists that are building special purpose computers (SPC). They have chosen a specific problem, like the molecular dynamics of a fluid, the Monte Carlo simulation of a spin glass or the transfer matrix algorithm for random conductivity, and now they build a computer specially tailored for this very purpose.

Why build a special purpose computer?

Isn't it much more comfortable to write a FORTRAN program and execute it on the next available general supercomputer (like a Cray XMP, an IBM 3090, a Cyber 205, etc.)? Several reasons have led people to do it the hard way:

First of all, general computers must usually be shared with many other users, so you will have access to a restricted number of hours, several hundred hours per year at best. If you build your own SPC it will calculate only for you every day of the year, 24 hours a day (except during blackouts).

There also is an important financial aspect: A grant proposal asking for 2×10^7 US dollars for a Cray 2 is likely to be rejected (unless perhaps you have

0378-4371/86/$03.50

already won a Nobel prize). The chances, however, to get 5×10^4 US dollars for tools and components to build an SPC are considerably better.

Finally there is speed! Supercomputers are famous for being very fast (that is what makes them price-efficient compared to smaller general computers). Now, however, there exist on the market highly integrated ultrafast chips. If, in addition, one arranges these chips inside the SPC in such a way that many of them work simultaneously (doing the same thing (parallelism), or doing different things (vectorization)) the SPC can be as fast or even faster than a supercomputer. Technologically speaking, this is the reason why SPC's are mainly a development of only the last couple of years.

All these reasons in favour of SPC's have to be counterbalanced by the fact that the building of an SPC involves an enormous amount of work in a field that usually has to be learned from scratch.

When is it worthwhile to build an SPC?

Since the building of an SPC is a non-negligible enterprise one should first consider carefully if one has chosen the right problem. There are certain general criteria:

–The problem one wants to solve should be sufficiently fundamental and sufficiently difficult that after two years – which is about the average time one needs to get results from an SPC – the scientific community is still interested.

–The algorithm one uses must be relatively simple, so that the SPC won't get too complex and its construction time too long.

–The method to solve the problem should involve a huge repetition of identical steps (typically 10^8 to 10^{10}) so that one can run the SPC for many days without intervening.

–The algorithm and the evaluation of the data should have had extensive previous testing on general computers and one should be so sure about the procedure that one is willing to fix in a definitive way what calculations one will perform. It is usually very difficult to change an SPC once it is built.

How to build an SPC?

Once one has chosen the algorithm for which one wants to build the SPC there are certain necessary conditions required before construction can start: One needs the financial support to buy the hardware components and electronic equipment (like an oscilloscope or a logic analyzer). An enthusiastic team willing to work hard must be formed. At least one person with a solid experience in electronics must belong to the team. Finally one has to find an electronics lab willing to put a reasonable infrastructure at your disposal.

For what follows, I will report some of the experience we have had in Saclay[1]) building an SPC to calculate the electrical conductivity of percolation clusters via the strip method[2]).

The first, and most creative step is the design of the architecture of the SPC. One has to try to decompose the algorithm into the largest possible number of elementary calculations that can be performed simultaneously. Some calculations are completely independent from the rest, like the generation of random numbers, and can therefore be executed by a processor of its own (MIMD = multiple instruction multiple data arrangement), so they are implemented on a separate electronic board.

For the part of the calculation that will be executed the most one has to construct a "pipeline". In the case of our Saclay machine the main calculation is the updating of a matrix R_{ij} through

$$R'_{ij} = R_{ij} - R_{11} - V_i * V_j, \tag{1}$$

where V_l is a vector and all operations are in 64 bit floating point arithmetic. In fig. 1 we see a simplified diagram of the data flow. Like on a car assembly line the first subtraction of eq. (1) is performed for one matrix element while the second subtraction of eq. (1) is done for another matrix element (MISD = multiple instruction single data arrangement). In order that such a pipeline can work every unit must be finished at the same time. So one has to investigate which chips are available on the market. It happens that the fastest 64 bit multipliers need 480 nsec per multiplication (WTL 1064 of Weitek) while a subtraction can be done in 120 nsec (WTL 1065). So one has to put four multipliers in parallel as shown in fig. 1 (SIMD = single instruction multiple data arrangement) and one can get a result every 120 nsec.

The design of the architecture is, however, far from being finished! In fig. 1 R_{ij} and V_i are stored in memories. One needs a device, the address generator, that determines the location of a given matrix element within the memory. Then one

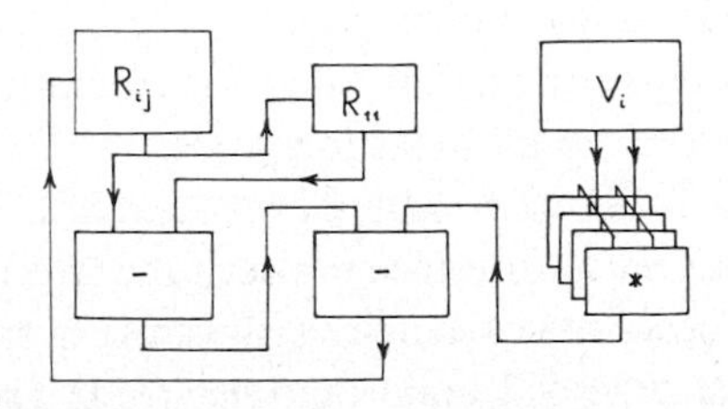

Fig. 1. Schematic diagram for an architecture to calculate eq. (1).

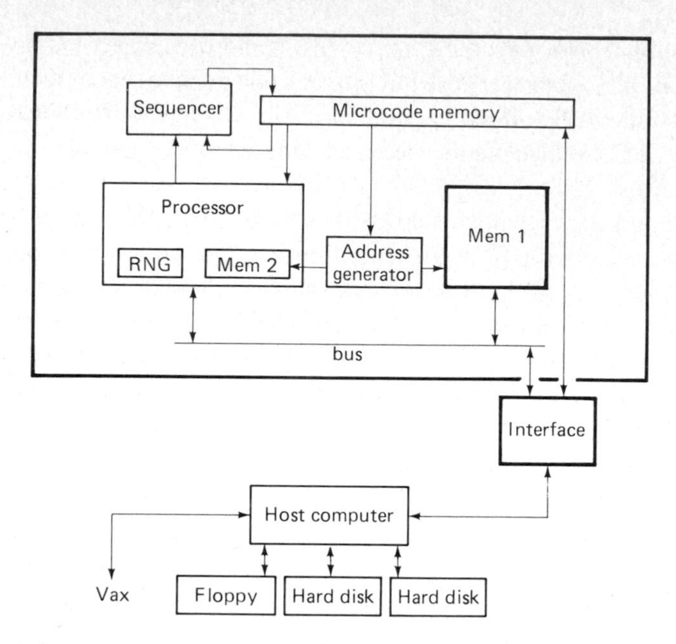

Fig. 2. General diagram of the Saclay machine[1]). The parts surrounded by heavy lines are the ones actually built.

needs somewhere to store the "vector-instructions", i.e., the sets of bits that tell each of the chips what function it has to perform. The ensemble of all these vector-instructions is called the microcode. Which instruction is to be applied and how often is managed by a "sequencer". Roughly speaking the sequencer makes the loops and jumps that we are accustomed to use in FORTRAN programs. All these units must have the same speed as the central processing unit (CPU) of fig. 1, i.e., 120 nsec.

Finally a connection to the external world must be established. One needs a host computer, for instance a PC, that will receive the data coming from the CPU, evaluate it and store it on a disk. The host computer need not be fast but must be reliable. From time to time it will make a complete backup of the state of the machine and make consistency checks: It is usually too complicated to use error correcting codes as those implemented on the big general computers, so if one finds an error during a calculation (due for example to a cosmic ray) it is restarted using the last backup as a starting point.

The host computer will not work with 64 bits and will, since it is slow, not be synchronous to the CPU, so an interface between the two is needed which is built on a separate board. The whole ensemble is sketched in fig. 2.

After the architecture is settled important electronic decisions must be taken: Which technology should be used to build the electronic boards: wirewrap or printed circuits, or should one even make integrated circuits? Also one has to choose all the chips out of the data books of the different manufacturers and to compare them one has to carefully study their specifications. Since about 200 different chips are needed this is a major effort.

After all these decisions are made, a detailed layout has to be made for each board. Each pin of each chip must be properly connected and, of course, the chips must fit on the board.

In order that the connections are not too long and do not cross each other too much the placing of the chips on the board and the routing of the connection lines between them has to be done very carefully. CAD (= computer assisted design) programs are commercially available and can be extremely useful for this purpose. CAD programs can write all information concerning the locations of chips and wires on tape. These tapes can be sent to a company that will make the boards for you within several weeks.

In the meantime you have ordered the chips from the respective manufacturers and built the backplane into which the boards can be placed. You put everything together and the SPC is ready for testing. The debugging will take you some more months before actual calculations can start.

Special purpose computers that have been built

Most of the about twenty SPC's that have been built by physicists in the last five years are related to statistical physics. Their construction time ranges from four months to four years, their price from 200 US dollars to 2×10^5 US dollars. About half of the SPC's built are restricted to very specific applications; the others are more general. In addition there are at least 30 projects for SPC's about which I will not talk. A partial list of the more specific existing SPC's that are related to statistical physics is given in table I.

Cellular automata have been the cheapest[3]) and the shortest times were needed to construct them[4]). One example for a cellular automaton SPC has been

TABLE I
Some SPC's built for statistical physics.

Place	Type	Completed	Ref.
M.I.T., USA	cellular automata	1982	3
E.N.S., France	cellular automata	1986	4
Santa Barbara, USA	Ising	1982	6
Delft, Netherlands	Ising	1982	10
Bell Labs, USA	Ising (spin glass)	1984	12
Delft, Netherlands	molecular dynamics	1983	16
Saclay, France	random conductivity	1986	1

presented by Vichniac during this conference[3]). Since cellular automata SPC's can visualize their configurations very nicely on a screen, many interesting pictures have been made with them. Quantitative results have, however, been quite rare up to now. This might change in the future if recent ideas by Pomeau and others[5]) to model hydrodynamic flow through cellular automata are exploited.

SPC's of the Ising type have been more successful in yielding precision data. The Santa Barbara machine[6]) held the world record for a while for the speed of simulation of an Ising model with 30 million updates per second. (This record was later broken again by general computers[7]).) It yielded precise values for the critical temperature and the dynamic critical exponent of the three dimensional Ising model making nearly 10^{12} updates[8]). However, more or less at the same time about equally precise estimates were obtained by other methods[9]). Also some incorrect results were produced for certain system sizes which were presumably due to correlations in the random number generator as was suggested for instance by calculations on the DISP[10]). The DISP is an SPC, slower but more versatile than the Santa Barbara machine. Several other calculations have already been performed on it including some that were reported on at this conference[11]).

Another successful Ising-type SPC is the spin glass machine[12]) built in Bell Labs. They found evidence for a finite transition temperature in the three-dimensional short range Ising spin glass[13]). The same result was also found by other authors at the same time using conventional computers[14]). The machine at Bell Labs was also used to study the random field Ising model[15]).

Much more complex than the machines described above are molecular dynamics SPC's since they involve real numbers instead of integers and since they involve long range instead of short range interactions. The first molecular dynamics SPC was built by Bakker[16]) in a pioneering effort using fixed point arithmetic. This machine was used to give some evidence that in two-dimensional melting no hexatic phase exists[17]) in agreement with work done on conventional computers[18]). Another SPC for molecular dynamics has just been completed in San Jose[18]).

Finally a number crunching SPC of equal order of difficulty has just been finished in Saclay[1]). It will be used to try to settle the question of the validity of the Alexander–Orbach conjecture for percolation clusters and to obtain accurate values for the dynamical critical exponents of percolation. Data are, however, not yet available.

Summarizing, we have described a new effort that has been made in the field of computational physics to overcome usual limitations of general supercomputers. It has to be left to the future to see if this development will survive in the long run or if a new generation of supercomputers will render it obsolete. Some more information about this field can be retrieved from older review articles[19]).

Acknowledgement

Many individuals helped me to get acquainted with this field through interesting discussions. I am particularly indebted to M. Hajjar, F. Hayot, J.M. Normand and D. Stauffer.

References

1) F. Hayot, H.J. Herrmann, J.M. Normand, P. Farthouart and M. Mur, J. Comp. Phys. **64** (1986) 380.
2) B. Derrida and J. Vannimenus, J. Phys. A **15** (1982) L557; H.J. Herrmann, B. Derrida and J. Vannimenus, Phys. Rev. B **30** (1984) 4080.
3) T. Toffoli, Physica **10D** (1984) 195.
4) D. d'Humieres, private communication.
5) Y. Pomeau, J. Phys. A **17** (1984) L415; U. Frisch, B. Hasslacher and Y. Pomeau, Phys. Rev. Lett. **56** (1986) 1505; N. Margolus, T. Toffoli and G. Vichniac, Phys. Rev. Lett. **56** (1986) 1694.
6) R.B. Pearson, J.L. Richardson and D. Toussaint, J. Comp. Phys. **51** (1983) 241.
7) S.F. Reddaway, D.M. Scott and K.A. Smith, Comp. Phys. Comm. **37** (1985) 351; H.J. Herrmann, J. Stat. Phys. **45** (1986) 145; Y. Okabe and M. Kikuchi, Statphys. 16 conference.
8) M.N. Barber, R.B. Pearson, D. Toussaint and J.L. Richardson, Phys. Rev. B **32** (1985) 1720; R.B. Pearson, J.L. Richardson and D. Toussaint, Phys. Rev. B **31** (1985) 4472.
9) G. Pawley, R.H. Swendsen, D.J. Wallace and K.G. Wilson, Phys. Rev. D **29** (1984) 4030; J. Adler, J. Phys. A **16** (1983) 3565; C. Kalle, J. Phys. A **17** (1985) L801; S. Wansleben and D.P. Landau, Statphys. 16 conference.
10) A. Hoogland, J. Spaa, B. Selman and A. Compagner, J. Comp. Phys. **51** (1983) 250.
11) H.W.J. Blöte, A. Compagner and A. Hoogland, Physica A, to be published; H.W.J. Blöte, A. Compagner, P.A.M. Cornelissen, A. Hoogland, F. Mallezie and C. Vanderzande, Physica **139A** (1986) 395.
12) J.H. Condon and A.T. Ogielski, Rev. Sci. Instr. **56** (1985) 1691.
13) A.T. Ogielski and I. Morgenstern, Phys. Rev. Lett. **54** (1985) 928; A.T. Ogielski, Phys. Rev. B **32** (1985) 7384.
14) R.N. Bhatt and A.P. Young, Phys. Rev. Lett. **54** (1985) 924; see also R.R.P. Singh and S. Chakravarty, Phys. Rev. Lett. **57** (1986) 245; and J.D. Reger and A. Zippelius, preprint.
15) A.T. Ogielski and D.A. Huse, Phys. Rev. Lett. **56** (1986) 1298.
16) A.F. Bakker, C. Bruin, F. van Dieren and H.J. Hilhorst, Phys. Lett. A **93** (1982) 67.
17) A.F. Bakker, C. Bruin and H.J. Hilhorst, Phys. Rev. Lett. **52** (1984) 449.
18) F.F. Abraham, Adv. in Phys. **35** (1986) 1.
19) H.J. Hilhorst, A.F. Bakker, C. Bruin, A. Compagner and A. Hoogland, J. Stat. Phys. **34** (1984) 987; H.J. Hilhorst and H.J. Herrmann, J. Phys. B Lett. **42** (1986) 52; Science **215** (1982) February issue; Phys. Today **36** (1983) May issue.

LIST OF CONTRIBUTORS

ANALYTIC SUBJECT INDEX